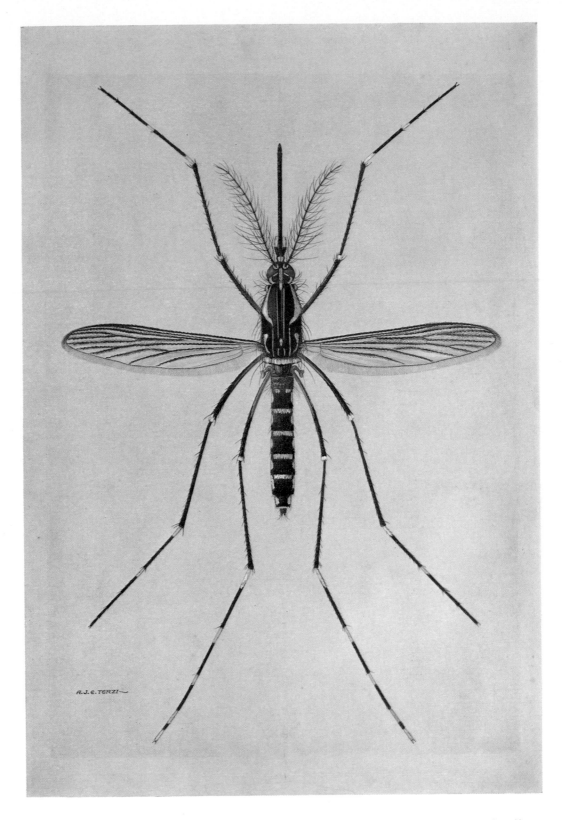

Plate 1 (Frontispiece). *Aedes (Stegomyia) aegypti*, female, Yellow Fever Mosquito (from Edwards) ($\times$38).

INSECTS AND OTHER ARTHROPODS OF MEDICAL IMPORTANCE

Edited by Kenneth G. V. Smith

With contributions by

T. Clay
B. H. Cogan
R. W. Crosskey
P. Freeman
M. S. K. Ghauri

D. J. Lewis
P. F. Mattingly
H. Oldroyd
A. C. Pont
W. H. Potts
D. R. Ragge

A. L. Rice
J. G. Sheals
F. G. A. M. Smit
K. G. V. Smith
I. H. H. Yarrow

With 12 plates (2 coloured) and 217 text-figures

The Trustees of the British Museum (Natural History)
London 1973

Publication number 720
ISBN 0 565 00720 3
British Museum (Natural History)
Cromwell Road
London SW7 5BD

FOREWORD

THE present work replaces *A Handbook for the Identification of Insects of Medical Importance* by John Smart, first issued in 1943. Smart's book went through four editions and proved very useful to medical entomologists and to students generally. The text was written during the Second World War to fill a need felt by the many medical entomologists employed in the tropical war zones of the Old World. Insects of the Americas were largely omitted.

The original book was, to a great extent, prepared entirely by Smart himself, adapted from the specialist literature, with chapters on fleas by Karl Jordan and on arachnids by R. J. Whittick. Many of the figures were from published works but the numerous new ones, mostly prepared by Arthur Smith, were all of the highest standard. Since 1943, however, the nature of medical entomology has changed considerably and fields that were of lesser importance at the time have now assumed greater prominence. An entirely new book seemed necessary with a rather different approach.

On considering the preparation of a new manuscript, it was obvious that, whilst retaining identification as the main purpose, it would be desirable to bring in more discussion on the biological and medical side. In 1943 it was still possible for one man to prepare the major part of the manuscript. This is no longer practicable and to give an adequate consideration of the rapidly changing scene in medical entomology with its complex zoonoses and disease-distribution problems, it was decided to invite specialists to contribute chapters on their own groups, under a general editor. I am glad to say that we received a most enthusiastic response from all the specialists we approached.

The work is now expanded to include the whole world and fuller treatment of groups of minor medical importance is given. The original book included keys to all stages of species of Anopheline mosquitoes known from the Old World, but in a group in which knowledge changes as rapidly as in the mosquitoes, such keys seem out of place nowadays. Instead, Dr Mattingly has prepared entirely new keys to mosquito genera on a world basis. A fuller treatment has been given to arthropods other than insects, and the title altered accordingly.

The work is intended primarily for identification and as a stepping-stone to the specialist literature, and to serve this end extensive bibliographies are given. In preparing the bibliographies, we have had in mind the needs of the student and of the younger medical entomologist working in the field, often isolated from major libraries. For this reason, many important references are cited even if they are not specifically mentioned in the text, and in addition, as far as possible all recent important works have been added right up to the time of going to press.

We have been very fortunate in being able to call on the services of Kenneth G. V. Smith to act as general editor. Not only has his extensive knowledge of Diptera been of great value, but it has been mainly through his drive and determination that the book has been published at all.

Paul Freeman
Keeper, Department of Entomology

CONTENTS

AUTHORS

Dr Theresa Clay Department of Entomology, British Museum (Natural History), London.

Mr Brian H. Cogan Department of Entomology, British Museum (Natural History), London.

Dr Roger W. Crosskey British Museum (Natural History), London; *formerly* Entomologist, Ministry of Health, Northern Nigeria and Commonwealth Institute of Entomology.

Dr Paul Freeman Department of Entomology, British Museum (Natural History), London.

Dr Mohammed S. K. Ghauri Commonwealth Institute of Entomology, c/o British Museum (Natural History), London.

Dr David J. Lewis External Staff, Medical Research Council, c/o Department of Entomology, British Museum (Natural History), London; Member of W.H.O. Advisory Panel for Parasitic Diseases; *formerly* Entomologist, Sudan Ministry of Health.

Dr Peter F. Mattingly Department of Entomology, British Museum (Natural History), London; Member of W.H.O. Advisory Panel for Parasitic Diseases; *formerly* entomologist, 7th Field Malaria Laboratory, Nigeria and Ghana, and Yellow Fever Research Institute, Lagos and consultant W.H.O. Filariaisis Unit, Rangoon.

Mr Harold Oldroyd Department of Entomology, British Museum (Natural History), London.

Mr Adrian C. Pont Department of Entomology, British Museum (Natural History), London.

Mr William H. Potts 24 Furze Road, High Salvington, Worthing, Sussex; *formerly* Chief Entomologist, East African Trypanosomiasis Research Organization, Tanganyika Territory.

Dr David R. Ragge Department of Entomology, British Museum (Natural History), London.

Dr Anthony L. Rice Department of Zoology, British Museum (Natural History), London.

Dr J. Gordon Sheals Department of Zoology, British Museum (Natural History), London.

Mr Frans G. A. M. Smit   Department of Entomology, British Museum (Natural History), London.

Mr Kenneth G. V. Smith   Department of Entomology, British Museum (Natural History), London.

Dr Ian H. H. Yarrow   Department of Entomology, British Museum (Natural History), London.

LIST OF PLATES

EDITOR'S PREFACE

IN a work of this sort it is inevitable that there will be some overlapping of subject matter, but I have been at pains to see that each chapter stands as an entity, although adequately cross-referenced. Further, as far as possible, authors have not been restricted in their approach or in the length of their contributions. Thus, where no recent comprehensive account of a group exists (e.g. Simuliidae), a fuller treatment is given than where such comprehensive studies are available (e.g. lice, cockroaches).

The chapters and sections are arranged approximately in order of medical importance of the group, and within a rough taxonomic framework for the Diptera. A full general index gives page reference to all mention of an animal or disease and a separate index is given to all authors cited which should facilitate use of the bibliographies. Finally a vector table is given, arranged under the insect orders rather than the usual arrangement by disease, and illustrated by distribution maps for some typical diseases.

The co-operation of the contributors has lightened the editorial task considerably, and I thank those of my colleagues who have shown so much interest in the work as a whole, over and above their own particular chapters, especially R. W. Crosskey, P. Freeman, D. J. Lewis and P. F. Mattingly.

Other colleagues who have freely given advice in their special fields include A. Hayes, A. M. Hutson, K. H. Hyatt, D. Macfarlane, K. G. McKenzie, H. Oldroyd, R. D. Pope, S. Prudhoe, A. L. Rice and R. I. Vane-Wright.

Illustrations are such a valuable feature of a work concerned primarily with identification and for those not prepared by the contributors we thank Dorothea Baker, M. Druckenbrod, Thelma Ford, Maureen Grogan, Alan Palmer, Arthur Smith and the late A. J. E. Terzi.

Peter Green and the photographic section are thanked for their careful photographic work in reproducing illustrations where originals were lost or otherwise unavailable.

Bernard Clifton and Pamela Gilbert of the entomological library were a constant help with bibliographic problems and Adrian Pont kindly helped with linguistic problems.

My wife, Vera Smith, carried out much emergency typing, checking and proof-reading.

Dr A. J. Duggan is thanked for discussion of Chagas' disease and for facilities provided at the Wellcome Museum of Medical Science, of which he is Director.

Professor G. S. Nelson and Dr B. R. Laurence of the London School of Hygiene and Tropical Medicine are thanked for information and comment.

Professor A. M. Fallis of the Department of Parasitology, University of Toronto is thanked for information on the transmission of *Leucocytozoon*.

Dr Mattingly is especially grateful to Dr Botha de Meillon and the staff of the Southeast Asia Mosquito Project and Dr Alan Stone of the Agricultural Research Service, U.S. Department of Agriculture, for much kindness and hospitality during his stay in Washington, for checking and commenting on his keys and for permission to reproduce a modified form of his paper in *Contributions of the American Entomological Institute*, vol. 7, part 4, 1971, as chapter 3a of the present work.

For the loan of, or permission to reproduce, certain illustrations and information the Trustees are grateful to the following individuals and institutions:

George Allen and Unwin Ltd for illustrations from *The African Trypanosomiases* (ed. H. W. Mulligan).

Dr H. Banziger of the Entomologisches Institut der Technischen Hochschule, Zürich.

The Bishop Museum Press, Honolulu, for figures from *Insects of Micronesia*.

The Commonwealth Agricultural Bureaux for figures from the *Bulletin of Entomological Research*.

The Company of Biologists Ltd for an illustration from the *Quarterly Journal of Microscopical Science*, vol. 14, 1931.

Dr J. R. Coura, Dr W. B. Petana and the Liverpool School of Tropical Medicine.

Dr H. R. L. Disney, formerly of the Helminthiasis Research Unit, Kumba, Cameroon.

Dr M. T. James and the United States Department of Agriculture for illustrations appearing in *The flies that cause myiasis in man*.

Messrs Raymond Lewis and Denys Dawnay of the Camera Press, London.

Methuen and Co. Ltd for illustrations from Imms' *A General Textbook of Entomology*.

Dr C. L. Pinango of the Department of Haematology, St Thomas's Hospital, London.

Professor A. A. Shtakelberg of the Soviet Academy of Sciences.

Mr S. A. Smith of the London School of Hygiene and Tropical Medicine.

Drs H. E. Whittingham and A. F. Rook and the *British Medical Journal*.

The World Health Organization, Geneva, for permission to include references to unpublished documents.

Kenneth G. V. Smith

August, 1971

1. INTRODUCTION

by Paul Freeman

GENERAL

INSECTS belong to the great group or Phylum of invertebrate animals called the Arthropoda. This group comprises at least 85 per cent of all known species of animals and includes such familiar forms as crabs, shrimps, spiders and centipedes, as well as the true insects (Insecta).

The Arthropoda have the body divided into separate rings or segments, each of which may bear jointed limbs. The whole of the body and limbs is covered by a cuticle, hardened areas of which form an exoskeleton, with flexible connecting membrane between the segments allowing movement. The cuticle contains chitin. The heart is dorsal and the body-cavity is a haemocoele; the central nervous system consists of a ganglionated ventral nerve cord, linked to a ganglion above the oesophagus, often referred to as the brain.

The Insecta (fig. 1) are the largest Class of the Arthropoda and adult insects possess the following characteristics:—

1. The body is divided into three regions: head, thorax and abdomen.
2. The head carries one pair of antennae only, one pair of mandibles and two pairs of maxillae, the second pair fused medially to form the labium.
3. The thorax carries three pairs of walking legs and usually one or two pairs of wings.
4. The abdomen has no walking appendages.
5. Respiration is by means of ramifying tubes or tracheae connected to the exterior by openings or spiracles along the sides of the body.

Larvae of the more advanced insects may be profoundly different from the adults and in particular, structures such as wings, legs, compound eyes, mouthparts and external genitalia are either entirely absent or considerably modified.

The other main groups of Arthropods, none of which possess wings, can be distinguished from Insecta as follows:—

1. *Crustacea* have two pairs of antennae and at least five pairs of legs; when the body segments are grouped, they are arranged in two regions only (e.g. lobsters and shrimps). Respiration is never by tracheae.
2. *Arachnida* have no antennae and four pairs of legs; the body segments are either grouped in two regions or are fused into an unsegmented whole. Respiration may be by tracheae or by 'lung-books'.
3. *Diplopoda* (Millipedes) have a single pair of antennae, the body trunk not differentiated into thorax and abdomen and each apparent segment carrying two pairs of legs and two pairs of spiracles; respiration is by tracheae.

1

4. *Chilopoda* (Centipedes) resemble Millipedes superficially but each segment is a true segment and thus carries only a single pair of legs and spiracles. The first pair of legs is modified to form poison claws.

More than three-quarters of a million insect species have so far been described and many more remain to be discovered. It is thought that the final figure for existing species may be nearly two million. They are amongst the most abundant living animals and have successfully colonized practically all terrestrial and freshwater ecological niches; a small number have succeeded in adapting themselves to life below the high tide mark and a very few to the open sea.

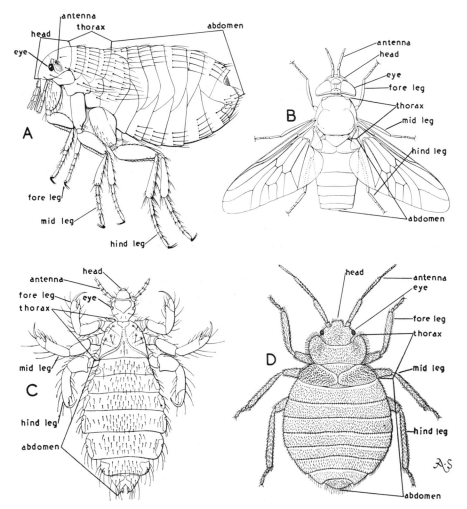

Fig. 1   A, Human flea (*Pulex irritans*); B, Horse-fly (*Chrysops*); C, Body louse (*Pediculus humanus* var. *corporis*); D, Bed-bug (*Cimex lectularius*) (various magnifications).

Insects are one of man's main competitors and many species impinge directly upon his activities. A considerable number are active transmitters of various diseases, of which malaria is perhaps the best known; others, such as the house-fly, are more passively involved.

In addition to the true insects, the Arachnida also include many pest species, a number of which may carry diseases. It is usual for books on medical entomology to cover these and a chapter has therefore been included to give some account of their structure and classification, with details of the groups of medical importance.

DEVELOPMENT AND LIFE HISTORY OF INSECTS

Most insects start life as eggs laid by adult females following mating with adult males. Occasionally the eggs hatch before being laid, so that the female appears to lay larvae. Other insects are able to lay unfertilized eggs that hatch normally (parthenogenesis) and in still other, much rarer cases, immature forms may reproduce (paedogenesis). On hatching, young insects may resemble the adults in general features, e.g., young cockroaches, or they may be very different, e.g., caterpillars and fly maggots. Young insects, with the possible exception of mayflies, never have functional wings, since these are only found in the adults.

As in the adult, the young insect is covered by a chitinous cuticle, with hardened parts and plates. The softer parts are provided with a loosely fitting and folded superficial epicuticle which allows room for growth. The more rigid parts such as mandibles and legs have no room for growth after their initial hardening. When the limit set by the extensibility of the epicuticle and the size of the harder parts, is reached, the inner layers of the old cuticle are dissolved from within and a new cuticle, a size larger, is laid down beneath the old one. After an appreciable thickness of new cuticle has been laid down, the old cuticle is ruptured along definite lines of weakness and the insect gently withdraws itself from the remnants of the old skin. The new skin in turn becomes extended to its limit and is cast off, until, after a series of such moults or ecdyses, maturity is reached and, normally, growth ceases.

The life cycle of an insect may be divided into three distinct phases: (1) the egg; (2) the growing stage, or larva; (3) the adult or imago, when it becomes sexually mature. In the higher Orders, where the difference in form and habit between larva and adult is great (as between a maggot and a blue bottle, fig. 6), there is an intermediate quiescent stage, called the pupa. In those Orders of insects in which the young resemble the adults in form and to some extent in habit, and in which this pupal stage is not present, the young insect is usually termed a nymph. Some entomologists do not use the term nymph, but refer to all as larvae. The changes of form through which an insect passes are collectively termed metamorphosis.

With a few rare exceptions, it is only in the adult stage that insects are sexually mature and capable of reproduction. With the exception of some of the very primitive insects, moulting ceases with sexual maturity. The young stages are usually voracious feeders, the adults are more concerned with reproduction and seeking new feeding grounds for their young.

The young stages of many insects live in places quite different from those in which the adults are found. Thus, mosquito larvae are aquatic; whilst many fly maggots live and grow in decaying flesh and other putrid matter. Invasion of quite different habitats by the young stages has become possible with the evolution of metamorphosis.

B

THE STRUCTURE OF INSECTS

General

Insects are invertebrate animals and thus do not possess an extensive internal skeleton to give support to the body and attachment to the muscles.   Instead, these functions, as in other Invertebrata, are mainly performed by the thickening and hardening of the integument, more particularly of the outer layer or cuticle.   The thickening is not uniform and continuous, otherwise complete rigidity would ensue, but takes the form of a series of plates, or sclerites, linked by thinner, membranous connective cuticle allowing flexion.   When great muscular effort, such as that involved in flight, is not needed, the cuticle may remain relatively soft, and rigidity of the body is then derived from the interplay of muscular tension on the cuticle on the one hand, and the internal pressure of the body fluid brought about by this tension on the other.   Such soft-bodied forms, which include many larvae, usually have the integument of the head thickened and strengthened giving a framework for the attachment of the muscles operating the mouth-parts.

The basis of insect cuticle is chitin, which is a fibrous nitrogenous polysaccharide. Chitin by itself is not a suitable covering for a terrestrial animal, although being tough and flexible it is well suited to provide hinges or joints between segments.   In most insects the greater part of the cuticle undergoes a hardening process called sclerotization, which is brought about by the deposition of sclerotin.   Sclerotin is an extremely strong skeletal substance and is formed by the addition of protein to the cuticle and by its subsequent tanning, which converts the protein to a cross-linked plastic.   Parts of the cuticle thus thickened and hardened are spoken of as being sclerotized and the whole integument is said to form an exoskeleton.

The cuticle extends to some internal organs that are of ectodermal origin; these include the proctodeal and stomodeal regions of the alimentary canal, the lower ducts and certain accessory organs of the reproductive systems of both sexes and the tracheae of the respiratory system.   In these parts, since they originate as invaginations of the ectoderm, the cuticular surface is a lining, i.e., it is internal in relation to the particular organ under discussion.   In some parts of the body, the integument becomes invaginated and hardened to form a rigid endoskeleton which supports certain organs and provides for the attachment of muscles.   The separate parts of such an endoskeleton are termed apodemes.

The surface of the integument may carry a variety of scales (which are flattened), hairs, spines and bristles.   Scales, hairs and bristles are alike in that they are articulated to the body surface and when one is broken off, the articulation can be distinctly seen in the form of a pit-like mark termed an alveolus.   These articulated hairs, etc., are also termed setae or macrotrichia.   The minute hairs on the wings of many insects and on the general body surface are termed microtrichia and are not articulated to the surface by an alveolus.   Spurs on the legs of many insects are large hairs differing in origin from the macrotrichia but similar to them in that they articulate in alveoli.   The sclerotized cuticle may be drawn out into spines, horns, knobs, etc., which naturally do not have alveoli at their bases; it may also be sculptured and coloured in such various ways as to give the insect a distinctive pattern. The cuticle may also be covered in part by areas of dusting, sometimes referred to as pollinosity or tomentum. These areas may appear grey and dull or there may be a shimmering effect with shifting lights, known as pruinosity.

The body is divided into three clearly distinguishable regions: the head, the thorax and the abdomen. Fundamentally, the whole insect consists of a series of segments or somites, which are not so obvious as in the earthworm for instance, but can be clearly seen in the abdomen. In the thorax and head the segmentation is indicated by the paired appendages. As a general guide, the integument of each individual segment consists of a sclerotized dorsal region called the tergum; a sclerotized ventral region called the sternum; and on each side a lateral region, the pleuron which may be sclerotized, as in the thorax, or soft, as in the abdominal segments. The terms tergite, sternite and pleurite refer to individual sclerites in these regions.

Head

The head consists of six fused segments and it contains the large supraoesophageal ganglion or 'brain'. The head carries the antennae, the large compound eyes, the simple eyes or ocelli (when present), and the mouthparts, that is feeding appendages grouped around the mouth.

The antennae consist of a series of segments. In primitive insects the antennae are thread-like as in the cockroach and are formed of numerous similar segments. The antennae in the various Orders of insects and within the Orders themselves are much modified in respect of the number, shape and size of the segments. The large compound eye is made up of large numbers of individual photo-sensitive elements called ommatidia; these are arranged side by side with their long axes orientated radially to the curved surface of the whole compound eye. The part of the surface of the eye corresponding to each ommatidium is referred to as a facet. The number of facets varies from as many as 28,000 to only a single one in some parasitic forms.

The sclerotized surface or exoskeleton of the head consists of several sclerites more or less welded together to form a hard head-capsule. The boundaries of these sclerites are not always clearly defined but in general terms the top of the head is termed the vertex and the surface behind this the occiput. Forward from the vertex is the frons which lies on the front of the head between the compound eyes and embraces the antennal sockets. Anterior to the frons is the clypeus to which is attached the labrum or upper lip, usually by a movable articulation. The lateral area below and posterior to each eye is termed the gena.

The mouthparts typically consist of a series of median and paired parts. Firstly, the labrum or upper lip is a simple plate capable of limited up and down movement; its ventral surface is the epipharynx or membranous roof of the mouth. Following this and clustered around the mouth are an anterior pair of jaws or mandibles and a posterior pair of jaws or maxillae with or without maxillary palps. The last appendage is the lower lip or labium which, although it appears to be a median structure, is in fact formed by the fusion of two appendages, or second maxillae, which are serially homologous with the maxillae. The labium normally bears a pair of labial palps. Arising at the base of the labium, and sometimes fused to its upper surface, is a median tongue-like process, the hypopharynx, having at its tip the opening of the salivary duct.

Primitively the mouthparts of insects are of a generalized chewing or biting type, of which those of the cockroach are a good example. However, there are many modifications and in some whole Orders the mouthparts have become modified in one way or another, for the purpose of piercing the integument of plants or animals so that the contained liquids may be sucked up, e.g., bugs (Hemiptera), lice (Phthiraptera) and fleas (Siphonaptera).

All the Diptera or true flies have the mouthparts modified for sucking up liquids or very finely divided solids, but the modifications are different throughout the Order. The females of the more primitive blood-sucking families, such as the biting midges, mosquitoes and horse-flies, possess both mandibles and maxillae, which are stylet-like for piercing or lacerating the skin of the host. The actual sucking tube for imbibing food is formed from the elongated labrum above and the elongated hypopharynx below. The labium either acts as a guide or in the case of the horse-fly assists with its spongy labella at the apex, in the soaking up of blood. Only the females of these groups suck blood.

The more advanced Diptera and the non-blood-sucking families have lost the mandibles and maxillae whilst retaining the sucking tube and labium with its spongy apical labella. The pair of palpi that remain are the maxillary palpi, those of the labium being modified into the spongy labella. Where the blood-sucking habit has been secondarily developed again, as in tsetse-flies and Hippoboscidae, it is the labium itself that is the main cutting and piercing organ. The labium is hard and trough-like and enters the wound along with the sucking tube. In this type, both sexes may take blood. In adult warble-flies the mouthparts are completely atrophied.

The structure of the mouthparts is of great importance in the major classification of insects but it is of only minor importance at species level. Specific or generic characters sometimes are to be found in the segmentation, for example, of the palpi. Mouthpart structure may be very important in relation to the function of insects as vectors of pathogenic organisms; further details are given in each chapter and a general review is given by Hocking (1971).

Thorax

The thorax is the locomotor part of the insect's body. It consists of three segments, referred to as pro-, meso- and metathorax. Each segment typically carries one pair of walking legs, whilst the meso- and metathorax each carry, in addition, one pair of wings. Absence of legs is rare in adult insects but it is the rule in larvae of Diptera and certain Coleoptera, for example. The absence of wings is thought to be a primitive character in the Apterygota (see below) but elsewhere it is a secondary feature when found in adult insects. No larvae or nymphs have functional wings.

The thoracic tergites are usually termed nota with the appropriate prefix pro-, meso- or metanotum; the lateral sclerites pro-, meso- or metapleuron; and the ventral sclerites pro-, meso- or metasternite. The plates of each segment are much subdivided and sometimes the subdivisions are used in classification. Names for these will be found in specialist monographs.

The thorax is seen at its simplest in the Thysanura (Bristle-tails) and in many larvae, where the segments differ little in size and proportions. With the development of wings the thorax becomes much more specialized. The meso- and metathorax become more or less intimately welded together and the resulting union may be so close that the limits of the two regions are almost indistinguishable. Where the wings are of about equal area these two segments are of equal size, but where the forewings are markedly the larger then the mesothorax has a corresponding greater development. Where the fore wings are small or not used in flight (Coleoptera) then the mesothorax is correspondingly reduced. The pronotum may be enlarged to form a shield, especially in Dictyoptera and Coleoptera; in many other Orders it is reduced to a narrow annular segment.

Insect legs are primarily for walking or running and this condition is well shown by the cockroach. They exhibit a wide range of adaptive modifications in the various Orders and families, for such functions as burrowing, jumping or seizing prey and so on. The proximal segment of the leg is usually small and is called the coxa; it is followed by another small segment called the trochanter. The principal leg segments are the femur, tibiae and tarsus and succeed one another in that order after the trochanter. There are usually strong spurs at the apex of the tibia. The tarsus is normally divided into subsegments, five being the basic number, of which the first is usually the longest and is often termed the basitarsus or metatarsus. The last or fifth tarsal segment bears the claws and in addition, paired pads or pulvilli and a single bristle or pad-like empodium, either or both of which may be absent. The tarsal segments may be reduced in number. The legs themselves are simply referred to as fore, middle and hind legs. When describing their features, the legs are, by convention, considered as projecting sideways at right angles to the body in a straight line; they are then treated as having ventral, dorsal, anterior and posterior surfaces.

Possession of wings is one of the most characteristic features of adult insects and accounts to a great extent for their dominance as a Class of animals. The wings are flat membranous structures, the membrane being supported by sclerotized hollow tubes called veins. The pattern of veins of a wing is termed the venation or neuration and presents characters of great systematic importance at all levels of classification. Unfortunately the various systems of nomenclature in use for the veins are confusing because the older systems were established uninfluenced by the modern concepts of evolution. The results have been that terminology of any particular group of authors was usually only applicable within the limits of the particular Order studied. The modern nomenclature is now largely based on the work of Comstock and Needham who first presented an evolutionary approach. However, in several groups, notably the Diptera, some knowledge of early systems is necessary to understand the literature. This is dealt with more fully in the chapter on the Diptera.

Both pairs of wings may be more or less equal in size but it is more normal for them to be unequal. In beetles and some other groups, the fore wings are hardened and when closed form a covering beneath which the membranous hind pair is folded on top of the abdomen, and here the hind pair of wings serve as the principal organs of flight. In Diptera the hind pair of wings are modified into sensory structures called halteres, knob-like organs with a stalk, which appear to function as gyroscopic organs and allow the fly to control rolling or yawing movements.

Some insects are without wings or have them reduced to small pads. This is particularly common in species, for example, lice and fleas, that are external parasites of warm-blooded vertebrates.

Abdomen

The insect abdomen is formed of a series of segments that are more equally developed than are those of other regions of the body. The tergites and sternites are usually simple, undivided plates, whilst the pleura are membranous. The abdomen consists basically of eleven segments but the apical ones may be so much modified in connection with the external genitalia that only as few as four may be recognizable at first glance in the higher Orders. In some higher Diptera the tergites may be so strongly developed as to encircle the sternites completely so that they are hidden from view; this allows the abdomen to expand after a large meal, or when eggs are developing. The abdomen contains the organs of digestion, reproduction, circulation and excretion.

The anus and reproductive organs are at the apex of the abdomen. The male usually possesses a median intromittent organ and a pair of claspers which help to grip the female during copulation. The complexity of the external male genitalia varies widely and their structure is often of great importance in identification. In the female the external genitalia consist typically of three pairs of processes which form together an ovipositor or egg-laying organ. In bees and wasps this is modified to form a sting. In other groups the processes may be greatly reduced, so that, in the higher Diptera, for example, the ovipositor is formed from the attenuated apex of the abdomen itself.

Internal Structure

The internal organs of an insect need hardly be considered here since only exceptionally are they used in identification. However, it is worthy of note that the general body cavity is a haemocoele and the blood contained in it is usually a colourless fluid. The internal organs, the respiratory tracheae and the muscles all lie in the haemocoele, bathed in the blood. In some insects the blood plasma contains visible pigments and, although little is known of the chemical structure of many of these, haemoglobin has long been known from the plasma of the larvae of both some Chironomid midges and also of the Horse Bot-fly (Diptera). There is no true vascular system but a longitudinal dorsal vessel or 'heart' is present, which pulsates and, by means of a series of orifices and valves, keeps the blood circulating in the haemocoele.

The alimentary, nervous and excretory systems may be omitted from consideration here, beyond noting that the excretory organs (Malpighian tubules) open into the beginning of the hind gut and hence their products reach the exterior in the faeces. There is thus no excretory opening in the body wall of the insect.

It has already been mentioned that the reproductive organs may possess important taxonomic characters in their external genitalia. Some of the internal parts of the female reproductive system, being of ectodermal origin, have strongly sclerotized linings. These parts, especially the spermathecae, of which there may be one, two or three, can easily be seen in macerated specimens (see Section on Collection and

Preservation), and, in some groups of insects, such as Sand-flies and Fleas, have been used for separating species.

The respiratory system consists of a network of internal air-tubes known as tracheae, which ramify throughout the body and the appendages. Their finest branches, termed tracheoles, penetrate the tissues of the various organs. The air enters the tracheae through paired segmental openings or spiracles, placed in the pleural regions of the thorax and abdomen. There are typically two pairs of spiracles in the thorax—the mesothoracic and metathoracic spiracles, though the anterior one is often called the prothoracic because of its forward position. The abdomen may carry as many as eight pairs of spiracles but this number is often reduced. The tracheae are of ecto-dermal origin and are lined with cuticle which carries a thickened spiral thread serving to keep them distended. When the tracheae are filled with air, they present a silvery appearance. The spiral thread may become detached when dissections are made and cause some questioning until it is recognized.

CLASSIFICATION AND NOMENCLATURE

Insects are a Class of the Phylum Arthropoda, divided into two Subclasses, Apterygota and Pterygota. In the classification recognized here, there are twenty-eight Orders mostly placed in the second Subclass.

Subclass I: Apterygota. Here are included wingless insects of primitive structure with virtually no metamorphosis. The young resemble the adults in all essential features except sexual maturity and size. They mostly go through many moults and may continue moulting after reaching sexual maturity. Their wingless condition is considered to be primitive, i.e., at no time in their evolutionary history have they ever possessed wings. Some modern authorities give Orders 2–4 Class status.

Order 1. Thysanura   (Silver Fish, Bristle-tails, Fire Brats).
  ,,   2. Diplura   (some Bristle-tails).
  ,,   3. Protura   (microscopic soil insects, no common name).
  ,,   4. Collembola   (Spring-tails).

Subclass II: Pterygota. This includes all winged insects and those wingless ones, such as lice and fleas, that are considered to be secondarily apterous, i.e., winged forms have occurred at some point in their evolutionary history. Their metamorphosis is varied, but it is rarely only slight or wanting.

Division 1. **Exopterygota** (or **Hemimetabola**). The young, usually termed nymphs, bear a close resemblance to their parents, with the wings appearing in the later stages of nymphal life as pads developing externally on the thorax (hence the name Exopterygota). There may be a change of habitat when the adult stage is assumed but there is rarely a resting stage or pupa. The changes that take place when the nymph becomes an adult are usually sufficient to be called a partial metamorphosis.

Order 5. Ephemeroptera (Mayflies).
  ,,   6. Odonata   (Dragonflies).

,,	7.	Dictyoptera	(Cockroaches and Mantids).
,,	8.	Isoptera	(Termites).
,,	9.	Plecoptera	(Stoneflies).
,,	10.	Grylloblattodea	(small rare group, no common name).
,,	11.	Dermaptera	(Earwigs).
,,	12.	Phasmida	(Stick and Leaf-insects).
,,	13.	Orthoptera	(Crickets and Grasshoppers).
,,	14.	Embioptera	(Web-spinners).
,,	15.	Zoraptera	(small rare group, no common name).
,,	16.	Psocoptera	(Psocids and Book-lice).
,,	17.	Thysanoptera	(Thrips).
,,	18.	Phthiraptera	(Lice).
,,	19.	Hemiptera	(Plant Bugs, Bed Bugs, Plant-lice, Scale Insects, etc.).

Division 2. **Endopterygota** (or **Holometabola**). The young, termed larvae, bear little resemblance to their parents and the change to adult is complex and always includes a resting stage or pupa during which the insect does not feed and is apparently quiescent, although great internal changes are taking place. In this group the wings develop internally in the larva (hence Endopterygota) and do not become visible externally until the pupal stage. The changes that take place are said to be a complete metamorphosis.

Order	20.	Neuroptera	(Alderflies, Lace-wings, Ant-lions, etc.).
,,	21.	Mecoptera	(Scorpionflies).
,,	22.	Trichoptera	(Caddisflies).
,,	23.	Lepidoptera	(Butterflies and Moths).
,,	24.	Diptera	(Two-winged Flies, Gnats, etc.).
,,	25.	Siphonaptera	(Fleas).
,,	26.	Hymenoptera	(Bees, Wasps, Ants, etc.).
,,	27.	Coleoptera	(Beetles).
,,	28.	Strepsiptera	(rare insect parasites, no common name).

Nomenclature

The modern system of naming and classifying animals dates from the 10th Edition of Linnaeus' *Systema Naturae* (1758), in which appeared not only the first complete and orderly grouping of animals, but also a new system of nomenclature. It was Linnaeus who first devised the method of substituting single specific names for the descriptive phrases which up till then had been employed in combination with the words that are now known as generic names. Linnaeus recognized six 'Classes' of Animals, the fifth being the Insecta, the definition of which allowed the inclusion of a great many creatures no longer called insects, except popularly, such as spiders, mites, crabs and centipedes. His Class Insecta was divided into seven Orders each containing a number of genera of which each in its turn included numerous species.

With the accumulation of more knowledge and the great increase in the numbers of species described, new series of categories have at various times been introduced. The

Table 1 THE GROUPS OF INSECTS AND ARACHNIDS OF MEDICAL IMPORTANCE

Class INSECTA
Subclass Pterygota

Division Exopterygota

Order Dictyoptera — Cockroaches and Mantids.
Cockroaches spoil food. They can act as the intermediate hosts of certain helminths that sometimes occur in man.

Order Hemiptera — Bugs, Bed Bugs.
Mainly important because of the direct irritation from bites of blood-sucking species. Neotropical Cone-nose Bugs known to include vectors of Trypanosomes (Chagas' disease, etc.)

Order Phthiraptera — Lice, including the Head and Body Louse and the Pubic Louse or 'Crab'.
Commonly called vermin, give rise to irritation and, indirectly, to skin infections. Vectors of Typhus, Trench Fever and Louse-borne Relapsing Fever.

Division Endopterygota

Order Coleoptera — Beetles.
Rare cases of Canthariasis (invasion of the alimentary canal). Blister-Beetles. They can act as intermediate hosts of certain helminthes that sometimes occur in man. Some larvae cause urticaria.

Order Lepidoptera — Butterflies and Moths.
Rare cases of Scholechiasis (caterpillars in the alimentary canal). Certain caterpillars possess hairs which cause urtication on contact with the skin or mucous membranes. Some Moths feed on eye secretions, mainly of cattle; one sucks blood.

Order Hymenoptera — Wasps, Bees and Ants.
Venomous stings and bites.

Order Diptera — Flies and Gnats, Mosquitoes, Black-flies, Sand-flies, Midges, Horse-flies, Clegs, Forest-flies, Louse-flies, House-flies, Blow-flies, Blue bottle, Greenbottles, Hover-flies, Eye-flies, Warble-flies, Horse-bots, Tsetse-flies, etc. The larvae of many flies are commonly known as maggots.
Biting flies may cause direct irritation, but their most important role is the transmission of disease, e.g. Malaria, Dengue, Yellow Fever, Filariasis (Elephantiasis), Leishmaniasis, Trypanosomiasis (Sleeping Sickness, etc.). The infestation of tissues or organs by the larvae of Diptera is called Myiasis.

Order Siphonaptera — Fleas, including the Jigger or Chigoe.
Direct irritation. Plague transmission.

Class ARACHNIDA

Scorpions, Spiders, Ticks and Mites.

Scorpions and some spiders are venomous. Some mites are endo-parasites, others and all ticks are blood-sucking ectoparasites. Some ticks and mites are vectors of disease.

Textbooks of general and medical entomology are listed at the end of this chapter.

older 'Classes' are now termed Phyla, the Class Insecta of Linnaeus becoming the Phylum Arthropoda, and including four large Classes—Crustacea, Myriapoda, Insecta and Arachnida, with some smaller ones of no medical importance. In this present book we now recognize twenty-eight Orders instead of the seven recognized by Linnaeus. In addition between Order and Genus there have come into use additional grouping such as Sub-Order, Division, Superfamily, Family, Subfamily, Tribe. For all practical purposes the most important groups are Order, Family, Genus and Species.

Properly cited, the name of an insect should consist of the genus name followed by the specific (or trivial) name which may be followed by the name of its author, either in full, or abbreviated if there is no chance of confusion, with no comma between, e.g., *Anopheles maculipennis* Meigen, *Musca domestica* L. The genus always has a capital initial letter, the species always a small one. If the date of description is cited, this should follow and be separated by a comma. The author's name (especially with the date) affords a clue to the publication in which the original name and description appeared. If the insect is placed in a genus other than the original one then the author's name (and date) are placed in brackets, thus: *Anopheles claviger* (Meigen, 1804). The author's name is often omitted for well-known insects; if a name is quoted frequently in a work, then the author may be given on the first citation and omitted afterwards.

The dates of publication are of great importance since the name to be used is the name under which the insect was first described (Law of Priority). It often happens that an insect, especially a common one, has been named and described independently several times over by different authors. In such cases only the earliest name is valid and all the others are synonyms of it. Where the same name has been used inadvertently more than once, these are homonyms. To avoid confusion, the later (or junior) homonym is suppressed and either a new name or the earliest synonym used.

So many difficulties have arisen in practice that an International Code of Zoological Nomenclature has been drawn up by a Commission set up by the International Congresses of Zoology. The latest edition was revised and adopted by the XVth Congress of Zoology held in London in 1958. It was published in 1961 and amendments were adopted by the XVIth Congress of Zoology at Washington in 1963. Anyone proposing to pursue studies in the field of systematic Zoology should acquaint himself with this Code.

The formation of the scientific names of animals is subject to certain rules and recommendations, laid down in the Code. They should be words which are either Latin or Latinized, or considered and treated as such in cases where they are not of classical origin. If derived from the names of people then a specific name should be in the genitive case, i.e., end in -*i* for a man and -*ae* for a woman; specific names always begin with a small letter regardless of their origin, they should also agree in gender with the generic name. Diacritic marks (accents etc.) should never be retained.

So far as higher groups are concerned, family names always end in '-idae', subfamily names in '-inae', superfamily names in '-oidea'. The ending '-ini' signifies a tribe, which is of lower rank than a subfamily. The majority of the Orders of insects have names ending in '-ptera', e.g., Diptera.

One of the more important results of the strict application of the Code is the not infrequent suppression of well-known names of common animals in favour of earlier described names of which they are found to be synonyms. Examples are the supression of *Anopheles mauritianus* Grandpré and Charmoy, 1901 in favour of *Anopheles coustani* Laveran, 1900, a synonymy settled in 1932. A similar case is that of the suppression of the well-known name *Anopheles fuliginosus* Giles, 1900 in favour of *Anopheles annularis* Wulp, 1884 of which *fuliginosus* was recognized to be a synonym only in 1932. It is possible to apply to the Commission on Zoological Nomenclature to retain the later name in cases where great inconvenience or confusion would result from a change of name and so nowadays, there are fewer changes of this type.

Difficulties with homonyms arise less frequently and when they occur they are usually quickly noticed and dealt with by renaming the species. Where homonyms were first described both in the same genus ('primary homonyms') the later name is suppressed for all time. However, where they occur by the bringing together of two species by an acceptance of wider generic limits ('secondary homonyms') the later name may be revived if the large genus is split up again by another author.

This brief outline is sufficient to indicate in broad outline some of the principles and pitfalls of nomenclatorial practice. Further information may be obtained from the published Code, from Mayr, Linsley and Usinger (1953), Blackwelder (1967) and Mayr (1969).

Variation, which is the essence of evolution, presents the taxonomist with many difficult problems concerning the limits of specific, subspecific and individual variation. Opinions concerning these limits are bound to differ but, pending their thorough investigation in the fields of genetics, cytology etc., these opinions must rest upon characteristics of the insects that are easily observed such as morphology and distribution on the earth's surface. The modern Code only recognizes two categories at the lowest level, that is, species and subspecies, the latter normally having a definite geographical distribution. Varieties and aberrations are not accorded any nomenclatorial status. Subspecific rank is indicated by the use of a trinominal, thus: *Anopheles aitkenii stantoni* Puri. It will often be found that authors differ in the rank accorded a species or subspecies, some regarding particular subspecies as of specific rank and *vice versa*.

There are also certain rules to be followed when describing a species as new. The formation of the actual name has been mentioned above, also the importance of ensuring that it has not been used before in the same genus. All the specimens available to the author are referred to as the 'type series', and the description should be based on all of them. A typical one, in good condition, and of the sex with the most distinctive characters, is then selected and labelled 'Holotype'. This specimen is the name-bearing specimen if it is found later that the series consists of more than one species. The holotype should receive special mention at the end of the description with its full collecting data and the Museum or collection where it is deposited. If this is not done, all the type series are accorded equal rank as 'syntypes' and a later author is at liberty to make a selection of a name-bearing specimen, now termed a 'lectotype', though he must publish such a selection. When a holotype has been correctly designated the remainder of the series are termed 'paratypes'; when a lectotype has been subsequently designated the remainder of the series become paralectotypes.   In addition, mention should be made of the difference between the new species and its close allies.   Wherever possible, dichotomous keys should be included, especially when more than one species is described.

ZOOGEOGRAPHY

High mountain ranges, oceans and deserts provide barriers to the dispersal of terrestrial animals; insects seem to be especially affected by desert areas.   The limits imposed by these factors and by climatic and other conditions, have resulted in the development of the faunas of the larger land masses upon lines so characteristic that zoologists have been able to define six main faunistic Regions (fig. 2).

The Palaearctic Region is approximately the whole of the Old World north of a line running through the Saharan and Middle Eastern deserts and along the Himalayan chain to the China Sea by way of the Yang-tse-Kiang River.   The Nearctic Region is roughly America north of Mexico.   These two Regions show many points of similarity, and there are numerous species common to both; they are frequently given the comprehensive term Holarctic.

The remainder of Africa and Madagascar form the Ethiopian Region, which is now sometimes called 'Africa south of the Sahara' or subsaharan Africa.   The Oriental Region comprises all the tropical eastern lands and islands, except for Australia, New Guinea and certain neighbouring islands, which, together with New Zealand and the Pacific Islands, are known collectively as the Australasian Region.

The Regions are based largely on the distribution of mammals and birds and have not always been found to apply quite so convincingly to insects.   They were first worked out by Sclater (1858) and adopted by A. R. Wallace (1876) in his great work on animal distribution and the map in fig. 2 shows the Regions as defined by him with their major Sub-Regions.

There will seldom be any doubt as to the Region into which any given country falls except along the boundaries of the faunistic Regions.   In these areas there is inevitable infiltration of elements from one Region to another and care must be taken in using faunistic keys to species.   Particular places where difficulties occur are in the Middle East where the Palaearctic and Oriental Regions meet the Ethiopian; in China, where Palaearctic and Oriental Regions meet; in Mexico, where there is overlap between Nearctic and Neotropical Regions.

Probably one of the most confusing areas is the division between Oriental and Australasian Regions (see Scrivenor et al., 1944 and Mayr 1944).   The classical dividing line here is 'Wallace's Line' which runs through the Macassar and Lombok Straits and thus places Celebes and Timor etc., in the Australasian Region.   Lee and Woodhill (1944) suggest that the dividing line is immediately west of Molucca, Ceram, Wetter and Timor, i.e., more to the east, but it seems that the line varies greatly from group to group within the insects.   The dividing line between the Palaearctic and Oriental Regions in China is discussed by Chang (1965).

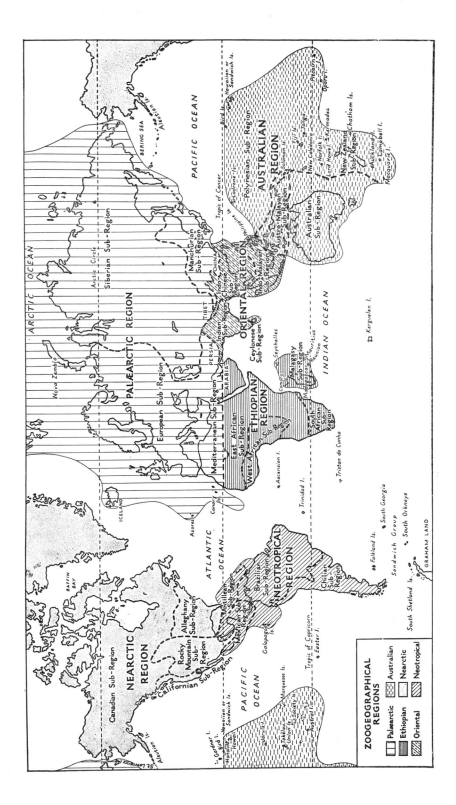

Fig. 2 Zoogeographical Regions, from Bartholomew, Eagle Clark and Grimshaw's Atlas of Zoogeography (Bartholomew's Physical Atlas, vol. 5, 1911).

METHODS OF COLLECTING AND PRESERVING INSECTS

These notes have been compiled for reference rather than with the idea of providing comprehensive instructions. No attempt has been made to list in detail the special apparatus that the medical entomologist will require if he is carrying out intensive mosquito surveys, etc., but works on this are given separately at the end of the general references to this chapter, the books by Oldroyd (1970) and Southwood (1966) being especially useful. Some techniques for specific insects are described in the relevant chapters.

The procedure for collecting and preserving adults of the larger insects of medical importance is essentially to kill, pin and dry the specimens and then to store them in such a way that they will be protected from mites, mould and other pests of the entomological collection, such as museum beetles. Adult mosquitoes are also dealt with in this way, but the smaller biting flies (midges, sand-flies, etc.) are more often placed in a fluid medium. Larval stages, eggs and pupae, fleas, lice and the majority of other ectoparasitic insects, and all Arachnida are also always preserved in fluid media.

The importance of labelling specimens cannot be over-emphasized. A specimen is of little value from the scientific point of view if it is without data telling where and when it was captured. In giving localities on labels it is helpful if some feature is mentioned that is marked on a map commonly in use, e.g., a town, hill or river. An indication of the type of terrain should be given if possible and also the height in hilly country. Insects collected *in coitu* should always be kept carefully together and the fact noted in writing on the label.

NETS. The usual type nowadays has a light folding aluminium frame fitted with a bag made of white organdie that should be deeper than it is wide. Collectors of Lepidoptera prefer the bags to be made of black, terylene, mosquito netting but the mesh of these is too coarse for smaller insects of medical importance and small insects are not easily seen against a black background. Most collectors do not use a long handle except in special circumstances. Nets for use in water are best made from a fine nylon material or else from bolting silk, which is extremely expensive.

KILLING BOTTLE. The familiar entomologists' killing bottle uses potassium or sodium cyanide as the lethal agent. For larger insects a bottle of 300 cm³ or 10 fluid ounces capacity is needed but for small insects a corked tube of 2·5 cm diameter has advantages. The cork, bung or stopper must be close-fitting and the cyanide must be of a coarse commercial grade for the best results. Crushed cyanide is placed in a layer 0·5 cm deep at the bottom of the bottle or tube. This is covered with dry plaster of paris—again of a coarse commercial grade, and the two mixed thoroughly. Another 0·5 cm layer of dry plaster is followed by a 0·5 cm layer of plaster, again, coarse commercial, mixed with water to a thick cream. Bubbles that appear during setting should be burst and the cavities filled up. The bottles should be left, uncorked, in a safe, airy place overnight and allowed to 'sweat' and dry before use. Subsequent 'sweating' can be counteracted by putting discs of blotting or filter-paper on top of the hardened plaster of paris.

It cannot be too strongly emphasized that potassium and sodium cyanide are extremely lethal to vertebrates as well as to insects and that great precautions must be taken during preparation and use of these killing bottles. Especially during the drying process, they must be placed where unauthorized people, particularly children, cannot gain access to them. Great precautions must also be taken in the destruction of old bottles, which, even though insects may hardly die in them, are still highly lethal to humans.

Other, safer, killing agents frequently used by entomologists include ethyl acetate, tetrachlorethane, carbon tetrachloride, benzine and chloroform. The last two render the insects very stiff. A simple killing bottle consists of a layer of plain, set and dried plaster of paris at the bottom of the bottle or tube on to which a few drops of one of these fluids is placed from time to time. This requires drying out after some use because the plaster becomes saturated with moisture from the insects' bodies. Insects can also be killed by placing a drop of one of these fluids on the cork of a collecting tube while it is momentarily removed. If the killing bottle is made of a transparent plastic, this is sometimes attacked by the fluids, especially ethyl acetate, and rendered non-transparent.

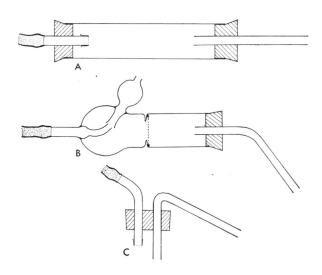

Fig. 3   Insect aspirators or 'pooters'. A, Straight sucking type; B, [blow-type; C, angled sucking type (collecting tube removed).

SUCTION-TUBES, EXHAUSTERS OR ASPIRATORS. These (fig. 3) are simple pieces of apparatus, sometimes termed a 'pooter', used for sucking up small insects from leaves, stones, window-panes etc., or out of the net. They are particularly useful for collecting mosquitoes, etc., on domestic, farm or other premises. Suction is usually applied by the mouth, but a rubber bulb fitted with suitable valves may be used. There are two usual forms of the apparatus.

The first (fig. 3A) consists of a wide-bore glass or plastic tube of a suitable length, which may be anything from 7–18 cm, with a cork or rubber bung at either end, through which is inserted a smaller glass tube. A piece of gauze is fastened over the tube to which suction is applied, to keep the insects in the wide tube and prevent them getting into the operator's mouth. A piece of rubber tubing is attached to this tube, often with a mouthpiece at the other end. The insects are sucked in through the tube at the other end.

In the second form a bottle or solid-bottomed tube is used and both tubes pass in through the cork (fig. 3C). The inlet tube is bent at an angle to suit the operator and the end in the bottle should reach near the bottom. The suction tube should be short and end just below the cork.

It is worth bearing in mind that some of the insects aspirated may be species involved in cases of myiasis. Hurd (1954) records a case of infestation of the sinuses by Coleoptera, Lepidoptera and mites after a prolonged use of the aspirator. To obviate this danger a blow-pooter (fig. 3B) has recently been designed which is much more pleasant to use, especially in forensic work when collecting from cadavers.

The best way of dealing with the delicate insects caught in suction tubes and bottles is to stupefy them by drawing in ethyl acetate fumes. This, however, may attack plastic tubes, in which case chloroform or tobacco smoke are preferable. Care must of course be taken not to suck up by the mouth any dangerous killing agent.

Sucking tubes on the ends of sticks can be used for collecting insects from walls or ceilings. When this is done, it is of course necessary to lengthen the suction tube and not the inlet tube since this would greatly increase the amount of air that would have to be aspirated.

For general collecting a variety of items are necessary such as pill-boxes (often made completely of plastic nowadays), corked glass tubes of various sizes, fine-pointed forceps, pipettes and a pocket lens.

BIBLIOGRAPHY

General

ALICATA, J. E. 1969. *Parasites of man and animals in Hawaii.* 190 pp. Basel and New York.

ALLINGTON, H. V. & ALLINGTON, R. R. 1954. Insect Bites. *J. Amer. med. Ass.* **155**: 240–247.

ASKEW, R. R. 1971. *Parasitic Insects.* 316 pp. London.

BARTHOLOMEW, J. G., CLARKE, W. E. & GRIMSHAW, P. H. 1911. Atlas of Zoogeography. *Bartholomew's Physical Atlas* 5.

BEKLEMISHEV, V. N. 1958. Identification des insectes nuisibles à la santé de l'homme. Moscow [In Russian].

BLACKWELDER, R. E. 1967. *Taxonomy.* 698 pp. New York, London and Sydney.

BRUES, C. T. & MELANDER, A. L. 1954. Classification of Insects. 2nd Edn. 917 pp. Cambridge, Mass.

BULLETIN ANALITIQUE d'ENTOMOLOGIE MÉDICALE ET VÉTÉRINAIRE. Paris [titles only].

BUSVINE, J. R. 1966. *Insects and hygiene* (2nd Edn.). 467 pp. London.

CASTELLANI, A. & CHALMERS, A. J. 1913. *Manual of tropical medicine.* 1747 pp. London.

CHANDLER, A. C. 1949. *Introduction to parasitology with special reference to the parasites of man.* 756 pp. New York and London.

CHANG, S. M. 1965. A discussion on the line of demarcation of the Palaearctic and Oriental regions east of Chinling based on the knowledge of the distribution of some agricultural insects. *Acta ent. sin.* **14** (4): 411–419 [in Chinese with English summary].

CHAPMAN, R. F. 1969. *The Insects. Structure and Function.* 819 pp. London.

CHOW, C. Y., HUANG, T. C. & YUE, T. F. 1950. Bibliography of Chinese Arthropods of medical and veterinary importance. *Q. Jl Taiwan Mus.* **3**: 157–185.

CHU, H. F. 1949. *How to know the immature insects.* 234 pp. Duboque.

CROWSON, R. A. 1970. *Classification and Biology.* 350 pp. London.

C.S.I.R.O. 1970. *The Insects of Australia.* 1029 pp. Melbourne.

CUSHING, E. C. 1957. *History of Entomology in World War II.* 117 pp. Washington.

DARLINGTON, P. J. 1957. *Zoogeography: the geographical distribution of animals.* 675 pp. New York and London.

DARLINGTON, P. J. 1965. *Biogeography of the southern end of the world.* 236 pp. Cambridge, Mass.

FERRIS, G. F. 1928. *Principles of Systematic entomology.* 169 pp. New York and London.

FURMANN, D. P. & CATTS, E. P. 1970. *Manual of Medical Entomology,* 3rd edn. 163 pp. Palo Alto.

GORDON, R. M. & LAVOIPIERRE, M. M.-J. 1962. *Entomology for students of medicine.* 333 pp. Oxford.

GRASSÉ, P. P. 1951. *Traité de zoologie.* 10. 1940 pp. Paris.

GRESSITT, J. L. 1958. Zoogeography of insects. *A. Rev. Ent.* **3**: 207–230.

HAESELBARTH, E., SEGERMAN, J. & ZUMPT, F. 1966. *The arthropod parasites of vertebrates in Africa south of the Sahara,* **3**. 283 pp. Johannesburg.

HERMS, W. B. 1961. *Medical entomology,* 5th edn. (revised M. T. James). 616 pp. New York.

HOCKING, B. 1971. Blood-sucking behaviour of terrestrial arthropods. *A. Rev. Ent.* **16**: 1–26.

HORSFALL, W. R. 1962. *Medical Entomology.* 467 pp. New York.

HULL, T. G. 1963. *Diseases transmitted from animals to man,* 5th edn. 967 pp. Springfield.

IMMS, A. D. 1957 (reprinted 1964). *A general textbook of entomology,* 9th edn. (revised O. W. Richards & R. G. Davies). 886 pp. London.

INTERNATIONAL CODE OF ZOOLOGICAL NOMENCLATURE, 2nd edn. 1963. 176 pp. London.

JAMNBACK, H. 1969. Blood-sucking flies and other outdoor nuisance arthropods of New York State. *Mus. Bull. N.Y. St. Mus. Sci. Serv.* Memoir **19**. 90 pp.

JENKINS, D. W. 1964. *Pathogens, parasites, and predators of medically important arthropods.* *Bull. Wld. Hlth Org.* **30**. Suppl. 150 pp.

KERRICH, G. J., MEIKLE, R. D. & TEBBLE, N. (Eds). 1967. Bibliography of key works for the identification of the British Fauna and Flora. *Systematics Association* Pubn No. **1** (3rd edn). 186 pp. London. [Insects pp. 57–96, Arachnids 97–104.]

LAPAGE, G. 1962. *Mönnings Veterinary Helminthology and Entomology,* 5th edn. 600 pp. London.

LECLERQ, M. 1969. *Entomological parasitology.* 158 pp. London, New York, etc.

LEE, D. J. & WOODHILL, A. R. 1944. Some new records and new synonymy of Australia species of *Anopheles* (Dipt., Culicidae). *Proc. Linn. Soc. N.S.W.* **69**: 67–72.

MACKIE, T. T., HUNTER, G. W. & WORTH, C. B. 1954. *A manual of tropical medicine,* 2nd edn. 907 pp. Philadelphia and London.

MANI, M. S. 1968. *General entomology.* 501 pp. Calcutta, Bombay and New Delhi.

MANSON-BAHR, P. H. 1968. *Manson's tropical diseases,* 16th edn. 1131 pp. London.

MARIANI, M. 1952. *Compendio di entomologia medica.* 201 pp. Palermo.

MARTINI, E. 1946. *Lehrbuch der medizinischen Entomologie.* 633 pp. Jena.

MATHESON, R. 1950. *Medical entomology,* 2nd edn. 612 pp. Ithaca, New York and London.

MAYR, E. 1942. *Systematics and the origin of species.* 334 pp.

MAYR, E. 1944. Wallace's Line in the light of recent zoogeographic studies. *Q. Rev. Biol.* **19**: 1–14.

MAYR, E., LINSLEY, E. G. & USINGER, R. L. 1953. *Methods and principles of systematic zoology,* 2nd edn. 328 pp. New York.

MAYR, E. 1969. *Principles of systematic zoology.* 428 pp. New York.

MIHALYI, F. 1963. Human- und veterinärmedizinische Bedeutung und Erfolg der entomologischen Forschungen in Ungarn. *Folia ent. Hung.* **16**: 387–400.

MUNRO, J. W. 1966. *Pests of stored products.* 234 pp. London.

NEVEU-LEMAIRE, M. 1938. *Traité d'entomologie médicale et vétérinaire.* 1339 pp. Paris.

PATTON, W. S. & EVANS, A. W. 1929. *Insects, ticks, mites and venomous animals.* 1. 786 pp. Liverpool.

PAVLOVSKI, E. N. 1946–1948. Manual of parasitology of man with discussion of the theory of the vectors of transmissible diseases. *Izd. Akad. Nauk S.S.S.R.* 2 vols, 1022 pp. Moscow [In Russian].

C

PETERSON, A. 1957. *Larvae of insects*, 3rd edn. 2 parts. 416 pp. Columbus.

REVIEW OF APPLIED ENTOMOLOGY, SERIES B. London. [A monthly review of medical and veterinary entomological literature including Arachnida, published by the Commonwealth Institute of Entomology, 56 Queen's Gate, London, S.W.7.]

RILEY, W. A. & JOHANNSEN, O. A. 1932. *Medical entomology*. 476 pp. New York and London.

SCLATER, P. L. 1858. On the general geographical distribution of the members of the class Aves. *Proc. Linn. Soc. Lond.* **2**: 130–145.

SCRIVENOR, J. B., *et al.* 1944. A discussion on the biogeographic division of the Australian archipelago, with criticisms of the Wallace and Weber lines and of any other dividing lines and with an attempt to obtain uniformity in the names used for the divisions. *Proc. Linn. Soc. Lond.* **1941–2**: 120–165.

SEN, S. K. & FLETCHER, T. B. 1962. *Veterinary entomology and acarology for India.* 668 pp. New Delhi.

SMART, J. 1965. *A handbook for the identification of insects of medical importance*, 4th edn. 303 pp. London.

SNODGRASS, R. E. 1935. *Principles of insect morphology.* 667 pp. London and New York.

TORRE-BUENO, J. R. 1937. *A glossary of entomology.* 336 pp. + suppl. A, 36 pp. New York.

TROPICAL DISEASES BULLETIN. London. [Titles, abstracts and reviews.]

WALLACE, A. R. 1876. *The geographical distribution of animals.* 2 vols, 503 + 607 pp. London.

WIGGLESWORTH, V. B. 1965. *The principles of insect physiology*, 6th edn. 741 pp. London.

WIGGLESWORTH, V. B. 1964. *The life of insects.* 360 pp. London.

ZOOLOGICAL RECORD, London. [An annual international bibliography of zoological literature published by the Zoological Society of London and the Commonwealth Institute of Entomology, London.]

Collecting and preserving

ANON. 1972. *Instructions for Collectors* No. 4a. Insects, 5th edn. B.M.(N.H.), London.

BIERNE, B. P. 1955. Collecting, preparing and preserving insects. *Science Service Entomology Division, Canada Department of Agriculture.* Pubn **932**. Ottawa.

ELTRINGHAM, H. 1930. *Histological and illustrative methods for entomologists.* 139 pp. Oxford.

GURNEY, A. B., KRAMER, J. P. & STEYSKAL, G. C. 1964. Some techniques for the preparation, study, and storage in microvials of insect genitalia. *Ann. ent. Soc. Am.* **57**: 240–242.

HAMMOND, H. E. 1960. The preservation of lepidopterous larvae using the inflation and heat-drying technique. *J. Lepid. Soc.* **14**: 67–78.

HOOD, J. D. 1940. *Microscopical whole mounts of insects*, 2nd mimeographed edn. Cornell University.

HURD, P. D. 1954. 'Myiasis' resulting from the use of the aspirator method in the collection of insects. *Science, N.Y.* **119**: 814–815.

OLDROYD, H. 1970. *Collecting, preserving and studying insects*, 2nd edn. 336 pp. London.

OMAN, P. W. & CUSHING, A. D. 1948. Collection and preservation of insects. *Misc. Publs U.S. Dep. Agric.* **601**. 42 pp.

SABROSKY, C. W. 1966. Mounting insects from alcohol. *Bull. ent. Soc. Am.* **12**: 349.

SMITH, G. E., BREELAND, S. G. & PICKARD, E. 1965. The Malaise trap—a survey tool in medical entomology. *Mosquito News* **25**: 398–400.

SMITH, K. G. V. (In press.) Insect whole mounts. IN Gray, P. (Ed.). *Encyclopaedia of Microscopy and Microtechnique.* New York.

SOUTHWOOD, T. R. E. 1966. *Ecological methods.* 391 pp. London.

2. DIPTERA—INTRODUCTION
(Flies, Gnats, Midges, etc.)

by Paul Freeman

THE Diptera can be distinguished from most other insects by the possession of only one pair of functional wings: immediately behind the wings arises a pair of club-shaped organs called halteres which are derived from the posterior pair of wings and control stability in flight (fig. 11). The great majority of Diptera are quite harmless, but the habit of sucking blood is found in a number of families and amongst these are important vectors of disease-causing organisms. Others, by virtue of their feeding habits may convey pathogenic organisms from decaying organic matter to food intended for human consumption.

The metamorphosis of the Diptera is of the complete type. The larval stages of many species live in decaying organic matter and may be accidentally swallowed in food and some may feed upon human blood or parasitize man as well as other vertebrate animals (myiasis, see Chapter 6).

Keys are given below (pages 30–36) for the identification of the families of adult Diptera. However, experience will soon enable the important groups (families, etc.) of blood-sucking and other medically important species to be recognized at sight. Notes on methods of collecting and preserving are given at the end of chapter 1.

STRUCTURE, INCLUDING WING VENATION

In adult Diptera (fig. 1) the body is clearly differentiated into the three main divisions— head, thorax and abdomen. The head is often large in relation to the size of the body and much of its surface is taken up by the compound eyes. Frequently, especially in the males, the compound eyes are so large that they meet in front, obliterating the strip of the head capsule that usually separates them. In this condition the head is said to be holoptic, in contrast to the dichoptic condition when the eyes are distinctly separated. On the top of the head there may or may not be three simple eyes (ocelli), which, when present, appear as shining, bead-like spots slightly raised above the surrounding area (fig. 8). Dipterous mouthparts are of a sucking type. In the blood-sucking species, the mouthparts are stylet-like and adapted for piercing the skin of the host (see page 35). In non-blood-sucking species, fluids and finely divided solids such as pollen grains are drawn up through the labium, which may be sponge-like.

The visible part of the thorax from above consists very largely of the mesothorax, the pro- and metathorax being correspondingly reduced in size. The semicircular posterior part of the mesothorax is called the scutellum and the larger anterior section or mesonotum may be further subdivided by a transverse suture (figs 10 and 11). The normal three pairs of legs are present and do not ordinarily present any great modifications (fig. 5).

The wings (fig. 4) of Diptera show features of great taxonomic importance. They are membranous and do not ordinarily bear large hairs (macrotrichia), or scales on the membrane. The wing membrane is, however, usually covered, at least in part, by the minute hairs termed microtrichia. The wing veins themselves and the wing margin carry macrotrichia to a greater or lesser extent. In Mosquitoes (family Culic-cidae) the wing veins always carry rows of flattened macrotrichia, or scales.

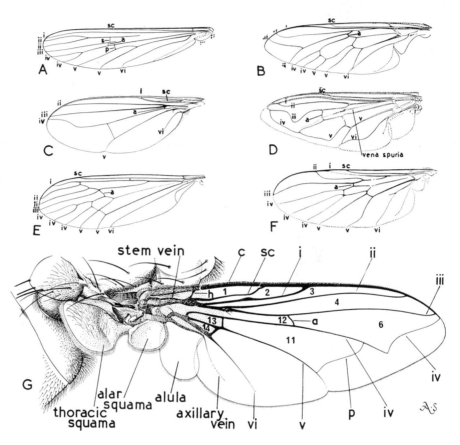

Fig. 4   Wings of Diptera showing some types of venation (any scales or markings omitted). A, *Anopheles maculipennis* (Culicidae); B, *Tabanus sudeticus* (Tabanidae); C, Doli-chopodidae; D, *Eristalis tenax* (Syrphidae); E, *Trichocera hiemalis* (Trichoceridae); F, *Sylvicola* (=*Anisopus*) *fenestralis* (Anisopodidae); G, *Calliphora vicina* (=*erythro-cephala*) (Calliphoridae) (various magnifications).

The most important feature of the wings, for purposes of identification and classifi-cation, is the *wing-venation*: i.e. the pattern of the *wing-veins*. This is quite astonish-ingly constant within one species, and often genera, tribes and families can be recognized immediately by some apparently insignificant detail. Even the small curvatures of different sections of a vein are characteristic, to the trained eye, and it is these that give an appearance of reality to a good drawing. When the venation is used for the purpose of identification, the smallest differences should be noted; sometimes they are individual aberrations, but this cannot be assumed.

The veins originate from branching *tracheae*, two of which run into the developing wing-bud. The veins thus fall into two groups, an anterior and a posterior: these are linked by the *anterior-* or *small cross-vein* (r-m), and this is a most useful landmark when interpreting a wing.

Table 2 VENATIONAL NOMENCLATURE

Fig. 4	Comstock-Needham Tillyard's modification	Verrall (1909: 57) (two systems)
VEINS		
C	C	costal and ambient vein
Sc	$\left\{ \begin{array}{l} Sc_1 \\ Sc_2 \end{array} \right\}$	mediastinal (auxiliary)
i	R_1	subcostal (first longitudinal)
ii	$\left. \begin{array}{l} R_{2+3} \\ R_{4+5} \end{array} \right\} R_s$ = praefurca	radial (second longitudinal)
iii		cubital (third longitudinal)
iv	M_{1+2}	discal (fourth longitudinal)
v	M_{3+4}	upper $\left. \right\}$ postical (fifth longitudinal)
	$\left\{ \begin{array}{l} Cu_1 \\ Cu_2 \\ 1A \end{array} \right.$	lower
vi		(shown as a fold, but not named)
		anal (sixth longitudinal)
axillary	ax	axillary
CROSS-VEINS		
h	h	humeral
a	r-m	discal (middle)
p	im	postical
—	base of M_3	lower (small)
—	m-cu	base of upper postical
CELLS		
1	1	costal
2	2	subcostal
3	3	marginal
4	4	first submarginal
—	5	second submarginal
6	6	first posterior
—	7	second posterior
—	8	third posterior
—	9	fourth posterior
—	10	fifth posterior
11	11	discal
12	12	first basal
13	13	second basal
14	14	anal
—	15	axillary

The membrane of the wing is pleated, like a fan, thereby achieving stiffness, and in principle the veins run along the creases, so that the veins are alternatively *convex* (i.e. on a ridge) and *concave* (i.e. in a valley).   This effect is more pronounced in the basal half of the wing, where it helps in identifying the main stems, but nearer the apex and hind margin the membrane is flatter, and the veins more branched.

If the veins are to be used in classification they must be given names, letters or numbers.   Three different systems are in use among dipterists:

(1) *The Schiner System.*   This is the oldest, and gives arbitrary names—radial, mediastinal, cubital, etc.—to the main stems, including their principal branches if any. In England this system was used by Verrall in *British Flies* (1901, 1909), along with the next system.

(2) *The 'longitudinal' system.*   Attempts to replace the arbitrary names of the Schiner system with a logical system of numbers: '1st longitudinal', '2nd longitudinal', etc.; or '1st vein', '2nd vein', etc.   The numbering starts with the first convex vein, ignoring the first concave vein, which is often interrupted, or even absent.

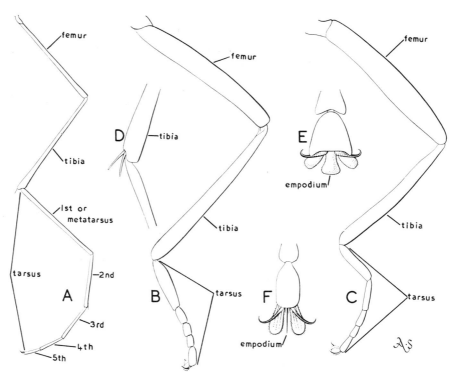

Fig. 5   Legs of Diptera.   A, Mosquito; B, Horse-fly (*Tabanus*); C, Flesh-fly (*Sarcophaga*); D, tibial-tarsal joint of hind leg of *Pangonia* (Tabanidae); E, foot of *Tabanus*; F, foot of robber-fly (Asilidae) (various magnifications).

(3) *The Comstock-Needham system.*   Introduced at the beginning of the century by Comstock and Needham, this system was intended to rationalize the naming of the veins in all Orders of insects.   All wings are seen as variants of a basic pattern, shown in fig. 4.   Unfortunately two of the names—subcostal and cubital—were also used in the Schiner system, but for different veins.

All three systems work equally well for a fairly complete venation, such as that of the Tabanid shown in fig. 4B, though even then, the conversion from one system to another is not easy, as Table 2 shows.   It is in the groups where the venation is greatly reduced that most trouble arises—what to call the veins that are left.

The user of this book will need to learn all three systems, since workers in different groups of medically important insects are very conservative in adhering to one or the other.

The veins divide the wing-membrane into areas known as *cells*, which are said to be 'open' if they reach to the wing-margin, and 'closed' if they do not.   The Comstock-Needham system names each cell after the vein which forms its anterior border—simple in theory, but difficult in practice, when the identity of the vein itself may be debatable. For the cells the old Schiner system remains the best.   At the base of the wing there are nearly always two closed cells—the 1st and 2nd *basal cells*—and beyond these, in the centre of the wing, the *discal cell*, one of the most valuable landmarks.   The other cells can be given names by derivation from fig. 4, even in wings with greatly reduced venation.

THE BIOLOGY OF DIPTERA

Diptera are holometabolous insects, with four stages: egg, larva, pupa and adult.   The larvae are always completely different in structure from the adult, with quite different requirements in habitat and food.   A few Diptera—chiefly Cecidomyiidae—practise paedogenesis, i.e. reproduce from larvae without the intervention of an adult stage; and a few others, tsetse-flies and Pupipara (Hippoboscidae, Streblidae, Nycteribiidae) have abolished the free larval stage, retaining the larva in a 'uterus' until it is ready to pupate.   With these exceptions Diptera live two lives in succession, as a larva and as an adult, in either of which they may become of medical importance if they attack man, or transmit disease to him in other ways.   Few Diptera are medically important both as larvae *and* as adults.

Larvae of Diptera nearly all require a fairly high degree of humidity, and therefore live in habitats that are moist to varying degrees.   Some are truly aquatic, in the sense of living continuously in water, without access to atmospheric air: Simuliidae and Chironomidae are the outstanding examples, together with Blepharoceridae, Deutero-phlebiidae and a few members of other families.   Mosquito larvae are generally thought of as aquatic, but they breathe atmospheric air through a siphon and—with occasional exceptions—pass their lives in the surface film.   Others which can survive immersion for a limited time, but not indefinitely, include Tabanidae, Stratiomyidae, Ephydridae and water-living Syrphidae.   Those few larvae that are able to obtain air from sub-merged plants are no more aquatic than a skin-diver.

If the complete absence of external spiracles is taken as indicating a fully aquatic larva, this category must include certain members of otherwise terrestrial families: e.g. *Ceroplatus* (Mycetophilidae), which lives on spray-soaked rocks, and in similar thin films of water.   The larvae of *Phlebotomus*, the adults of which are often associated with very arid regions, require a very high humidity approaching saturation, and this is produced by condensation of moisture in the deep recesses in which they live.

The larvae of many Nematocera and Brachycera and most Cyclorrhapha are terrestrial, but seldom live exposed to the open air like caterpillars. They live and feed in some soft medium, usually moist and decaying. Soil-living larvae feed on decaying vegetable matter, including fungus—e.g. Bibionidae, Mycetophilidae—but may some-

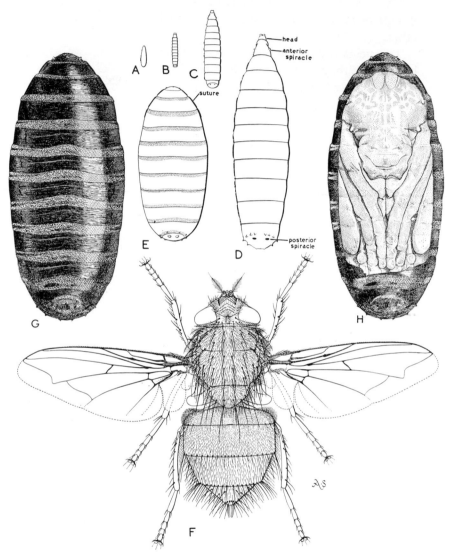

Fig. 6　The common Bluebottle, *Calliphora vicina* (=*erythrocephala*). A, egg; B, first instar larva; D, second instar larva; D, third instar larva; E, puparium; F, adult or imago (×5); G, enlarged and detailed view of puparium; H, puparium with portion removed showing the pupa inside.

times attack living plants—Tipulidae (Leather jackets)—or prey on other soil animals—e.g. Asilidae. Larvae of *Chrysops* (Tabanidae) live in mud beneath shallow water, and are vegetarian, in contrast to the carnivorous larvae of *Tabanus* and *Haematopota*. Larvae of the muscoid families (house-flies and blow-flies) choose media where decay has produced a high concentration of organic matter, mainly of vegetable origin in the case of house-flies and animal origin in blow-flies. The carnivorous and parasitic habits of screw-worms, bots and warbles have probably evolved from the latter.

All Dipterous larvae are legless (apodous), though many have pseudopods of various types, all being secondary adaptations. The larvae of the more primitive groups (Nematocera) have a well developed, sclerotised head with opposable mandibles. The head is reduced in the more advanced forms, culminating in the headless maggots of the Cyclorrhapha in which the mandibles are present as mouth-hooks that work side by side in a vertical plane. Further details of the structure of larvae, with special reference to identification are given in Chapter 6.

Pupae of the Nematocera are usually free, that is, the appendages are not stuck down and there is no protecting case (except in Simulidae where there is a cocoon). The pupae of mosquitoes are exceptional in being active swimmers. The higher groups show a progression towards the so-called coarctate pupa of the Cyclorrhapha in which the pupa is enclosed in a case termed a puparium, formed from the hardened last larval skin (fig. 6G, H). The adult fly escapes by forcing off a circular cap at the head end.

The biology of adult flies is concerned almost entirely with mating and egg-laying. The adult life has probably been reduced to its shortest in the Chironomidae, the adults of which emerge from freshwater ponds, lakes, rivers, etc. in great numbers; they seldom take food, and mate, oviposit and die within a day or two. To be able to do this, the larva must store up food, not only for its own needs, but also for those of the adult. Such a life-history confines the species to one type of habitat.

Most other Diptera have exploited the mobility of the adult, which enables them to occupy a habitat different from that of the larva. The mosquitoes are a particular example of this. Early control methods were concentrated on the larval habitat, such as altering the flow of water, cutting down or planting shade trees, and oiling the surface to kill the larvae. More modern control methods are directed against the adult flies in their resting places between blood-meals. Control of tsetse is partly directed against the pupation-sites in the ground, and partly against the resting-places of the adult.

For the purposes of medical entomology, the double life of the flies is important both among those of medical importance and those that are merely a nuisance. To underline this division between larva and adult we have dealt with them in different sections of the book.

CLASSIFICATION OF THE DIPTERA

The Diptera are divided into three sub-orders on characters shown by adults, larvae and pupae. Each sub-order contains a number of families and in the following account families of medical or domestic importance are printed in capitals; some other families are mentioned but many of the lesser families are omitted.

Sub-order NEMATOCERA

Antennae of adult many-segmented, often longer than head and thorax, most segments alike (fig. 7A, D, G); the antenna never bears a style or arista; palpi usually four or five segmented and with some exceptions (e.g. mosquitoes) tend to hang down. The larvae have well-developed heads and opposable biting mandibles. The pupae are free and the adult escapes by a dorsal longitudinal slit in the skin of the pupal thorax.

Families: TIPULIDAE, PHLEBOTOMIDAE, PSYCHODIDAE, CULICIDAE, CHIRONOMIDAE, CERATOPOGONIDAE, SIMULIIDAE, ANISOPODIDAE, Bibionidae, Scatopsidae, Mycetophilidae, Sciaridae, Cecidomyiidae.

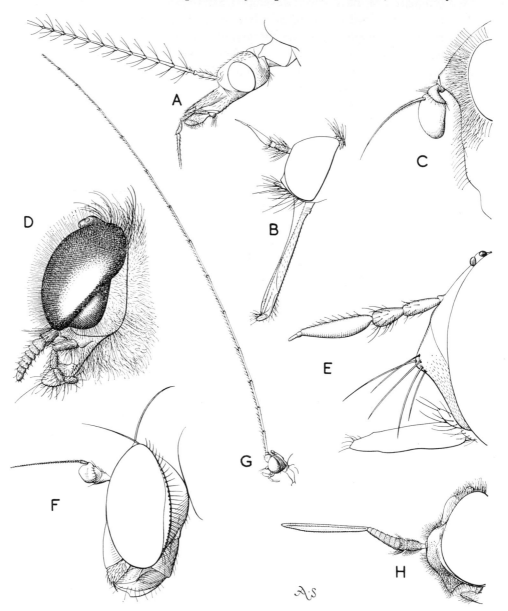

Fig. 7   Heads of Diptera. A, Tipulidae; B, Empididae; C, Syrphidae; D, Bibionidae; E, Asilidae; F, Dolichopodidae; G, Mycetophilidae; H, Stratiomyidae (various magnifications).

The Nematocera are essentially soft-bodied flies with a more normal insect type of antenna, in some families bushy in the males.   They include midges, gnats, mosquitoes, 'daddy-long-legs', etc.

Sub-order BRACHYCERA

Antennae shorter than the thorax, of variable structure, generally of three definite segments, the terminal one elongate and showing traces of annulation or else bearing a terminal style or arista (fig. 7B, E, F, H); palpi with one or two segments and projecting forwards. The larvae have incomplete, usually retractile, heads and vertically biting mandibles. The pupae are free or very rarely enclosed in the last larval skin.

Families: STRATIOMYIDAE, RHAGIONIDAE, TABANIDAE, Scenopinidae, Asilidae, Bombyliidae, Empididae, Dolichopodidae.

Many of the Brachycera are large and conspicuous insects, often rather rare and seldom seen by ordinary people. They may be of fearsome aspect and sometimes possess a 'beak' (fig. 7B) causing them to appear potentially dangerous if captured, and a few of these may, rarely and accidentally, actually bite human beings. These, however, are of no medical importance, it being only the Tabanidae and Rhagionidae that habitually attack man and other mammals to feed on blood. The brachycerous antenna shows an intermediate state between Nematocera and Cyclorrhapha.

Sub-order CYCLORRHAPHA

Antennae three-segmented with an arista or style carried dorsally on the last segment (fig. 7C); palpi with a single segment. The larvae are headless maggots with sclerotized mouth hooks working side by side in the vertical plane. The pupae are formed within the last larval skin which is sloughed from the pupa without being discarded and then forms a hardened case or puparium around the pupa proper (figs 6G, H).

Series ASCHIZA

Families: PHORIDAE, SYRPHIDAE and certain other small families.

Series SCHIZOPHORA

Section *Acalypterae:*
Families: DROSOPHILIDAE, CHLOROPIDAE, SEPSIDAE, PIOPHILIDAE, GASTEROPHILIDAE and many other small or medically unimportant families.
Section *Calypterae:*
Families: MUSCIDAE, ANTHOMYIIDAE, GLOSSINIDAE, CALLIPHORIDAE, SARCOPHAGIDAE, OESTRIDAE, Tachinidae.
Section *Pupipara:*
HIPPOBOSCIDAE, Streblidae, Nycteribiidae.

The Cyclorrhapha is a large group, consisting mainly of the insects normally called 'flies'. Many are closely similar to one another and their classification can be very difficult; the antennae, larvae and pupae show the highest stages of evolution in the Order.

The flies of the Series Schizophora are remarkable for the possession of an eversible sac or ptilinum, which protrudes from their heads and with which they push open the puparium and work their way through the soil beneath which they may have pupated. To allow of the extrusion of this bladder-like organ a flap falls forward in the centre of the face. During the hardening process, the ptilinum is withdrawn and ceases to have any function but its position is indicated by the frontal or ptilinal suture, which is horse-shoe shaped and encloses the antennae, ending at a variable distance from the mouth margin (fig. 9B).

KEY TO FAMILIES OF DIPTERA—ADULTS

This key is intended to assist in the recognition of adults of families of Diptera of medical importance. Some other families are mentioned, but most families of no medical importance are simply eliminated at appropriate places in the key. A simplified key to known blood-sucking families is given after the main key.

1   Antennae composed of two basal segments (of which the first may be reduced) and a flagellum of at least six, and usually more, similar segments (fig. 7A, D, G). Palpi with several segments, drooping. Anal cell of wing open, almost never narrowed towards wing margin (fig. 4A, E, F) .   .   .   .   NEMATOCERA   2

−   Antennae composed of two basal segments and a third compound segment formed by fusion of the elements of the flagellum. In some Brachycera the elements of the flagellum can still be seen (fig. 98C, D). Palpi with 1–3 segments, often held forwards. Anal cell narrowed towards wing margin, usually closed and often very much reduced towards base of wing (figs 4B, 97)   11

2   Dorsum of thorax with a V-shaped suture (fig. 8A), female with a horny ovipositor. Crane-flies, daddy-long-legs, winter gnats, etc.   .   .   .   .   .
      .   .   *Several families of which only Tipulidae are of slight medical importance*

−   Dorsum of thorax without a V-shaped suture, female without a horny ovipositor   .   3

3   Wings with fine net-like folds as well as the usual veins   .   .   .   .
    .   .   .   .   .   *Blepharoceridae etc., of no known medical importance*

−   Wings without fine net-like folds   4

4   Ocelli present (fig. 8C). March-flies, gall midges, fungus midges .   .   *Several families of which only Anisopodidae and Scatopsidae are of slight medical importance*

−   Ocelli absent   5

5   Costal vein passing round and beyond apex of wing; wings often hairy or scaly .   .   6

−   Costal vein ending before apex of wing; wings usually appearing bare   .   .   9

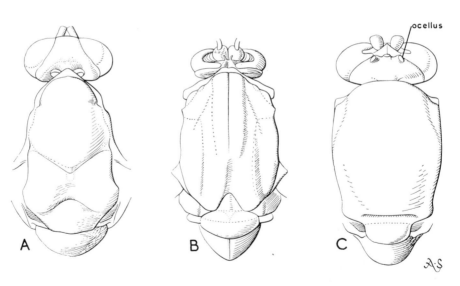

Fig. 8   Heads and thoraces of Nematocera. A, Tipulidae; B, Chironomidae; D, Mycetophilidae. Note the V-shaped suture in A; the longitudinal groove in B; the absence of a groove or suture in A; the presence of ocelli in C and their absence in A and B (various magnifications).

6   Wings broadly ovate or pointed, often short, never held flat over back when at rest, no cross veins except sometimes near base so that venation consists mainly of longitudinal veins, with few forks; Sc very short, weak, ending free. Second antennal segment not enlarged. Small flies with densely hairy body and wings .   7

–   Wings long and narrow, held flat over back when at rest; Sc ending in costa beyond middle of wing. Second antennal segment enlarged   8

7   Mouthparts adapted for blood-sucking, mandibles and maxillae of females elongated for piercing; labellum elongate and not bulbous; radius with 5 branches (fig. 9F) (Sand-flies) . . . . **PHLEBOTOMIDAE**

–   Mouthparts not adapted for blood-sucking except in a very few species (*Horaiella* and *Sycorax*) in which the maxillae are not known to be long, and the radius has 4 branches (R_2 and R_3 fused into a single vein) (fig. 93F) (Moth-flies) . . . . **PSYCHODIDAE**

8   Wing margin with nine terminations of long veins or their branches; mouthparts in form of long sucking proboscis (Mosquitoes) . . . . **CULICIDAE**

–   Wing margin with less than nine such terminations or else mouthparts not in form of long sucking proboscis . . . *Small families and gall midges mostly of no known medical importance (Chaoboridae* and *Dixidae*, see Chapter 3a)

9   Wings broad, body thick-set, antennae shorter than thorax and composed of almost cylindrical segments closely set together and never plumose (figs 81, 82) (Black-flies) . . . . **SIMULIIDAE**

–   Not with this combination of characters   10

10   Wings narrow and in rest held either apart or roof-like at the sides of the body; legs thin, fore-legs often particularly long; most posterior part of thorax (postnotum) with a longitudinal groove; mouthparts reduced and incapable of biting (figs 8B, 96) non-biting midges . . . . . . **CHIRONOMIDAE**

–   Wings broader, held flat on top of each other in rest, often spotted in blood-sucking species (fig. 94A); postnotum without groove; mostly minute insects with piercing and sucking mouthparts (Biting Midges) . . **CERATOPOGONIDAE**

11   Last tarsal segment bearing three pads and a pair of claws (fig. 5E) . . . . Part of BRACHYCERA   12

–   Last tarsal segment bearing two pads, with or without a normal bristle-like empodium, and with a pair of claws (fig. 5F) . . . . *Part of* BRACHY-CERA (*Scenopinidae* etc.), *of no known medical importance* and CYCLORRHAPHA   15

12   Antennae variously shaped but never with a bristle-like style or arista, the distal part shows rings or annuli (fig. 98C, D)   13

–   Antennae usually with a terminal bristle-like style or arista, when absent the distal part never shows rings or annuli (figs 7C, 101)   14

13   Squamae of wings large (Horse-flies, Clegs) (fig. 100) . . . **TABANIDAE**

–   Squamae small . . . . *Of no known medical importance*

14   Squamae not exceptionally large and lobate; venation normal (fig. 101) (Snipe-flies) . . . . **RHAGIONIDAE**

–   Squamae exceptionally large and lobate, or if small, then venation shows a net-like arrangement towards the tip of the wing . *Of no known medical importance*

15   Somewhat crab-like or spider-like flies with a leathery or horny integument, some winged. Ectoparasites of birds or mammals which occasionally find their way on to man (fig. 134B, C) . . . . *PUPIPARA*   16

–   Not like this   17

16   Parasites of bats. Either completely wingless, long-legged and spider-like with a small head sunk into a groove on the dorsal surface of the thorax (Nycteribiidae) or wings present, although sometimes reduced, head anterior on thorax, legs generally short and stout (Streblidae) . . **NYCTERIBIIDAE & STREBLIDAE**

- Parasites of other mammals and birds. Wings fully developed, reduced, absent or caducous, head anterior on thorax, legs short and stout (fig. 134B, C) . . . . **HIPPOBOSCIDAE**

17 Ptilinal (frontal) suture short, semi-circular, indistinct or absent (fig. 9A); front of head between eyes uniformly sclerotized without differentiation of a median stripe . . . . *ASCHIZA*   18

- Ptilinal (frontal) suture distinct, long and horse-shoe shaped (fig. 9B); front of head between eyes differentiated to form a median stripe . . *SCHIZOPHORA*   20

18 Wing veins markedly thickened towards fore-margin of wing; remaining veins all run markedly parallel to each other; small flies of a characteristic bristly appearance (fig. 102) . . . . **PHORIDAE**

- Not like this 19

19 Wings with false vein (*vena spuria*) (fig. 4D) and a tendency for the veins to form an internal margin round the wing; often brightly coloured and bare, though some are bee-like (Hover-flies) . . . . **SYRPHIDAE**

- Not like this . . . . *Of no known medical importance*

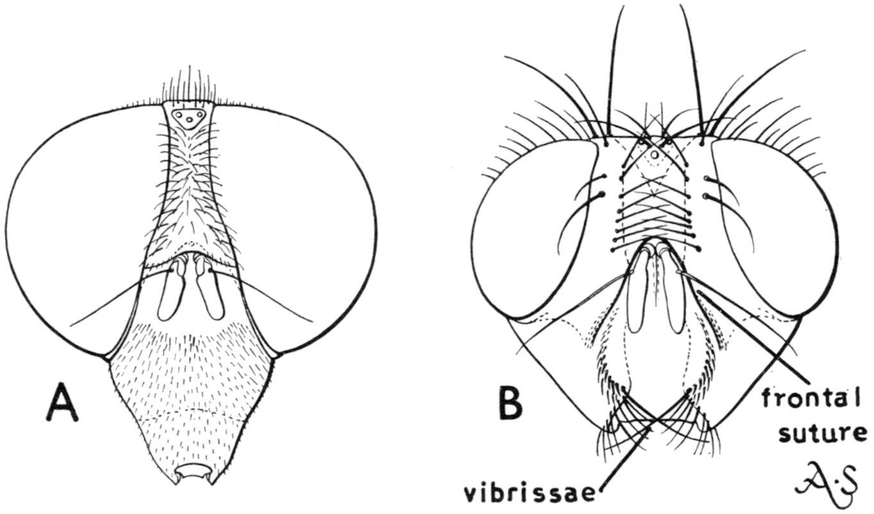

Fig. 9   Heads of A, a Syrphid (Aschiza); B, a Calliphorid (Schizophora) to show the presence of a frontal suture in the latter and its absence in the former. Note also the vibrissae. The antennae of both are typical of the Cyclorrhapha.

20 Second antennal segment without a longitudinal cleft on its upper outer edge (fig. 10D); thorax usually without a distinct transverse suture (fig. 10B), posterior calli usually undeveloped except in *Gasterophilus*; squamae usually small. Mostly small flies with eyes well separated in both sexes. A large group of many families, few of medical importance . . . . *ACALYPTERAE*   21

- Second antennal segment with a distinct longitudinal cleft on its upper outer edge (fig. 10C); thorax usually with a distinct transverse suture (fig. 10A), posterior calli developed; squamae usually conspicuous (fig. 4G). Mostly medium to larger flies, males frequently with eyes approximated or touching . *CALYPTERAE*   31

21 Brown, furry flies about 15 mm. long, with vestigial mouthparts sunk in a tiny oral pit; all transverse veins confined to basal half of wing (fig. 134A) (Horse Bot-flies) . . . . **GASTEROPHILIDAE**

- Not like this 22

22 Costa complete 23
- Costa with at least one break 25

23 Palpi vestigial, abdomen usually waisted anteriorly, apical wing spot may be present
    .   .   .   .   **SEPSIDAE**

  – Palpi well developed   24

24 Tibiae usually with a pre-apical dorsal bristle and the last tarsal segment broadened.
    Posterior vertical bristles convergent and three to four fronto-orbital bristles
    present.   Vein 6 reaches the hind margin of the wing (Seaweed-flies)   .   .
    .   .   .   .   **COELOPIDAE**

  – Without the above combination of characters   *Families of no known medical importance*

25 Anal cell and sixth longitudinal vein absent, vibrissae lacking   .   .   .   26

  – Anal cell and sixth vein present, vibrissae normally present   28

26 Costa with only one break, ocellar triangle large and plate-like, frons without strong
    bristles.   Subcostal vein vestigial (fig. 133C) (Eye-flies)   .   **CHLOROPIDAE**

  – Costa with one or two breaks (fig. 133A), other characters not like this   .   .   .   27

27 Costa with two breaks; mouth opening usually large; hairs on the arista, when present,
    only on the upper surface   .   .   .   .   .   .   **EPHYDRIDAE**

  – Not with this combination of characters   .   *Families of no known medical importance*

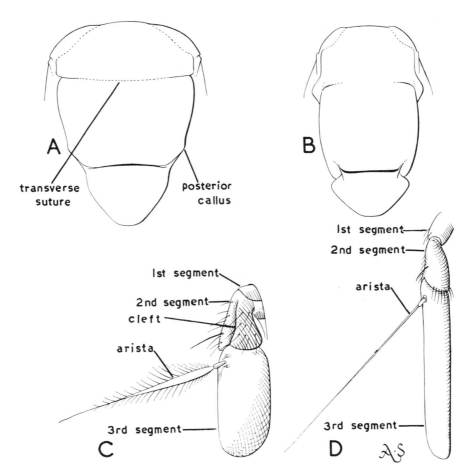

Fig. 10   A, Thorax of a Calypterate fly in dorsal view; note the complete transverse suture
and the presence of postalar calli; B, thorax of an Acalypterate fly; note the incomplete
transverse suture and the absence of posterior calli; C, antenna of a Calyptrate fly;
note cleft on 2nd segment; D, antenna of an Acalypterate fly; note the absence of a
cleft on the 2nd segment (various magnifications).

28 Anterior frons without strong bristles; posterior vertical bristles divergent. Sixth longitudinal vein not reaching the wing margin. Small, metallic dark blue or black flies (fig. 133D) . . . . **PIOPHILIDAE**

  – Not with this combination of characters 29

29 Hind metatarsus usually incrassate; no sternopleural bristles. Mouth opening wide; arista often long, with short pubescence . . . **SPHAEROCERIDAE**

  – Not with this combination of characters 30

30 Costa with two breaks (fig. 133A). Tibiae with a pre-apical bristle; arista usually branched, with at least one ventral hair. Mesopleuron without bristles . . . . **DROSOPHILIDAE**

  – Not with this combination of characters *Families of no known medical importance*

31 Hypopleuron (meropleuron) without a row of bristles below the spiracle . . 32

  – Hypopleuron (meropleuron) with one of more vertical series of bristles below the spiracle (fig. 11A) 34

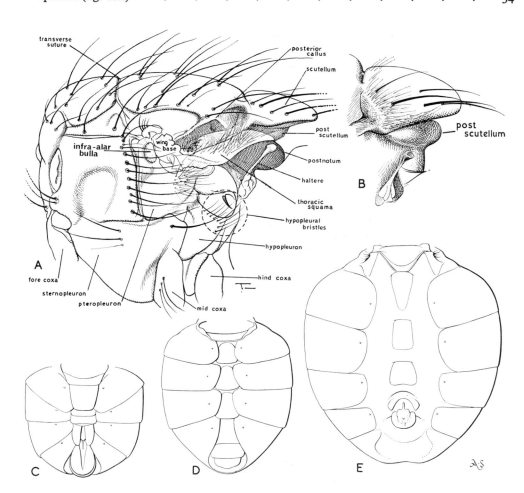

Fig. 11 A, Thorax of Calliphorid fly in lateral view; note presence of hypopleural bristles and poorly developed postscutellum; the infra-alar bulla can be seen just anterior to the wing base; B, scutellar area of thorax of a Tachinid fly showing the well-developed postscutellum; C, abdomen of a Calliphorid fly in ventral view, note sternites, especially the second visible, one overlapping tergites; D & E, abdomens of two Tachinid flies viewed from beneath; note tergites overlapping the sternites in the first and not overlapping at all in the second (A, after Senior-White, Aubertin & Smart).

32 Apical cell not narrowed at the margin, no cruciate frontal bristles; eyes broadly separated in both sexes. Thoracic squama shorter than, or equal to, the alar squama. Scutellar suture interrupted medially   .   . *Not of known medical importance*

–   Fourth longitudinal vein often bending forward to narrow the apical cell at the margin; if the apical cell is not narrowed, then eyes are not widely separated, or else frons with cruciate bristles.   Scutellar suture complete   33

33 Arista plumose with feathered lateral hairs; proboscis needle-like, porrect and at rest completely ensheathed by the long slender palpi. Prosternum membranous (Tsetse-flies)   .   .   .   . **GLOSSINIDAE**

–   Hairs of arista not feathered; proboscis, if elongate and porrect, not ensheathed by the palpi; prosternal plate developed (House-flies, etc.)   .   . **MUSCIDAE**

34 Mouth parts vestigial (fig. 135) (Bot and Warble-flies) .   .   . **OESTRIDAE**
–   Mouth parts functional and well developed   35

35 Post-scutellum prominent and strongly convex (fig. 11B).   The plates forming the dorsal surface of the abdomen (tergites) completely overlapping the plates forming the ventral surface (sternites) (fig. 11D, E)   .   .   . **TACHINIDAE**

–   Post-scutellum not prominent (fig. 11A).   At least the second visible sternite overlapping the tergites (fig. 11C) (Flesh-flies, Bluebottles, etc.)   .   .   .   .
    .   .   .   **CALLIPHORIDAE** (*including the* **SARCOPHAGINAE**)

KEY TO THE FAMILIES OF BLOOD-SUCKING FLIES

(modified from Smart)

This is a simplified key for determination to families of flies known to be blood-suckers. The key should **not** be used for flies that are not **known** to suck the blood of man or animals.

1 'Abnormal' flies with leathery bodies flattened dorso-ventrally.   Seldom taken in flight, but usually crawling on their host or other accidental carrier (Keds, Louse-flies, Tick-flies) (fig. 134B, C)   .   .   .   .   . **HIPPOBOSCIDAE**

–   'Normal' flies, gnats or midges   2

2 Antennae consisting of a series of very similar segments apart from the basal two which may be very small and not noticeable on a casual examination   .   .   .   3

–   Antennae otherwise   6

3 Antennae compact (fig. 82); thick-set, slightly humpbacked flies (fig. 81) (Black-flies) .   .   .   . **SIMULIIDAE**

–   Antennae otherwise   4

4 Proboscis elongate (fig. 12), wings with venation as in fig. 4A (Mosquitoes, Gnats) .
    .   .   .   . **CULICIDAE**

–   Proboscis not exceptionally elongated   5

5 Small moth-like flies (fig. 90A) (Sand-flies)   .   . **PHLEBOTOMIDAE**
–   Relatively bare and usually very small flies with the wings frequently mottled (figs 94A, 95) (Biting Midges)   .   .   .   . **CERATOPOGONIDAE**

6 Wing venation more complex (fig. 97) with at least nine terminations of veins or branches of veins reaching or approaching the wing margin; arista usually wanting (fig. 98C, D)   7

–   Wing venation less complex (figs 4B, G, 105) with seven or fewer veins reaching or approaching the wing margin; antennae with an arista on the dorsal surface of the third, terminal, segment   8

D

7 Squamae large, antennae without a style or bristle (figs 98C, D, 99A) (Horse-flies, Clegs, etc.) . . . . **TABANIDAE**

– Squamae smaller, antennae with a terminal style or bristle (fig. 101) (Snipe-flies) . . . . **RHAGIONIDAE**

8 Antennal arista plumose with feathered lateral hairs; proboscis needle-like and porrect ensheathed at rest by the long slender palpi (figs 106A, 120E) (Tsetse-flies) . . . . **GLOSSINIDAE**

– Hairs of arista not feathered; proboscis if elongate and porrect, not ensheathed by the palpi (fig. 129) (Stable-flies, etc.) . . . . . **MUSCIDAE**

BIBLIOGRAPHY

BRAUNS, A. 1954. *Terricole Dipterenlarven* and *Puppen terricole Dipterenlarven*, 2 vols. 179 and 156 pp. Berlin.

BRUES, C. T. & MELANDER, A. L. 1954. 2nd edn. Classification of Insects. 917 pp. Cambridge, Mass. [Diptera including larvae pp. 305–538].

COLE, F. R. 1969. *The Flies of Western North America.* 693 pp. Berkeley and Los Angeles.

CURRAN, C. H. 1965. 2nd edn. *The Families and Genera of North American Diptera.* 515 pp. New York.

EDWARDS, F. W., OLDROYD, H. & SMART, J. 1939. *British Bloodsucking Flies.* 156 pp. B.M. (N.H.) London.

GREENBERG, B. 1971. *Flies and disease* 1. 856 pp. Princeton.

HAYES, W. P. 1938–1939. A bibliography of keys for the Identification of Immature Insects. Part 1. Diptera. *Ent. News* 49: 246–251; 50: 5–10, 76–82.

HENDEL, F., KÜKENTHAL, W. & KRUMBACH, T. 1937. *Handbuch der Zoologie*, Bd. 4, Hft. 2. *Diptera.* 1730–1998 pp. Berlin and Leipzig.

HENNIG, W. 1948–1952. *Die Larvenformen der Dipteren.* 3 vols. Berlin (reprinted 1968).

IMMS, A. D. 1957 (reprinted 1964). (9th edn. by Richards, O. W. and Davies, R. G.) *A General Textbook of Entomology.* 866 pp. London.

JOHANNSEN, O. A. 1933–1937. Aquatic Diptera, parts 1–5. *Bull. Cornell Univ. agric. Exp. Stn.* **164**: 71 pp; **177**: 62 pp; **205**: 84 pp; **210**: 80 pp.

LEWIS, D. J. 1958. Some Diptera of medical interest in the Sudan Republic. *Trans. R. ent. Soc. Lond.* **110**: 81–98.

LINDNER, E. (Ed.). 1925–1949. *Die Fliegen der Palaearktischen Region.* Bd. **1** Handbuch. 422 pp. Stuttgart. [Parts on individual families still appearing.]

MALLOCH, J. R. 1917. A preliminary classification of Diptera exclusive of Pupipara, based upon larvae and pupal characters, with keys to imagines in certain families. Part 1. *Bull. Ill. St. Lab. nat. Hist.* **12**: 161–407.

OLDROYD, H. 1964. *The Natural History of Flies.* 324 pp. London.

—— 1971. (2nd edn.) Diptera: 1. Introduction and Key to families. 104 pp. *Handbk Ident. Br. Insects* 9 (1). London.

OLDROYD, H. & SMITH, K. G. V. 1967. Diptera. In Kerrich, G. J. *et al.* Bibliography of key works for the Identification of the British Fauna and Flora. *Publs. Syst Ass.* **1**. (3rd edn). 186 pp. London. [Diptera pp. 82–96].

PAPAVERO, N. (Ed.). 1966–. *A Catalogue of the Diptera of the Americas South of the United States.* São Paulo [contains bibliography of all Neotropical taxonomic works on Diptera, still appearing].

SÉGUY, E. 1950. *La Biologie des Diptères.* 609 pp. Paris.

STONE, A., SABROSKY, C. W., WIRTH, W. W., FOOTE, R. H. & COULSON, J. R. 1965. A Catalog of the Diptera of America North of Mexico. *U.S. Dept. Agric. Handb.* 276. 1696 pp. Washington. [Contains bibliography to all Nearctic taxonomic works.]

ZUMPT, F. 1965. *Myiasis in man and animals in the Old World.* 267 pp. London.

3a. CULICIDAE
(Mosquitoes)
by P. F. Mattingly

Some authors regard Dixidae and Chaoboridae as subfamilies of Culicidae, but in the present work the world catalogue of mosquitoes (Stone, *et al.* 1959) is followed and these groups are given family status. The Dixidae and Chaoboridae are of no medical importance and are distinguished by the characters given in the key to families in Chapter 2. True mosquitoes may be distinguished at once by the long proboscis which projects forward with the palpi. The subfamilies of Culicidae now recognized are included in the generic keys below.

DISEASE RELATIONS AND BIONOMICS

Disease relations

Setting aside mechanical transmission and phoresy, mosquitoes serve as vectors of three groups of human pathogens. These are Haemosporidia (four species of human malaria parasite and an occasional simian malaria parasite, all belonging to the genus *Plasmodium*), two or more species of Filarioidea belonging to the genera *Brugia* and *Wuchereria*, and the arboviruses, more than 40 of which have been recovered from man while a further 40 are believed to infect man on the evidence of antibody surveys.

The diseases to which these give rise are discussed as ecological systems by Mattingly (1969). Monographs or reviews dealing with particular groups of diseases include Edeson & Wilson (1964), Garnham (1966), Reeves (1965), Russell *et al.* (1963), Taylor (1967), Berge *et al.* (1970) and Ward & Scanlon (1970). For epidemiology and epidemiological sampling methods see Detinova (1968), Muirhead-Thomson (1968), and Pampana (1969). For control methods see Busvine (1966).

Human malaria is transmitted exclusively by *Anopheles* spp. Vectors include *An. labranchiae* Falleroni, *sacharovi* Favre, *sergentii* (Theobald), *superpictus* Grassi and *pharoensis* Theobald in parts of the Mediterranean area, *An. funestus* Giles, *moucheti* Evans, *nili* (Theobald) and members of the *gambiae* Giles complex in the Ethiopian Region, *An. stephensi* Liston, *fluviatilis* James and *pulcherrimus* Theobald in western Asia, *An. culicifacies* Giles in India and Ceylon, *An. maculatus* Theobald, *sundaicus* (Rodenwaldt) and members of the *An. barbirostris* Van der Wulp, *hyrcanus* (Pallas), *umbrosus* (Theobald), *leucosphyrus* Dönitz, *minimus* Theobald and *annularis* Van der Wulp groups in southern Asia, members of the *An. punctulatus* complex in Melanesia and *An. pseudopunctipennis* Theobald, *bellator* Dyar & Knab, *cruzii* Dyar & Knab, *darlingi* Root, *aquasalis* Curry, *albimanus* Wiedemann, *albitarsis* Lynch Arribálzaga and *nuneztovari* Gabaldon in Central and South America. Other species may be locally important and in general a list such as the above can present only a very crude picture since the importance of a particular vector in a given area will vary with the prevailing ecological conditions (see, e.g. Reid in Ward & Scanlon, 1970). Complicating factors include the development of insecticide resistance among a number of important vectors (World Health Organisation, 1970) and the habit of some others of feeding and resting outside houses where they fail to make contact with the insecticide. (For behavioural aspects see Mattingly, 1962b.) Vectors of Wuchererian filariasis include several of the major malaria vectors and, in addition, the highly domestic *Culex pipiens fatigans* Wiedemann throughout the tropics except in the Pacific area where the nocturnal periodicity of the parasite is suppressed and day-biting vectors, particularly *Aedes* (*Stegomyia*) *polynesiensis* Marks, take over. Vectors of Brugian filariasis include *Anopheles* and *Mansonia* spp. (see Wharton, 1962, Edeson & Wilson, 1964, Reid, 1968).

Viruses known or believed to infect man have been recovered from more than 150 species of mosquitoes belonging to 14 different genera (*Aedeomyia, Aedes, Anopheles, Culex, Culiseta, Deinocerites, Eretmapodites, Haemagogus, Limatus, Mansonia, Psorophora, Sabethes, Trichoprosopon, Wyeomyia*). (See Chapter 19.)

Bionomics

The egg stage, though of cardinal ecological importance, has been somewhat neglected. The nearest to a general review will be found in Mattingly (*Riv. Parassit.*, 1971). Oviposition sites (or larval breeding places) may be roughly classified as follows:

A. Running water habitats (stream edges, mainly anophelines).
B. Still water habitats.

    1. Ground water habitats.
        (a) Permanent (anophelines and culicines other than *Aedes* and aedine genera).
        (b) Temporary (especially *Aedes*).
        (c) Specialized (crab holes, rock holes, wells, soak aways) (various culicines, a few anophelines in wells).

    2. Container habitats (artefacts, tree holes, cut, split or bored bamboos, leaf axils, plant pitchers, flower petals, bracts, spathes, fallen leaves or rachids, fruits and husks, snail shells, cup fungi etc.) (*Toxorhynchites, Anopheles* subgenus *Kerteszia*, sabethine genera, some aedine genera, some subgenera of *Aedes*, individual species and species groups of *Anopheles* and various culicine genera).

Larval bionomics are reviewed in the various taxonomic monographs listed in the references, notably Hopkins (1952), and by Laird (1956). Since the introduction of residual insecticides a quarter of a century ago adult bionomics have received increasing attention. Coverage in taxonomic monographs is not always adequate. Exceptions are Gillies & De Meillon (1968) and Reid (1968). Among general reviews Bates (1949) and Muirhead-Thomson (1951) are classics, the former recently republished. Horsfall (1955) is a compilation useful mainly for references to earlier literature. Marshall (1938) has enjoyed great popularity as an elementary introduction and has also been recently republished. Natvig (1948) contains much information regarding northern species. Clements (1963) covers a much wider spectrum than the title might suggest. Mattingly (*Riv. Parassit.*, 1971) summarizes a wide range of topics and has an extensive and up-to-date bibliography. Publications on bionomics of individual vectors include Christophers (1960a), Jachowski (1954) and De Meillon *et al.* (1967). Wharton (1962) covers various aspects of Old World *Mansonia* spp. Literature on neotropical species is scattered. Galindo *et al.* (1950), Trapido & Galindo (1957) and Aitken *et. al.*, 1969 will serve as an introduction.

The role of diel rhythms in cyclical behaviour has received much attention but there is no general review. Mattingly (*Trans. 13th internat. Congr. Ent.*, in press) reviews some ecological implications. Haddow (1955) discusses techniques for recording and analysing biting cycles. More fundamental work has been concerned chiefly with the oviposition cycle. Gillett (1962) is a key paper for the modus operandi of the internal 'clock'. More recently this type of study has been extended to the diel activity rhythm (Taylor & Jones, 1969).

Interest in mosquito genetics stemmed mainly from the development of insecticide resistance but many other aspects of mosquito biology are now involved. The field as a whole is well covered by Wright & Pal (1967).

GEOGRAPHICAL DISTRIBUTION

Outline distributions are given in the Synoptic Catalog (Stone *et al.* 1959). For further details see the various regional monographs listed below. Mattingly (1962a) discusses mosquito zoogeography in general. For zoogeography of the Pacific area see Belkin (1962, 1968).

KEYS TO SUBFAMILIES AND GENERA OF CULICIDAE

Introduction

The suprageneric and generic classification adopted here follow closely the Synoptic Catalog of the Mosquitoes of the World (Stone *et al.*, 1959) and the various supplements (Stone, 1961, 1963, 1967, 1970). Changes in generic nomenclature arising from the publication of the Catalog include the substitution of *Mansonia* for *Taeniorhynchus* and *Culiseta* for *Theobaldia*, bringing New and Old World practice into line, the substitution of *Toxorhynchites* for *Megarhinus* and *Malaya* for *Harpagomyia*, the suppression of the diaeresis in *Aëdes*, *Aëdeomyia* (formerly *Aëdomyia*) and *Paraëdes* (Christophers, 1960b) and the inclusion of the last named as a subgenus of *Aëdes* (Mattingly, 1958). The only new generic name to appear since the publication of the Catalog is *Galindomyia* (Stone & Barreto, 1969). *Mimomyia*, previously treated as a subgenus of *Ficalbia*, is here treated, in combination with subgenera *Etorleptiomyia* and *Ravenalites*, as a separate genus. Ronderos & Bachmann (1963a) proposed to treat *Mansonia* and *Coquillettidia* as separate genera and they have been followed by Stone (1967, 1970) and others. I cannot accept this and they are here retained in the single genus *Mansonia*.

It will be seen that the treatment adopted here, as always with mosquitoes since the early days, is conservative. Inevitably, therefore, difficulties arise in connection with occasional aberrant species. In order to avoid split, or unduly prolix, couplets I have preferred, in nearly every case, to deal with these in the Notes to the Keys.   The latter are consequently to be regarded as very much a part of the keys themselves and should be constantly borne in mind.   A bibliography of key works is included for the benefit of those who may wish to carry out identification at the subgeneric or species level.

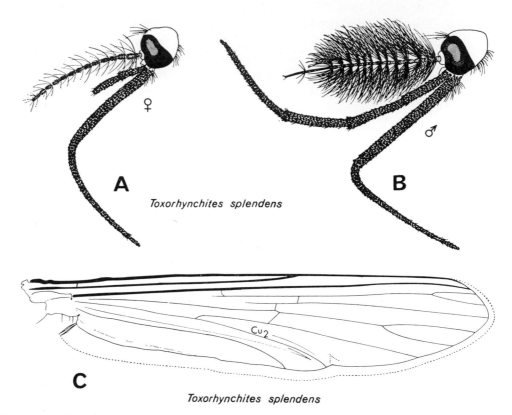

Toxorhynchites splendens

Toxorhynchites splendens

Fig. 12   *Toxorhynchites splendens*.   A, female head; B, male head; C, wing.

I. Female Adults

1   Proboscis long, strongly attenuated and recurved with prominent setulae confined to base; posterior edge of wing emarginated just beyond tip of vein Cu2 (fig. 12) (Subfamily Toxorhynchitinae)   .   .   .   ***TOXORHYNCHITES*** Theobald[5]

−   Proboscis and wing otherwise   2

2   Abdomen with sterna (and usually also terga) wholly or largely devoid of scales (Subfamily Anophelinae)   3

−   Abdominal terga and sterna with dense, uniform covering of scales (Subfamily Culicinae)   5

3   Veins Cu1 and M, distad of the cross vein, wavy (fig. 13); all wing scales dark; Moluccas, Melanesia and northern Australia only   .   .   ***BIRONELLA*** Theobald

−   These veins very rarely wavy; if so then wing with conspicuous pattern of light and dark scales[1]   4

4   Scutellum trilobed with bristles in three distinct groups (fig. 14A); posterior pronotal bristles present (fig. 15); New World tropics only   .   .   .   ***CHAGASIA*** Cruz

Scutellum smoothly rounded with bristles more or less evenly distributed (fig. 14B); posterior pronotal bristles absent   .   .   .   .   .   ***ANOPHELES*** Meigen

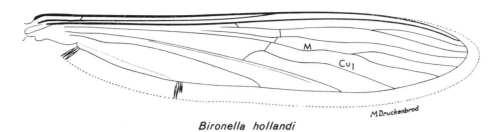

Bironella hollandi

M.Druckenbrod

Fig. 13 *Bironella hollandi* wing.

5   Tip of proboscis swollen, upturned and hairy (fig. 16); Old World tropics only
                                                                    MALAYA Leicester[2,3,5]
–   Proboscis sometimes with tip swollen, otherwise unmodified   6
6   Scutum with double median longitudinal stripe of broad, flat, usually white or
      silvery scales; spiracular bristles present, postspiraculars absent (see fig.
      15); South-east Asia and New Guinea only   .   .   .   .   *TOPOMYIA* Leicester[2,3,5]
–   Without this combination of characters   7
7   Squama and postnotum bare; vein 1A reaching wing margin at most very slightly
      beyond base of fork of vein Cu (fig. 17A)   8
–   Squama fringed at least in part (fig. 25A,B,D) or postnotum with bristles (fig. 14C, D) or
      vein 1A reaching wing margin well beyond the base fork of vein Cu (fig. 24) or
      with any combination of these characters   10
8   Outstanding scales on outer half of wing field with emarginated tips (fig. 17B); Old
      World tropics only   .   .   .   .   *HODGESIA* Theobald[2]
–   Wing scales otherwise; distribution various   9
9   Pleuron with conspicuous vertical stripe of broad, silvery scales extending from the
      prealar area to mid coxa (fig. 18); wing membrane with microtrichia visible under
      magnification of about ×50; South-east Asia only   .   .   *ZEUGNOMYIA* Leicester[2]
–   Pleural ornamentation otherwise; microtrichia minute, visible only under high
      magnification   .   .   .   .   .   .   *URANOTAENIA* Lynch Arribalzaga[2]
10 Spiracular area with scales or one or more bristles (fig. 15)   11
–   Spiracular area bare   18

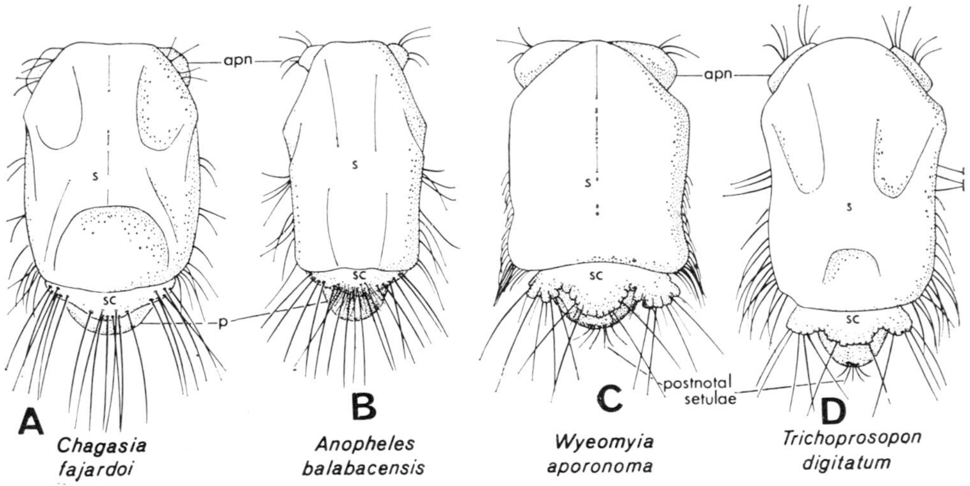

A
Chagasia
fajardoi

B
Anopheles
balabacensis

C
Wyeomyia
aporonoma

D
Trichoprosopon
digitatum

Fig. 14   Thorax in general view.   A, *Chagasia fajardoi*; B, *Anopheles balabacensis*; C, *Wyeomyia aporonoma*; D, *Trichoprosopon digitatum* apn, anterior pronotum; p, postnotum; s, scutum; sc, scutellum.

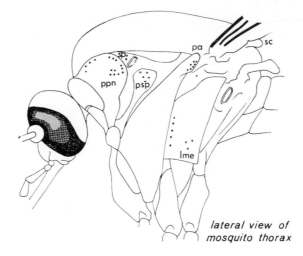

*lateral view of
mosquito thorax*

Fig. 15   Generalized mosquito thorax in lateral view showing setae used in the keys.   lme,
lower mesepimeral; pa, prealar; ppn, posterior pronotal; psp, spiracular.

11   At most one or two prealar bristles present (see fig. 15); stem vein and base of subcosta without bristles (see fig. 17).   Oriental and Australasian Regions and far eastern
Palaearctic only .   .   .   . ***TRIPTEROIDES*** Giles[5,6]

 –   Prealar bristles more numerous; stem vein, at least, with bristles or genera confined
to New World   12

12   Postspiracular bristles absent; abdomen blunt tipped; prealar bristles relatively
numerous; bristles present dorsally on stem vein and, usually, also ventrally at
base of subcosta; Nearctic Region and Old World only   .   .   ***CULISETA*** Felt[4]

 –   Without this combination of characters; New World species only   13

13   Postspiracular bristles present; postnotum bare; abdomen pointed at tip
PSOROPHORA Robineau-Desvoidy

 –   Postspiracular bristles absent; postnotum usually with a tuft of setulae (fig. 14C, D);
abdomen blunt tipped   14

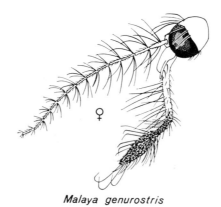

Malaya genurostris

Fig. 16   *Malaya genurostris*, female head.

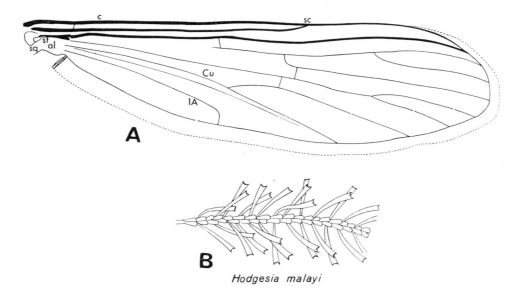

Fig. 17 *Hodgesia malayi.* A, wing; B, outstanding scales from outer half of wing field. al, alula; c, costa; sc, subcosta; sq, squama; st, stem vein.

14 Spiracular area with broad scales only, without bristles; proboscis shorter than fore femur; scutal scales with gold and purple metallic reflections; pleura with abundant golden and silvery scales; hind tarsus with only one claw . **LIMATUS** Theobald

  - Spiracular area with one or more small bristles (fig. 15); proboscis and thoracic ornamentation various; hind tarsus with two claws as usual 15

15 Antenna at most about half the length of the proboscis, usually less; proboscis long and slender, at least one-sixth as long again as fore femur (as in fig. 27c); scutum covered with broad, flat scales; scutellum with silvery scales at least on part of mid lobe; one or more tarsi nearly always with white markings on one side at least . . . . **PHONIOMYIA** Theobald[7]

  - Without this combination of characters 16

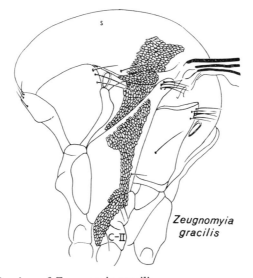

Fig. 18 Thorax in side view of *Zeugnomyia gracilis.*

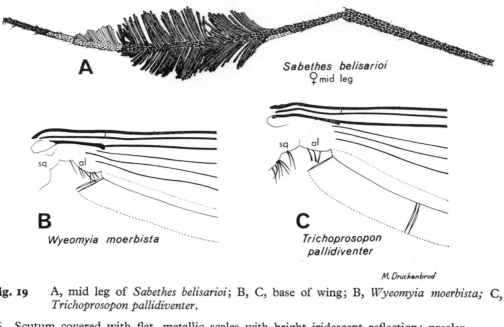

Fig. 19   A, mid leg of *Sabethes belisarioi*; B, C, base of wing; B, *Wyeomyia moerbista*; C, *Trichoprosopon pallidiventer*.

16   Scutum covered with flat, metallic scales with bright iridescent reflection; prealar bristles absent (see fig. 15); anterior pronotal lobes large, almost touching in mid line; one or more tarsi often with conspicuous 'paddles' of erect scales (fig. 19A)
                                                    SABETHES Robineau-Desvoidy[8]

–   Without this combination of characters; tarsi never with 'paddles'   .   .   .   17

17   Anterior pronotal lobes large and closely approximated (fig. 14C); squama bare or with one to three bristles or hairlike scales arising from the upper portion near the alula (fig. 19B); clypeus sometimes with scales, never with setulae . **WYEOMYIA** Theobald

–   Pronotal lobes smaller, well separated (fig. 14D) or squama with bristles more numerous or arising from lower portion (fig. 19C) or clypeus with conspicuous setulae or with any combination of these characters . **TRICHOPROSOPON** Theobald

18   Antenna short, thick, tapering, basal flagellomere with a prominent scale tuft; mid and hind femora with large tufts of suberect scales (fig. 20A, C)
                                                         AEDEOMYIA Theobald

–   Antenna and femora otherwise   19

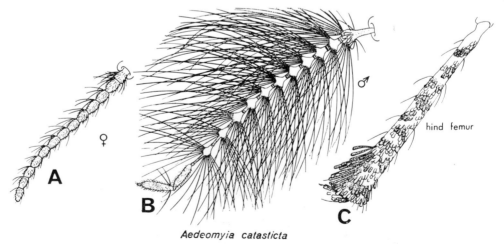

Fig. 20   *Aedomyia catasticta*.   A, female antenna; B, male antenna; C, hind femur.

19 Antenna with flagellomeres short and thick and verticillary hairs unusually short;
    vertex and occiput with numerous short hairs (fig. 21A, B); New Zealand only
                                                                  OPIFEX Hutton

–   Head quite otherwise   20

20 Postnotum with a small patch of setulae (as in fig. 14C, D) or scutum covered with
    flat scales with bright metallic reflection, usually both; southern and eastern Asia
    only .   .   .   .   **HEIZMANNIA** Ludlow[9]

–   Postnotum without setulae or distribution otherwise   21

21 Scutum covered with broad, flat scales with bright metallic reflection; anterior
    pronotal lobes unusually large; New World tropics and subtropics only
                                                              HAEMAGOGUS Williston

–   Scutal ornamentation otherwise; anterior pronotal lobes not unusually large   .   .   22

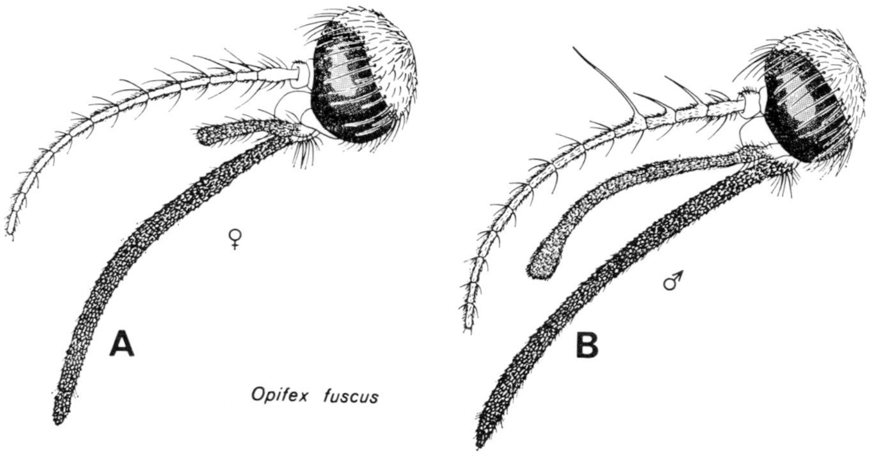

Fig. 21   Head of *Opifex fuscus*: A, female; B, male.

22 Antenna unusually long, exceeding the proboscis by about the length of the last four
    flagellomeres or more; first flagellomere greatly elongated, three or more times as
    long as the terminal flagellomere (fig. 22A); proboscis not noticeably swollen
    apically; New World only   .   .   .   .   **DEINOCERITES** Theobald

–   Without this combination of characters   23

23 Antenna longer than proboscis with all flagellomeres markedly elongated; apical
    flagellomere at least half as long as the basal one (fig. 22C); proboscis distinctly
    swollen apically; scutum without conspicuous ornamentation; postspiracular
    bristles absent; tarsi unbanded; New World tropics only
                                                      GALINDOMYIA Stone & Barreto[10]

–   Without this combination of characters   24

24 First fore and mid tarsomere distinctly longer than the other four together; fourth
    tarsomere shorter than fifth, only a little longer than broad (fig. 26C); (post-
    spiracular bristles absent; all claws simple)   .   .   **ORTHOPODOMYIA** Theobald

–   Proportions of these tarsomeres otherwise   25

25 Postspiracular bristles present or fore tarsal claws toothed or both .   .   .   26

–   Postspiracular bristles absent; all tarsal claws simple   30

26 Paratergite broad and bare; postnotum usually with a group of setulae (as in fig.
    14C, D); back of head, pleura and posterolateral corners of abdominal terga with
    broad, silvery scales; tropical Africa only   .   .   **ERETMAPODITES** Theobald

–   Without this combination of characters   27

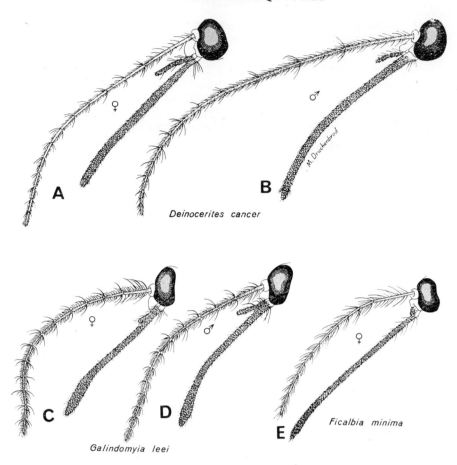

Fig. 22   Antenna, palp and proboscis: A, B, *Deinocerites cancer*: A, female; B, male.   C, D,
Galindomyia leei: C, female; D, male.   E, *Ficalbia minima*, female.

27   Decumbent scales of vertex broad, flat; postspiracular and lower mesepimeral bristles
     present and proboscis curved and laterally compressed (fig. 23A) or postspiraculars
     absent and palpus half length of proboscis or more; southern Asia, Japan and
     Melanesia only   .   .   .   .   **ARMIGERES** Theobald[11]
–   Without this combination of characters   28

28   Upper surface of wing with all or most scales very broad, many often asymmetrical
     (fig. 24); all tarsal claws simple; decumbent scales of vertex narrow
                                    MANSONIA Blanchard (part)[12]
–   Without this combination of characters   29

29   Squama bare or with at most six short hairs (fig. 25A); alula with broad scales; posterior
     pronotum without scales; pleural scales restricted to four small white spots; those
     on sternopleuron and mesepimeron sometimes fused; hind tarsomeres II–IV with
     conspicuous white basal bands; South-east Asia only   .   .   **UDAYA** Thurman
–   Without this combination of characters   .   .   .   .   .   **AEDES** Meigen[13]

30   Small or very small species; alula with fringe of narrow scales; antenna with first
     flagellomere at least twice as long as fifth, usually longer (fig. 22E); palps less than a
     fifth of length of proboscis; scutellum with narrow scales only; wing with anterior
     fork cell less than twice as long as its stem; Old World tropics only
                                    FICALBIA Theobald[14]
–   Without this combination of characters   31

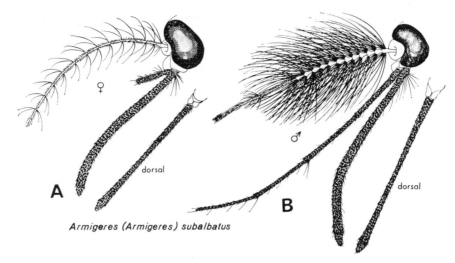

Fig. 23 *Armigeres* (*Armigeres*) *subalbatus*, antenna, palp and proboscis: A, female, B, male.

31 Alula bare or with flat, decumbent scales (fig. 25B, C); Old World tropics only
                                                    MIMOMYIA Theobald[14,15]
 – Alula with narrow fringe scales (fig. 25D) **32**
32 Hind tarsal claws very small and inconspicuous; all tarsi with well-developed pulvilli
    (fig. 26A, B) . . . . ***CULEX*** Linnaeus
 – Hind tarsal claws not unusually small; pulvilli absent . ***MANSONIA*** Blanchard (part)[16]

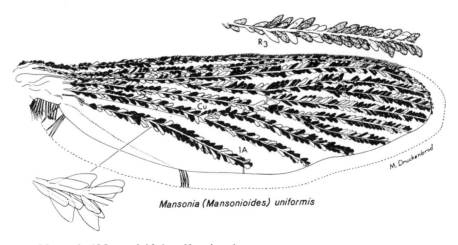

Fig. 24 *Mansonia* (*Mansonioides*) *uniformis*, wing.

II. Male Adults[17]

1 Proboscis long, strongly attentuated and recurved with prominent setulae confined to
    base; posterior edge of wing emarginated just beyond tip of vein Cu2; palps of
    the same order of length as the proboscis (fig. 12B, C) (Subfamily Toxorhynchitinae)
                                                TOXORHYNCHITES Theobald[5]
 – Proboscis and wing otherwise; palps various **2**

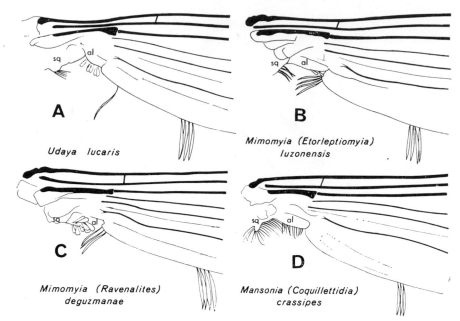

Fig. 25   Base of wing: A, *Udaya lucaris*; B, *Mimomyia (Etorleptiomyia) luzonensis*; C, *Mimomyia (Ravenalites) deguzmanae*; D, *Mansonia (Coquillettidia) crassipes*.

2   Abdominal sterna wholly or largely bare; scutellum evenly rounded, not trilobed (fig. 14B); fore tarsus with a single claw only   3
–   Without this combination of characters   4

3   Veins Cu1 and M, distad of the cross vein, wavy (fig. 13); all wing scales dark; palps various; Moluccas, Melanesia and northern Australia only   **BIRONELLA** Theobald
–   These veins wavy only in a few species with brightly ornamented wing[1]; palps at least three-quarters of the length of the proboscis, usually longer **ANOPHELES** Meigen

4   Abdomen devoid of scales; fore tarsi with paired claws; New World only
CHAGASIA Cruz
–   Abdomen densely scaly   5

5   Proboscis strongly modified, as in the female (fig. 16); Old World tropics only
MALAYA Leicester[2,3,5]
–   Proboscis otherwise   6

6   Scutum with double median stripe of broad, flat scales; spiracular bristles present; postspiraculars absent; palps minute, as in female; South-east Asia and New Guinea only .   .   .   .   **TOPOMYIA** Leicester[2,3,5]
–   Without this combination of characters   7

7   Squama bare; vein 1A reaching wing margin at most very slightly beyond base of fork of vein Cu (fig. 17A); palps minute, as in female   8
–   Squama fringed, at least in part (fig. 25A, B, D), or vein 1A reaching wing margin well beyond this or both; palps various   10

8   Outstanding scales on distal half of wing emarginated at tips (fig. 17B); antenna as in female, non-plumose and with all flagellomeres, including the last two, subequal
HODGESIA Theobald[2]
–   Wing scales otherwise; antennae various   9

9   Pleuron with conspicuous vertical stripe of broad, silvery scales (fig. 18); one fore and one mid claw toothed .   .   .   .   .   .   **ZEUGNOMYIA** Leicester[2]
    Pleuron otherwise; fore and usually also mid claws both simple
URANOTAENIA Lynch Arribalzaga[2,18]

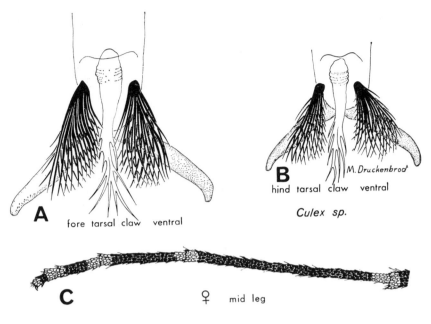

Fig. 26   Female tarsi: A, B, *Culex*; A, fore; B, hind; C, *Orthopodomyia anopheloides*, mid.

10   Spiracular area with scales or one or more bristles (fig. 15) 11
 –   Spiracular area bare 18

11   Prealar bristles numerous (fig. 15); postspiracular bristles absent; bristles present
     dorsally on stem vein and usually also ventrally at base of subcosta (see fig. 17); palps
     at most slightly shorter than proboscis; Nearctic Region and Old World only
                                                                         CULISETA Felt[4]
 –   Without this combination of characters and distribution 12

12   Old World only . . . . **TRIPTEROIDES** Giles[5,6]
 –   New World only 13

13   Postspiracular bristles present; postnotum bare; palps longer than proboscis
                                                         PSOROPHORA Robineau-Desvoidy
 –   Postspiracular bristles absent; postnotum normally with a tuft of setulae (fig. 14C, D);
     palps often only a quarter of the length of the proboscis or less . . . 14

14   Proboscis shorter than antenna, and with a conspicuous scale tuft at tip (fig. 27A),
     or with an abrupt flexure beyond half-way (fig. 27B); palps minute as in female;
     scutal scales with gold and purple metallic reflection . . **LIMATUS** Theobald
 –   Proboscis otherwise; palps and scutal ornamentation various 15

15   Palps about a quarter of the length of the proboscis or less; antenna at most about
     half the length of the proboscis, usually less; proboscis long and slender, at least
     about one-sixth as long again as fore femur (fig. 27C); scutum covered with broad,
     flat, often metallic scales; scutellum with silvery scales on part of mid lobe at
     least; one or more tarsi nearly always with white markings . **PHONIOMYIA** Theobald[19]
 –   Without this combination of characters 16

16   Palps less than a quarter of the length of the proboscis; scutum with broad scales
     with bright, iridescent, metallic reflection; one or more pairs of legs often with
     conspicuous 'paddles' (as in fig. 19A); prealar bristles absent (see fig. 4); anterior
     pronotal lobes very large, almost touching in mid line   **SABETHES** Robineau-Desvoidy[20]
 –   Without this combination of characters; legs never with 'paddles' 17

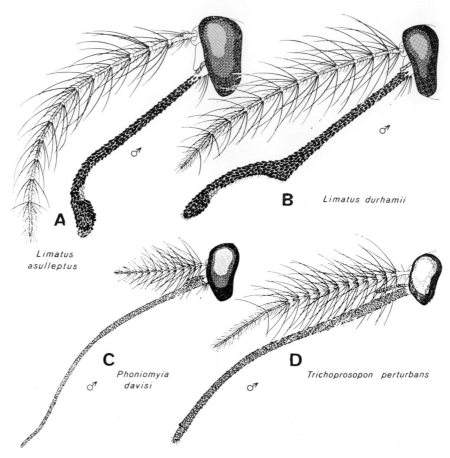

Fig. 27 Male antenna, palp and proboscis: A, *Limatus asulleptus*; B, *Limatus durhamii*; C, *Phoniomyia davisi*; D, *Trichoprosopon perturbans*.

17 Palps usually more than half the length of the proboscis; if not then either scutum with narrow scales or proboscis long and slender and antenna much more than half the length of the proboscis with the two terminal flagellomeres greatly elongated (fig. 27D) . . . . . . **TRICHOPROSOPON** Theobald[21]

  – Palps at most about a quarter of the length of the proboscis, usually less; scutum covered with broad, flat scales; antennae various but never with the 2 terminal flagellomeres greatly elongated . . . . **WYEOMYIA** Theobald[19,20]

18 Antenna with the 2 terminal flagellomeres markedly thickened (fig. 20B); hind femur with a large apical scale tuft (fig. 20C); palps very short, as in female **AEDEOMYIA** Theobald

  – Antenna and hind femur otherwise; palps various 19

19 Antenna with flagellomeres 2–4 each with a stout dorsal spine; back of head with numerous short hairs; palps with apex clavate (fig. 21B); New Zealand only **OPIFEX** Hutton

  – Antenna, palps and back of head quite otherwise 20

20 Palps at most about one-fifth of the length of the proboscis; scutum covered with broad, flat scales with bright, metallic reflection or postnotum with a group of setulae (as in fig. 14C, D) or both; antenna usually with verticillary hairs short as in female (fig. 28A); southern and eastern Asia only . . . **HEIZMANNIA** Ludlow[22]

  – Without this combination of characters 21

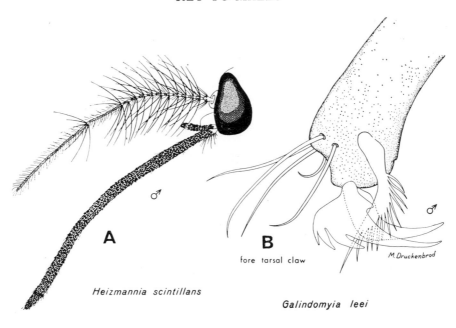

Fig. 28   A, male antenna, palp and proboscis of *Heizmannia scintillans*; B, male fore tarsal claws of *Galindomyia leei*.

21 Antenna with basal flagellomere (and sometimes also some succeeding flagellomeres) greatly elongated, at least 3 times as long as the terminal flagellomere; verticillary hairs short as in female; proboscis not noticeably swollen apically (fig. 22B); New World only .   .   .   .   **DEINOCERITES** Theobald

–   Antenna otherwise   22

22 Palps very short as in female; antennal flagellum with verticillary hairs short and scanty; apical flagellomere at least half as long as the basal one; proboscis distinctly swollen apically (fig. 22D); inner claw of fore tarsus with 2–5 teeth arising from the convex surface (fig. 28B); other claws all simple; New World tropics only **GALINDOMYIA** Stone & Barreto[10,23]

–   Without this combination of characters   23

23 Scutum covered with bright, metallic, iridescent scales; anterior pronotal lobes unusually large (see fig. 14); palps at most two-thirds of the length of the proboscis, often much less; New World tropics and subtropics only **HAEMAGOGUS** Williston

–   Without this combination of characters   24

24 Postspiracular bristle or bristles present   25
–   Postspiracular area without bristles   29

25 Paratergite broad and bare; postnotum usually with a group of setulae (as in fig. 14C, D); back of head, pleura and posterolateral corners of abdominal terga with conspicuous patches of broad, silvery scales; tropical Africa only **ERETMAPODITES** Theobald

–   Without this combination of characters   26

26 Decumbent scales of vertex broad, flat; postspiracular and lower mesepimeral bristles present; acrostichals and dorsocentrals absent; proboscis curved and laterally compressed (fig. 23B); southern Asia, Japan and Melanesia only **ARMIGERES** Theobald (part)[11]

–   Without this combination of characters   27

E

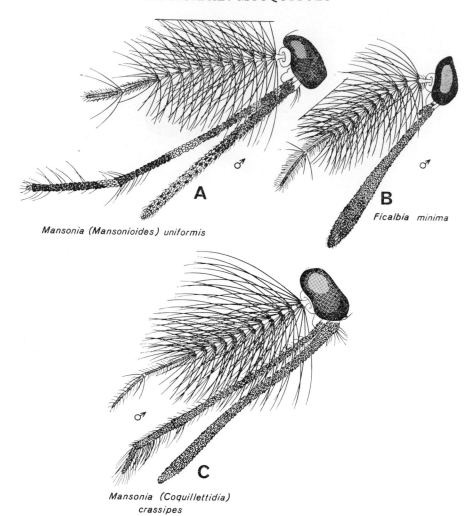

A

Mansonia (Mansonioides) uniformis

B

Ficalbia minima

C

Mansonia (Coquillettidia)
crassipes

Fig. 29 Male palps and proboscis: A, *Mansonia (Mansonioides) uniformis*; B, *Ficalbia minima*; C, *Mansonia (Coquillettidia) crassipes*.

27 Upper surface of wing with all or most scales very broad, many often asymmetrical (fig. 24); species confined to New World or if occurring in Old World then with apical segment of palp greatly reduced (fig. 29A) . **MANSONIA** Blanchard (part)[24]

  — Wing seldom with scales of this type and then only in Old World species; the latter with apical and subapical segments of palps either both well developed or both greatly reduced 28

28 Palps long, slender, almost hairless, the 2 terminal joints together more than half the length of the shaft; squama bare or with at most six short hairs; alula with broad scales; posterior pronotum without scales; hind tarsomeres II–IV with conspicuous white basal bands; South-east Asia only . . **UDAYA** Thurman

  — Without this combination of characters . . . . . **AEDES** Meigen[25]

29 Alula bare or with flat, decumbent scales (fig. 25B, C); Old World tropics only **MIMOMYIA** Theobald[14,15]

  — Alula with narrow fringe scales (fig. 25D) 30

30 Proboscis greatly swollen on distal third or more (fig. 29B); Old World tropics only **FICALBIA** Theobald[14]

  — Proboscis at most slightly swollen apically 31

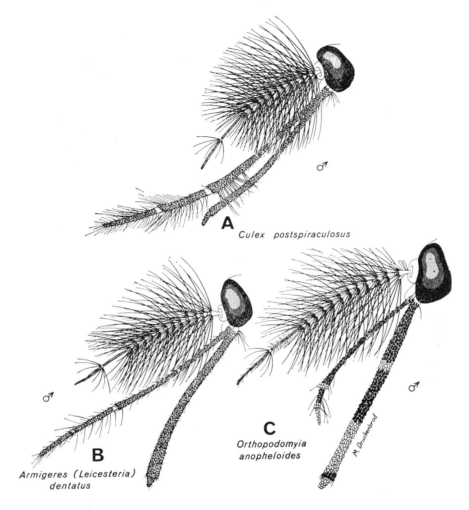

Fig. 30   Male palps and proboscis: A, *Culex postspiraculosus*; B, *Armigeres* (*Leicesteria*) *dentatus*; C, *Orthopodomyia anopheloides*.

31 Pulvilli present (best seen on hind legs); hind tarsal claws unusually small (as fig. 26B); palps various, in some cases very much shorter than proboscis . **CULEX** Linnaeus

–   Pulvilli absent; hind claws not unusually small; palps always at least as long or almost as long as proboscis   32

32 Palps longer than proboscis, the terminal segment not much shorter than the subterminal; subterminal segment and tip of shaft with numerous long hairs (fig. 29C)
MANSONIA Blanchard (part)[26]

–   Palps otherwise   33

33 Palps with the last two segments not greatly reduced (fig. 30B); southern Asia only
ARMIGERES Theobald (part)[27]

–   Palps with the apical segment (and sometimes also the subapical) greatly reduced (fig. 30C)   .   .   .   .   **ORTHOPODOMYIA** Theobald[28]

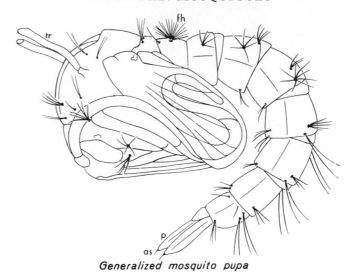

Generalized mosquito pupa

Fig. 31   Generalized mosquito pupa: as, apical paddle seta; fh, float hair; p, paddle; tr, trumpet.

III. Pupae

The pupa of genus **GALINDOMYIA** is undescribed.

1   Segment x with a conspicuous branched hair; seta 9 on segment VIII greatly reduced; paddles without apical seta (fig. 32A) (Subfamily Toxorhynchitinae)

TOXORHYNCHITES Theobald[5]

–   Segment x without setae; seta 9–VIII and paddles various   2

2   Paddles nearly always with an accessory seta arising anterior to and in line with the apical seta (fig. 32B, C); if not then seta 9 on segments IV–VII in the form of a short, stout, dark spine arising from the extreme posterior corner of the segment (as in fig. 32B) and trumpets short, flared and split nearly to base (Subfamily Anophelinae)[29] 3

–   Paddles with accessory seta absent or, if present, arising level with and laterad of the apical seta; seta 9, usually, and trumpets almost always, otherwise (Subfamily Culicinae) 4

3   Seta 2-III–VII a short, stout, dark spine (fig. 32C); New World tropics only

CHAGASIA Cruz

–   Seta 2 on these segments otherwise .   .   .   .   **ANOPHELES** Meigen
BIRONELLA Theobald

4   Trumpets modified for insertion into subaqueous plant tissues (fig. 33A, B); float hair suppressed . . . . **MANSONIA** Blanchard[30]

–   Trumpets otherwise or float hair well developed or both   5

5   Trumpets with a hinged tragus (fig. 33C) .   .   .   .   **HODGESIA** Theobald

–   Trumpets otherwise   6

6   Paddles small, usually more or less pointed, without apical seta (fig. 34) (Tribe Sabethini) 7

–   Paddles otherwise, nearly always with apical seta[31]   14

7   Old World only   8

–   New World only   10

8   Trumpets subcylindrical with inner and outer walls widely separated (fig. 33D, E) . 9

–   Trumpets subconical or with inner and outer walls closely apposed or both (fig. 33F, G) . . . . **TRIPTEROIDES** Giles[32]

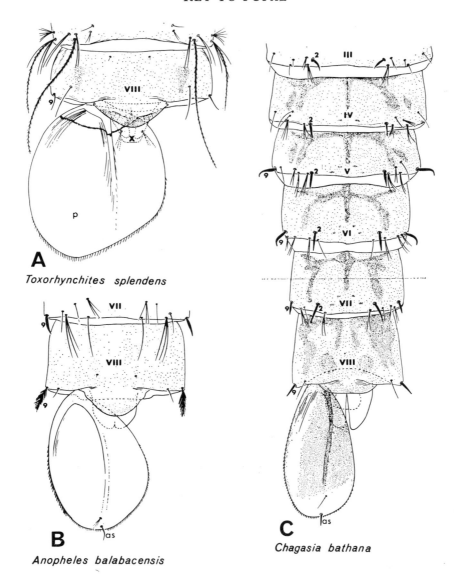

A *Toxorhynchites splendens*

B *Anopheles balabacensis*

C *Chagasia bathana*

Fig. 32 Terminal segments of pupa: A, *Toxorhynchites splendens*; B, *Anopheles balabacensis*; C, *Chagasia bathana*.

9   Seta 6-VII relatively well developed, arising well cephalad of seta 9-VII (fig. 35A)

MALAYA Leicester

–   Seta 6-VII usually less well developed and arising close to and laterad of seta 9-VII
  (fig. 35B)   .   .   .   .   *TOPOMYIA* Leicester[33]

10 Paddles short and broad, much shorter than seta 9-VIII, with edges sometimes serrated but tips entirely bare; posterior border of segment VIII deeply excavated;
  trumpets narrow, cylindrical or subconical (figs 34D, 36A)   .   *LIMATUS* Theobald

–   Without this combination of characters   11

11 Trumpets slender, tubular, narrowing basally, not or only slightly expanded at apex
  (fig. 36B); paddles broad with apex pointed and usually spiculate (fig. 34E)

PHONIOMYIA Theobald[34]

–   Trumpets broader, usually conical or beaker-shaped with basal portion expanded
  and inner lining well separated (fig. 36C, D)   12

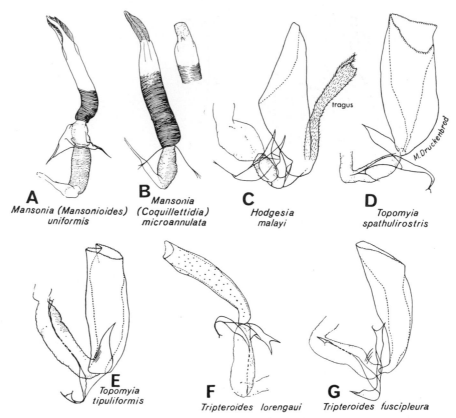

Fig. 33   Pupal trumpets: A, *Mansonia (Mansonioides) uniformis*; B, *Mansonia (Coquillettidia)
microannulata*; C, *Hodgesia malayi*; D, *Topomyia spathulirostris*; E, *Topomyia tipuli-
formis*; F, *Tripteroides lorengaui*; G, *Tripteroides fuscipleura*.

12   Either with long, delicate fringe on both borders of paddle (fig. 37A) or with genital sac
       sunk in a deep embrasure (fig. 37B)   .   .   ***WYEOMYIA*** Theobald (part)[35]

 –   Paddles without such a fringe; posterior border of segment VIII with at most a shallow
       excavation (fig. 37C, D)   13

13   Seta 5-VI longer, usually much longer, than segment VII; seta 9-VI variously developed
       but always conspicuous (fig. 37C)   .   .   ***SABETHES*** Robineau-Desvoidy[36]
                                                               WYEOMYIA Theobald (part)[35]

 –   Seta 5-VI less strongly developed, often shorter than segment VII; seta 9-VI very small
       and inconspicuous (fig. 37D)   .   .   .   .   ***TRICHOPROSOPON*** Theobald[36]
                                                               WYEOMYIA Theobald (part)[35]

14   Paddles smooth on both borders with apex convex; apical paddle seta at least two-
       thirds as long as paddle; seta 9-VIII long, single, simple (fig. 38A); New World
       only   .   .   .   .   ***DEINOCERITES*** Theobald

 –   Without this combination of characters   15

15   Paddle smooth on both borders, deeply cleft at apex; apical seta at least half as long
       as paddle; seta 5-IV-VI with long, frayed median branch and short lateral branches
       (fig. 38B)   .   .   .   .   ***AEDEOMYIA*** Theobald

 –   Paddles and seta 5-IV-VI otherwise   16

16   Seta 9-VII very short, single, simple as on anterior segments (fig. 38C); New Zealand
       only   .   .   .   .   ***OPIFEX*** Hutton

 –   Seta 9-VII more strongly developed than on anterior segments or distribution other-
       wise or both   17

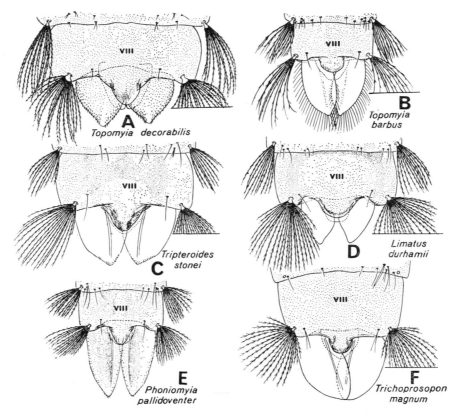

Fig. 34 Sabethine pupal paddles: A, *Topomyia decorabilis*; B, *Topomyia barbus*; C, *Tripteroides stonei*; D, *Limatus durhamii*; E, *Phoniomyia pallidoventer*; F, *Trichoprosopon magnum*.

17   Trumpets at least about ten times as long as their breadth at half way, usually more; paddles narrow or very narrow, not or only very slightly inflated on inner aspect; fringe, usually of irregular spicules, on both borders; apical seta minute or absent (fig. 39); Old World tropics only .   .   .   .   ***MIMOMYIA*** Theobald[14,15,37]

–   Without this combination of characters   18

18   Paddles with long, delicate fringe on both borders; float hair arising unusually near mid line, tending to point forwards in mounted specimens; seta 1-II long, stout, single or bifid; seta 5-III-VII in each case longer than the following segment (fig. 40); South-east Asia only .   .   .   .   .   ***ZEUGNOMYIA*** Leicester

–   Without this combination of characters   19

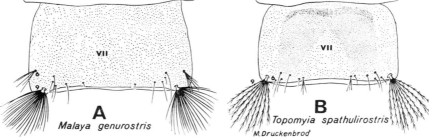

Fig. 35 Segment VII of pupal abdomen: A, *Malaya genurostris*; B, *Topomyia spathulirostris*.

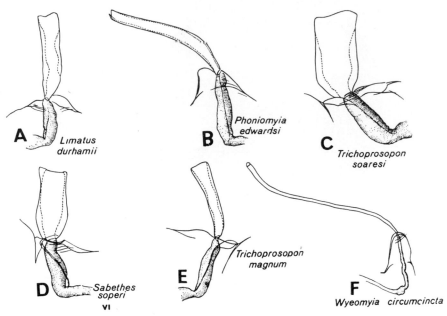

Fig. 36   Sabethine pupal trumpets: A, *Limatus durhamii*; B, *Phoniomyia edwardsi*; C, *Tricho-prosopon soaresi*; D, *Sabethes soperi*; E, *Trichoprosopon magnum*; F, *Wyeomyia circum-cincta*.

19   Paddles with inner half deeply excavated towards base, usually much broader than outer half; segment IX usually with a pair of small setulae; paddles fringed or toothed on both borders, usually extensively so (fig. 41)

            URANOTAENIA Lynch Arribalzaga[38]

–   Without this combination of characters   20

20   Paddles small, with long, delicate fringe on both borders; apical paddle seta long and stout; seta 8-C much longer and stouter than 9-C (fig. 42A, B); tropical Africa only

            ERETMAPODITES Theobald[39]

–   Without this combination of characters and distribution .   .   .   21

21   Paddles with long, delicate fringe on both borders; seta 6-VI very strongly developed, longer and much stouter than 5-VI (nearly always spinose or multibranched, often subplumose); seta 9-VI a minute, colourless setula (fig. 43A, B); southern Asia, China, Japan and Melanesia only .   .   .   .   *ARMIGERES* Theobald[40]

–   Paddles without such a fringe or segment VI otherwise; distribution various .   22

22   Paddles with long, delicate fringe on both borders; mid rib of paddle very poorly developed, barely visible, if at all, even towards base; South-east Asia only (fig. 44A) .   .   .   .   *UDAYA* Thurman

–   Paddles various, if with long fringe then mid rib always strongly developed, conspicuous (fig. 44B–D)   23

23   Paddles with long, delicate fringe on both borders, oval, usually more or less pointed, never indented at tip; seta 9-VI minute, colourless (fig. 44D) or if not then either seta 5-II long, stout, dark (fig. 45A) or seta 5 on all segments very short (fig. 45B); southern Asia only .   .   .   .   *HEIZMANNIA* Ludlow[41]

–   Without this combination of characters   24

24   Trumpets at least seven times as long as their breadth at half way, the pinna occupying at least half the length, meatus entirely tracheoid or almost so; seta 9-VIII very feebly developed; paddle edge strongly serrated on outer half, inner half smooth (fig. 48); Old World tropics only .   .   .   .   *FICALBIA* Theobald[14]

–   Without this combination of characters   25

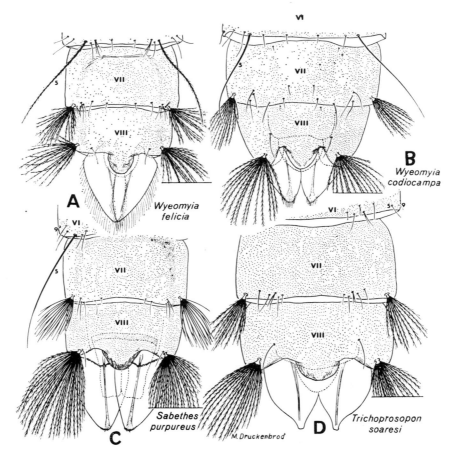

Fig. 37   Posterior segments of pupal abdomen: A, *Wyeomyia felicia*; B, *Wyeomyia codiocampa*; C, *Sabethes purpureus*; D, *Trichoprosopon soaresi*.

25   Paddles more or less rectangular with thickened basal portion of outer edge some-
      times spiculate but whole border otherwise smooth, hyaline; accessory paddle seta
      absent; apical paddle seta very short; seta 9-VI–VIII long, stout, plumose, on VIII
      about half the length of the paddle or more (fig. 49) **ORTHOPODOMYIA** Theobald[42]

 – Without this combination of characters **26**

26   Trumpets with tubular portion occupying most of the length and with rudimentary
      basal tracheation at most; seta 9-VIII arising from the posterolateral corner of the
      segment, not, or only very slightly, displaced anteriorly; either with posterior
      corner of abdominal segment IV toothed or with prominent ventral lobes on
      posterior border of segment VIII, partly covering the bases of the paddles, or with
      accessory paddle seta present (fig. 50); New World only
                                 PSOROPHORA Robineau-Desvoidy[43]

 – Without this combination of characters **27**

27   Seta 8-C arising anterior or at most slightly posterior to base of trumpet, very much
      anterior to 9-C; trumpet nearly always with rudimentary basal tracheation at most;
      seta 9-VIII rarely arising cephalad of the posterior border of the segment (fig. 51) .   **28**

 – Seta 8-C arising level with or posterior to the base of the trumpet, more nearly level
      with 9-C; trumpets frequently with extensive subbasal tracheation; position of
      seta 9-VIII various (fig. 53) **29**

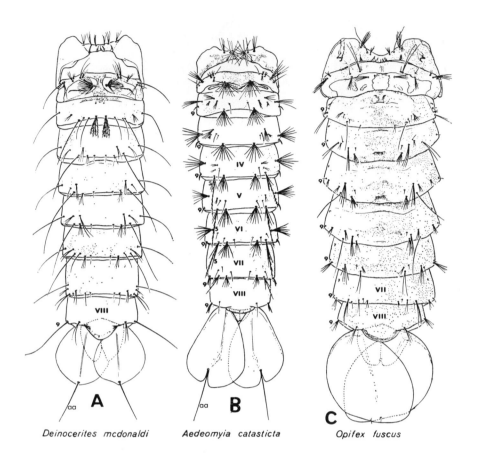

Deinocerites mcdonaldi   Aedeomyia catasticta   Opifex fuscus

Fig. 38   Pupal abdomen: A, *Deinocerites mcdonaldi*; B, *Aedeomyia catasticta*; C, *Opifex fuscus*.

28   Setae 8-C and 9-C poorly developed; setae 5-II and 5-III very feebly developed, not
     or barely reaching on to the following segment; either with seta 5-VII as long as or
     longer than the following segment or with mid rib of paddle deeply pigmented and
     seta 5-IV–VI shorter than the following segment; seta 9-III–VI minute; seta 9-VIII
     with four or more branches, about half the length of the paddle or more (figs 51,
     52A); New World tropics and subtropics only   .   .   **HAEMAGOGUS** Williston[44]

–   Without this combination of characters or distribution otherwise   .   **AEDES** Meigen[45]

29   Trumpets with well-developed subbasal tracheation or seta 9-VIII arising well
     cephalad of the posterior border of the segment, usually both (figs 53A, B; 54A, B, D)
                                                                              CULEX Linnaeus[45]

–   Trumpets with rudimentary basal tracheation at most; seta 9-VIII always arising
     from the posterior border of the segment (fig. 53C)   .   .   **CULISETA** Felt[4,46]

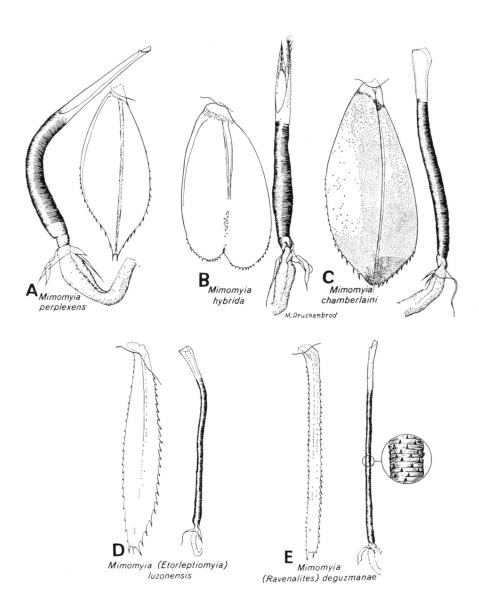

Fig. 39 Pupal trumpets and paddles, genus *Mimomyia*: A, *M. perplexens*; B, *M. hybrida*; C, *M. chamberlaini*; D, *M. (Etorleptiomyia) luzonensis*; E, *M. (Ravenalites) deguzmanae.*

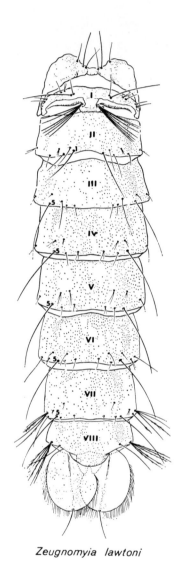

Zeugnomyia lawtoni

Fig. 40   Pupal abdomen: *Zeugnomyia lawtoni*.

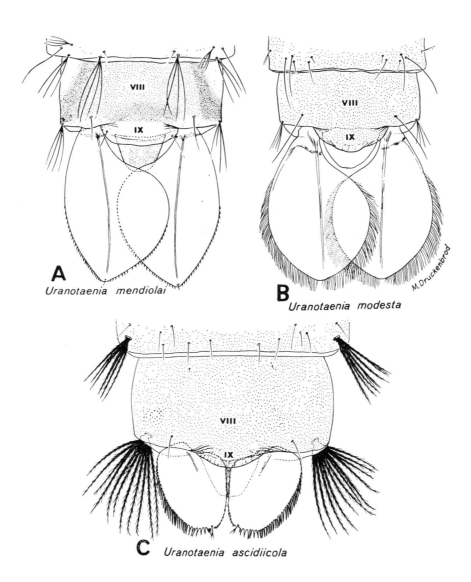

Fig. 41   Terminal segments of pupa, genus *Uranotaenia*: A, *U. mendiolai*; B, *U. modesta*: C, *U. ascidiicola*.

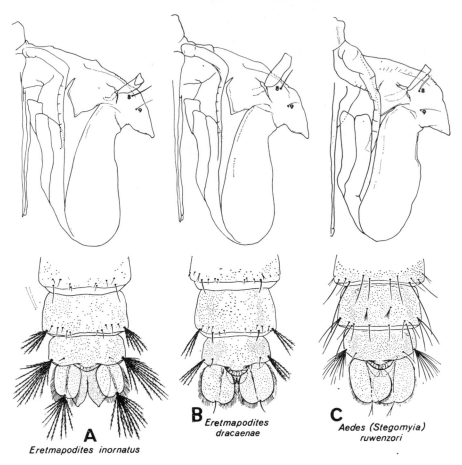

Fig. 42   Terminal segments and cephalothorax of pupa: A, *Eretmapodites inornatus*; B, *Eretmapodites dracaenae*; C, *Aedes (Stegomyia) ruwenzori*.

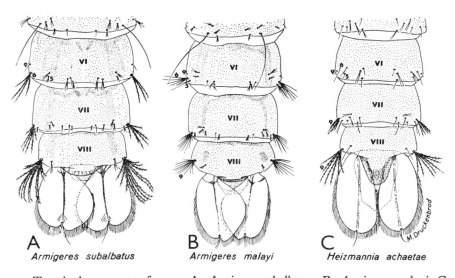

Fig. 43   Terminal segments of pupa: A, *Armigeres subalbatus*; B, *Armigeres malayi*; C, *Heizmannia achaetae*.

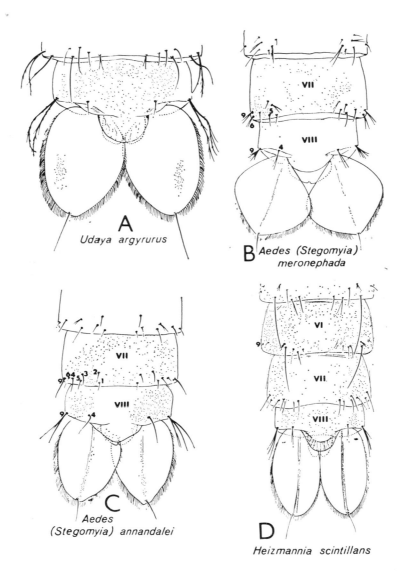

Fig. 44 Terminal segments of pupa: A, *Udaya argyrurus;* B, *Aedes (Stegomyia) meronephada;* C, *Aedes (Stegomyia) annandalei;* D, *Heizmannia scintillans.*

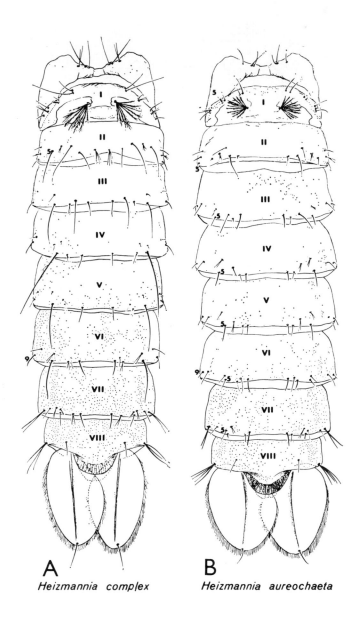

A
Heizmannia complex

B
Heizmannia aureochaeta

Fig. 45   Pupal abdomen, genus *Heizmannia*: A, *H. complex*; B, *H. aureochaeta.*

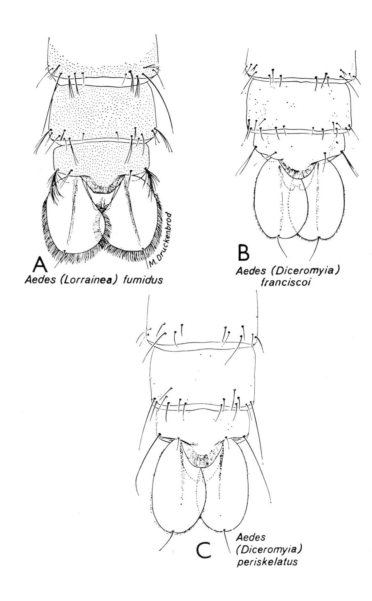

Fig. 46   Terminal segments of pupa: A, *Aedes (Lorrainea) fumidus*; B, *Ae. (Diceromyia) franciscoi*; C, *Ae. (Diceromyia) periskelatus*.

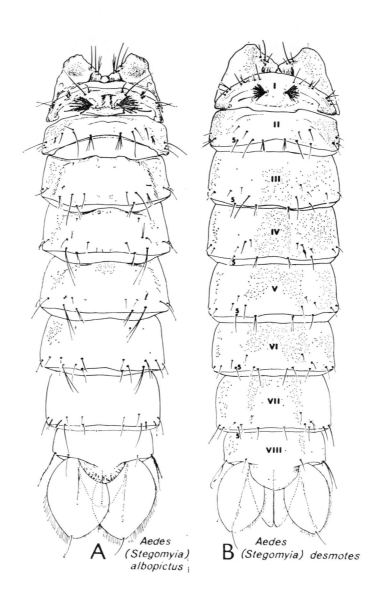

Fig. 47   Pupal abdomen, *Aedes* subgenus *Stegomyia*: A, *Ae. albopictus*; B, *Ae. desmotes*.

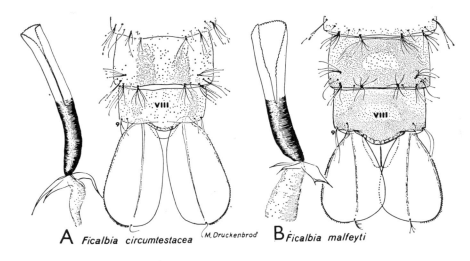

Fig. 48   Pupal trumpet and terminal segments, genus *Ficalbia*: A, *F. circumtestacea*; B, *F. malfeyti*.

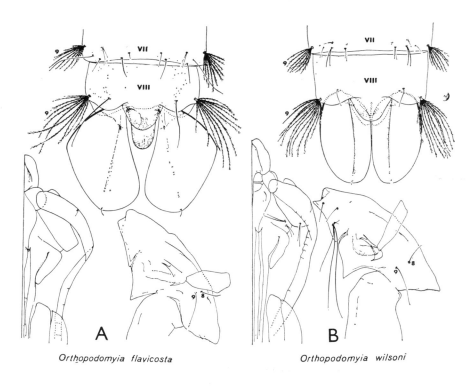

Fig. 49   Pupal cephalothorax and terminal segments, genus *Orthopodomyia*: A, *O. flavicosta*; B, *O. wilsoni*.

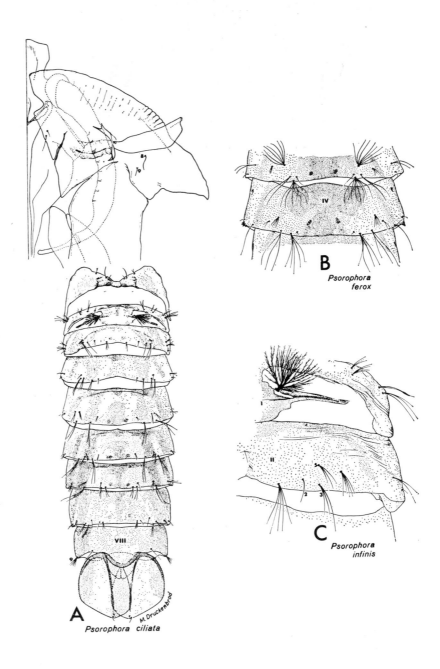

Fig. 50   Pupa, genus *Psorophora*: A, cephalothorax and abdomen of *Ps. ciliata*; B, segment IV of abdomen of *Ps. ferox*; C, segments I-II of *Ps. infinis*.

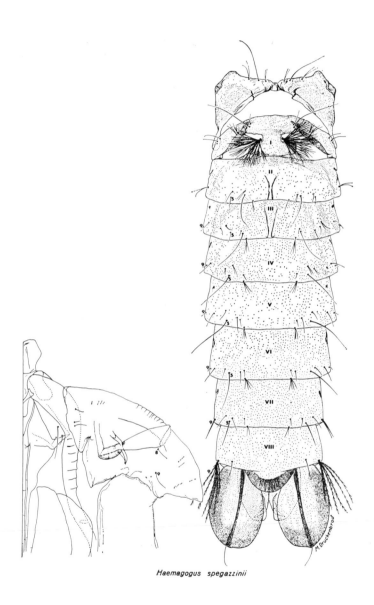

Haemagogus spegazzinii

Fig. 51 : Pupal cephalothorax and abdomen of *Haemogogus spegazzinii*.

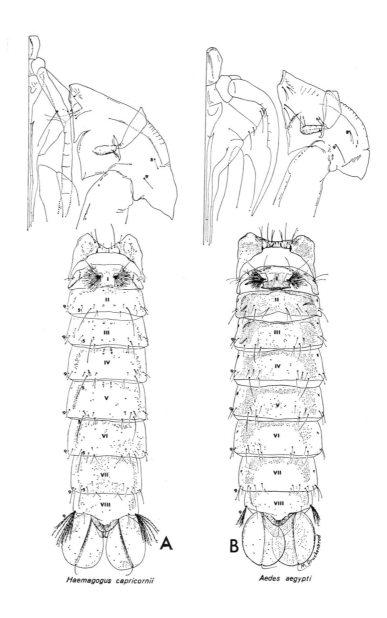

Fig. 52   Pupal cephalothorax and abdomen: A, *Haemogogus capricornii*; B, *Aedes aegypti*.

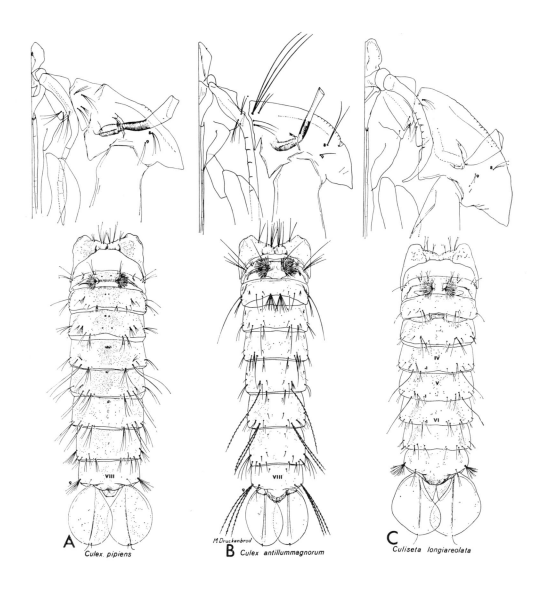

Fig. 53   Pupal cephalothorax and abdomen: A, *Culex pipiens*; B, *Culex antillummagnorum*; C *Culiseta longiareolata*.

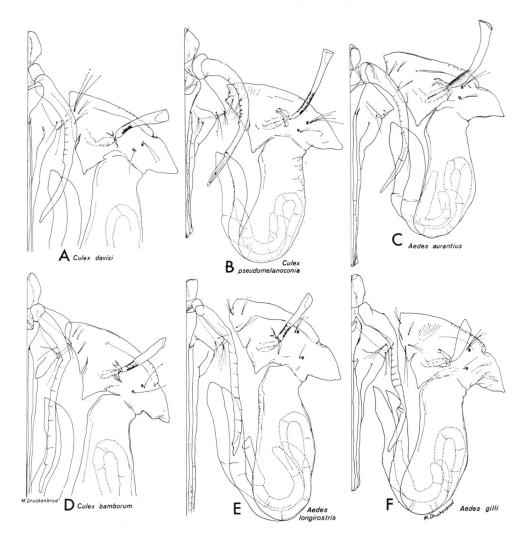

Fig. 54   Pupal cephalothorax: A, *Culex davisi*; B, *Culex bamborum*; C, *Culex pseudomelanoconia*;
C, *Aedes aurantius*; D, *Culex bamborum*; E, *Aedes longirostris*; F, *Aedes gilli*.

IV. Fourth Stage Larvae

The following lettering is used throughout the key to larvae:

- A   Antenna
- C   Head
- M   Mesothorax
- P   Prothorax
- PT Pecten tooth
- T   Metathorax

Other lettering is explained in the accompanying legends.

The larva of genus *GALINDOMYIA* is undescribed.

1   Respiratory siphon absent; seta 1 usually palmate on most abdominal segments
    (fig. 55) (Subfamily Anophelinae)   2

–   Siphon present; seta 1 never palmate.   4

2   Anterior flap of spiracular apparatus produced into a long, spinelike process, ventral
    valves with fringe of fine hairs; palmate hairs characteristically shaped (fig. 55A);
    New World tropics only   .   .   .   .   .   .   ***CHAGASIA*** Cruz

–   Spiracular apparatus and palmate hairs otherwise   3

3   Inner clypeal setae close together; seta 1 of mesothorax palmate (fig. 56); northern
    Australia, Melanesia and Moluccas only .   .   .   .   ***BIRONELLA*** Theobald

–   Without this combination of characters and distribution   .   ***ANOPHELES*** Meigen

4   Mouthbrushes with about ten flattened, non-pectinate blades; antenna with setae 2-A
    and 3-A arising basad of seta 1-A; comb and pecten absent (fig. 57) (Subfamily
    Toxorhynchitinae)   .   .   .   .   .   ***TOXORHYNCHITES*** Theobald[5]

–   Mouthbrushes with numerous hairs; antenna with setae 2-A and 3-A distad of 1-A;
    comb almost always present; pecten present or absent (Subfamily Culicinae)   .   5

5   Ventral brush with at most two pairs of setae, usually only one (fig. 58); New World
    only[47].   6

–   Ventral brush with three pairs of setae or more or species confined to Old World   .   10

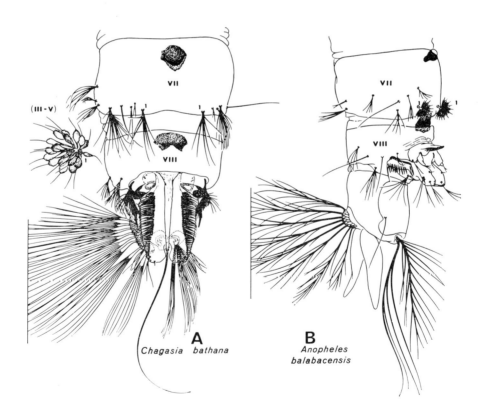

Fig. 55   Terminal segments of anopheline larvae: A, *Chagasia bathana*; B, *Anopheles balabacensis*.

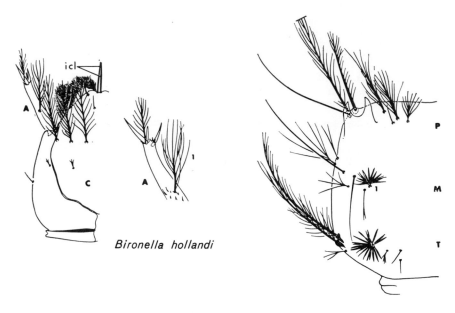

Fig. 56   Larval head and thorax of *Bironella hollandi*; icl, inner clypeal setae.

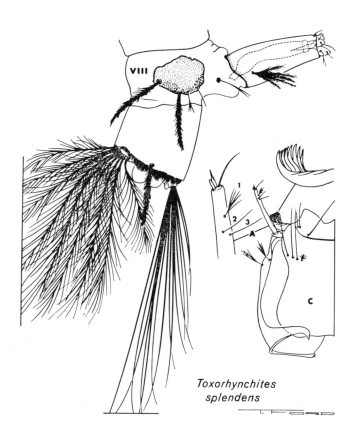

Fig. 57   Larval head and terminal segments of *Toxorhynchites splendens*.

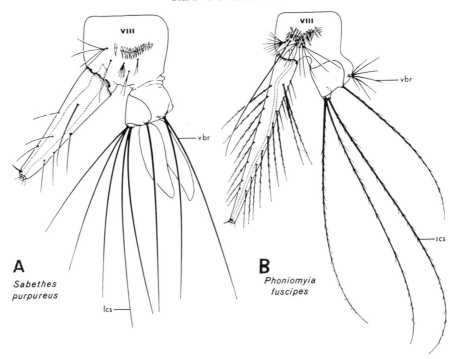

Fig. 58 Terminal segments of New World sabethine larvae: A, *Sabethes purpureus*: B, *Phoniomyia fuscipes*; vbr, ventral brush; lcs., lower caudal seta.

6　Setae of ventral brush as long, or almost as long, as lower caudal setae; siphon relatively slender, at least about 3.5 times as long as saddle; comb teeth in a single row or with at most three or four detached (fig. 58A) .　***SABETHES*** Robineau-Desvoidy
　　WYEOMYIA Theobald (part)[48]

-　Setae of ventral brush much shorter than lower caudal setae or siphon short and stout or comb teeth in at least two complete rows or with any combination of these　.　7

7　Siphon long, slender, strongly tapering, about five times as long as its breadth at base or more, with numerous long, unbranched setae dorsally and ventrally; comb teeth in two or more rows (fig. 58B)　.　.　.　***PHONIOMYIA*** Theobald
　　WYEOMYIA Theobald (part)[49]

-　Siphon otherwise or comb teeth in a single, regular row (sometimes arising from a sclerotized plate) or both　.　.　.　.　.　.　.　.　.　.　8

8　Siphon short, stout, at most about three times as long as its breadth at base with several branched setae dorsally and ventrally; head setae 4-C, 5-C and 6-C single; maxillary 'horns' not developed; comb with about 4–7 teeth in a single row, not arising from a sclerotized plate (fig. 59A) .　.　.　.　.　***LIMATUS*** Theobald

-　Without this combination of characters; maxillary 'horns' present in some species (fig. 59B)　.　.　.　.　.　.　.　.　.　.　.　.　9

9　Mandible greatly enlarged (fig. 60A) or maxilla with a large 'horn' (fig. 60B); siphon with a dense midventral row of setae extending for almost the whole length (fig. 59B) or setae of ventral brush at least three times as long as saddle
　　TRICHOPROSOPON Theobald

-　Mandibles never thus; maxilla seldom with conspicuous 'horn', if so then ventral brush or siphon otherwise　.　.　.　.　.　***WYEOMYIA*** Theobald (part)[50]

10　Ventral brush with a single pair of setae (one or two small, supplementary hairs present in occasional individuals); antenna short, without articulated apical segment; siphon with two or (usually) more subdorsal setae as well as various ventral or subventral setae (fig. 61)　.　.　.　.　.　.　.　.　.　11

-　Without this combination of characters[51]　.　.　.　.　.　.　.　13

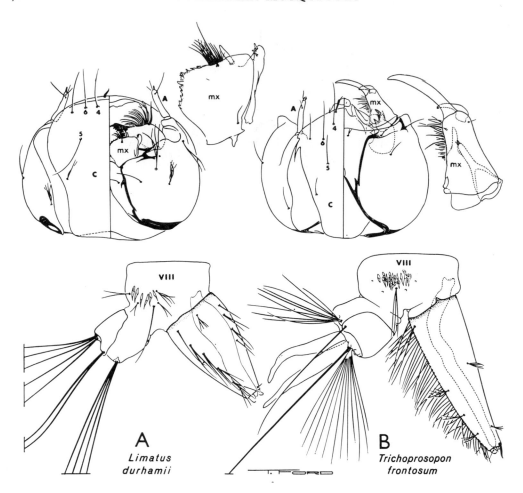

Fig. 59   Larval head and terminal segments: A, *Limatus durhamii*; B, *Trichoprosopon frontosum*; mx, maxilla.

11   Prothoracic setae 5 and 6 large fan-shaped tufts arising from a common tubercle; comb usually a patch of teeth in two or more rows (fig. 61); seta 6 of mesothorax and seta 7 of metathorax never stout spines; tropical Africa, southern Asia and Melanesia only   12

  –   Prothoracic setae otherwise; seta 6 of mesothorax and/or seta 7 of metathorax often a stout spine; comb teeth in a single row, sometimes arising from a sclerotized plate, occasionally absent or reduced to a single tooth (fig. 63), or distribution otherwise   .   .   .   .   ***TRIPTEROIDES*** Giles[52,53]

12   Either abdominal segments IV–VI (at least) with one or more pairs of stellate setae with numerous short, stiff branches (figs 61B, 64A) or maxillae with conspicuous 'horns' (fig. 64B) or siphon at least six times as long as saddle (fig. 64C); southern and eastern Asia and Melanesia only   .   .   .   .   ***TOPOMYIA*** Leicester[53]

  –   Setae of this kind never present; maxillae never with 'horns'; siphon at most about four times as long as saddle (fig. 61A); Old World tropics from Africa to eastern Asia and Melanesia   .   .   .   .   ***MALAYA*** Leicester

13   Siphon modified for piercing plant tissues, with sclerotized saw-toothed process at tip (fig. 65A)   .   .   .   .   ***MANSONIA*** Blanchard

  –   Siphon not so modified or, if so, without any saw-toothed process (fig. 65B)[54]   .   . 14

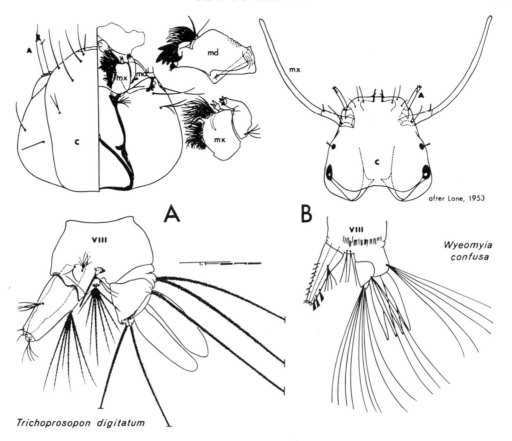

Fig. 60 Larval head and terminal segments: A, *Trichoprosopon digitatum*; B, *Wyeomyia confusa*; md, mandible; mx, maxilla.

14 Antenna broad, flattened; some thoracic setae enormously long, others stellate; tip
  of siphon with paired hooks and branched setae (fig. 66)   ***AEDEOMYIA*** Theobald
– Antenna, thoracic setae and siphon otherwise   15

15 Siphon with a single pair of subventral setae; metathoracic setae 9–12 very short,
  unbranched (fig. 67); New Zealand only   .   .   .   ***OPIFEX*** Hutton
– Without this combination of characters or distribution otherwise .   .   .   16

16 Siphon with a single pair of subventral setae arising at not more than one-fifth of the
  distance from base to apex; comb a single row of at most twenty teeth; siphon less
  than twice as long as saddle (fig. 68); Old World tropics only .   .   17
– Without this combination of characters or distribution otherwise[55]   .   .   18

17 Head seta 5 arising almost directly behind 6 which is single; head seta 4 nearly as
  long as 5; pecten with at least three teeth, usually more (fig. 68A) ***HODGESIA*** Theobald
– Head setae otherwise; pecten with at most two teeth (fig. 68B)   ***FICALBIA*** Theobald[56]

18 Distal portion of antenna freely articulated (fig. 69); siphon with a single pair of
  subventral setae; pecten with at most four teeth on either side, often fewer; ventral
  brush with 2–4 pairs of setae, one or two supernumerary setae occasionally present
  in addition; Old World tropics only   .   .   .   ***MIMOMYIA*** Theobald[14,15,54]
– Antenna never thus; siphon often otherwise; ventral brush with at least three pairs
  of setae, often with five or more pairs   19

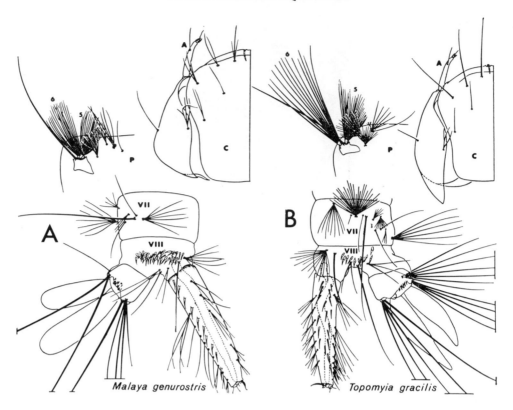

Fig. 61　　Larval head, prothorax and terminal segments: A, *Malaya genurostris*; B, *Topomyia gracilis*.

19　Pecten absent; siphon very short and broad with a single pair of subventral setae arising beyond half way; antennal seta minute (fig. 70A); southern Asia, Japan and Melanesia only　.　　.　　.　　.　　.　　.　　. ***ARMIGERES*** Theobald[57]

－　Without this combination of characters and distribution　.　　.　　.　　. 20

20　Pecten absent; antennal seta arising on basal half, with four or more branches; head setae 5 and 6 long and branched; siphon at least about 2.5 times as long as its breadth at base, often much longer, with a single pair of subventral setae; ventral brush with six pairs of setae or more (fig. 70B)　***ORTHOPODOMYIA*** Theobald[57]

－　Without this combination of characters　.　　.　　.　　.　　.　　. 21

21　Head with a pair of conspicuous lateral pouches; siphon with a pair of large sub-ventral setae and two pairs of smaller setae distal to this, one subnormal, the other subventral; saddle poorly developed; a small accessory sclerotized plate usually present basad of the ventral brush (fig. 71A); New World only

DEINOCERITES Theobald

－　Without this combination of characters　.　　.　　.　　.　　.　　.　　. 22

22　Siphon with a single pair of subventral setae arising at not less than a quarter of the distance from base to apex, usually more; (a pair of minute subdorsal setae also usually present near tip) (figs 73–76)[58]　.　　.　　.　　.　　.　　.　　. 23

－　Siphon with subventral setae more numerous or (in one or two neotropical species) entirely absent or if with a single pair of such setae then these arising at about one-fifth of the distance from base to apex or less (figs 71B, 80A)　.　　.　　.　　. 30

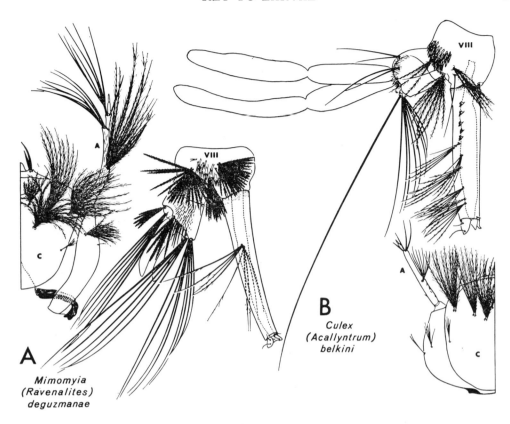

Fig. 62   Larval head and terminal segments: A, *Mimomyia (Ravenalites) deguzmanae*; B, *Culex (Acallyntrum) belkini*.

23   Maxillary suture absent or incomplete, not reaching posterior tentorial pit (fig. 72A); head seta 5 or 6 or both often flattened, barbed, spinelike; comb often arising from a large sclerotized plate (fig. 73A)   .   ***URANOTAENIA*** Lynch Arribalzaga[57,59]

  -   Maxillary suture well developed, extending from the level of the mentum to the posterior tentorial pit (fig. 72B); head setae 5 and 6 sometimes single and barbed but never spinelike; comb plate, if present, smaller (fig. 73B)   24

24   Antennal seta and head setae 4–7 small, delicate, inconspicuous; comb teeth never in a regular row; siphon at most about 3.5 times as long as its breadth at base, usually much shorter; pecten with at most seven teeth, usually fewer (sometimes absent); ventral brush with four (rarely five) pairs of stout, strongly plumose setae, some of them usually single (fig. 74A, C); Ethiopian Region only
        ERETMAPODITES Theobald[60]

  -   Without this combination of characters and distribution   25

25   Comb with at most ten teeth, in a single row; head seta 5 single, 4 and 6 shorter than 5 with two or more delicate branches, 7 single and much longer than any of these (fig. 75); South-east Asia only   .   .   .   .   .   ***UDAYA*** Thurman

  -   Comb various; head setae otherwise; distribution various   26

26   Comb teeth in a single row; saddle incomplete with strongly developed spines along the distal edge; head seta 7 slender, delicate, single or bifid, 5 and 6 long, single, somewhat stouter than 7 but not conspicuously thickened (fig. 76A); South-east Asia only   .   .   .   .   ***ZEUGNOMYIA*** Leciester[61]

  -   Without this combination of characters and distribution   27

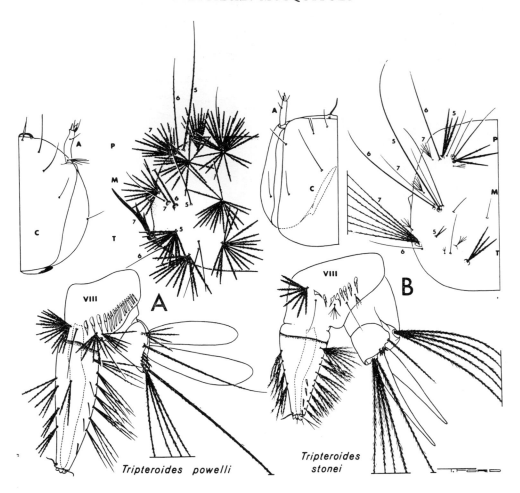

Fig. 63   Head, thorax and terminal segments of larva, genus *Tripteroides*: A, *T. powelli;* B, *T. stonei.*

27   Head seta 4 large and conspicuous, 6 markedly anterior to 5 and 7, 7 with at least
         five branches, usually more; stellate setae absent; thoracic integument devoid of
         spicules; comb teeth never fused at base or arising from a sclerotized plate; pecten
         teeth with secondary denticles, if any, confined to base; saddle incomplete (fig.
         76B); southern and eastern Asia only   .   .   .   **HEIZMANNIA** Ludlow[62]
  –   Without this combination of characters and distribution   28

28   Comb teeth in a single, regular row; anal segment completely ringed by saddle, the
         latter pierced in the mid line by the proximal setae of the ventral brush which
         forms a midventral row extending almost to the base of the anal segment (fig. 78A, B);
         New World only .   .   .   .   .   **PSOROPHORA** Robineau-Desvoidy[63]
  –   Without this combination of characters and distribution   29

29   Antenna short, smooth or almost so; antennal seta very small, single or bifid, rarely
         trifid; head setae 5 and 6 slender, single or bifid, 6 markedly anterior to 7; saddle
         incomplete; ventral brush arising from a sclerotized boss (fig. 79A); New World
         tropics and subtropics only   .   .   .   .   .   **HAEMAGOGUS** Williston[64]
                                                                            AEDES Meigen (part)
  –   Without this combination of characters and distribution   .   .   **AEDES** Meigen (part)

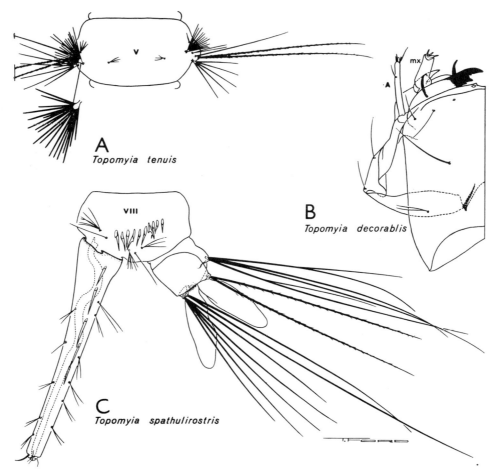

Fig. 64   Genus *Topomyia*: A, segment V of larval abdomen of *T. tenuis*; B, larval head of *T. decorabilis*; C, terminal segments of larva of *T. spathulirostris*; mx, maxilla.

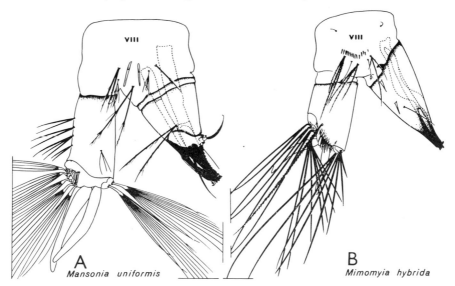

Fig. 65   Terminal segments of larva: A, *Mansonia uniformis*; B, *Mimomyia hybrida*.

G

30 Siphon with a single pair of subventral setae arising near base (with or without a midventral row of setae beyond this) (figs 71B, 80A) . . . *CULISETA* Felt[55]

– Siphon with subventral setae usually well removed from mid line (fig. 80B), occasionally forming a midventral row (fig. 80C) but never with a separate pair arising near base . . . . *CULEX* Linnaeus

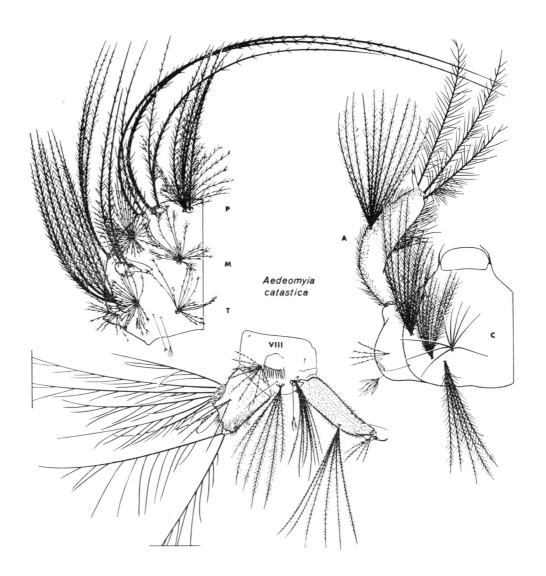

Fig. 66 Larval head, thorax and terminal segments of *Aedeomyia catasticta*.

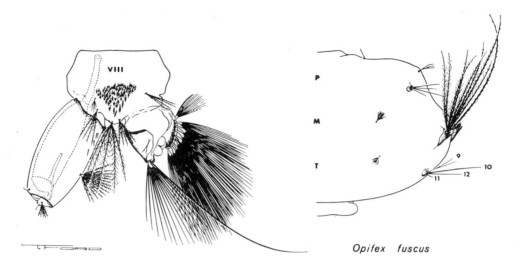

Fig. 67   Larval thorax and terminal segments of *Opifex fuscus*.

Notes on the Keys

1. Outside the genus *Bironella* this condition is found only in certain members of the *Anopheles leucosphyrus* Dönitz complex.

2. Genera *Malaya, Hodgesia, Uranotaenia, Zeugnomyia* and, in part, *Topomyia* are characterized by the fact that vein 1A turns down abruptly to reach the wing margin before or at most very slightly beyond the base of the fork of vein Cu (fig. 17A). This is a very distinctive character shared only by *Aedes* subgenus *Cancraedes* and the males of a few *Culex*. In some *Limatus* spp. there is a tendency for the tip of vein 1A to turn down abruptly but this seems always to take place well beyond the base of the fork of vein Cu. In *Zeugnomyia* the tip of vein 1A turns down abruptly but sometimes reaches the wing margin slightly beyond the base of the fork of vein Cu. Doubtful specimens are easily recognized by the vertical silvery stripe extending from the prealar area down to the mid coxa (fig. 18).

3. *Topomyia* and some species of the closely related genus *Malaya* (formerly *Harpagomyia*) are very similarly ornamented, having a conspicuous longitudinal white or silvery stripe on the scutum (occasionally brownish in *Topomyia*). They can, however, always be distinguished with ease by the proboscis. This is strongly swollen towards the tip in some *Topomyia* but never shows the extensive modifications, associated with feeding on the regurgitations of ants, which are found in *Malaya*.

4. Formerly called *Theobaldia* by workers in the Old World.

5. *Tripteroides, Malaya* and *Topomyia*, together with the New World genera in couplets 14–17, form the tribe Sabethini of the subfamily Culicinae. The other subfamilies of Culicidae are the Anophelinae (*Anopheles, Bironella, Chagasia*) and the Toxorhynchitinae with the single genus *Toxorhynchites* (*Megarhinus* of earlier authors).

6. It has been proposed to place the New Zealand species *T. argyropus* (Walker) in a monotypic genus *Maorigoeldia* but this is retained here as a subgenus of *Tripteroides*.

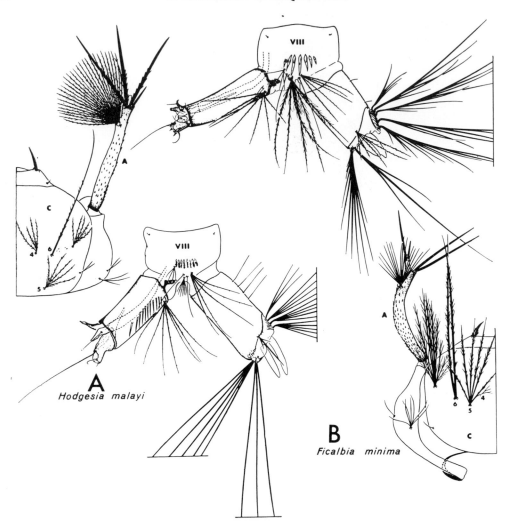

Fig. 68   Larval head and terminal segments: A, *Hodgesia malayi*; B, *Ficalbia minima*.

7. One species of *Phoniomyia* has the tarsi entirely dark in a proportion of individuals.
    The very short antenna, of the order of a third of the length of the proboscis, distin-
    guishes it from other New World sabethines with comparable proboscis.   Some *Tricho-
    prosopon* or a few *Wyeomyia* spp. with unusually long proboscis might be confused but
    these can be recognized by the entirely dark scutellum or tarsi or both.   A few *Tricho-
    prosopon* spp. with white tarsal markings have a peacock blue, rather than silvery, scale
    patch on the mid lobe of the scutellum.

8. A few *Trichoprosopon* spp. have scutal scales with dull, bluish lustre but they are never
    brightly iridescent.   A few *Wyeomyia* spp. with bright metallic scutal scaling might run
    down here but these are distinguished by the presence of prealar bristles.

9. Two species lacking postnotal bristles have been placed in a separate genus, *Mattinglyia*.
    I prefer to treat this as a subgenus of *Heizmannia* and am describing an annectant species
    elsewhere.

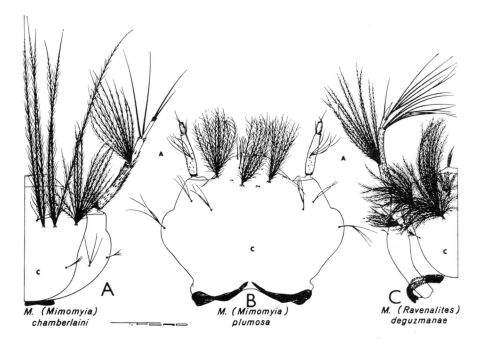

A — M. (Mimomyia) chamberlaini

B — M. (Mimomyia) plumosa

C — M. (Ravenalites) deguzmanae

Fig. 69   Larval head, genus *Mimomyia*: A, *M. (Mimomyia) chamberlaini*; B, *M. (M.) plumosa*; C, *M. (Ravenalites) deguzmanae*.

10.   Known only from a single species with undescribed early stages and uncertain affinities.

11.   *Armigeres* differs markedly from *Mansonia* in the broad scaled vertex and from *Udaya* in the heavily scaled pleuron.   *Aedes* is distinguished, except from *Armigeres* s. str., by the presence of postspiracular bristles.   *Aedes* subgenus *Alanstonea* resembles *Armigeres* s. str. closely in general facies but differs in having no lower mesepimeral bristle.   All other *Aedes* lack the curved proboscis.   Another character distinguishing many *Aedes* is the presence of acrostichal or dorsocentral bristles.

12.   Some authors recognize *Mansonia* and *Coquillettidia* as distinct genera, each with two subgenera.   I prefer to include all four subgenera in the genus *Mansonia*.   Apart from a single species all the subgenera, except *Coquillettidia*, run to the present couplet. *Coquillettidia* and one species of subgenus *Rhynchotaenia* run to couplet 32.

13.   *Culex postspiraculosus* Lee, from Australia, would also run down here.   It is the only known *Culex* in which postspiracular bristles occur otherwise than as an occasional aberration.   It is recognizable as a *Culex* by the well-developed pulvilli (see fig. 26A). One species of *Aedes*, formerly placed in a separate genus, *Ayurakitia*, lacks postspiracular bristles but runs down correctly, via couplet 25, to the present couplet since it has toothed fore and mid tarsal claws and palps less than a quarter of the length of the proboscis.   See also note 27.

14.   The genus *Ficalbia* is currently held to include four subgenera.   In my view, however, the nominotypical subgenus differs too widely from the others for this treatment to be acceptable.   I prefer, therefore, to treat *Ficalbia* s. str. as a separate genus, referring the other three subgenera to genus *Mimomyia*.   Diagnostic characters are given in the keys.

15.   One species extends beyond the tropic, in the extreme eastern part of its range, as far north as Okinawa.   Two others extend some distance south of the tropic in Queensland.

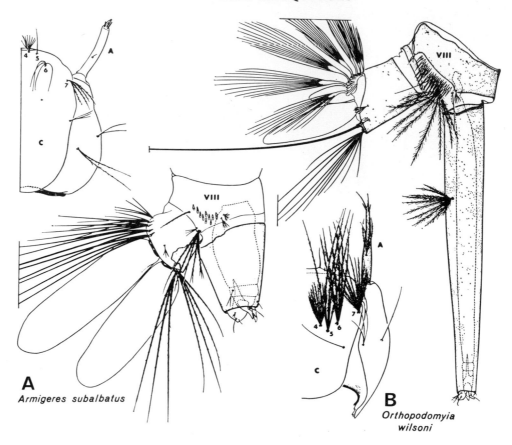

Fig. 70   Larval head and terminal segments: A, *Armigeres subalbatus*; B, *Orthopodomyia wilsoni*.

16.   Subgenus *Coquillettidia* and one species of subgenus *Rhynchotaenia* lacking postspiracular bristles (see note 12).   One species of *Armigeres*, with simple claws, would also run down here but this can be immediately recognized by the flat scaled vertex, palps more than half as long as proboscis and the curved proboscis (as in fig. 23).

17.   Based on external characters only.

18.   Males of *Aedes* subgenus *Cancraedes*, from southern Asia, would also run down here. They differ from *Uranotaenia* in having postspiracular bristles and two or more lower mesepimerals (one or none in *Uranotaenia*).

19.   A few species of *Wyeomyia*, with unusually long proboscis, might run down here but all of these have either the scutellum or the tarsi entirely dark scaled or both.   One species of *Phoniomyia* has the tarsi entirely dark in a proportion of individuals but this is recognizable from any *Wyeomyia* with a comparable proboscis by the very short antenna (not more than about one-third of the length of the proboscis).

20.   One or two species of *Wyeomyia* with bright metallic scutal scaling might run down here but these can be recognized by the presence of prealar bristles.

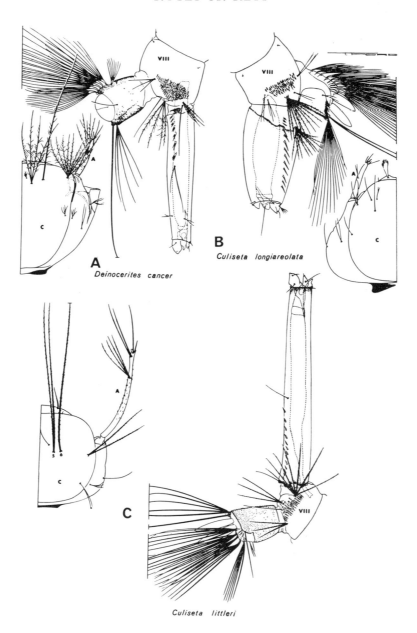

Fig. 71   Larval head and terminal segments: A, *Deinocerites cancer*; B, *Culiseta longiareolata*; C, *Culistea littleri*.

21.   The very long male palps will separate most *Trichoprosopon* spp. from other New World sabethines.   Some of the few *Trichoprosopon* spp. with short male palps have narrow scutal scales which are completely diagnostic.   The others have the two terminal antennal flagellomeres greatly elongated, together about ten times as long as the antepenultimate flagellomere or more (fig. 27D).   No *Trichoprosopon* spp. have bright, metallic scutal scaling though a few have scutal scales with dull bluish reflection.

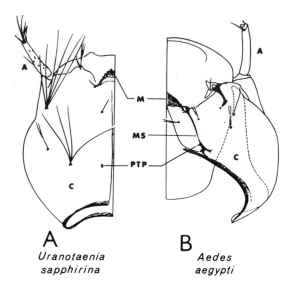

A
*Uranotaenia
sapphirina*

B
*Aedes
aegypti*

Fig. 72   Larval head in ventral view: A, *Uranotaenia sapphirina*; B, *Aedes aegypti*; *M*, mentum; MS, maxillary suture; PTP, posterior tentorial pit.

22.   See note 9.   Males of subgenus *Mattinglyia* differ from those of the nominotypical subgenus in having quite strongly plumose antennae and in lacking postnotal setae.

23.   Some *Haemagogus* have short male palps and reduced flagellar setae but their antenna in no way resembles that of *Galindomyia*.   It is subplumose with all flagellomeres except the last very short.

24.   All except subgenus *Coquillettidia* and one species of subgenus *Rhynchotaenia* (see notes 16 and 27).

25.   The Australian *Culex postspiraculosus* would run down here (see note 13).   It can be recognized by the long, upturned, tapering palps (fig. 30A).

26.   Subgenus *Coquillettidia* only (see note 24).

27.   Subgenus *Leicesteria* only. Individuals of *Mansonia* (*Rhynchotaenia*) *arribalzagai* Theobald lacking postspiracular bristles would also run down here but this species and subgenus are found only in the New World tropics.   *Aedes* (*Kompia*) *purpureipes* Aitken, which is almost unique among *Aedes* in lacking postspiracular bristles, would also run down here but this species is found only in the United States and Mexico.

28.   *Aedes* subgenus *Ayurakitia* would run down here (see note 13).   It includes only one species and is known only from Thailand.   It differs from *Orthopodomyia* in many details of ornamentation among them the restriction of the pleural scaling to four small silvery spots.

29.   In some *Bironella* spp. the accessory paddle seta is absent or arises level with the apical paddle seta.   These can at once be recognized as anopheline by the character of seta 9 and the trumpets.   Among non-anopheline genera only *Aedeomyia* (fig. 38B) has seta 9 approximately as in *Anopheles*.   One species of *Uranotaenia* has a trumpet of anopheline type but this is recognizable by the other characters given.

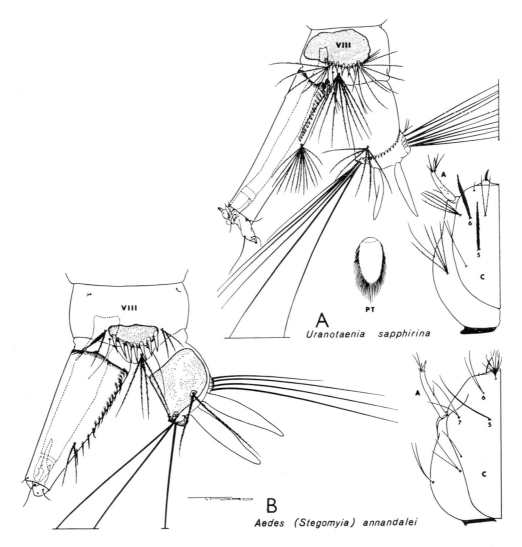

Fig. 73   Larval head and terminal segments: A, *Uranotaenia sapphirina*; B, *Aedes* (*Stegomyia*)
annandalei.

30.   Some *Mimomyia* have a rather similar trumpet but, apart from the African *M. perplexens*,
these have the float hair well developed. *M. perplexens* has the trumpet modified in quite
a different way from any *Mansonia* sp. (fig. 39A). Pupae of subgenus *Coquillettidia*,
and apparently also *Rhynchotaenia*, have the tips of the trumpets equipped with backwardly directed barbs which prevent them being withdrawn. The tips break off short
when the pupa rises to the surface at the time of emergence. They are consequently
seldom seen in cast skins.

31.   Absent only in some *Mimomyia* with a highly characteristic type of paddle.

32.   Complete separation of *Tripteroides* pupae from those of other Old World sabethines is
not at present possible but almost all can be recognized by the combination of characters
given in the key.

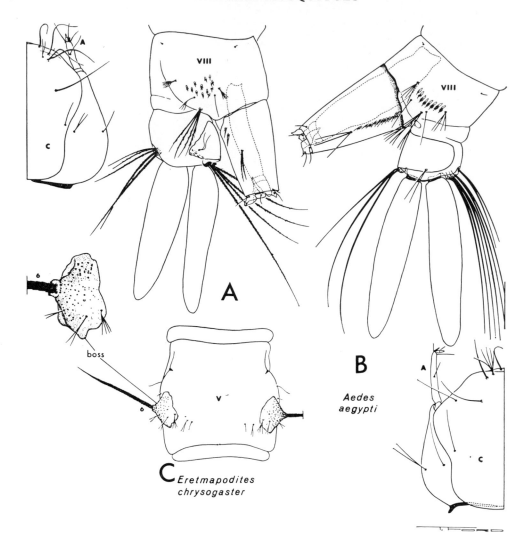

Fig. 74 A, B, larval head and terminal segments: A, *Eretmapodites chrysogaster*; B, *Aedes aegypti*; C, segment V of larval abdomen of *Eretmapodites chrysogaster*.

33. All but two of the known *Topomyia* pupae have seta 6-VII as in fig. 35. Both the others have it as in *Malaya* but differ from that genus in having all or most of the paddle surface spiculate (fig. 34A).

34. *Trichoprosopon magnum* (fig. 36E) has longer trumpets than are usual in the genus but differs from all known *Phoniomyia* in having them strongly expanded at the tip and from known *Limatus* spp. in the large paddles and relatively shallow excavation of the posterior border segment VIII (fig. 34F). *Wyeomyia circumcincta* would key here with *Phoniomyia* but has the trumpets much longer than any known in that genus (fig. 36F).

35. Pupae of *Wyeomyia* cannot be entirely separated from those of *Sabethes* or *Trichoprosopon* on currently available material and descriptions.

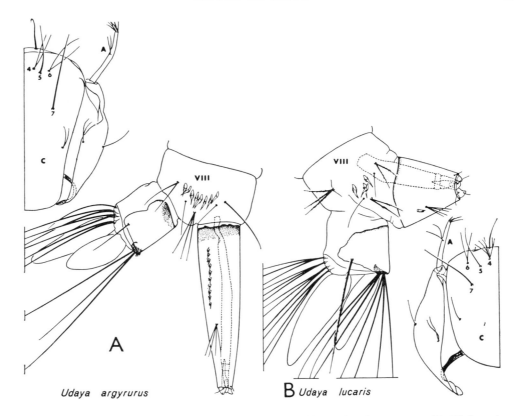

Fig. 75   Larval head and terminal segments, genus *Udaya*: A, *U. argyrurus*; B, *U. lucaris*.

36.   Some *Sabethes* spp. have seta 9-VI more strongly developed, others less strongly developed, than in the figure but it is always more conspicuous than in any known *Trichoprosopon*.   All those *Sabethes* spp. with feebly developed seta 9-VI which are known to me have seta 5-VI more strongly developed than in any known *Trichoprosopon* except perhaps *Tr. magnum* with its distinctive trumpet (fig. 36E and see note 34).

37.   Subgenera *Etorleptiomyia* and *Ravenalites* of genus *Mimomyia* have highly distinctive pupae unlikely to be confused with any others (fig. 39D, E).   The resemblance of some pupae of subgenus *Mimomyia* to those of *Mansonia* has already been mentioned (note 30).   A few *Uranotaenia* resemble *Mimomyia* superficially in the shape of the paddle and the long trumpets but can be recognized by the excavation of the basal part of the inner half of the paddle and the presence of paired setulae on abdominal segment IX (fig. 41A).   The African *M. splendens* Theobald is unique in having a fringe of long, hairlike spicules on both borders of the paddle but it should run down correctly.

38.   The characteristic paddle shape is usually conspicuous, least so in one or two aberrant species resembling the Sabethini but differing from these in the presence of an apical paddle seta (fig. 41C).   A few *Aedes* (*Lorrainea*) and *Armigeres* s. str. have a similar type of paddle but with a delicate fringe on both borders which is very rare in *Uranotaenia* (figs 43A, B; 46A).   The few *Uranotaenia* with such a fringe can be recognized from both the above genera by the reduction of seta 9-VIII and the paddle seta (fig. 41B).   One or two *Haemagogus* spp. have a somewhat similar paddle but are easily recognized by the combination of characters given in couplet 27.   Paired setulae are present on abdominal segment IX in most *Uranotaenia* (fig. 41A) but absent in a few atypical ones (fig. 41B, C).

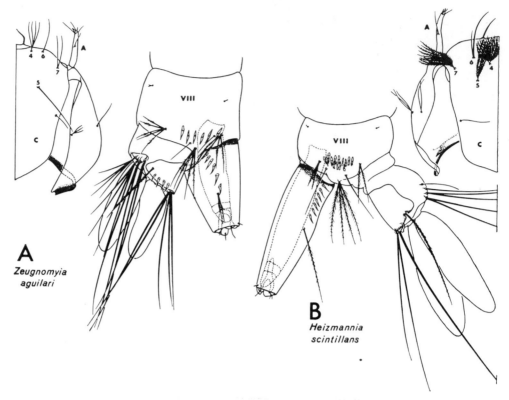

Fig. 76   Larval head and terminal segments: A, *Zeugnomyia aguilari*; B, *Heizmannia scintillans*

39. The only species which might be confused are a few members of the *Aedes* (*Stegomyia*) *africanus* Theobald group (fig. 42C) but these have the paddle fringe and apical seta less strongly developed and differ in various other characters among them the cephalothoracic setae as indicated in the key.

40. *Armigeres malayi* (fig. 43B) has seta 6-VI less strongly developed than in other species but still much better developed than 5-VI.   One species of *Heizmannia* (fig. 43C) would run down here but this differs from *Arm. malayi* in the much more strongly developed seta 9-VIII and from other *Armigeres* in having seta 6-VI much less strongly developed.

41. *Aedes* (*Lorrainea*) spp. have a similar paddle fringe but are recognizable by the charac-teristic paddle shape (fig. 46A and see note 38).   *Ae.* (*Diceromyia*) spp. differ in the in-dented tip of the paddle (fig. 46B) except for *Ae. periskelatus* which has the apical paddle seta longer and stouter than in any *Heizmannia* (fig. 46C) as do some *Stegomyia* (fig. 47A). A few other *Stegomyia* present difficulties but can be recognized by having seta 5 on most segments intermediate in type (fig. 47B), less reduced than in *H. aureochaeta* (fig. 45B), more so than in other *Heizmannia* (figs 44D, 45A).

42. Some *Aedes*, particularly *Ae.* (*Finlaya*) spp., have a similar paddle but differ either in the much longer apical paddle seta or in the reduction of seta 9-VII–VIII or both.   The relative position of setae 8-C and 9-C also furnishes an absolute distinction from *Aedes* (see fig. 52B).

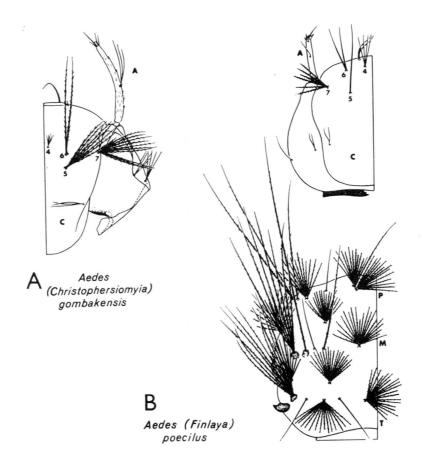

A
*Aedes
(Christophersiomyia)
gombakensis*

B
*Aedes (Finlaya)
poecilus*

Fig. 77   A, larval head of *Aedes* (*Christophersiomyia*) *gombakensis*; B, larval head and thorax of *Aedes* (*Finlaya*) *poecilus*.

43.   Subgenera *Psorophora* and *Janthinosoma* are distinguished from other genera by the lobed posterior border of segment VIII and/or the toothed posterolateral corner of segment IV.   Most species of subgenus *Grabhamia* are distinguished, except from *Culex* and one species of *Culiseta* by the presence of an accessory paddle seta.   *Culex* are readily distinguished by the tracheated trumpet or anterior displacement of seta 9-VIII, usually both.   The only New World *Culiseta* with an accessory paddle seta, *C. melanura* Coquillett, is distinguished by having the trumpet cleft nearly to base, i.e. with very short tubular portion.   *Aedes atropalpus* (Coquillett) is said sometimes to have an accessory paddle seta but this species can be recognized by the short, broad paddle, almost as broad as, or broader than, long with apex flattened and more or less indented.   In *Grabhamia* the paddle is distinctly longer than broad or has a rounded or pointed apex, usually both.   One species of *Grabhamia* from the Caribbean area lacks the accessory paddle seta and would run to *Aedes* in couplet 28.   It can be recognized by having seta 2-II well inside 3-II and seta 5-II almost directly anterior to 3-II (fig. 50c).

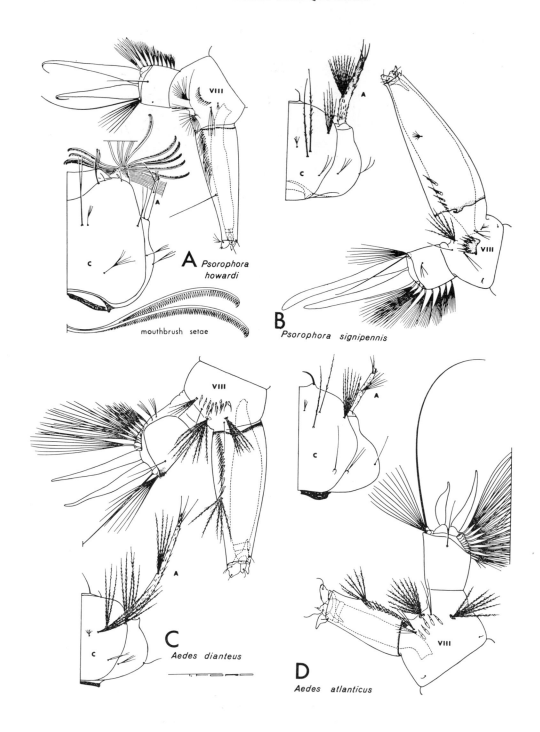

mouthbrush setae

A *Psorophora howardi*

B *Psorophora signipennis*

C *Aedes dianteus*

D *Aedes atlanticus*

Fig. 78   Larval head and terminal segments: A, *Psorophora howardii* Coquillett; B, *Psorophora signipennis* (Coquillett); C, *Aedes diantaeus* Howard, Dyar & Knab; D, *Aedes atlanticus* Dyar & Knab.

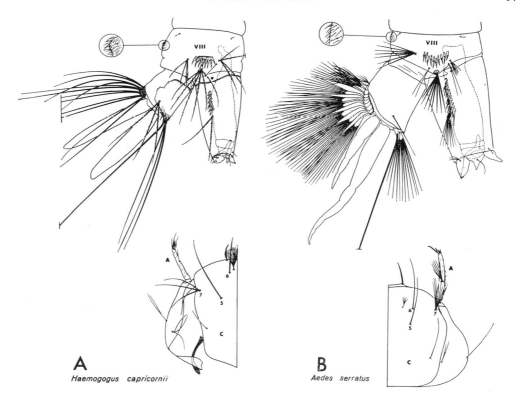

Fig. 79   Larval head and terminal segments: A, *Haemogogus capricornii*; B, *Aedes serratus*.

44.   The only *Haemagogus* sp. known to me which might not run down here is *H. chalco-spilans* Dyar.   It is recognizable from any known *Aedes* or *Psorophora* by the combination of very short setae 5-II and 5-III, setae 5-IV and 5-V only a little longer than the following segments, seta 9-III–VI minute, seta 9-VII with several branches, 9-VIII about half the length of the paddle, paddle with apex pointed and short, branched apical paddle seta. A number of *Aedes* otherwise resembling *Haemagogus* have seta 5 on one or more abdominal segments as long as the two following segments together.   In *Haemagogus* seta 5 is never more than a little longer than the following segment.   *Aedes aegypti* is at once distinguished by the well developed seta 9-III–VI (fig. 52B).

45.   The relative position of setae 8-C and 9-C appears to be almost completely diagnostic as between *Aedes* and *Culex*.   In the former however seta 8-C occasionally arises somewhat posterior to the trumpet (fig. 52B) while in the latter it arises farther forward in some species than in others.   In comparing pupal skins, therefore, the specimens must be carefully oriented.   A few small New World *Culex* apparently have these setae as in *Aedes* but these have trumpets with subbasal tracheation of a kind unknown in the New World aedine genera (fig. 53B).   In the great majority of *Culex* seta 9-VIII arises well cephalad of the posterior border of the segment (fig. 53A).   The few species known to me in which this is not so have slender trumpets with extensive tracheation of a kind almost unknown in *Aedes* (fig. 54A, B, D).   They are all small or very small species.   The few *Aedes* with comparable tracheation are all Old World species and are either large or very large species or have seta 8-C much farther forward than in any Old World

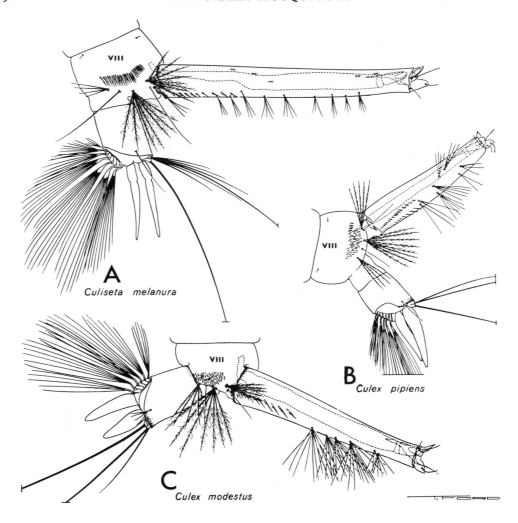

Fig. 80   Terminal segments of larva: A, *Culiseta melanura*; B, *Culex pipiens*; C, *Culex modestus*.

Culex (fig. 54C, E).   In contrast to this seta 9-VIII rarely arises cephalad of the posterior border of the segment in *Aedes*.   All the species in which it is known to do so have trumpets with poorly developed basal tracheation and seta 8-C arising well forward (fig. 54F, see also note 43).

46.   The relative position of setae 8-C and 9-C appears to be absolutely diagnostic as between *Aedes* and *Culiseta* (figs 52B, 53C).   In the latter seta 8-C always arises far back.   Additional partial characters are the abnormally long seta 1-IV–VI in most *Culiseta* (but not the tropical African species) and the presence of an accessory paddle seta in about half the known species.   These characters should not, however, be needed for diagnosis.

47.   Two species of *Sabethes* have a pair of small, supernumary setae in addition to the single pair of well-developed setae by which the ventral brush is represented in all other known sabethines.

48. Most *Wyeomyia* spp. have the setae of the ventral brush short (often stellate). In some they are longer, though still much shorter than the lower caudal seta. All but about three of the remaining species can be recognized by the arrangement of the comb teeth as indicated in the key. Some *Sabethes* spp. have a pair of dorsal, chitinous hooks on abdominal segment VII. These are unknown in any other New World genus.

49. A few *Trichoprosopon* spp. have the siphon moderately long, slender, tapering with several short subdorsal setae. In these, however, the ventral setae form a long row of branched tufts quite unlike anything found in *Phoniomyia*. Two or three *Wyeomyia* spp. would run down here and are best separated at the species level.

50. None of the few species of *Wyeomyia* with the siphon as in fig. 59B has a conspicuous maxillary horn. Two species which might run down to the first half of the present couplet are *W. confusa* (fig. 60B) and *W. occulta* Bonne-Wepster & Bonne. The former, however, has an enormously elongated maxilla, quite unlike that of any other species in either genus, while the latter differs from any known *Trichoprosopon* lacking a mid-ventral row of setae on the siphon in having a comb of numerous scales in a patch three rows deep. An additional character serving to distinguish most, though not quite all, species of *Trichoprosopon* with the siphon otherwise than in fig. 59B is the very small number of comb teeth (five or less; not less than ten in *Wyeomyia* spp. with maxillary horn).

51. Two African *Culex* spp. lack the ventral brush entirely. In *Mimomyia* subgenus *Ravenalites* it comprises two pairs of setae only. Some members of this subgenus have a superficial resemblance to *Tripteroides* but they can be recognized at once by the articulated terminal joint of the antenna and the presence of only one pair of subventral setae on the siphon (fig. 62A). In *Culex (Acallyntrum)* spp. from Melanesia the ventral brush is represented by a single pair of setae. Larvae of these species differ from those of the sabethine genera in lacking dorsal or subdorsal setae on the siphon (fig. 62B). In addition the larvae of the sabethine genera almost always have the antenna and antennal seta much reduced while in many the thorax and abdomen are covered with large stellate setae (fig. 63A), a condition not met with in *Acallyntrum*.

52. Some *Topomyia* have the comb reduced to a single row or almost so. These, however, can at once be separated from *Tripteroides*, as can all other *Topomyia* and *Malaya*, by the large fan-shaped setae 5 and 6 on the prothorax (fig. 61). A few *Topomyia* spp. have well-developed stellate setae on the thorax and/or abdomen but these are never so strongly developed as in many *Tripteroides* (figs 61B, 63A). The modified mesothoracic seta 6 and/or metathoracic seta 7 (fig. 63A), when present as they are in the majority of species, are completely diagnostic of *Tripteroides*. *T. (Maorigoeldia) argyropus* (Walker) is unique in the genus in having a comb with numerous scales in a patch but this species is confined to New Zealand.

53. Two of the three known species of *Topomyia* with strongly developed maxillary horns lack stellate setae on the abdomen. They are, however, readily distinguished from *Malaya* in which, so far as is known, maxillary horns never occur. Horns of this kind are found in various Australasian *Tripteroides* but not in either of the known Australasian *Topomyia* (one undescribed). They are found in two Philippines *Tripteroides* (one undescribed). One species of Topomyia (*T. spathulirostris*) lacks maxillary horns and has the stellate setae poorly developed. It can, however, be easily recognized by the siphon which is much longer than in any *Malaya* (fig. 64C).

54. A few African, South-east Asian and Melanesian *Mimomyia* have a piercing siphon (fig. 65B) but this is much less elaborately modified than in *Mansonia* and there should be no danger of confusion.

H

55.   Most *Culiseta* have a pair of subventral setae arising at or near the base of the siphon and are without other siphonal setae, but only three of them are found in the tropics. Two of these are recognizable by the fact that the comb teeth are arranged in a patch (fig. 71B) and the third by the presence of a midventral row of setae on the siphon.

56.   An additional character is the presence in most *Hodgesia* larvae of several long, very delicate setae on the anal segment anterior to the barred area of the ventral brush (fig. 68A).   These are easily broken off, however, and when this happens their bases are virtually impossible to detect.   One or two short, delicate setae anterior to the ventral brush are also sometimes seen in *Ficalbia*.

57.   The only culicine genera consistently lacking a pecten are *Armigeres* and *Orthopodomyia*. Among the other genera remaining to be keyed this condition is found in whole or in part in two species of *Culex*, two of *Uranotaenia* and some half dozen *Eretmapodites*. The latter are confined to the Ethiopean Region where *Armigeres* is absent and *Orthopodomyia* very rare.   They differ from *Orthopodomyia* in many characters including the reduced antennal seta, head setae and ventral brush (fig. 74A).   Of the two *Culex* spp. *C. moucheti* Evans lacks the ventral brush entirely while *C. dispectus* Bram has four pairs of subventral tufts on the siphon.   The two *Uranotaenia* spp. both occur outside the known geographical range of *Armigeres* and *Orthopodomyia*.   One (*U. browni* Mattingly from the Seychelles) has the siphon densely covered with long spicules.   The other (*U. colocasiae* Edwards from Fiji) has the ventral brush much as in *Eretmapodites*.

58.   One species of *Culiseta*, known only from southern Australia, would run down here instead of to the second half of the couplet.   It is easily recognized from any other species with which it might be confused by the long siphon, complete saddle and long head setae 5 and 6 and antennae (fig. 71C).

59.   New World *Uranotaenia* larvae are recognizable also by the more or less thickened, barbed, spinelike head setae 5 and 6, very large comb plate and apically fringed pecten scales.   Many of the Old World species also exhibit some or all of these characters but some do not and for these the only absolutely diagnostic character is the absence or extreme reduction of the maxillary suture (fig. 72A).

60.   The only species likely to be confused are some *Aedes* subgenus *Stegomyia*.   These are easily recognized by the long, slender, non-plumose setae of the ventral brush (fig. 74B).   Most of them also differ in having the comb teeth in a single row and in the more numerous pecten teeth.   A character which is completely diagnostic for *Eretmapodites*, when present, is the occurrence of rugose, sclerotized bosses at the bases of the principal lateral abdominal setae (fig. 74C).

61.   Among the remaining South-east Asian genera with a single pair of subventral setae on the siphon *Heizmannia* is distinguished by the much more strongly developed head seta 7 (and usually also 5) and less strongly spiculate saddle edge (fig. 76B).   Many *Aedes* differ in having head seta 7 (at least) more strongly developed or the comb teeth in more than one row.   Those which do not differ in either of these characters, e.g. most *Stegomyia*, are distinguished by having the saddle edge finely spiculate at most while some also differ in having the comb teeth arising from a sclerotized plate.   One South-east Asian *Finlaya* (*Aedes harperi* Knight from the Philippines) seems from the published description to be distinguishable only at the species level.

62.   Larvae of this genus closely resemble those of some *Aedes*.   They often have a characteristic head seta 6 with two unequal branches (fig. 76B).   This, when present, is absolutely

diagnostic except from *Aedes* subgenus *Christophersiomyia*, easily recognized by the much smaller head seta 4 (fig. 77A) and one or two *Aedes* (*Finlaya*) spp. differing in other characters given in the key. *Ae.* (*Christophersiomyia*) spp., like many other *Aedes*, are also distinguished by the fact that head seta 6 arises only very slightly, if at all, anterior to 7. *Aedes* spp. resembling *Heizmannia* in the well-developed head seta 4 and the position of head seta 6, e.g. many *Ae.* (*Stegomyia*) spp., can be recognized by the feebly developed head seta 7 (fig. 73B). Other characters serving to separate a few doubtful species are the presence of stellate setae on thorax and abdomen (fig. 77B) or the possession of a spiculated thoracic and abdominal integument.

63. Larvae of subgenus *Psorophora* s. str. have the mouthbrushes modified for predation (fig. 78A) and are unlikely to be mistaken for anything else. Those of the other subgenera show some resemblances to *Aedes*. However, most New World *Aedes* have the comb teeth in two or more rows while those which do not either have the saddle incomplete (fig. 78C) or the ventral brush confined to that part of the anal segment which is not ringed by the saddle (fig. 78D). None of them have the comb arising from a sclerotized plate as in many *Psorophora* (fig. 78B). Some *Haemagogus* have the comb arising from a small sclerotized plate but these all have the saddle incomplete and lack the anterior extension of the ventral brush. They are also distinguished by the much reduced antenna (fig. 79A).

64. Most New World *Aedes* are distinguishable from *Haemagogus* either by the longer or more strongly spiculate antenna, position and branching of the head setae, complete saddle or absence of lateral sclerotizations from the barred area at the base of the ventral brush (fig. 79B) or by a combination of these. There remain a number of *Aedes*, particularly in the subgenera *Finlaya* and *Howardina*, which are best separated at the species level.

BIBLIOGRAPHY

As a source of references to the taxonomic literature the Synoptic Catalog (Stone *et al.*, 1959) and Supplements (Stone, 1961–70) are indispensable. They can be kept up to date with the aid of one of the monthly bibliographies. The *Bulletin Analytique d'Entomologie Médicale et Vétérinaire*, published by the Office de la Recherche Scientifique et Technique Outre-mer at Bondy, Seine-et-Oise, is the best. As a general introduction to the fauna of the various regions, and a source of further references, the following are recommended.

Palaearctic Region

Gutsevich *et al.* (1970), Guy (1959), Kramar (1958), La Casse & Yamaguti (1950), Mattingly & Knight (1956), Mihályi (1963), Mohrig (1969), Natvig (1948), Rioux (1958), Senevet & Andarelli (1959), University of Maryland (1963).

Ethiopian Region

Edwards (1941), Gillies & De Meillon (1968), Grjebine (1966), Haddow *et al.* (1951), Hamon (1963), Hopkins (1952), Leeson (1958), Mattingly (1952, 1953), Mattingly & Brown (1955), Mattingly & Lips (1953), Muspratt (1955, 1956), Ovazza *et al.* (1956), Van Someren *et al.* (1955).

Oriental Region

Barraud (1934), Bonne-Wepster (1954), Borel (1930), Bram (1967a), Christophers (1933), Delfinado (1966), Lien (1968), Macdonald (1957, 1960), Mattingly (1965), Reid (1968), Reuben (1969), Thurman (1959), Tyson (1970), Wharton (1962).

Australian Region (including Micronesia)

Belkin (1962, 1968), Bohart (1956), Dobrotworsky (1965), Huang (1968a, b), Marks (1954), O'Gower (1958), Ramalingham & Belkin (1965), Sirivanakarn (1968), Steffan (1966, 1968), Van den Assem & Bonne-Wepster (1964).

Nearctic Region

Barr (1958), Beckel (1954), Carpenter (1968), Carpenter & La Casse (1955), Darsie (1951), Gjullin *et al.* (1961), King *et al.* (1960), Ross & Horsfall (1965), Smith (1969), Steward & McWade (1961), Vockeroth (1954), Zavortink (1970).

Neotropical Region

Belkin *et al.* (1970), Belkin & Hogue (1959), Berlin (1969a, b), Bram (1967), Correa & Ramalho (1956), Cova-Garcia (1961), Cova-Garcia *et al.* (1966), Foote (1954), Forattini (1962, 1965a, b), Galindo *et al.* (1954), Garcia & Ronderos (1962), Guedes *et al.* (1965), Lane (1953), Ronderos & Bachmann (1963b), Schick (1970).

General

Certain of the above overlap two or more adjacent regions, notably Mattingly & Knight, 1956 (Palaearctic, Oriental and Ethiopian), University of Maryland, 1963 (Palaearctic and Oriental), Bonne-Wepster, 1954 (Oriental and Australian), Belkin & Hogue, 1959, Bram, 1967b, Foote, 1954, Galindo *et al.*, 1954, Zavortink, 1970 (Neotropical and Nearctic). Recent taxonomic works with world coverage are Maslov (1967) and Zavortink (1968). Foote & Cook (1959) is useful as a general survey of vector species.

BIBLIOGRAPHY

AITKEN, T. H. G., SPENCE, L., JONKERS, A. H. & DOWNS, W. G. 1969. A 10-year survey of Trinidadian arthropods for natural virus infections (1953–1963). *J. med. Ent.* **6**: 207–215.

BARR, A. R. 1958. The mosquitoes of Minnesota (Diptera: Culicidae: Culicinae). *Tech. Bull. Univ. Minn. Agric. Exp. Stn*, **228**. 154 pp.

BARRAUD, P. J. 1934. *The Fauna of British India.* Diptera, 5 (CULICIDAE: Megarhinini and Culicini). London: Taylor & Francis. 463 pp.

BATES, M. 1949. *The Natural History of Mosquitoes.* 378 pp. New York. (Republished by Harper Torchbooks, 1965).

BECKEL, W. E. 1954. The identification of adult female *Aedes* mosquitoes (Diptera, Culicidae) of the black-legged group taken in the field at Churchill, Manitoba. *Can. J. Zool.*, **32**: 324–330.

BELKIN, J. N. 1962. *The Mosquitoes of the South Pacific (Diptera, Culicidae).* Berkeley & Los Angeles: Univ. Calif. Press. 2 Vols. 1,020 pp.

—— 1968. Mosquito studies (Diptera, Culicidae) VII. The Culicidae of New Zealand. *Contr. Am. ent. Inst.* **3** (1): 1–182.

BELKIN, J. N., HEINEMANN, S. J. & PAGE, W. A. 1970. Mosquito studies (Diptera, Culicidae) XXI. The Culicidae of Jamaica. *Contr. Am. ent. Inst.* **6** (1): 1–458.

BELKIN, J. N. & HOGUE, C. L. 1959. A review of the crabhole mosquitoes of the genus *Deinocerites* (Diptera, Culicidae). *Univ. Calif. Publs Ent.* **14**: 411–458.

BERGE, T. O., SHOPE, R. E. & WORK, T. H. 1970. Catalogue of arthropod-borne viruses of the World. *Am. J. trop. Med. Hyg.* **19**: 1082–1160.

BERLIN, O. G. W. 1969a. Mosquito studies (Diptera, Culicidae) XII. A revision of the neotropical subgenus *Howardina* of *Aedes. Contr. Am. ent. Inst.* **4** (2): 1–190.

—— 1969b. Mosquito studies (Diptera, Culicidae) XVIII. The subgenus *Micraedes* of *Culex. Contr. Am. ent. Inst.* **5** (1): 21–63.

BOHART, R. M. 1956. *Insects of Micronesia* **12** (1). *Diptera: Culicidae.* Honolulu: Bishop Museum. 85 pp.

BONNE-WEPSTER, J. 1954. Synopsis of a hundred common non-anopheline mosquitoes of the Greater and Lesser Sundas, the Moluccas and New Guinea. *Publs R. trop. Inst. Amsterdam* **111**. 147 pp.

BOREL, E. 1930. Les moustiques de la Cochinchine et du Sud-Annam. *Monogr. Colln Soc. Path. exot.* **3**. 423 pp.

BRAM, R. A. 1967a. Contributions to the mosquito fauna of Southeast Asia II. The genus *Culex* in Thailand (Diptera: Culicidae). *Contr. Am. ent. Inst.* **2** (1): 1–296.

—— 1967b. Classification of *Culex* subgenus *Culex* in the New World (Diptera: Culicidae). *Proc. U.S. natn. Mus.* **120**: 1–120.

BUSVINE, J. R. 1966. *Insects and Hygiene.* 2nd Edn. 467 pp. London.

CARPENTER, S. J. 1968. Review of recent literature on mosquitoes of North America. *Calif. Vector Views* **15**: 71–98.

CARPENTER, S. J. & LA CASSE, W. J. 1955. *Mosquitoes of North America.* 360 pp. Berkeley & Los Angeles: Univ. Calif. Press.

CHRISTOPHERS, S. R. 1933. *The Fauna of British India. Diptera* **4** (CULICIDAE: *Anophelini*). London: Taylor & Francis. 371 pp.

—— 1960a. *Aedes aegypti (L.), the Yellow Fever Mosquito.* 739 pp. Cambridge Univ. Press.

—— 1960b. The generic name *Aedes. Trans. R. Soc. trop. Med. Hyg.* **54**: 407–408.

CLEMENTS, A. N. 1963. *The Physiology of Mosquitoes.* 393 pp. London.

CORREA, R. R. & RAMALHO, G. R. 1956. Revisão de *Phoniomyia* Theobald, 1903 (Diptera, Culicidae, Sabethini). *Folia clin. biol.* **25**: 1–176.

COVA-GARCIA, P. 1961. *Notas sobre los Anofelinos de Venezuela y su Identificación.* 2nd Edn. 213 pp. Caracas: Editora Grafas.

COVA-GARCIA, P., SUTIL, E. & RAUSSEO, J. A. 1966. *Mosquitos (Culicinos) de Venezuela.* 2 Vols. 823 pp. Caracas: Ministerio de Sanidad.

DARSIE, R. F. 1951. Pupae of the culicine mosquitoes of the Northeastern United States (Diptera, Culicidae, Culicini). *Mem. Univ. Cornell agric. Exp. Stn.* **304**. 67 pp.

DELFINADO, M. 1966. The culicine mosquitoes of the Philippines, Tribe Culicini (Diptera, Culicidae). *Mem. Am. ent. Inst.* **7**. 252 pp.

DE MEILLON, B. *et al.* 1969. Papers on *Culex pipiens fatigans. Bull. Wld Hlth Org.* **36** (1): 1–100 & 163–180.

DETINOVA, T. S. 1968. Age structure of insect populations of medical importance. *A. Rev. Ent.* **13**: 427–450.

DOBROTWORSKY, N. V. 1965. *The Mosquitoes of Victoria.* 237 pp. Melbourne.

EDESON, J. F. B. & WILSON, T. 1964. The epidemiology of filariasis due to *Wuchereria bancrofti* and *Brugia malayi. A. Rev. Ent.* **9**: 245–268.

EDWARDS, F. W. 1941. *Mosquitoes of the Ethiopian Region III.* Culicine adults and pupae. 409 pp. London: Br. Mus. (Nat. Hist.).

FOOTE, R. H. 1954. The larvae and pupae of the mosquitoes belonging to the *Culex* subgenera *Melanoconion* and *Mochlostyrax. Tech. Bull. U.S. Dep. Agric.* **1091**. 126 pp.

FOOTE, R. H. & COOK, D. R. 1959. Mosquitoes of medical importance. *Agric. Handb. Forest Serv. U.S.* **152**. 156 pp.

FORATTINI, O. P. 1962. *Entomologia Médica. I. Parte Geral, Diptera, Anophelini.* 662 pp. Univ. de São Paulo.

—— 1965a. *Entomologia Médica. II. Culicini: Culex, Aedes e Psorophora.* 506 pp. Univ. de São Paulo.

FORATTINI, O. P.　1965b.　*Entomologia Médica. III. Culicini: Haemagogus, Mansonia, Culiseta. Sabethini. Toxorhynchitini.　Arboviroses. Filariose bancroftiana. Genética.*　416 pp.　Univ. de São Paulo.

GALINDO, P., TRAPIDO, H. & CARPENTER, S. J.　1950.　Observations on diurnal forest mosquitoes in relation to sylvan yellow fever in Panama.　*Am. J. trop. Med.* **30**: 533–574.

GALINDO, P., BLANTON, F. S. & PEYTON, E. L.　1954.　A revision of the *Uranotaenia* of Panama with notes on other American species of the genus.　*Ann. ent. Soc. Am.* **47**: 107–177.

GARCIA, M. & RONDEROS, R. A.　1962.　Mosquitos de la Republica Argentina I. Tribu Anophelini (Diptera-Culicidae-Culicinae).　*Anu. Comis. Invest. Cient. Prov. B. Aires* **3**: 105–164.

GARNHAM, P. C. C.　1966.　*Malaria Parasites and other Haemosporidia.*　1114 pp.　Oxford.

GILLETT, J. D.　1962.　Contributions to the oviposition-cycle by the individual mosquitoes in a population.　*J. Insect Physiol.* **8**: 665–681.

GILLIES, M. T. & DE MEILLON, B.　1968.　*The Anophelinae of Africa South of the Sahara.*　Publs S. Afr. Inst. med. Res. **54**.　343 pp.

GJULLIN, C. M., SAILER, R. I., STONE, A. & TRAVIS, B. V.　1961.　The mosquitoes of Alaska.　*U.S. Dep. Agric. Hdbk* **182**.　98 pp.

GRJEBINE, A.　1966.　*Biologie et Taxonomie des Anophelinae de Madagascar et des Iles Voisines.*　Doctoral thesis.　Université de Dijon.　487 pp.

GUEDES, A. S., SOUZA, M. A., MACIEL, C. S. & XAVIER, S. H.　1965.　Catálogo illustrado dos mosquitos da coleção do Instituto Nacional de Endemias Rurais I. Género *Psorophora* Robineau-Desvoidy, 1827.　*Revta bras. Malar. Doenç. trop.* **17**: 3–24.

GUTSEVICH, A. V., MONCHADSKII, A. S. & SHTAKEL'BERG, A. A.　1970.　Fauna of U.S.S.R. Vol. III. No. 4. Mosquitoes. Family Culicidae.　Leningrad: Zoological Institute.　384 pp.　[In Russian.]

GUY, Y.　1959.　Les Anophèles du Maroc.　*Mém. Soc. Sci. nat. phys. Maroc. Zool.*, n.s. **7**: 1–235.

HADDOW, A. J.　1955.　Studies of the biting habits of African mosquitoes. An appraisal of methods employed, with special reference to the twenty-four-hour catch.　*Bull. ent. Res.* **45**: 199–242.

HADDOW, A. J., VAN SOMEREN, E. C. C., LUMSDEN, W. H. R., HARPER, J. O. & GILLETT, J. D.　1951.　The mosquitoes of Bwamba County, Uganda, VIII. Records of occurrence, behaviour and habitat.　*Bull. ent. Res.* **42**: 207–238.

HAMON, J.　1963.　Les moustiques anthrophiles de la région de Bobo-Dioulasso (République de Haute-Volta).　*Annls Soc. ent. Fr.* **132**: 84–144.

HOPKINS, G. H. E.　1952.　*Mosquitoes of the Ethiopian Region I. Larval Bionomics of Mosquitoes and Taxonomy of Culicine Larvae.*　2nd Edn.　355 pp.　London: Br. Mus. (Nat. Hist.).

HORSFALL, W. R.　1955.　*Mosquitoes: their Bionomics and Relation to Disease.*　723 pp.　London.

HUANG, Y. -M.　1968a.　A new subgenus of *Aedes* (Diptera, Culicidae) and illustrated key to the subgenera of *Aedes* of the Papuan Subregion.　*J. med. Ent.* **5**: 169–188.

—— 1968b.　*Aedes* (*Verrallina*) of the Papuan Subregion (Diptera: Culicidae).　*Pacif. Insects Monogr.* **17**: 1–73.

JACHOWSKI, L. A.　1954.　Filariasis in American Samoa V. Bionomics of the principal vector, *Aedes polynesiensis* Marks.　*Am. J. Hyg.* **60**: 186–203.

KING, W. V., BRADLEY, G. H., SMITH, C. N. & MCDUFFIE, W. C.　1960.　A handbook of the mosquitoes of the southeastern United States.　*U.S. Dep. Agric. Handbk* **173**.　188 pp.

KRAMAR, J.　1958.　*Fauna of Czechoslovakia.* 13. *Biting Mosquitoes-Culicinae.*　286 pp.　Prague: Csechoslovenské Akad. Ved.　[In Czech.]

LA CASSE, W. J. & YAMAGUTI, S.　1950.　*Mosquito Fauna of Japan and Korea.*　3rd Edn.　487 pp.　Kyoto: Office of Surgeon, 8th Army H.Q.

LAIRD, M.　1956.　Studies of mosquitoes and fresh water ecology in the South Pacific.　*Bull. R. Soc. N.Z.* **6**.　213 pp.

LANE, J.　1953.　*Neotropical Culicidae.*　1112 pp.　Univ. de São Paulo.　2 Vols.

LEESON, H. S.　1958.　An annotated catalogue of the culicine mosquitoes of the Federation of Rhodesia and Nyasaland and neighbouring countries, together with locality records from Southern Rhodesia.　*Trans. R. ent. Soc. Lond.* **110**: 21–51.

LIEN, J. C. 1968. New species of mosquitoes from Taiwan (Diptera: Culicidae) Part V. Three new subspecies of *Aedes* and seven new species of *Culex*. *Trop. Med., Nagasaki* **10**: 217–262.

MACDONALD, W. W. 1957. Malaysian Parasites XVI. An interim review of the non-anopheline mosquitoes of Malaya. *Stud. Inst. med. Res. Malaya* **28**: 1–34.

—— 1960. Malaysian Parasites XXXVIII. On the systematics and ecology of *Armigeres* subgenus *Leicesteria* (Diptera, Culicidae). *Stud. Inst. med. Res. Malaya* **29**: 110–153.

MACDONALD, W. W. & TRAUB, R. 1960. Malaysian Parasites XXXVII. An introduction to the ecology of the mosquitoes of the lowland dipterocarp forest of Selangor, Malaya. *Stud. Inst. med. Res. Malaya* **29**: 79–109.

MARKS, E. N. 1954. A review of the *Aedes scutellaris* subgroup with a study of variation in *Aedes pseudoscutellaris* (Theobald) (Diptera: Culicidae). *Bull. Br. Mus. nat. Hist. Ent.* **3**: 349–414.

MARSHALL, J. F. 1938. *The British Mosquitoes.* London: Br. Mus. (Nat. Hist.). 341 pp. [Republished by Johnson Reprint Company, New York in 1966.]

MASLOV, A. V. 1967. Bloodsucking mosquitoes of the sub-tribe Culisetina (Diptera, Culicidae) of the world fauna. *Opred. Faune SSSR.* **93**. 182 pp. [In Russian.]

MATTINGLY, P. F. 1952. The subgenus *Stegomyia* (Diptera: Culicidae) in the Ethiopian Region. I. A preliminary study of the distribution of species occurring in the West African Sub-region with notes on taxonomy and bionomics. *Bull. Br. Mus. nat. Hist. Ent.* **2**: 235–304.

—— 1953. The subgenus *Stegomyia* (Diptera: Culicidae) in the Ethiopian Region. II. Distribution of species confined to the East and South African Sub-region. *Bull. Br. Mus. nat. Hist. Ent.* **3**: 1–65.

—— 1958. A revision of *Paraëdes* Edwards and *Cancraëdes* Edwards (Diptera: Culicidae). *Proc. R. ent. Soc. Lond., B* **27**: 76–83.

—— 1962a. Towards a zoogeography of the mosquitoes. *Publs Syst. Ass.* **4**: 17–36.

—— 1962b. Mosquito behaviour in relation to disease eradication programmes. *A. Rev. Ent.* **7**: 419–436.

—— 1965. *The Culicine Mosquitoes of the Indomalayan Area VI. Genus Aedes Meigen subgenus Stegomyia Theobald (Groups A, B and D).* 67 pp. London: Br. Mus. (Nat. Hist.).

—— 1969. *The Biology of Mosquito-borne Disease.* 184 pp. London.

—— 1971. Ecological aspects of mosquito evolution. *Riv. Parassit.* **13**: 31–65.

—— 1971. Cyclical behaviour in mosquitoes. *Trans. 13th internat. Congr. Ent., Moscow* **1**: 416–417.

MATTINGLY, P. F. & BROWN, E. S. 1955. The mosquitoes (Diptera: Culicidae) of the Seychelles. *Bull. ent. Res.* **46**: 69–110.

MATTINGLY, P. F. & KNIGHT, K. L. 1956. The mosquitoes of Arabia. I. *Bull. Br. Mus. nat. Hist.* (Ent.), **4**: 91–141.

MATTINGLY, P. F. & LIPS, M. 1953. Notes on the Culicini of the Katanga (Diptera, Culicidae). *Revue Zool. Bot. afr.* **47**: 311–343 and **48**: 49–72.

MIHALYI, F. 1963. *Biting Mosquitoes of Hungary.* 229 pp. Budapest: Akademiai Kiado. [In Hungarian.]

MOHRIG, W. 1969. Die Culiciden Deutschlands. *Parasit. SchrReihe* **18**: 1–260.

MUIRHEAD–THOMSON, R. C. 1951. *Mosquito Behaviour in Relation to Malaria Transmission and Control in the Tropics.* 219 pp. London.

—— 1968. *Ecology of Insect Vector Populations.* 174 pp. London and New York.

MUSPRATT, J. 1955. Research on South African Culicini (Diptera, Culicidae) III. A check list of the species with notes on taxonomy, bionomics and identification. *J. ent. Soc. sth. Afr.* **18**: 149–207.

—— 1956. The *Stegomyia* mosquitoes of South Africa and some neighbouring territories. *Mem. ent. Soc. sth. Afr.* No. **4**. 138 pp.

NATVIG, L. R. 1948. Contributions to the knowledge of the Danish and Fennoscandian mosquitoes. *Norsk. Ent. Tidsskr.* Suppl. I. 567 pp.

O'GOWER, A. K. 1958. The mosquitoes of North Western Australia. *Serv. Publs Dep. Hlth Aust. Sch. publ. Hlth trop. Med.* No. **7**. 46 pp.

OVAZZA, M., HAMON, J. & NERI, P. 1956. Contribution à l'étude des diptères vulnérants de l'Empire d'Ethiopie. I. Culicidae. *Bull. Soc. Path. exot.* **49**: 151–182.

PAMPANA, E. J. 1969. *A Textbook of Malaria Eradication*. 2nd Edn. 593 pp. London.

RAMALINGHAM, S. & BELKIN, J. N. 1965. Mosquito studies (Diptera, Culicidae) III. Two new *Aedes* from Tonga and Samoa. IV. The mosquitoes of the Robinson-Peabody Museum of Salem expedition to the Southwest Pacific, 1956. *Contr. Am. ent. Inst.* **1** (4): 1–34.

REEVES, W. C. 1965. Ecology of mosquitoes in relation to arboviruses. *A. Rev. Ent.* **10**: 25–46.

REID, J. A. 1968. Anopheline mosquitoes of Malaya and Borneo. *Stud. Inst. med. Res. Malaya* **31**. 520 pp.

REUBEN, R. 1969. A redescription of *Culex vishnui* Theo., with notes on *C. pseudovishnui* Colless and *C. tritaeniorhynchus* Giles from southern India. *Bull. ent. Res.* **58**: 643–652.

RIOUX, J. A. 1958. Les Culicides du 'Midi' Méditerranéen. *Encyclopédie Entomologique*. 303 pp. Paris.

RONDEROS, R. A. & BACHMANN, A. O. 1963a. A proposito del complejo *Mansonia* (Diptera, Culicidae). *Revta Soc. ent. argent.* **25**: 43–51.

——, —— 1963b. Mansoniini neotropicales I (Diptera- Culicidae). *Revta Soc. ent. argent.* **26**: 57–65.

ROSS, H. H. & HORSFALL, W. R. 1965. A synopsis of the mosquitoes of Illinois. *Illinois nat. Hist. Surv. biol. Notes* **52**. 50 pp.

RUSSELL, P. F., WEST, L. S., MANWELL, R. D. & MACDONALD, G. 1963. *Practical Malariology*. 2nd Edn. 750 pp. London.

SCHICK, R. X. 1970. Mosquito studies XX. The *terrens* group of *Aedes* (*Finlaya*). *Contr. Am. ent. Inst.* **5** (3): 1–158.

SENEVET, G. & ANDARELLI, L. 1959. Les moustiques de l'Afrique du Nord et du bassin méditerranéen. *Encyclopédie Entomologique*. **37**. 383 pp Paris.

SIRIVANAKARN, S. 1968. The *Culex* subgenus *Lophoceraomyia* in New Guinea and Bismarck Archipelago (Diptera: Culicidae). *Pacif. Insects Monogr.* **17**: 75–186.

SMART, J. 1965. *A Handbook for the Identification of Insects of Medical Importance*. 303 pp. 4th Edn. London: Br. Mus. (Nat. Hist.).

SMITH, M. E. 1969. The *Aedes* mosquitoes of New England (Diptera, Culicidae) II. Larvae. *Can. Ent.* **101**: 41–51.

STEFFAN, W. A. 1966. A checklist and review of the mosquitoes of the Papuan Subregion. *J. med. Ent.* **3**: 179–237.

STEWARD, C. C. & McWADE, J. W. 1961. The mosquitoes of Ontario (Diptera: Culicidae) with keys to the species and notes on distribution. *Proc. ent. Soc. Ont.* **91**: 121–188.

STONE, A. 1961. A synoptic catalog of the mosquitoes of the world, Supplement I (Diptera: Culicidae). *Proc. ent. Soc. Wash.* **63**: 29–52.

—— 1963. A synoptic catalog of the mosquitoes of the world, Supplement II (Diptera, Culicidae). *Proc. ent. Soc. Wash.* **65**: 117–140.

—— 1967. A synoptic catalog of the mosquitoes of the world, Supplement III (Diptera, Culicidae). *Proc. ent. Soc. Wash.* **69**: 197–224.

—— 1970. A synoptic catalog of the mosquitoes of the world, Supplement IV (Diptera, Culicidae). *Proc. ent. Soc. Wash.* **72**: 137–171.

STONE, A. & BARRETO, P. 1969. A new genus and species of mosquito from Colombia, *Galindomyia leei* (Diptera, Culicidae, Culicini). *J. med. Ent.* **6**: 143–146.

STONE, A., KNIGHT, K. L. & STARCKE, H. 1959. *A Synoptic Catalog of the Mosquitoes of the World*. Thomas Say Foundation. Vol. **6**. Washington, D.C.: Entomological Society of America. 358 pp.

TAYLOR, B. & JONES, M. D. R. 1969. The circadian rhythm of flight activity in the mosquito *Aedes aegypti* (L.); the phase-setting effects of light-on and light-off. *J. exp. Biol.* **51**: 59–70.

TAYLOR, R. M. 1967. *Catalogue of Arthropod-borne Viruses of the World.* Public Health Service Publication No. **1760.** 898 pp. Washington, D.C., U.S. Govt. Printing Office.

THURMAN, E. B. 1959. A contribution to a revision of the Culicidae of Northern Thailand. *Bull. Md agric. Exp. Stn.* A– **100.** 180 pp.

TRAPIDO, H. & GALINDO, P. 1957. Mosquitoes associated with sylvan yellow fever near Almirante, Panama. *Am. J. trop. Med. Hyg.* **6**: 114–144.

TYSON, W. H. 1970. Contributions to the mosquito fauna of Southeast Asia VII. Genus *Aedeomyia* Theobald in Southeast Asia, VIII. Genus *Aedes*, subgenus *Mucidus* Theobald in Southeast Asia. *Contr. Am. ent. Inst.* **6** (2): 1–80.

UNIVERSITY OF MARYLAND. 1963. *Index Catalogue to Russian, Central and Eastern European, and Chinese Literature in Medical Entomology. I. Diptera.* College Park, Md: Dept of Zoology. 243 pp.

VAN DEN ASSEM, J. & BONNE-WEPSTER, J. 1964. New Guinea Culicidae, a synopsis of vectors, pests and common species. *Zool. Bijdr.* **6.** 136 pp.

VAN SOMEREN, E. C. C., TEESDALE, C. & FURLONG, M. 1955. The mosquitoes of the Kenya coast; records of occurrence, behaviour and habitat. *Bull. ent. Res.* **46**: 463–493.

VOCKEROTH, J. R. 1954. Notes on the identities and distributions of *Aedes* species of northern Canada with a key to the females (Diptera: Culicidae). *Can. Ent.* **86**: 241–255.

WARD, R. A. & SCANLON, J. E. (Eds). 1970. Conference on anopheline biology and malaria eradication. *Misc. Publs ent. Soc. Am.* **7**: 1–196.

WHARTON, R. H. 1962. The biology of *Mansonia* mosquitoes in relation to the transmission of filariasis in Malaya. *Bull. Inst. med. Res. Malaya* **11.** 114 pp.

WORLD HEALTH ORGANIZATION. 1970. Insecticide resistance and vector control. *Tech. Rep. Ser. Wld Hlth Org.* **443.** 279 pp.

WRIGHT, J. W. & PAL, R. (Eds). 1967. Genetics of Insect Vectors of Disease. 794 pp. Amsterdam, London and New York.

ZAVORTINK, T. J. 1968. Mosquito studies (Diptera: Culicidae) VIII. A prodrome of the genus *Orthopodomyia*. *Contr. Am. ent. Inst.* **3** (2): 1–221.

—— 1970. The treehole *Anopheles* of the New World. *Contr. Am. ent. Inst.* **5** (2): 1–35

3b. SIMULIIDAE
(Black-flies, German: Kriebelmücken)
by R. W. Crosskey

THE Simuliidae are small stout-bodied midges generally known as black-flies, although not all of them are black and some Neotropical species are even predominantly yellow or orange in colour. The family is nearly cosmopolitan, black-flies being found almost anywhere if there are suitable rivers and streams for the developmental stages; some forms have colonized remote oceanic islands such as Crozet and St. Helena. Slightly more than one thousand species are known, but new species are frequently discovered, especially in the tropics.

A few black-fly species cannot bite, but the great majority of them bite and suck the blood of warm-blooded vertebrates. As a consequence of this feeding habit black-flies are able to transmit several pathogenic organisms, and some species are incriminated as vectors of skin-inhabiting filarial nematodes of the genus *Onchocerca* Diesing in mammals (including man) and of protozoan blood parasites of the genus *Leucocytozoon* Ziemann among birds; they are also considered to be natural vectors of avian trypanosomes (Bennett, 1961). Viruses have been isolated from some ornithophilic black-flies and simuliids are reputed to be involved in the transmission of myxomatosis virus among Australian rabbits (Mykytowycz, 1957), but at present there is no evidence that simuliids are involved in the transmission of arboviruses in man; experimental mechanical transmission of virus among laboratory mice has, however, been demonstrated (Austin, 1967) with *Austrosimulium ungulatum* Tonnoir (a New Zealand man-biting simuliid) as transmitting agent.

The simuliids include the only known vectors of human onchocerciasis, and herein lies the particular importance of these insects in medical entomology. This disease is caused by *Onchocerca volvulus* (Leuckart) and is prevalent in tropical Africa and in parts of the tropical Americas. There is good evidence that bovine onchocerciasis caused by *Onchocerca gutturosa* Neumann is transmitted in England by *Simulium ornatum* Meigen, and it is possible—though evidence is lacking at present—that simuliids of the genus *Austrosimulium* Tonnoir can be vectors of *O. gibsoni* Cleland & Johnson among cattle in Queensland. Among the avian filariae it is known that *Splendidofilaria fallisensis* (Anderson), a parasite of wild and domestic ducks, is transmitted by two species of black-fly (Anderson, 1968) and it is extremely probable that other avian filariae have simuliid vectors.

Simuliids are the vectors of *Leucocytozoon* (as this genus is now re-defined by Bennett *et al.*, 1965) and eight species of these avian blood-inhabiting parasites, seven in North America and one in Britain, have been shown fairly conclusively to have one main species of vector. It appears now that each species of *Leucocytozoon* parasitizes one particular family of birds, but that some species of simuliids are capable of transmitting (since complete sporogony takes place in them) more than one species of *Leucocytozoon*;

only *L. fringillinarum* occurs in more than one bird family. These parasites occur in both domestic and wild birds, and diseases of turkeys and ducks, caused by *L. smithi* and *L. simondi* respectively, can be commercially important in North America.

Apart from their rôle as transmitters of human, bovine and avian diseases the black-flies can have very serious and harmful effects on man and domestic animals because of their severe biting attacks, especially at times when mass outbreaks occur or when seasonal abundance is at its peak. They may then have a deleterious economic effect on, for instance, the lumbering and tourist industries, on the health and milk-yield of live-stock, or on the egg-laying of poultry. Man-biting simuliids are important pests in places as different as New York State, the Himalayas, and the Marquesas Islands, and some zoophilic species can be serious cattle pests in the Canadian prairies, in the Danube valley, and on the Queensland cattle-ranches (in the worst outbreaks in these areas livestock can die through suffocation by myriads of flies as well as by toxaemias resulting from the bites).

Table 3 summarizes present knowledge on the rôle of Simuliidae in the transmission or probable transmission of vector-borne pathogens.

FAMILY RECOGNITION AND CHARACTERISTICS

The family Simuliidae is very homogeneous and can be quickly and easily recognized in both adult and early stages by the features shown in the accompanying figures.

Adult flies have a short body form with rather short broad wings (fig. 81). The eyes are very large and occupy most of the head: in the male they meet above the antennae so that the frons is obliterated (except in some rare primitive forms) and the upper eye facets are enlarged and sharply demarcated from the smaller lower facets (fig. 82B), but in the female they are separated above the antennae by a broad frons and have small facets of uniform size (fig. 82A). The antennae are short and stout and have from 9–12 segments of rather uniform size (in the great majority of forms the antennae are 11-segmented); they are similar in both sexes and are devoid of long sensory whorls of hair (cf. mosquitoes and ceratopogonids which possess such long hairs). Ocelli are absent. Both sexes have a short thick downwardly directed proboscis and a pair of long slender 5-segmented maxillary palps (fig. 82A, B). The mouthparts of the female are adapted for blood-sucking, the labrum having small apical teeth which stretch the host's skin during biting and the mandibles and maxillae (fig. 82C, E) having small apical teeth which can penetrate the host; male flies do not bite and their maxillae and mandibles lack cutting teeth, and some black-flies also have non-biting females in which the teeth are atrophied. The arrangement of the mouthparts of the female is shown in simplified cross-section in fig. 82D. The labella are large, the pharynx unarmed, and the dorsal part of the hypopharynx (cibarium) has paired basal cornuae; in many species the cibarium is armed between the cornuae with minute teeth of uncertain function (fig. 82F). The thorax is short and deep with a strongly convex scutum (fig. 85A), and the ventral part of the mesothorax (katepisternum) is sharply differentiated from the remainder of the thorax by an impressed groove (mesepisternal sulcus) (except in the very rare North American genus *Parasimulium* which lacks the sulcus). The sides of the thorax have a characteristic membranous area (pleural membrane) just in front of the wing base. The wings (fig. 84A) are without scales (cf. mosquitoes) and have only the anterior veins strong and conspicuous; there is a characteristic forked submedian fold between veins M_2 and Cu_1, and vein Cu_2 has a sigmoid curvature as figured (except in the South American genus *Gigantodax* in which it is straight). The legs in all simuliids are rather short and stout compared to those of other Nematocera, and both sexes of many species have a rounded lobe (calcipala) on the inner side of the apex of the hind basitarsus and a deep transverse groove (pedisulcus) on the second hind tarsal segment (fig. 84B); the claws are either simple or have one basal tooth developed to varying degrees in different species. The abdomen has the first tergite modified into a prominent flap (the so-called

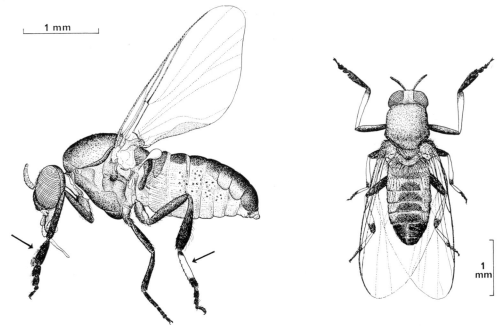

1 mm

1 mm

Fig. 81   Adult female black-flies: left, *Simulium damnosum*, lateral view; right, dorsal view of typical species in resting position (arrows on left fig. indicate the dorsal hair-crest of the fore tarsus and the white hind basitarsal band characteristic of *damnosum*. **Right** hand fig. shows the resting position of the wings closed over the abdomen and the manner in which the fore legs are often extended well forwards in life.)

basal scale) bearing a long hair-fringe (fig. 85A), but the remainder of the abdomen is largely membranous—especially in the female in which it is capable of great distension to accommodate the blood-meal; the tergites are small except on the terminal segments and the sternites rudimentary (the female lacking sternites completely before the seventh segment). The genitalia of the male form a small compact hypopygium which shows features of great importance in taxonomy; the female terminalia include a characteristic sclerotized Y-shaped genital rod and there is a single, usually subspherical, spermatheca.

Black-fly *larvae* and *pupae* are perhaps the most characteristic organisms in running water. The *larva* (fig. 83A) is distinguished by having a mid-ventral proleg just behind the well-formed head and by the conspicuous posterior circlet of serially arranged hooklets on the end of the body. The head, with very rare exceptions, bears a pair of large cephalic fans which open as hemispherical basket-like 'mouth-brushes' (fig. 83B); the concave inner edges of the fan rays bear each a single row of minute spicules lying in the same plane as the ray (and used in life for filtration of food from the water). The head appendages comprise a large anterior labrum bearing dense close-set setae, a

Table 3 Summary of present knowledge of the rôle of Simuliidae as vectors of pathogenic organisms (experimental transmissions and infections included).

[Note: The generic name *Simulium* is used in its broad sense and is abbreviated as S. Names of the authors (describers) of the parasites and simuliids are omitted for convenience and if required should be found from standard parasitological and entomological works. References cited can be found in the appropriate part of the classified bibliography at the end of this chapter.]

Pathogenic organism		Simuliid host	Vertebrate host	Geo-graphical area	Status of present knowledge or evidence for transmission (with main references)
Group	Species				
Nematode	*Mansonella ozzardi*	*Simulium amazonicum*	Man	Brazil	Natural transmission reported (Cerqueira, 1959)
	Onchocerca gutturosa	*Simulium ornatum*	Cattle	England	Natural transmission presumed. Development of parasite to infective stage in wild flies (Steward, 1937; Eichler, 1971).
	Onchocerca volvulus	*Simulium damnosum, S. neavei, S. woodi, S. metallicum, S. ochraceum*	Man	Tropical Africa and tropical Americas	Natural transmission confirmed beyond reasonable doubt (Blacklock, 1926, and extensive subsequent literature: see bibliography).
	"	*Simulium callidum, S. exiguum S. haematopotum, S. veracruzanum*	Man	Guatemala	Development of parasite to infective stage in experimental flies (Dalmat, 1955).
	"	*Simulium exiguum*	Man	Venezuela	Development of 'Venezuela strain' of parasite (but not of W. African strains) to infective stage in experimental flies (Duke, 1970).
	"	*Simulium vorax*	Man	Tanzania	Development of parasite to infective stage in experimental flies (Wegesa, 1967).

Pathogenic organism		Simuliid host	Vertebrate host	Geographical area	Status of present knowledge or evidence for transmission (with main references)
Group	Species				
Nematode (cont.)	Onchocerca volvulus (cont.)	Simulium dukei [='aureosimile' in error]	? man	Cameroons	Infective stage of parasite found in wild flies (Duke, 1962).
	"	Simulium damnosum	Chimpanzee	Cameroons	Experimental transmission from man with infective stage of parasite developed in experimentally fed flies (Duke, 1962).
	Splendidofilaria fallisensis [=Ornithofilaria f.]	Simulium rugglesi, S. anatinum	Ducks (domestic & wild)	Canada	Natural transmission confirmed (Anderson, 1956, 1968).
	Unidentified filaria (probably Onchocerca gutturosa)	Simulium exiguum	? cattle	Guatemala	Advanced 'sausage' stage of parasite in wild flies (Gibson & Dalmat, 1952).
	Unidentified filaria 'A'	Simulium neavei	[Unknown]	Uganda	Infective stage of parasite found in wild flies (Nelson & Pester, 1962).
	Unidentified filariae 'D' and 'E'	Simulium damnosum	[Unknown]	Cameroons	Infective stages of parasites found in wild flies (Duke, 1967).
	Unidentified filariae 'Types I–IV'	Simulium damnosum	[Unknown]	Liberia	Infective stages of parasites found in wild flies (Garms & Voelker, 1969).
	Unidentified filaria	Simulium bovis	[Unknown]	Nigeria	Infective stage of parasite found in wild flies (Crosskey, 1957).
	Unidentified filaria	Simulium griseicolle	[Unknown]	Nigeria	Infective stage of parasite found in wild flies (Crosskey & Crosskey, 1958).
Protozoan	Crithidia simuliae	Simulium colombaschense [='columbaczense' in error]	[Unknown]	Yugoslavia	Trypanosomids described from gut of wild flies (Georgewitch, 1909).

Pathogenic organism		Simuliid host	Vertebrate host	Geographical area	Status of present knowledge or evidence for transmission (with main references)
Group	Species				
Protozoan	*Leucocytozoon berestneffi* (prob. good sp. distinct from *sakharoffi*	*Prosimulium decemarticulatum*, *Simulium (Eusimulium) aureum* group	Corvidae	Canada	Schizogony in experimental flies (Khan & Fallis, 1971)
	Leucocytozoon bonasae	*Simulium (Eusimulium)* spp. (*aureum*-group & *latipes*-group)*	Grouse (Tetraonidae)	Canada	Natural and experimental transmission (Fallis & Bennett, 1958).
	Leucocytozoon danilewskyi	*Simulium (Eusimulium)* spp. (*aureum*-group & *latipes*-group)*	Owls (Strigidae)	Canada	Sporogony in flies and experimental transmission (Bennett et al. 1965).
	Leucocytozoon dubreuli [syn. *L. mirandae*]	*Simulium (Eusimulium)* spp. (*aureum*-group)*	Thrushes (Turdidae)	Canada	Sporozoites of parasite developed in experimental flies (Fallis & Bennett, 1962).
	Leucocytozoon fringillinarum	*Simulium (Eusimulium)* spp. (*aureum*-group & *latipes*-group)*	Passerine birds (mainly Fringillidae)	Canada	Sporogony in flies and experimental infections (Fallis & Bennett, 1962, 1966).
	Leucocytozoon sakharoffi	*Simulium angustitarse*†	Rook (Corvidae)	England	Natural transmission presumed from sporogony in wild & experimental flies and experimental infections (Baker, 1970).
	Leucocytozoon simondi	*Simulium rugglesi, S. anatinum*	Ducks (Anatidae)	Canada & U.S.A.	Natural transmission & experimental infections (Anderson et al., 1962; Fallis & Bennett, 1966).
	Leucocytozoon smithi	*Simulium meridionale* [syn. *occidentale*], *S. jenningsi*	Turkey (Meleagrididae)	Canada & U.S.A.	Natural transmission & experimental infections (Skidmore, 1932; Anderson & DeFoliart, 1961).

Pathogenic organism		Simuliid host	Vertebrate host	Geographical area	Status of present knowledge or evidence for transmission (with main references)
Group	Species				
Protozoan (cont.)	*Trypanosoma avium* (or *avium*-complex)	Ornithophilic simuliids, several spp.	several species of birds of different families	Canada	Natural transmission presumed. Experimental infections with trypanosomes passaged through flies (Bennett, 1961).
Virus	Myxomatosis virus	*Simulium melatum*	Rabbits	Australia	Natural transmission presumed. Experimental infections from wild flies (Mykytowycz, 1957).
	Whataroa (M78)	*Austrosimulium ungulatum*	Mice	New Zealand	Experimental mechanical transmission (Austin, 1967).
	Semliki Forest (SFV)	*Austrosimulium ungulatum*	Mice	New Zealand	Experimental mechanical transmission (Austin, 1967).
	Eastern encephalitis (EEV)	*Simulium meridionale*, *S. johannseni*	Turkeys	U.S.A.	Isolations from flies fed on experimentally exposed and brooder-house birds, *meridionale* suspected natural vector (Anderson et al., 1961).
	Unidentified virus	*Simulium meridionale*	—	U.S.A.	One isolation (DeFoliart et al., 1969).
	Unidentified virus	Ornithophilic simuliids	—	U.S.A. (Wisconsin)	One isolation (ex 15588 specimens of three simuliid species) (DeFoliart & Hanson, 1968).

*Host simuliids recorded in literature as *S. aureum* and *S. latipes*, but these species belong to taxonomically difficult groups and it appears probable that the North American species involved are not the true *aureum* and *latipes* of Europe. The recorded hosts in Canada are therefore referred to as *aureum*-group and *latipes*-group species. For nomenclatural reasons the *latipes*-group has recently been renamed as *vernum*-group.

†It is now known that the British species hitherto called *S. angustiarse* is specifically distinct from the true *angustiarse* Lundström and should be called *S. lundstromi* Enderlein.

I

thickly haired hypopharynx, a pair of large toothed mandibles, a pair of haired maxillae each with a shaft-like single-segmented maxillary palp (fig. 83c), and a pair of slender antennae of not more than four segments (though secondary subdivision of segments may occur).   The venter of the head-capsule has a broad apically-toothed median projection (hypostomium), and usually has a posteromedian unsclerotized area (postgenal cleft) of which the shape is of great importance in taxonomy.   The head is partially pigmented in most forms, and may have dark spots surrounded by paler areas (positive pattern) or pale areas surrounded by dark areas (negative pattern); it has a pair of minute but conspicuous black eye-spots on each side.   The elongated larval body is typically rather swollen in the thoracic region and posteriorly on the abdomen, and the cuticle is bare or has only very minute hairs or scales (which normally are very inconspicuous).

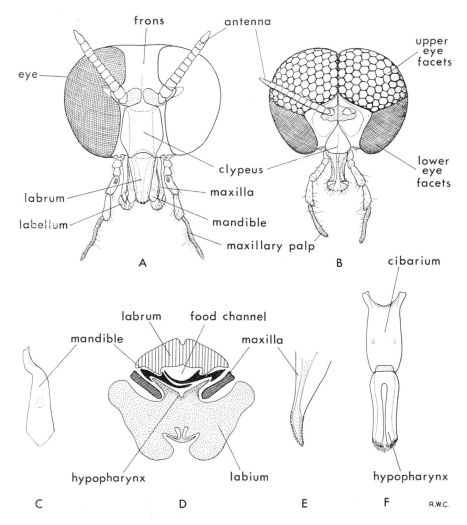

Fig. 82   Head structures of adult Simuliidae: A, head of female *Simulium*, facial view; B, head of male *Simulium*, facial view: C, mandible of biting female; D, cross-section of middle of proboscis of female *Simulium*, slightly schematic; E, maxilla of biting female; F, cibarium and hypopharynx of female. (Fig. B drawn from Oriental species with relatively few rows of very large upper facets, and antenna shown on one side only.)

The rectum has extrusible colourless gills (fig. 83A), believed to have an osmoregulatory function, and larvae in collections are often found with these extended. An X-shaped anal sclerite is present between the anus and the posterior circlet in nearly all larvae (the sclerite is Y-shaped in the aberrant Holarctic forms which lack cephalic fans and absent in one species).

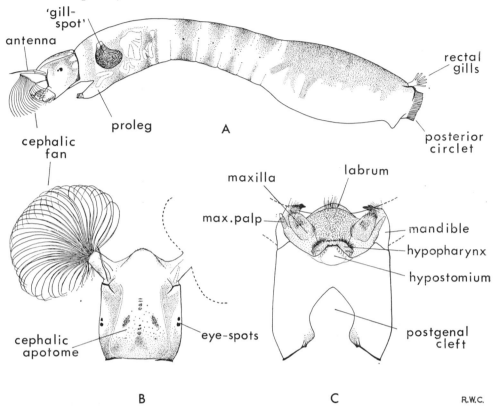

Fig. 83   Larva of Simuliidae: A, mature larva of *Simulium* sp., showing 'gill-spot' (see text), lateral view; B, larval head in dorsal view of typical species with cephalic fans (fan shown one side only); C, ventral view of head and mouthparts, cephalic fans omitted. (The ventral papilla just in front of the posterior circlet shown in fig. A is absent in many species.)

The *pupa* is immobile and is recognized by being invested in a pocket-like (fig. 86C) or shoe-shaped (fig. 86D) cocoon, and by bearing a pair of variously branched or filamentous gills anteriorly on the thorax (fig. 86C). The pupal abdomen bears small hooks for locking the pupa in the cocoon, and sometimes has a pair of strong recurved terminal hooks. Some species have a feeble weakly-woven cocoon and some very rare forms have no cocoon at all. The form of the pupal gill and the arrangement of abdominal hooks (onchotaxy) are rather constant in a species and of great taxonomic importance.

The *egg* is smooth-shelled, eccentric ovoid in shape (fig. 86A) and has no special adaptations (cf. mosquito eggs); it is white when laid but darkens during embryogenesis.

Size is fairly constant in each species, though there can be some seasonal variation and high-altitude populations can be larger than lowland ones. Adult size is best indicated by wing-length, which ranges from about 1.5 mm up to 6 mm; mature larvae range in length from about 4.5 to 12 mm and eggs from about 0.1 to 0.4 mm.

Larval Simuliidae, like many other larval Diptera, have giant polytene chromosomes especially well developed in the salivary gland cells. The banding patterns and other micromorphological characters of these chromosomes are of great importance for unmasking complexes of cryptic species (cytotaxonomy). There are normally three pairs of chromosomes.

LIFE-HISTORY, ECOLOGY AND BEHAVIOUR

The Simuliidae are aquatic in the early stages, and nearly always require running water for their development. Most of them need a slow or fast steady current, but a few African species are capable of occasional development in completely stagnant places. They are found in all types of watercourse from the merest temporary trickles to the greatest rivers, though each species tends to prefer one particular type of water habitat.

The eggs are usually laid in clusters numbering from about 150 to 600 or more which are stuck on to grasses (fig. 89B), rock-surfaces or other submerged substrates, but some black-flies fly over the water and scatter the eggs loosely and singly. Oviposition most often takes place in the early evening, and often many gravid females lay the eggs communally. Embryonic development can take only a day or two in the tropics, but egg diapause occurs in many temperate species and in these life in the egg may last several months; overwintering can occur in this way in the northern Holarctic regions.

Larval life is completed in only a few days or a week or two in warm tropical regions, but is often very long in temperate species which overwinter in the larval stage; these larvae growing slowly through the winter often attain a greater size and produce larger adult flies than those which complete their development quickly, and the simuliids of the northern temperate regions are in general conspicuously larger species than those of the warm lowland tropics. Larvae have from six to eight instars during their growth (though there are few reliable observations). Larvae are often found in great abundance, especially in the fast-water loving species, and they attach themselves to grasses, dead leaves, trailing roots, stones and smooth rock-surfaces submerged by the water, using the hooklets of the posterior circlet to latch themselves on to a pad of silk spun by the salivary glands and adhered to the substrate. The larval proleg has a similar hook circlet, though much smaller, to that at the end of the abdomen and larvae can move about with looping movements by alternately attaching and releasing the two hook-circlets; larvae can also release themselves into the current on silk threads for re-attachment further downstream and, if necessary, can regain an upstream position by 'climbing' back up the threads. Larvae normally orientate themselves in an out-stretched position in the direction of the current with the head downstream (fig. 89A). They are filter-feeders and use their cephalic fans to strain fine particulate matter from the water, ingesting the filtrate and using the organic element in it (such as diatoms and vegetable detritus) as food; the cephalic fans are held open in the current and closed at intervals for cleaning with the mandibles (they do *not* beat to create a current as used to be wrongly stated). Black-fly larvae are sometimes predaceous and may ingest small chironomid larvae or indulge in cannibalism, swallowing the very young instars of their own kind. There are a few aberrant forms (species of *Gymnopais* & *Twinnia*) in which the cephalic fans are absent, feeding then being accomplished by scraping algal matter from the substrate with the labrum and chisel-like teeth of the hypostomium (the function of the hypostomium in normal filtering forms is not clear, but it is apparently not involved in feeding and may perhaps have the function of cutting extruded silk during movement and cocoon formation).

At maturity the larva undergoes what is physiologically the larval-pupal moult, but does not immediately slough off the last larval skin. It continues to masquerade as a larva to all appearances but is actually the prepupa or pharate pupa recognized by the presence of a much-darkened 'gill-spot' (fig. 83A) on the thorax (the future pupal gill) and by the dorsal abdominal hooks of the future pupa showing through the cuticle. At this stage the black-fly is popularly termed the 'mature larva' or 'gill-spot larva' and for all practical purposes may still be regarded as the larva. The short pharate stage is followed by pupation, at which time the 'gill-spot larva' may seek a more sheltered position prior to weaving the cocoon in which it pupates (though many cascade-inhabiting or grass-attaching forms pupate on the larval sites so that both larvae and pupae are found commonly together). Construction of the cocoon, from silk produced by the salivary glands, normally takes about 40–60 minutes, and once pupation within the cocoon has occurred the pupa lies inactive until the adult fly is ready to emerge. Pupae present a streamlined surface to the current by orientating themselves so that the respiratory gills and open end of the cocoon point downstream. Pupal life is always short, normally from two to six days; the duration of the pupal stage appears to be independent of temperature (cf. mosquitoes), and there is some evidence of a regulatory mechanism which ensures only diurnal emergence of the adult flies. Eclosion of the adult is rapid, requiring only very few minutes; flies may take wing instantly on reaching the water surface, which they do either by crawling upon some object that is not entirely submerged or by releasing themselves into the water and rising in a bubble of gas. The empty cocoons, left behind by emerging flies, are often useful evidence of the presence of simuliids in a stream.

The larvae and pupae of some remarkable African Simuliidae live attached to nymphal mayflies, river-crabs and prawns in an obligate relationship. In these phoretic associations, as they are usually called, the eggs appear never to be laid on the phoretic partner, but the larvae (except for the very early instars) and the pupae are never found elsewhere; attachment to the arthropod partner takes place as early as possible during larval life (an idea once current that only older larvae are found in phoretic associations is now abandoned, as recent work has revealed first instar larvae of several species on their arthropod partners). Nothing is known of how the very young black-fly larvae locate the phoretic partner. The ecological significance of the association is far from clear: it appears to enable some of these simuliids to exist in stony torrents that are subject to rapid flooding and rolling of the boulders, where normal free-living forms are scarce; but other phoretic forms occur commonly in normal streams with stable substrates, and plentiful free-living simuliids, where the biological advantage is less obvious.

Adult male black-flies cannot bite and only the females bite and suck blood. Not all species are blood-suckers, however, some having the teeth of the female mouthparts atrophied and non-functional. The great majority of species require a blood feed before each batch of eggs can mature (i.e. they are anautogenous), but non-blood-sucking forms are able to mature each egg-batch without a blood meal (autogenous forms); this is not a hard and fast distinction, as some blood-sucking species with fully developed biting mouthparts are sometimes autogenous for the first oviposition. Blood-sucking black-flies feed on warm-blooded vertebrates, and though some of them are widely zoophilic the majority show a distinct preference for feeding either on birds (ornithophilic forms) or on mammals (mammalophilic forms). Feeding on man (anthropophily) is an incidental form of feeding by species that are widely zoophilic, and no species is exclusively anthropophilic. The ornithophilic species, which probably predominate in the Simuliidae as a whole, nearly all possess a strong tooth at the base of the claw (fig. 85B) in the biting female which is thought to help the fly to grip the feather-barbs of the host bird and push amongst the overlapping feathers to reach the skin; the truly mammalophilic species, on the other hand, have a simple untoothed claw (fig. 85C), but there are intermediate conditions and *Simulium damnosum*, for example, which feeds both on avian and mammalian blood has claws with a small peg-like basal tooth.

The biting mechanism of blood-sucking female Simuliidae is of special significance in relation to the uptake and transmission of skin-inhabiting onchocercal helminths. Black-flies are 'pool-feeders', imbibing blood from the oozing reservoir which results from the laceration of the host's skin by the fly's mouthparts. During biting the skin is first stretched by the labrum and its apical teeth, cut by the saw-toothed apices of the mandibles working towards each other, and then pierced by the toothed maxillae and the hypopharynx; later the labrum appears to assist in opening the wound. The mandibles, maxillae and hypopharynx enter the host's tissues and may penetrate to a depth of some 120–150 microns, relatively deep probing in relation to the short length of the black-fly proboscis. Blood is imbibed through a canal ('food channel' in fig. 82D) enclosed anteriorly by the labrum and posteriorly by the anterior one of the paired overlapping mandibles, and it is through this channel (which communicates directly with the pharynx and alimentary system of the fly) that blood- and skin-inhabiting parasites gain access to their simuliid vectors. Uptake of blood is rather slow and complete engorgement usually takes at least 4–6 minutes, though much more prolonged feeds (up to a quarter of an hour or more) are frequent; the slow engorgement process is closely related to the likelihood of skin-borne microfilariae being imbibed from the dermal wound, a likelihood increased by the fact that simuliids rasp at the skin instead of piercing directly to the capillaries in the manner of mosquitoes or tsetse-flies. Blood-feeding is very determined in simuliids if flies are feeding on the favoured host, and the flies are dislodged with difficulty once blood-sucking has started; furthermore they never normally make interrupted feeds. As a result of these biting habits the black-flies include efficient intermediate hosts of blood-borne and skin-borne pathogens of birds and mammals, but play little or no part in the mechanical transmission of organisms.

Flies take rather more than their own body-weight of blood at each feed, and a blood-meal is normally taken every few days (after each oviposition and to provide nourishment for the succeeding egg-batch). But black-flies are not as dependent upon blood as is often supposed, and females of blood-sucking forms often feed on nectar, storing a colourless fluid in the crop which appears always to contain sugars (male flies, too, if they feed at all are assumed to derive nutriment from flowers). There are a few reliable records of female flies feeding on invertebrate hosts.

Black-flies bite almost entirely by day (though some crepuscular and even nocturnal biting is recorded for some species) and out-of-doors, and simuliid biting behaviour is therefore in contrast to that of many malarial mosquitoes (which are often nocturnal and endophilic). Location of the hosts is probably largely visual, and biting occurs mainly on specially favoured areas of the host's body that differ between species. Biting is much influenced by weather conditions, and often occurs far from the breeding sites of the flies. The proportion of old and young flies in the biting population varies with season and time of day, older parous flies (that have previously fed and laid eggs) forming a higher proportion of the feeding flies at—for example—the end of a wet season or at a special time of day than the younger recently emerged nulliparous flies (that have not taken their first blood-meal or laid eggs); the factors governing the physiological age structure of the biting populations are poorly understood, but are important in medical entomology as it is only the parous flies that have previously taken a blood-meal which can harbour pathogenic organisms at the infective stage ready for transmission. There is no simple rule-of-thumb for reliably distinguishing parous and nulliparous black-flies, but typically they differ as follows: when the abdomen is dissected the nulliparous female has abundant fat-body (with which it has been endowed from the larval stage), shows no follicular relics in the ovarioles and all the ovarioles of the ovary are tightly packed, and there are never relict unlaid eggs; the parous fly normally has little or no fat-body, the ovary tends to contain loose ovarioles in which follicular relics are evident, and often there are some fully developed eggs left over from a previous oviposition—the relict eggs providing conclusive proof that the fly has already oviposited and strong presumptive evidence that it has had a previous blood-meal.

Some black-flies mate on stones by the water almost immediately after emergence, but most appear to mate on the wing in small swarms not far from the breeding streams (the males finding the females visually). Virtually nothing is known of the mating behaviour of tropical forms, but swarming behaviour probably occurs in many of these as in temperate species. Sperm passes to the female in a small hollow bag (spermatophore) formed during copulation. A few species (subarctic) are parthenogenetic.

All black-flies are fully winged and females can disperse far from their emergence sites, flight ranges of many miles having been authenticated for some species; in Africa flies often bite in open bush far from any rivers or streams. Air movements and birds possibly aid dispersal, but nothing is known positively on this; simuliids have, however, been trapped in the air at various altitudes up to 1530 m (about 5000 ft) (Glick, 1939) and it is likely that flies can reinfest arid areas by being carried on air currents. Adult flies and their biting nuisance are often seasonal, and are normally absent in the northern winter or in severe dry-season conditions, biting then being largely confined to spring and early summer in Europe and North America or to the wet-season in Africa (though *Simulium ochraceum* bites predominantly in the dry-season in southern Mexico); exceptionally severe outbreaks of some large-river species occur periodically, for reasons not yet satisfactorily explained.

Little is known about adult longevity in nature, but female flies can live for a few weeks and some two to three weeks is perhaps normal; marking experiments have, however, shown that females of *Simulium metallicum*, an important onchocerciasis vector species, can live up to at least 85 days, and that several weeks is not unusual (Dalmat, 1955). The reappearance of black-flies after long periods of drought may, some workers think, be due to aestivation of the adult females; if proved this will imply a possible longevity of several months under certain conditions.

Distribution of Simuliidae is governed mainly by the existence of suitable water-courses and therefore by the geology and topography that determine the hydrological characteristics of the river systems. Mountainous areas are often richer in species than the lowlands, and Simuliidae occur up to very high altitudes, being known to about 4700 m (15 275 ft) a.s.l. in the Chilean Andes and up to 4520 m (14 700 ft) and 4030 m, (13 100 ft) on isolated mountains of East Africa (Mt. Kenya and Ruwenzori respectively). Some very high altitude species have atrophied mandibles and evidently cannot suck blood. Man-biting species seldom occur much above an altitude of some 1530 m (5000 ft) above sea-level.

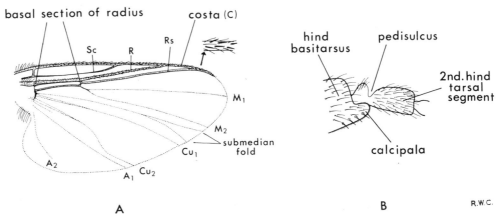

Fig. 84   A, wing of *Simulium*; B, part of hind leg of *Simulium* showing the calcipala and pedisulcus.

TAXONOMY AND IDENTIFICATION

Smart's (1945) classification of the world Simuliidae is now much outdated because of the intensive taxonomic work, and the discovery and description of many new forms, that have taken place in the past twenty-five years, but no completely comprehensive classification has been proposed to supersede it. At present there is no fully agreed arrangement of the genera of black-flies into tribes and subfamilies, and the satisfactory definition of these higher categories—and even of the genera themselves—is very difficult when the whole world fauna is taken into account. The black-flies show almost continuous transition of form from the least specialized to the most advanced, making it hard to categorize them really well and to get complete agreement among specialists on the ranking of definable groups of species; modern classifications attempt to take into account the characters of the larval and pupal stages as well as the adult flies, but the existence of many species (especially phoretic and island forms) with aberrant immature stages, and even of some with normal juveniles but aberrant adults, only adds to the taxonomist's difficulties—and so far there is no evidence that the new cyto-taxonomic techniques are of value above the specific level. It is not necessary, however, to consider the higher classification, or most of the genera, in the present work, as only a few genera contain species having any medical importance; the reader interested in the higher classification of world forms should consult the recent work of Rubtsov (1959–1964), Stone (1964) and Crosskey (1969).

There are about a dozen genera of Simuliidae, providing that the genus *Simulium* is treated in a broad 'lumpers' sense, as is here considered best (see later in this section), although over 70 generic names have been proposed in the family.   Only three of the genera contain man-biting species and have any medical importance, *Prosimulium*, *Austrosimulium* and *Simulium*.   The world genera here recognized are briefly summarized in Table 4, and the three containing man-biting species can be differentiated by the following key.

KEY TO THE GENERA OF SIMULIIDAE CONTAINING MAN-BITING SPECIES*

[Note: this key should only be used for the generic identification of adult flies caught biting man: the simplified key characters will not necessarily hold true for all non-anthropophilic species included in the genera.   The key excludes from coverage the American 'buffalo-gnat', *Cnephia pecuarum* (Riley), as this once notorious simuliid is no longer a man-biting pest.]

1   Hind leg with a rounded lobe (calcipala) at the inner apex of the basitarsus, and with a dorsal groove (pedisulcus) near the base of the second tarsal segment (fig. 84B). Costa of the wing with two types of vestiture, short strong spiniform setae mixed with fine hairs.   Wing vein *Rs* running straight to wing margin without dividing (fig. 84A) 2

–   Hind leg without calcipala or pedisulcus.   Costa of wing with fine hair-like vestiture only.   Wing vein *Rs* forked towards its apex (the two branches sometimes only slightly splayed apart or even subparallel, careful examination required). (North America and northern Eurasia) . . . . ***PROSIMULIUM*** Roubaud

2   Antenna with eleven segments.   (Nearly cosmopolitan, but not in New Zealand) ***SIMULIUM*** Latreille s. l.

–   Antenna with ten segments.   (Australia and New Zealand only) ***AUSTROSIMULIUM*** Tonnoir

*Specimens of Simuliidae collected whilst feeding on man in any part of the tropical regions of the world may be expected to run to *Simulium* in the key; any that do not should be submitted to a specialist.

Genus PROSIMULIUM Roubaud
Taxonomists are not agreed on the limits of this genus but the man-biting species belong only to the typical element in the genus, having the following main characteristics which distinguish them from *Simulium*: mesepisternal sulcus wide and shallow; costa with hairs only; vein *Rs* forked into two terminal branches; hind leg without calcipala or pedisulcus; pupa with feeble irregular cocoon and a pair of long strong terminal hooks.

The most important man-biting species are *Prosimulium mixtum* Syme & Davies and its allies in eastern Canada and New York State.

Genus AUSTROSIMULIUM Tonnoir
It has become taxonomic tradition to treat *Austrosimulium* as a very distinct genus of Simuliidae but in fact its characters appear to ally it very closely with *Simulium* s.l., from which it differs by having 10-segmented antennae (very rarely only 9 segments),

Table 4 Summary of the world genera of Simuliidae (in alphabetical order).

Genus*	Distribution	No. of Species	Anthropophilic habit
Afrosimulium Crosskey	southern Africa	1	No
Austrosimulium Tonnoir	Australia, Tasmania, New Zealand, Campbell Island	18	Yes, in a few species
Crozetia Davies	Crozet Islands	1	No
Gigantodax Enderlein	Andean Mountains from Central America to Tierra del Fuego	21†	No
Gymnopais Stone	subarctic parts of North America and Eurasia	5†	No
Metacnephia Crosskey	Holarctic regions, including North Africa	about 35	No
Parasimulium Malloch	western United States	2	No
Prosimulium Roubaud (including *Cnephia* Enderlein and allied forms)	Holarctic regions, also a few species in Africa, Australia, Central and South America	about 125	Yes, in a few Holarctic species
Simulium Latreille s.l.	worldwide (but excluding New Zealand and Hawaii)	about 810	Yes, in some species from all regions
Tlalocomyia Wygodzinsky & Diaz Najera	Mexico	1	No
Twinnia Stone & Jamnback	northern Holarctic regions	5	No

*The genera *Paraustrosimulium* Wygodzinsky & Coscarón and *Lutzsimulium* d'Andretta, containing one and two species repectively from South America, are omitted as their status is very obscure. Some additional small South American genera are currently being described (Wygodzinsky & Coscarón, *in prep.*).
†Additional undescribed species of these genera are known.

hair-like vestiture only on R_1, more than one apical spinule on the styles of the male genitalia, and by having as a rule some complex multiramous hooklets on the terminal abdominal pupal segments, traces of an annular sclerite on the larva in front of the posterior circlet and by having forwardly directed sclerotized arms from the larval anal sclerite. However, all of these features (with the possible exception of multiramous pupal hooklets) can be found somewhere in the genus *Simulium* when the whole world fauna is considered, and the two genera are not very clearly separable.

Austrosimulium is confined to Australia, Tasmania, New Zealand, and Campbell Island. The only species biting man at all regularly are *A. pestilens* Mackerras & Mackerras in Queensland, *A. bancrofti* Taylor in south-western Australia and *A. ungulatum* Tonnoir in New Zealand.

Genus SIMULIUM Latreille

Definition. Black-flies with the following characters in combination: Antenna with 11 segments (very rare exceptions with 10 or 12 segments in Oriental region). Pleural membrane bare or haired. Mesepisternal sulcus deep and narrow, virtually complete anteriorly. Costa and apical part of R_1 with spiniform setae as well as hairs (except in one Pacific island species with hair only on R_1). Vein Rs undivided, vein Cu_2 sinuous, basal section of radius bare or haired. Calcipala large (except in a few South American species), pedisulcus present. Style of male genitalia with one apical spinule (more in a few species). Pupa with discrete cocoon and without long abdominal tail hooks. Larval cuticle often with small setae or scales. [For more complete diagnosis see Crosskey (1969)].

Rubtsov (1959–1964) treats *Simulium* as a large number of separate genera, and restricts the use of the name *Simulium* to a relatively small number of Holarctic species, but this course has not found general acceptance among other specialists on the Simuliidae. The majority of taxonomists prefer to retain *Simulium* in its old broad sense (although most now recognize many named subgenera within the genus, which need not be considered in the present work). In the following discussion and elsewhere in this book the name *Simulium* is used in its old wide sense, as there is no doubt that the retention of a broad generic concept is to the mutual benefit of the taxonomist, the medical entomologist and veterinarian, and especially to all workers concerned with Simuliidae in tropical medicine.

The world fauna of *Simulium* s.l. includes about 35 subgenera and 810 species (the latter comprising some 300 Palaearctic, 60 Nearctic, 200 Neotropical, 110 Ethiopian, 80 Oriental and 60 Australasian species). Some species are now known to have populations with different cytological characteristics although their external morphology is homogeneous, and these may be complexes of sibling species; in *Simulium damnosum*, for example, seventeen different cytotypes have been demonstrated in populations from different parts of Africa, and the chromosome differences which are now being found in many black-fly species will, if accepted as evidence of specificity, imply many more species than at present. The existence of cytologically separable siblings, which lack tangible and well correlated macromorphological features for their recognition, poses potential problems of nomenclature in important vectors such as the *S. damnosum*-complex; it will be best for the chromosomally distinguishable forms to be called by

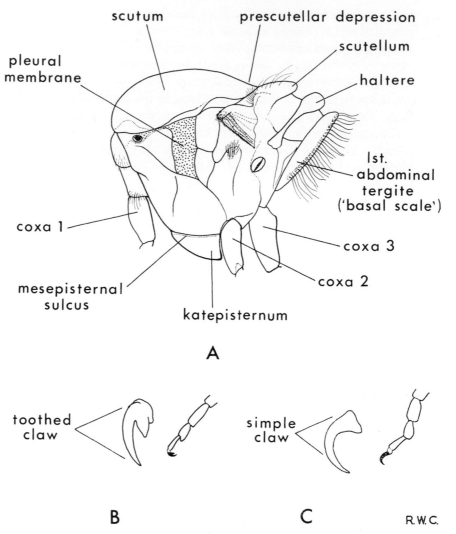

scutum

prescutellar depression

scutellum

haltere

pleural membrane

1st. abdominal tergite ('basal scale')

coxa 1

coxa 3

coxa 2

mesepisternal sulcus

katepisternum

A

toothed claw

simple claw

B   C   R.W.C.

Fig. 85   A, side view of thorax of a typical black-fly; B, tarsal claw with large basal tooth, typical of bird-biting simuliids; C, tarsal claw lacking basal tooth, typical of mammal-biting simuliids.   (Only extreme base of wing shown in fig. A.)

vernacular names, at least until the problems are more thoroughly investigated and under-stood—then it may be justified to establish formal Latin names for them which are available in zoological nomenclature (as in similar complexes of mosquitoes).

The genus is inadequately covered by monographs with which the species can be identified, and there are none for most zoogeographical regions; even for the well worked fauna of North America no comprehensive taxonomic work exists.   The situation is better for the regions in which onchocerciasis occurs and *Simulium* is medic-ally important: Dalmat (1955) and Vargas & Diaz Najera (1957) have monographed the species of Guatemala and Mexico, and their works are invaluable for the identification of the Central American fauna; Freeman & de Meillon (1953) have monographed the species of the Ethiopian zoogeographical region, and, although more species have been described since its publication, this work is excellent for identification of most of the tropical

African species in the adult and pupal stages.   The keys of Crosskey (1960), though dealing mainly with the West African fauna, to a large extent form a larval supplement to Freeman & de Meillon (*op. cit.*).

Even with the best works identification to species is often not easy, depending at times on minute features of the male or female genitalia or of the larval head or vestiture for which slides must be prepared, but pupae of African species are frequently easy to identify by the form of the gills.   Specific identifications should be made or confirmed by a specialist, and the medical entomologist or field worker (unless concerned with onchocerciasis) need normally only recognize the genus as such, for which the following simple rule-of-thumb is sufficient: *Any adult black-fly having a calcipala and pedisulcus on the hind leg and 11-segmented antenna belongs to Simulium.*

The only *Simulium* of serious medical importance are those habitually or frequently biting man in areas where onchocerciasis is endemic, and the following keys are given to *help* in identification of these anthropophilic species; but it must be kept in mind that a few other species come to man but very rarely bite, and that other species could be encountered that are not yet known to be at all anthropophilic (some species not habitually anthropophilic will feed on man when their numbers reach exceptionally high levels).   The keys to adults include the known onchocerciasis vectors and also the other man-biting species found in the regions where the disease is endemic, as the latter are of potential importance although not yet incriminated as certain vectors; the keys to the early stages (given for Africa only) attempt merely to distinguish the vectors from the harmless species.

KEY TO MAN-BITING FEMALES OF *SIMULIUM* IN TROPICAL AFRICA

[Note: the geographical areas cited are those in which man-biting is known, *not* the full range of species distribution. All included species except *damnosum* and *neavei* bite man only rarely and atypically, or very locally, or at very low frequency.]

1   Fore tarsus broad and flattened, with conspicuous dorsal hair-crest (fig. 81).   Abdomen with recumbent silvery or silver-yellow hairs arranged in clumps on the sides, and with the posterior tergites bare and shining black (except for some minute erect dark hairs).   Legs black with characteristic and very conspicuous creamy-white band on hind basitarsus. Medium-sized species, wing-length 2.0–2.4 mm. (Commonest man-biting species, West to East Africa, Congo Basin, Ethiopia, Sudan: see accompanying map)   .   .   .   .   .   . ***damnosum*** Theobald

–   Fore tarsus not conspicuously dilated and without obvious hair-crest.   Abdomen with rather evenly distributed recumbent hairing, present on posterior tergites and not aggregated into definite separated clumps laterally.   Legs usually either unicolorous black or extensively pale.   Size varied   2

2   Large blackish species with silvery-yellow to golden or partly black hair vesture on thorax and abdomen.   Scutum without trace of pattern.   Wing-length 2.6–3.6 mm.   3

–   Very small greyish species (blackish in *wellmanni*) with silvery hair vesture on thorax and abdomen.   Scutum with three fine dark longitudinal lines, best seen as fly is turned.   Wing-length 1.3–2.3 mm.   6

3   Abdomen with rather evenly distributed silvery-yellow to golden recumbent hairing, sometimes coppery medially, base of abdomen not strikingly contrasting with remainder.   Ground colour of basal part of hind basitarsi paler than remainder of hind legs (inconspicuously so in *neavei*). (Eastern Congo and East Africa)   .   4

—   Base of abdomen thickly covered with silvery-yellow to golden hairing and strikingly
     contrasting in colour with remainder of abdomen, which appears partly or entirely
     black with black hairing. Hind basitarsi entirely black like the rest of the legs.
     (Cameroons)   5

4   Abdominal hairing uniformly silvery-yellow to golden.   Paler part of hind basitarsus
     inconspicuous, not forming a definite pale band. (Congo, Uganda, formerly Kenya
     also)   .   .   .   .   ***neavei*** Roubaud

—   Abdominal hairing not unicolorous, distinctly coppery or bronze coloured on middle
     segments (fourth tergite and sometimes fifth also) which therefore contrast in colour
     with silvery-yellow vestiture on other segments. Hind basitarsus with definite
     pale yellow or reddish-yellow pale band in ground colour of basal half or two-thirds.
     (Tanzania)   .   .   .   .   ***woodi*** de Meillon

5   Very black species with hair of frons, clypeus, prescutellar area of scutum, and scu-
     tellum black or bronze-black.   Abdomen almost entirely black-haired, except for the
     contrasting yellow-haired basal two segments.   (Forested parts of Cameroons)
                                                    dukei Lewis, Disney & Crosskey

—   Not such conspicuously black species, hair of frons, clypeus, all of scutum and much
     of scutellum silvery-yellow to pale golden.   Abdomen with conspicuous yellow hair
     on sides of segments 3–6 (middle part), in addition to thick yellow hairing of basal
     two segments, and bronze-black to black haired only on terminal segments and
     mid-dorsally on segments 3–6.   (Mainly savanna parts of Cameroons)
                                                    ovazzae Grenier & Mouchet

6   Claws with large basal tooth (fig. 85B)*   7
—   Claws simple, without basal tooth (fig. 85C)*   8

7   Basal section of radius haired.* Pleural membrane haired. (Mainly northern savanna
     areas of West Africa)   .   .   .   .   .   .   ***adersi*** Pomeroy
—   Basal section of radius bare.   Pleural membrane bare.   (Mainly Sudan, also Nigeria)
                                                    griseicolle Becker

8   Legs and antennae entirely black.   (Angola).   .   .   .   ***wellmanni*** Roubaud
—   Legs extensively pale or with at least a pale hind basitarsal band; antennae with first
     two segments reddish-yellow to dark reddish and conspicuously paler than remainder
     of antennae   9

9   Legs mainly pale reddish-yellow, only darkened brown to blackish-brown on tarsi and
     apices of tibiae, sometimes faintly on mid femora.   (Mainly Nigeria)   ***bovis*** de Meillon
—   Legs dark brown or blackish-brown, only yellowish on basal two-thirds of hind basi-
     tarsi and sometimes paler brownish on bases of tibiae. (Congo and Zambia)
                                                    albivirgulatum Wanson & Henrard

          * High magnification is required to see these features adequately and slide preparations
          are useful.

KEY TO PUPAE FOR RECOGNITION OF AFRICAN VECTOR *SIMULIUM*

1   Pupae attached to crabs   .   .   .   .   .   ***neavei, woodi,*** and allied species
     [For further identification see Lewis & Hanney (1965).]
—   Pupae not attached to crabs   2

2   Pupal gill as in fig. 86E, composed of inflated thin-walled tubular branches and having
     some resemblance to bunch of bananas; branches arranged as pair of large basal arms
     and between them three curved outer branches and six weaker inner branches, the
     latter arising as three pairs from short common stems   .   ***damnosum*** Theobald
—   Pupal gill not as in fig. 86E, of immensely varied form, but if (very rarely) branches in
     form of enlarged thin-walled tubes then not arranged as above
                                                    Harmless species, none proven vectors

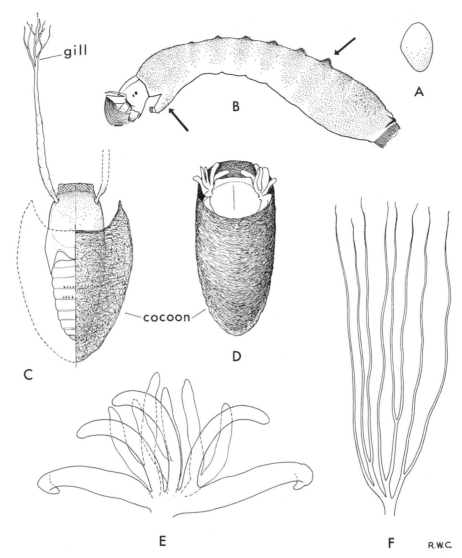

Fig. 86 Early stages of Simuliidae: A, typical black-fly egg, showing eccentric shape; B, larva of *Simulium damnosum*, side view; C, typical pupa and cocoon of *Simulium*; D, pupa and cocoon of *Simulium damnosum*, dorsal view; E, pupal gill of *Simulium damnosum*; F, pupal gill of *Simulium neavei*. (Arrows on fig. B point to the abdominal tubercles and the setae on the proleg which identify *S. damnosum* larvae. In fig. C the left half of the cocoon is cut away to show the pupal abdomen and its main hooks on the dorsal side.)

KEY TO LARVAE FOR RECOGNITION OF AFRICAN VECTOR *SIMULIUM*

[Note: key applies only to well-grown or mature larvae].

1   Larvae attached to crabs . . . . . . ***neavei, woodi,*** and allied species
     [For further identification see Lewis (1961) and Lewis & Hanney (1965).]
–   Larvae not attached to crabs 2

2 Larva with characteristic paired dorsolateral prominences, often in form of large conical tubercles, on the anterior abdominal segments (fig. 86B). Larval thorax (including proleg) and abdomen largely covered with conspicuous small black setae or scales, these aggregated on the dorsal tubercles and making them very obvious
damnosum Theobald

[Two species known from a single locality in Angola are similar.]

— Larva without dorsal abdominal tubercles, at most only slightly swollen in this position. Larval cuticle without extensive covering of setae, or if (a few species) abdomen and thorax both have setae then the proleg always bare   . *Harmless species, none proven vectors*

KEY TO MAN-BITING FEMALES OF *SIMULIUM* IN MEXICO, CENTRAL AMERICA AND NORTHERN SOUTH AMERICA

[Note: the localities given indicate where man-biting is most serious. A few species not included in the key may bite man very rarely.]

1 Scutum yellow or orange. Basal section of radius haired   2
— Scutum brownish-black or black in ground colour (narrowly orange-yellow on sides in *veracruzanum*). Basal section of radius bare   3

2 Legs mainly yellow, only brown on parts of tarsi and apically on hind femora and tibia. Scutum with a pattern of two curving silvery-grey pollinose longitudinal lines. Abdomen only yellow on first two segments. (Guatemala, Mexico)
callidum Dyar & Shannon
— Legs entirely dark brown. Scutum without a pattern of pollinose lines. Abdomen yellow on basal half (first four tergites) and black on remainder, the colour contrast very conspicuous. (Guatemala, Mexico)   .   .   .   .   *ochraceum* Walker

3 Abdomen entirely black. Scutellum brown or black. Ground colour of scutum all dark, usually with pattern also   4
— Abdomen with first two segments yellow. Scutellum yellow. Ground colour of scutum mainly black but orange-yellow on lateral margins and adjacent to humeri; median black area divided into three by a pair of broad and very boldly marked longitudinal stripes of silvery-grey pollinosity. (Guatemala) *veracruzanum* Vargas *et al.*

4 Hair of scutum all or mostly black. Scutellum without any pale hair. Legs dark brown or black except for conspicuous creamy-white band on each hind basitarsus .   5
— Hair of scutum entirely silvery-yellow to golden. Legs mainly pale yellow, only parts of tarsi and hind legs noticeably dark brown. Scutellum with conspicuous long soft recumbent yellow or golden hair on each side   6

5 Hair of scutum all black, very short. Scutum with pattern formed of three narrow stripes (outer pair convergent anteriorly) which appear black and darker than surrounding areas in some lights but reverse to appear silver-grey against a dark background as fly is turned into different lights. Medium-sized species, wing-length about 2.4 mm. (Guatemala, British Honduras, Venezuela) .   . *metallicum* Bellardi
— Scutal hair mainly black, but posterior angles of scutum and the prescutellar area with conspicuous soft yellow recumbent hair. Scutum with pattern formed of two brilliant subparallel silver pollinose stripes, their appearance more or less fixed. Small species, wing-length about 2 mm. (British Honduras, Costa Rica, Panama, Venezuela)   .   .   .   . *quadrivittatum* Loew

6 Hair of scutum arranged in small discrete clumps. Scutum without a pattern. Hind basitarsus entirely yellow. (Colombia, Mexico, Guatemala).   . *exiguum* Roubaud

  — Hair of scutum evenly distributed.   Scutum with very bold pattern, two broad parallel silver-grey pollinose stripes which merge posteriorly with each other and with similar grey pollinose side-margins leaving a median and two sublateral broad black vittae.   Hind basitarsus white on about basal two-thirds and black-brown on apical third.   (Northern South America)   .   .   .   .   .   ***haematopotum*** Malloch

MEDICAL IMPORTANCE AND VECTORS OF ONCHOCERCIASIS

The medical importance of black-flies arises from the direct effects of their bites, and—much more importantly—from their rôle in disease transmission.

Simuliidae as biting pests

The Simuliidae include some of the most intolerable and voracious pests that bite man. Some species are such a persistent menace at certain times that they can make large areas of land unpleasant to live or work in (and can thereby hinder economic development). The bites can be painful and give rise to swelling and inflammation of the badly bitten areas of the skin which can persist for several days; when the face is badly bitten the eyes may be partially closed by the facial swelling, and bleeding and irritation from the bites can in themselves be very unpleasant.   More severe reactions to the bites sometimes occur: in Central Europe, for example, the bites of *Simulium erythrocephalum* (De Geer) are a cause of dermatitis (Krstitsch & Zivkovitch, 1968), and in the United States *Simulium jenningsi* Malloch is reported to have caused allergic asthma (Brown & Bernton, 1970).   In the nineteenth century human deaths were reported in the lower Mississippi valley from mass biting of *Cnephia pecuarum* (Riley), the American 'buffalo-gnat' (a species which is not a serious pest since the building of the Mississippi levees); and the notorious Golubatz fly (*Simulium colombaschense* Fabricius = *S. columbaczense* Schönbauer) of the Danube in Central Europe is believed to have caused human fatalities during its bad eighteenth century outbreaks (from toxaemia due to the bites and indirectly while people were escaping the black-fly attacks or the livestock driven demented by these attacks).

Man-biting species are found in all zoogeographical regions but the number making serious attacks on man is fortunately small.   Fallis (1964) has listed one hundred species of simuliids known to bite man, but in most of these anthropophily is very atypical. Broadly speaking the onchocerciasis vector species (see below) are often not important biting pests as such, even *S. damnosum* often being no more than a minor irritant, but *S. ochraceum* and *S. metallicum* can bite avidly in immense numbers in Central America. The most vicious and troublesome man-biting simuliid pests are: *Prosimulium mixtum* and *S. venustum* Say in eastern Canada and New York State; *S. quadrivittatum* in Central America; *S. antillarum* Jennings in some Caribbean islands; *S. indicum* Becher in the lower Himalayas from India to Thailand; *S. jolyi* Roubaud in the New Hebrides; *S. buissoni* Edwards in the Marquesas Islands; and *S. amazonicum* Goeldi (or a related species) in the Amazon basin.

K

Simuliidae as disease vectors

The Simuliidae (unlike the mosquitoes, tsetse-flies and phlebotomids) do not transmit any protozoan blood parasites of man, and are not known to be involved in transmission of human virus infections.   Their medical importance as disease vectors lies in the ability of a few species of *Simulium* to transmit *Onchocerca volvulus* to man; this is the

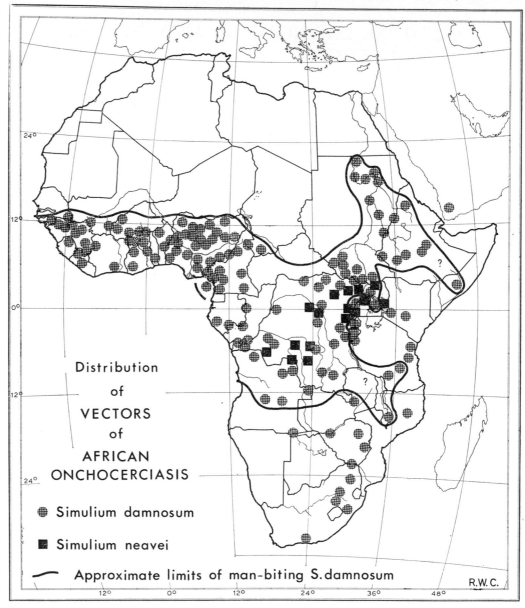

Fig. 87   Map showing the geographical distribution of *Simulium damnosum* and *S. neavei*, the chief vectors of African onchocerciasis. (The distribution of *S. woodi* is not shown as this localized species is only known to be a vector in one small focus in Tanzania.   *S. damnosum* might be a complex of sibling species (see text) and is not yet known to bite man in the Yemen where onchocerciasis occurs.   The map is based upon evidence available up to 1970.)

only pathogenic organism that is certainly transmitted to man by black-flies, but there is a strong possibility that the human filaria *Mansonella ozzardi* is transmitted by *Simulium amazonicum* Goeldi in Brazil (Cerqueira, 1959; Garnham & Walliker, 1965). On present evidence onchocerciasis caused by *O. volvulus* infection is not a zoonosis, but natural infections with this parasite have been recorded in two primates other than man, viz. in an *Ateles* monkey in Mexico (Caballero & Barrera, 1958) and in a gorilla in the Congo (Van den Berghe *et al.*, 1964); the chimpanzee has been experimentally infected

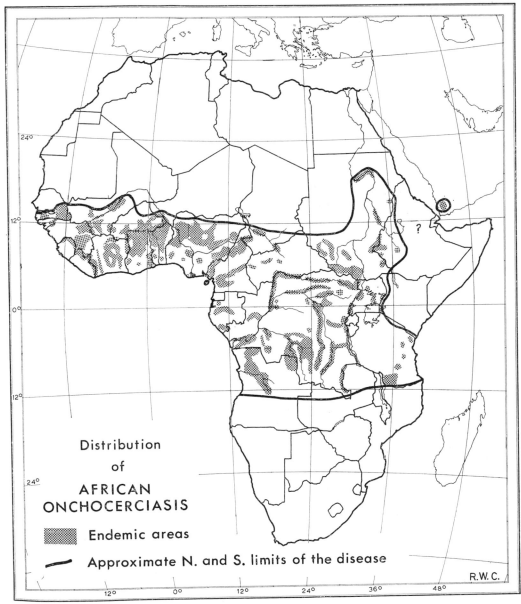

Fig. 88   Map showing the approximate geographical distribution of human onchocerciasis in Africa, the endemic areas shaded. (Map based upon evidence available up to 1970. The endemic areas shown along the rivers of the southern Congo basin, derived from Fig. 1, p. 7 of *W.H.O. Tech. Rep. Ser.*, No. 335, 1966, may be less extensive than indicated.)

(Duke, 1962). In Africa the areas most heavily infected with onchocerciasis in man lie in the savannas, outside the limited areas where these rain-forest primates occur, and the disease appears to be maintained in man solely by inter-human transmission (Nelson, 1965) through simuliid vectors. Earlier suspicions that wild and domestic African ruminant animals might act as reservoirs of *O. volvulus* have not been corroborated, and available evidence now indicates that the infections in cattle and antelopes so far discovered involve different species of *Onchocerca*.

Onchocerciasis is a filarial disease characterized in man by intense itching, lichenification and fissuring of the skin, the occurrence of subcutaneous nodules where the tissues are thin over the bones (as on the hips, ribs, knees or scalp) and, in the worst cases, of irreversible blindness. The slender adults of *O. volvulus* reach a few centimetres long in the male but up to 70 centimetres or more in the female and live in deep-seated tissues as well as in the superficial fibrous nodules (onchocercomata) just under the skin. The embryo worms (microfilariae), produced by the female viviparously and in enormous numbers, are found only in the skin (in contrast to those of the other important human filariases, viz. Bancroftian and Brugian filariasis and loiasis, in which the microfilariae occur in the blood); they have no known periodicity in the dermis. The microfilariae of *O. volvulus* are taken up by the *Simulium* intermediate hosts when they rasp the skin in order to obtain blood, and up to two hundred or more can be ingested at one blood-meal, although few of these (often not more than one or two) will complete their development to become infective forms that can be passed on to another human being; metamorphosis of the parasite within the fly involves change of form to a fat caudate 'sausage' stage embedded in the thoracic muscles of the fly and later to a relatively large infective (metacyclic) stage which is passed on from the mouthparts when the fly takes a subsequent feed, the metamorphic process within the *Simulium* vector usually taking about 10–13 days.

The disease is endemic in tropical Africa where it occurs sporadically in rural areas between about 15°N. and 13°S. latitudes and in localized foci in the tropical Americas (Mexico, Guatemala, Venezuela and Colombia, possibly also Surinam) (see figs 88 & 214); a small focus exists in Yemen, near the south-west tip of the Arabian peninsula. It has long been considered probable that New World onchocerciasis arose by importation of *O. volvulus* from Africa in infected negro slaves, but Duke (1970 : 430) considers this very unlikely: his recent cross-transmission experiments with African and American strains of the parasite strongly contraindicate such an origin for the New World infections. (The status of the name *Onchocerca caecutiens* (Brumpt), at one time applied to the New World parasite and for many years now considered a synonym of *O. volvulus* (Leuckart), becomes less sure in the light of Duke's findings, which appear to weaken the evidence for conspecificity of the American and African *Onchocerca* parasites of man.)

In Africa several million people are probably infected with the disease but incompleteness of clinical surveys prevents any accurate estimation at present (300,000 cases have been estimated for Northern Nigeria). Onchocercal blindness is especially serious in the savanna areas of West Africa where many villages have up to ten per cent of the inhabitants blind, or with serious eye lesions caused by *O. volvulus*. The exact causes of blindness are uncertain, but it is associated with the physical presence of microfilariae in or near the eyes. Some heavily infected communities in the West African forest areas have very little onchocercal blindness in comparison with equally infected communities in the savanna areas; the reasons for this remain obscure.   In West Africa onchocerciasis covers enormous diffuse areas, mostly in the savanna, but in East Africa it is largely restricted to small discrete foci; in the Congo basin it tends to occur along the courses of the great rivers where the population mostly lives.   The disease and at least one of the two chief vectors (*S. neavei* and *S. damnosum*) occur in Angola, Cameroun, Central African Republic, Chad, Congo (both republics), Dahomey, Ethiopia, Fernando Po, Gabon, Ghana, Guinea, Ivory Coast, Kenya (nearly eradicated), Liberia, Mali, Niger, Nigeria, Portuguese Guinea, Senegal, Sierra Leone, Sudan, Tanzania, Togo, Uganda, Upper Volta and Yemen.

American onchocerciasis (Robles' disease) is much more restricted: there are three small foci in Mexico (Oaxaca and Chiapas States), a main focus and two much smaller foci in Guatemala (the latter possibly continuous with a Mexican focus), and two diffuse foci in northern Venezuela; infection has also been found in a village in southern Colombia.   In Mexico and Guatemala about 50,000 people are infected, but there are no estimates for Venezuela.   The Mexican and Guatemalan foci occur in coffee-growing areas between about 460 m (1500 ft) and 1400 m (4500 ft) above sea-level on the Pacific side of the central mountains.   The restriction of the disease to such small foci is not easily explicable, as the chief vector in Mexico and Guatemala, *Simulium ochraceum*, occurs over a much wider area.   In Africa the vectors have a wider range than the disease (cf. figs 87 & 88) if *S. damnosum* is treated as one species (but it is probably a complex of species of which the typical member has a distribution more coincident with that of onchocerciasis); in West Africa the man-biting *S. damnosum* populations nearly all belong to the typical cytological form ('Nile form') and onchocerciasis seems to exist wherever *S. damnosum* is found.   In Africa there is no definite association of the disease with altitude (it occurs from sea-level to about 1550 m (5000 ft)).

The symptomatology of onchocerciasis is essentially similar in Africa and the New World, but there is a definite association in each area between the part of the body on which the vectors feed and the localization of microfilariae (and to a lesser extent of nodules).   In Mexico and Guatemala the chief vector (*S. ochraceum*) bites mainly on the upper parts of the body and these areas contain the highest microfilarial densities; in Africa the situation is reversed, both main vectors feeding primarily on the lower parts of the body where the highest worm loads tend to occur (the situation in Venezuela, where the vector *S. metallicum* bites the legs, is similar to Africa).

The biting habits of simuliids make them particularly well adapted to the uptake of the skin-borne *O. volvulus* microfilariae and for their subsequent transmission as worms in the infective stage.   Some species of *Simulium* are very efficient vectors, and there is increasing evidence that onchocerciasis transmission can be maintained by remarkably low vector densities (either naturally low as in the case of *Simulium woodi* which maintains the small Amani focus in Tanzania, or low because of vector control measures as in some *S. damnosum* foci).   The efficiency of *Simulium* species as *O. volvulus* vectors derives from five main aspects of their biology: (i) the 'pool-feeding' habit with its associated tearing of the skin and release of microfilariae into the well of blood from which imbibition takes place; (ii) the long engorgement time which enhances the likelihood of microfilariae being imbibed; (iii) the determined manner of feeding if man is the favoured host and unwillingness to be dislodged or make interrupted feeds; (iv) the need for blood every few days before the gonotrophic cycle can normally be initiated and eggs produced; and (v) the relatively long adult female life which can permit some worms from two or more intakes of microfilariae to reach the infective stage during the fly's lifetime. Superimposed upon these factors, and conducing to successful transmission, is the apparent response of the microfilariae to the bite of the fly: flies can imbibe in one blood-meal up to 200 or more microfilariae, a far greater number than could be present in the dermis in the relatively minute area damaged by the mouthparts, and a positive attraction of microfilariae to the site of the bite seems to occur.

It cannot be stated exactly how many species of *Simulium* can transmit *O. volvulus* in man, but the two tropical African species *S. damnosum* and *S. neavei*, and the two tropical American species *S. ochraceum* and *S. metallicum*, are without doubt the chief vectors responsible for most transmission. Some other species have been incriminated as minor or local vectors, or are regarded as potential vectors because *O. volvulus* can develop in them to the infective stage; but as some of these species bite man only infrequently or hesitantly, because he is not the most favoured source of blood, it is doubtful if they play any significant rôle in natural transmission.

Infectivity rates of *O. volvulus* in its vectors vary greatly in fly populations sampled at different times of season or day in any place, and at different places within an endemic area of onchocerciasis. Unless the physiological age structure of each vector population are known, and there is reliable assessment of biting rates, the infectivity rate contributes little to a knowledge of transmission (hence no indication is given here of the widely differing data on infectivity rates that have been published); the level of transmission needs to be assessed as a transmission index derived from the infective bites likely to be received at any time by an individual or community in a given time, and for *Simulium-Onchocerca* complexes little has yet been done on these lines.

The following is a brief *résumé* of the black-fly vectors of human onchocerciasis with pertinent notes on their biology.

Chief vectors:

S. damnosum. This is the most important and widespread African vector (fig. 87) and is responsible for transmission in West Africa, the Congo Basin, parts of Uganda and Tanzania, in Sudan and Ethiopia, and probably in the Yemen. It occurs in southern and south-central Africa but in these areas, as in parts of East Africa, it does not bite man. It is probably not a single species, as studies of the giant chromosomes of the larval salivary glands have shown that at least seventeen cytotypes exist in different parts of Africa; the chromosomal differences between these are at least partially correlated with either an exclusively zoophilic habit or with anthropophilic populations. The man-biting female of *damnosum* is easily recognized by the combination of conspicuous hair-crest on the front tarsi and white band on the hind basitarsi (fig. 81), the latter feature being very obvious to the naked eye on the living fly; the larva (fig. 86B) is also distinctive, having subconical protuberances on the dorsum of the middle segments (pair to each segment) and an almost complete covering of minute dark setae or slender scales. The immature stages occur characteristically in the rapids of small to very large rivers, the larvae and pupae being attached predominantly to submerged trailing grasses, on which most oviposition takes place; *damnosum* normally does not occur in tiny steams or temporary trickles. Adult flies disperse far from the breeding sites and can be found biting in open bush up to ten miles or more from their source; about 90 per cent of bites on man are on the legs. The species, or species-complex as it now appears, occurs in forested and savanna country and in the West African northern dry savannas is able to reappear after prolonged drying out of the breeding rivers (either, it is presumed—as firm evidence is lacking—because of aestivation of adult flies locally or because of air-borne reinfestation with flies from areas of perennial breeding farther south). In areas with sharply demarcated dry and wet seasons most transmission by *S. damnosum* is clearly seasonal, but the transmission peak is probably not at the time of maximum fly abundance in mid rains. There is evidence that flies take significantly longer to engorge on infected persons with advanced onchocercal skin lesions; in such cases the microfilarial densities are lower than in lightly infected cases showing no skin lesions, but there might be a similar transmission potential because of the prolonged engorgement time. Biting of wild and domestic birds and of mammals other than man occurs among anthropophilic *damnosum* in West Africa, and infections of female flies with unidentifiable filariae—probably of avian origin—are recorded.

The tiny black 'sweat-bees' of the genus *Trigona* (see Chapter 12) are sometimes mistaken by the layman for *S. damnosum* when they alight on the skin, because they have some superficial likeness (and are sometimes sent to the medical entomologist for identification on this account).

S. metallicum. This is the vector in Venezuela, and is considered to be a vector of minor importance in Mexico and Guatemala. In size, appearance (black species with white-banded hind basitarsi) and many of its habits it greatly resembles the African *S. damnosum*—for example it oviposits in large egg masses attached to trailing vegetation and bites man almost entirely on the legs (being particularly partial to the feet if these are exposed), though this preference is less marked in Guatemala than in Venezuela. The early stages occur in streams of very varied size, in Mexico and Guatemala often in large streams with much greater flow than those favoured by *ochraceum*, but in Venezuela usually in very small streams. It bites man avidly, and appears to be especially strongly anthropophilic in Venezuela, but in Mexico and Guatemala it is largely zoophilic with a preference for cows and horses; it feed occasionally on domestic birds. Female flies disperse farther from the breeding sites than *S. ochraceum* and *S. callidum* and a flight range up to at least nine miles is authenticated. On man it tends to feed nervously like *callidum* in Guatemala, but in Venezuela it is a determined biter which is not easily disturbed once blood-sucking has started (like *ochraceum* and *damnosum*); similarly in British Honduras, where there is no known onchocerciasis but *metallicum* is a serious pest, it is a determined feeder on man. The transmission potential of the species may vary therefore in different parts of the distribution range.

S. neavei. This is an important vector in the Congo and Uganda (fig. 87), and was formerly responsible for transmission in Kenya (from where it is now believed to be eradicated). It is a phoretic species in which the larvae and pupae live on river-crabs of the genus *Potamonautes*, usually in rather turbid streams and small rivers.

Several species of *Potamonautes* act as carriers, and together these have a wider distribution than *S. neavei*; the distribution of this vector, at least in East Africa, is therefore
not limited by the availability of suitable crabs (Williams, 1968). Larvae and pupae
occur mainly on the sides of the crabs and on the limb bases and eye-stalks (fig. 89c)
without any particular orientation. The oviposition sites are unknown and studies
are badly needed to locate them; it seems certain that eggs are not laid directly on the
crabs, and up to now the first instar larvae have not been found on the crabs either
(it is supposed that eggs might be laid on vegetation or inorganic substrates in parts of
the watercourses upstream of the stretches in which larvae occur on crabs). Adult
flies often bite in forest at higher altitudes than those at which the early stages are found,
and are frequently hard to find in areas where the juveniles are common. Man-biting
is mostly on the legs, but the preference is less positive than in *S. damnosum*. Little
is known about dispersal, seasonal abundance, or hosts other than man. Nelson &
Pester (1962) found infections with an unidentifiable filaria, in addition to *O. volvulus*,
among anthropophilic *S. neavei* in Uganda, which suggests that flies feed on alternative
animal hosts. Female flies are easily distinguished from *S. damnosum* by their all
black legs, larger size, and their much more yellowish appearance in life (caused by
the vestiture of golden-yellow hairing on the frons, scutum, and abdomen); the pupal
gill is of a quite different type from that of *damnosum*, having eight long slender filaments (fig. 86F).

S. ochraceum. This is the chief vector in Mexico and Guatemala and is widely
distributed in Central and northern South America. It breeds perennially in very
small streams or mere trickles of water that are often concealed under herbage and
fallen leaves and therefore hard to find; the oviposition habits are unusual, the eggs
being dropped a few at a time on to emergent floating vegetation. Flight range is
short, being limited by forest and mountain. Flies feed to a small extent on domestic
animals and birds, but in Guatemala and Mexico are avidly anthropophilic; biting on
man occurs largely on the head, neck and shoulders. Flies bite plentifully throughout
the year but are most abundant in the middle part of the weakly defined dry season
(December–January); there is conflicting evidence on whether this is the main period
of *O. volvulus* transmission, but it is generally agreed that *ochraceum* is the most efficient
vector in the Mexico-Guatemala foci. It is a very small species easily recognized by
its bright orange scutum and largely yellow abdomen (coloured illustration in Dalmat,
1955).

Other known or suspected vectors:

In addition to the four main vectors discussed above some other species are known
vectors that play a minor rôle sympatrically with one of the chief vectors or which alone
are responsible for transmission in some very restricted focus. A few more species
are probably minor vectors but are not definitely incriminated as such.

S. callidum. This is considered to be a minor vector in Mexico and Guatemala.
It breeds in a wide range of streams, but normally in larger watercourses than *ochraceum*,
and oviposits by sticking individual eggs or very small egg batches on to rock surfaces
that are just submerged. It is mainly a zoophilic species which feeds preferentially

on equine animals, occasionally on birds, but also in substantial numbers on man. Biting on man occurs mainly below waist level, and the flies are nervous feeders, apt to move about on the skin before feeding and easily disturbed during ingestion of blood

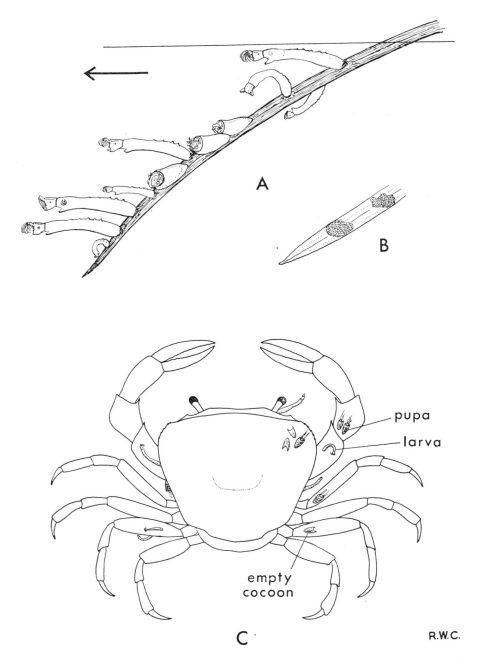

Fig. 89   A, larvae and pupae of *Simulium damnosum* on leaf-blade of grass, showing typical orientation of immature black-flies to the current (direction of flow as indicated by arrow); B, egg-masses of a *Simulium* species on tip of grass leaf-blade; C, larvae, pupae and empty cocoons of *S. neavei* attached to a river-crab. (Crab slightly schematic and drawing composed to show some typical attachment sites on eye-stalks, chelipeds, bases of walking legs, and sides of body.)

(unlike *ochraceum* which starts to feed very positively on man and is not readily disturbed during engorgement); hence *callidum* is a less efficient vector than *ochraceum*. It is a small species with yellowish-orange scutum and black abdomen (coloured illustration in Dalmat, 1955).

S. dukei.   This species bites man occasionally in the Cameroons forest and the early stages live in phoretic association with the river-prawn *Atya africana* and on another unidentified species of *Atya*.   Flies have been found infected with filariae indistinguishable from *O. volvulus* and *S. dukei* is considered to be a potential vector locally in the Cameroons (it was originally recorded as such under the name *S. aureosimile* because of misidentification of the fly involved) (Duke 1962).

S. exiguum.   This is the only anthropophilic *Simulium* found in the small human onchocerciasis focus in Colombia and is probably responsible for transmission there (Barreto *et al.*, 1970).   The species occurs from tropical South America to southern Mexico, and possibly plays a very minor rôle as a vector in Guatemala; though flies are sometimes infected with *O. volvulus* in Venezuela the species has a minimal rôle there as a vector (Duke, 1970).

S. vorax.   This species has been reported as a potential vector in Tanzania, where development of *O. volvulus* microfilariae to the infective stage has been demonstrated in flies fed on an infected man (Wegesa, 1967).

S. woodi.   This is a little-known phoretic species of the *S. neavei*-group responsible for transmission in the Amani area of north-east Tanzania (Raybould, 1967). Larvae and pupae occur on at least three species of crabs of the genus *Potamonautes* in very small heavily shaded streams, usually in low densities.   Flies bite throughout the year but there is some seasonal fluctuation in abundance; man is the only host animal known at present and flies bite almost solely on the legs, especially the ankles.

S. veracruzanum and **S. haematopotum.**   These two species have been reported as possible minor vectors in Guatemala, because *O. volvulus* developed in them to the infective stage (Gibson & Dalmat, 1952; Dalmat, 1955).

CONTROL

The natural enemies of black-flies include birds and predaceous flies which prey upon the adults, and fish which browse upon the larvae, but the most important are the microsporidian Protozoa which infest the abdomen of many black-fly larvae and mermithid nematodes which sometimes occur in the adults as well as the early stages. Infections with microsporidia can inhibit simuliid development and prevent pupation and mermithids can cause sterility in adult female flies.   Hopes have been entertained that microsporidia or mermithids might be used for control, but these are optimistic and up to now no means for the biological control of black-flies have been discovered.

Control still depends upon chemical insecticides, and there is an extensive North American literature on black-fly control by insecticides which is relevant to the control of onchocerciasis vector species.   Eggs and pupae of black-flies are immune, or almost so, to insecticides, and control operations are directed against larvae or adults; but spraying or fogging against adult flies usually has only a local temporary and palliative effect and all major control schemes are based on larvicides.   Black-fly larvae are highly susceptible to insecticides applied to rivers and streams where breeding occurs, and the flow enables larvicides to be effective over long stretches when applied at only a few places; with some insecticides low dosages of from 0.1–1.0 parts per million of stream discharge applied over fifteen or thirty minutes are sufficient to eliminate larvae, but as very few areas are sufficiently isolated for eradication to be practicable the larviciding usually has to be repeated at intervals. Reinfestation of cleared watercourses can be very rapid, though control can often be timed to coincide in its effects with the dry season or some other natural lull in adult fly activity which will obviate the need for further control measures until the following year.

The breeding preferences of the different pest species determine the methods of larviciding. Black-flies which are 'area' breeders in networks of innumerable tiny streams that it is difficult to reach or even to find from the ground (such as *Simulium venustum* and *Prosimulium mixtum* in Canada and New York State) are controlled by aerial spraying of larvicide (this having some adulticide effect also), but species which are 'linear' breeders in relatively few large rivers where accessibility is needed at only a few places (such as *S. arcticum* Malloch in Saskatchewan) can be effectively controlled by dosing from the ground. The chief African onchocerciasis vector, *Simulium damnosum*, comes in the latter category, and has been satisfactorily controlled by ground larviciding in parts of Uganda and Nigeria; but even with this species the vastness of the area requiring control is forcing the adoption of air-spraying in West Africa. *S. neavei* was, however, eradicated from its former small foci in Kenya by DDT applied from the ground. A control programme, mainly using Dieldrin, has been operated against *S. ochraceum* in the Chiapas focus of onchocerciasis in Mexico for many years, but the fact that this species breeds in thousands of tiny trickles and streams partially hidden by dead leaves and herbage in forested and mountainous country has made it impossible to find and dose enough breeding sites from the ground for control to be really successful. The difficulties encountered in Mexico and the new trials with air-spraying in West Africa emphasize that the problems of controlling the medically important black-flies are far from solved, and that there is little hope of eradicating the vector species.

COLLECTING, PRESERVING AND REARING BLACK-FLIES

Early stages are best preserved in 80 per cent alcohol, into which they can be placed straight from the streams; small wide-mouthed bottles with screw caps are the most practical containers for field collecting (the most useful for routine sampling being those with caps of same diameter as the bottle and total dimensions about 4.5 × 2.0 cm). If larvae are required with head fans and rectal gills extended they should be placed in 95 per cent alcohol. Larvae collected for cytological work should be fixed only in the manner recommended by the cytologist for whom material is required. When collecting pupae the undersides of stones, etc. should be searched, as pupae may be in more sheltered places than larvae; pupae are easily damaged, and are best removed from the substrate by freeing them with light pressure applied alternately to each side (rocking motion will free the cocoon which can then be safely picked up with forceps by the empty hind end without damage to the enclosed pupa). Empty cocoons (useful for some purposes) can usually be pinched off the substrate without tearing. Early stages of the *S. neavei*-group are collected by trapping the host crabs (some suitable traps are described by McMahon *et al.*, 1958). Larvae and pupae sent to a specialist for identification should be in alcohol, and should *never* be sent solely as slide-mounts; slide preparations of whole larvae are virtually useless.

Adults can be collected from their hosts while biting, occasionally swept, or trapped in fan-traps, light-traps, silhouette-traps, etc. They can be attracted in some cases with CO_2 or ether extracts from birds (Fallis & Smith, 1964). Adult flies can be preserved in alcohol, but should *always* be preserved as dry pinned specimens if identification by a specialist is required (unless he requests otherwise). For many purposes, especially for obtaining males, it is necessary to rear adult flies from pupae. This is easy, but can be frustrating in the tropics unless great care is exercised to keep material cool, not over-wet and free from mould. The following outline procedure is needed: the most blackened pupae only are collected, either on a piece of substrate or prised free from the substrate, and surplus water adhering to them is blotted off so that they are only just damp; pupae are then placed in a suitable vessel (e.g. large specimen tube) which contains a little just damp (on no account saturated) cotton wool or filter paper and kept in a cool place. Adult flies will emerge quite quickly (often in a few hours, normally at least within two days) and when emerged should be placed in completely dry tubes and left alive for some time to harden. The pupal pelt and cocoon possess important characters and each empty skin and cocoon left behind after the adult fly has emerged should be kept with its associated fly; for this purpose pupal skin and cocoon can be preserved dry and gummed to a piece of card mounted with the adult specimen. Adult and empty skin must not be associated unless it is known for certain which adult came from which skin (ideally therefore pupae used for adult emergence should be kept individually). The adult flies should be micro-pinned and staged, unless being preserved in alcohol or other fluid for some special purpose.

Useful accounts in more detail of collecting and rearing methods are given by Wood & Davies (1966) and Tarshis (1968); an excellent account for anyone reading Russian has been published by Rubtsov (1956).

Black-flies are recalcitrant laboratory insects and a self-perpetuating laboratory colony has not yet been obtained with any species. They can, however, be reared through under laboratory conditions from egg or larva to adult, and a variety of techniques has been described; as starting points see in particular Fredeen (1959), Hartley (1955), and references given by Wood & Davies (1966). For a method of rearing *Simulium damnosum* see Raybould (1967). With most species it has so far proved difficult or impossible to obtain mating of adults under laboratory conditions, and inducing females to take blood is also often difficult (though they can be kept alive for some time if offered sugar foods); but encouraging results have recently been obtained by Wenk & Raybould (1972) with one form of the *S. damnosum*-complex. Until techniques are evolved for inducing successful mating and feeding there is no possibility of a continuous self-maintaining laboratory colony (which is much needed for experimental work on black-fly transmission of pathogenic organisms).

BIBLIOGRAPHY

Information on this family is mostly contained in relatively small scattered papers and there are no comprehensive works except in the taxonomic field. To assist the medical entomologist the references given in the following selected bibliography have therefore been classified as follows:

    (a) Identification and taxonomy
    (b) Vectors of *Onchocerca volvulus* and other filariae
    (c) Simuliids and pathogens other than filariae
    (d) Morphology, biology and miscellaneous
    (e) Control
    (f) Collecting and rearing

(a) Identification and Taxonomy

(References mainly for areas where onchocerciasis vectors occur.)

CROSSKEY, R. W. 1960. A taxonomic study of the larvae of West African Simuliidae (Diptera: Nematocera) with comments on the morphology of the larval black-fly head. *Bull. Br. Mus. nat. Hist.* (Ent.) **10**: 1–74.

—— 1962. The identification of the larvae of African *Simulium*. *Bull. Wld Hlth Org.* **27**: 483–489.

—— 1965. The identification of African Simuliidae (Diptera) living in phoresis with nymphal Ephemeroptera, with special reference to *Simulium berneri* Freeman. *Proc. R. ent. Soc. Lond.* (B) **40**: 118–124.

—— 1969. A re-classification of the Simuliidae (Diptera) of Africa and its islands. *Bull. Br. Mus. nat. Hist.* (Ent.) Suppl. **14**: 1–195.

DALMAT, H. T. 1955. The black flies (Diptera, Simuliidae) of Guatemala and their rôle as vectors of onchocerciasis. *Smithson. misc. Collns* **125** (1): 1–425.

FREEMAN, P. & DE MEILLON, B. 1953. *Simuliidae of the Ethiopian Region.* London, Brit. Mus. (Nat. Hist.), 224 pp. [Includes keys to, and descriptions of, all tropical African species described up to 1952.]

LEWIS, D. J. 1961. The use of the larval cuticular pattern in classifying the *Simulium neavei* Roubaud complex (Diptera: Simuliidae). *Proc. R. ent. Soc. Lond.* (B) **30**: 107–111.

—— 1964. On the *Simulium bovis* complex (Diptera, Simuliidae). *Ann. Mag. nat. Hist.* (13) **7**: 449–455.

LEWIS, D. J. & HANNEY, P. W. 1965. On the *Simulium neavei* complex (Diptera: Simuliidae). *Proc. R. ent. Soc. Lond.* (B) **34**: 12–16.

RUBTSOV, I. A. 1959–1964. Simuliidae (Melusinidae) *in* Lindner, *Fliegen palaearkt. Reg.* **14**: 1–689.

SMART, J. 1945. The classification of the Simuliidae (Diptera). *Trans. R. ent. Soc. Lond.* **95**: 463–532. [Contains catalogue of world Simuliidae described up to 1944.]

STONE, A. 1964. Simuliidae and Thaumaleidae, *in* Guide to the insects of Connecticut, Part vi. The Diptera or true flies of Connecticut, 9th fascicle. *Bull. Conn. St. geol. nat. Hist. Surv.* **97**: 1–126.

VARGAS, L. & DIAZ NAJERA, A. 1957. Simúlidos mexicanos. *Revta Inst. Salubr. Enferm. trop., Méx.* **17**: 143–399. [Includes keys to Mexican species and useful for Central American forms generally.]

(b) Vectors of *Onchocerca volvulus* and other filariae

ANDERSON, R. C. 1956. The life cycle and seasonal transmission of *Ornithofilaria fallisensis* Anderson, a parasite of domestic and wild ducks. *Can. J. Zool.* **34**: 485–525.

ANDERSON, R. C. 1968. The simuliid vectors of *Splendidofilaria fallisensis* of ducks. *Can. J. Zool.* **46**: 610–611.

BAIN, ODILE. 1969. Morphologie des stades larvaires d'*Onchocerca volvulus* chez *Simulium damnosum* et redescription de la microfilaire. *Annls Parasit. hum. comp.* **44**: 69–82.

BAIN, O. & PHILIPPON, B. 1969. Recherche sur les larves de Nématodes Ascaridida trouvées chez *Simulium damnosum*. *Annls Parasit. hum. comp.* **44**: 147–156.

BARNLEY, G. R. & PRENTICE, M. A. 1958. *Simulium neavei* in Uganda. *E. Afr. med. J.* **35**: 475–485.

BARRETO, P., TRAPIDO, H. & LEE, V. H. 1970. Onchocerciasis in Colombia. Entomologic findings in the first observed focus. *Am. J. trop. Med. Hyg.* **19**: 837–841.

BEQUAERT, J. 1929. The insect carrier of *Onchocerca volvulus* in Liberia. *Trans. IVth Int. Congr. Ent.*, Ithaca (1928) **2**: 605–607.

BLACKLOCK, D. B. 1926. The development of *Onchocerca volvulus* in *Simulium damnosum*. *Ann. trop. Med. Parasit.* **20**: 1–48.

BROWNE, S. G. 1960. Observations on *Simulium neavei* Roubaud, with special reference to a focus of onchocerciasis in the Belgian Congo. *Bull. ent. Res.* **51**: 9–15.

BRUČENO IRAGORRY, L. & ORTIZ, I. 1957. Los simúlidos de Venezuela. (Importancia médica. Morfología y Sistemática. Distribución geográfica.) *Boln venez. Lab. clin.* **2**: 23–57.

BURTON, G. J. 1966. Observations on cocoon formation, the pupal stage, and emergence of the adult of *Simulium damnosum* Theobald in Ghana. *Ann. trop. Med. Parasit.* **60**: 48–56.

BURTON, G. J. & McRAE, T. M. 1965. Dam-spillway breeding of *Simulium damnosum* Theobald in northern Ghana. *Ann. trop. Med. Parasit.* **59**: 405–412.

CABALLERO, Y. C. E. & BARRERA, A. 1958. Estudios helmintologicos de la region oncocercosa de Mexico y de la Republica de Guatemala. Nematoda, IIa parte Filarioidea. V. Hallazgo de un nodulo oncocercoso en un mono cirana, *Ateles geoffroyi vellerosus* Gray, del Estado de Chiapas. *Revta lat.-am. Microbiol.* **1**: 79–94.

CERQUIERA, N. L. 1959. Sobre a transmissão da *Mansonella ozzardi*. *Revta bras. Med.* **1**: 885–914.

CRISP, G. 1956. *Simulium and onchocerciasis in the Northern Territories of the Gold Coast*. H. K. Lewis & Co. Ltd, London, xvi + 171 pp.

—— 1956. Observations on the distribution and biting habits of *Simulium damnosum* in the Gold Coast. *Ann. trop. Med. Parasit.* **50**: 444–450.

CROSSKEY, R. W. 1954. Infection of *Simulium damnosum* with *Onchocerca volvulus* during the wet season in Northern Nigeria. *Ann. trop. Med. Parasit.* **48**: 152–159.

—— 1955. Observations on the bionomics of adult *Simulium damnosum* Theobald (Diptera, Simuliidae) in Northern Nigeria. *Ann. trop. Med. Parasit.* **49**: 142–153.

—— 1956. The distribution of *Simulium damnosum* Theobald in Northern Nigeria. *Trans. R. Soc. trop. Med. Hyg.* **50**: 379–392.

—— 1957. Man-biting behaviour in *Simulium bovis* de Meillon in Northern Nigeria, and infection with developing filariae. *Ann. trop. Med. Parasit.* **51**: 80–86.

—— 1957. Further observations on infection of *Simulium damnosum* with *Onchocerca volvulus* in Northern Nigeria. *Trans. R. Soc. trop. Med. Hyg.* **51**: 541–548.

—— 1958. The body-weight in unfed *Simulium damnosum* Theobald, and its relation to the time of biting, the fat-body and age. *Ann. trop. Med. Parasit.* **52**: 149–157.

—— 1962. Observations on the uptake of human blood by *Simulium damnosum*: the engorgement time and size of the blood-meal. *Ann. trop. Med. Parasit.* **56**: 141–148.

CROSSKEY, R. W. & CROSSKEY, M. E. 1958. Filarial infection in *Simulium griseicolle* Becker. *Nature, Lond.* **181**: 713.

DALMAT, H. T. 1954. Ecology of simuliid vectors of onchocerciasis in Guatemala. *Am. Midl. Nat.* **52**: 175–196.

—— 1955. The black flies (Diptera, Simuliidae) of Guatemala and their rôle as vectors of onchocerciasis. *Smithson. misc. Collns* **125** (1): 1–425.

—— 1958. Biology and control of simuliid (Diptera) vectors of onchocerciasis in Central America. *Proc. 10th Int. Congr. Ent.*, Montreal (1956) **3**: 517–533.

DALMAT, H. T. & GIBSON, C. L. 1952. A study of flight ranges and longevity of blackflies (Diptera, Simuliidae) infected with *Onchocerca volvulus*. *Ann. ent. Soc. Am.* **45**: 605–612.

DAVIES, J. B. 1962. Egg-laying habits of *Simulium damnosum* Theobald and *Simulium medusaeforme* form *hargreavesi* Gibbins in Northern Nigeria. *Nature, Lond.* **196**: 149–150.

—— 1963. Further distribution records of the black-flies (Diptera: Simuliidae) of Nigeria with notes on the occurrence of *Simulium damnosum* Theo. in abnormal situations. *J. W. Afr. Sci. Ass.* **7**: 134–137.

DE LEÓN, J. R. 1957. Simuliid vectors of onchocerciasis in Guatemala. *Bull. Wld Hlth Org.* **16**: 523–529.

—— 1961. Contribución al conocimiento de la transmisión de la *Oncocerca* [sic] *volvulus* por los simúlidos de Guatemala. *Publnes Inst. Invest. cient. Univ. San Carlos de Guatemala,* No. 12, 53 pp.

DE LEÓN, J. R. & DUKE, B. O. L. 1966. Experimental studies on the transmission of Guatemalan and West African strains of *Onchocerca volvulus* by *Simulium ochraceum, S. metallicum* and *S. callidum*. *Trans. R. Soc. trop. Med. Hyg.* **60**: 735–752.

DISNEY, R. H. L. 1969. The timing of adult eclosion in blackflies (Dipt., Simuliidae) in West Cameroon. *Bull. ent. Res.* **59** (1968): 485–503.

—— 1970. The timing of the first blood meal in *Simulium damnosum* Theobald. *Ann. trop. Med. Parasit.* **64**: 123–128.

—— 1970. A note on variation within *Simulium damnosum* in the forest zone of West Cameroon. *Ann. trop. Med. Parasit.* **64**: 129–130.

DUKE, B. O. L. 1962. Experimental transmission of *Onchocerca volvulus* from man to chimpanzee. *Trans. R. Soc. trop. Med. Hyg.* **56**: 271.

—— 1962. *Simulium aureosimile* Pomeroy, a possible vector of *Onchocerca volvulus*. *Ann. trop. Med. Parasit.* **56**: 67–69. [This paper refers to the simuliid now known to be *S. dukei*, see text.]

—— 1962. Studies on factors influencing the transmission of onchocerciasis. I.—The survival rate of *Simulium damnosum* under laboratory conditions and the effect upon it of *Onchocerca volvulus*. *Ann. trop. Med. Parasit.* **56**: 130–135.

—— 1962. Studies on factors influencing the transmission of onchocerciasis. II.—The intake of *Onchocerca volvulus* microfilarae by *Simulium damnosum* and the survival of the parasites in the fly under laboratory conditions. *Ann. trop. Med. Parasit.* **56**: 255–263.

—— 1966. *Onchocerca-Simulium* complexes. III.—The survival of *Simulium damnosum* after high intakes of microfilariae of incompatible strains of *Onchocerca volvulus*, and the survival of the parasites in the fly. *Ann. trop. Med. Parasit.* **60**: 495–500.

—— 1967. Infective filaria larvae, other than *Onchocerca volvulus*, in *Simulium damnosum*. *Ann. trop. Med. Parasit.* **61**: 200–205.

—— 1967. *Onchocerca-Simulium* complexes. IV.—Transmission of a variant of the forest strain of *Onchocerca volvulus*. *Ann. trop. Med. Parasit.* **61**: 326–331.

—— 1968. Studies on factors influencing the transmission of onchocerciasis. IV.—The biting-cycles, infective biting density and transmission potential of 'forest' *Simulium damnosum*. *Ann. trop. Med. Parasit.* **62**: 95–106.

—— 1968. Studies on factors influencing the transmission of onchocerciasis. V.—The stages of *Onchocerca volvulus* in wild 'forest' *Simulium damnosum*, the fate of the parasites in the fly, and the age-distribution of the biting population. *Ann. trop. Med. Parasit.* **62**: 107–116.

—— 1968. Studies on factors influencing the transmission of onchocerciasis. VI.—The infective biting potential of *Simulium damnosum* in different bioclimatic zones and its influence on the transmission potential. *Ann. trop. Med. Parasit.* **62**: 164–170.

—— 1970. *Onchocerca-Simulium* complexes. VI.—Experimental studies on the transmission of Venezuelan and West African strains of *Onchocerca volvulus* by *Simulium metallicum* and *S. exiguum* in Venezuela. *Ann. trop. Med. Parasit.* **64**: 421–431.

DUKE, B. O. L. & BEESLEY, W. N. 1958. The vertical distribution of *Simulium damnosum* bites on the human body. *Ann. trop. Med. Parasit.* **52**: 274–281.

DUKE, B. O. L. & LEWIS, D. J.   1964.   Studies on factors influencing the transmission of oncho-
    cerciasis.   III.—Observations on the effect of the peritrophic membrane in limiting the
    development of *Onchocerca volvulus* microfilariae in *Simulium damnosum*.   *Ann. trop. Med.
    Parasit.* **58**: 83–88.

DUKE, B. O. L., LEWIS, D. J. & MOORE, P. J.   1969.   *Onchocerca-Simulium* complexes.   I.—
    Transmission of forest and Sudan-savanna strains of *Onchocerca volvulus*, from Cameroon,
    by *Simulium damnosum* from various West African bioclimatic zones.   *Ann. trop. Med.
    Parasit.* **60**: 318–336.

DUKE, B. O. L., MOORE, P. J. & DE LEÓN, J. R.   1967.   *Onchocerca-Simulium* complexes.   V.—The
    intake and subsequent fate of microfilariae of a Guatemalan strain of *Onchocerca volvulus*
    in forest and Sudan-savanna forms of West African *Simulium damnosum*.   *Ann. trop. Med.
    Parasit.* **61**: 332–337.

DUNBAR, R. W.   1966.   Four sibling species included in *Simulium damnosum* Theobald (Diptera:
    Simuliidae) from Uganda.   *Nature, Lond.* **209**: 597–599.

—— 1969.   Nine cytological segregates in the *Simulium damnosum* complex (Diptera: Simuliidae).
    Bull. Wld Hlth Org. **40**: 974–979.

EICHLER, D. A.   1971.   Studies on *Onchocerca gutturosa* (Neumann, 1910) and its development
    in *Simulium ornatum* (Meigen, 1818).   II.—Behaviour of *S. ornatum* in relation to the trans-
    mission of *O. gutturosa*.   II.—*J. Helminth.* **45**: 259–270.

EICHLER, D. A. & NELSON, G. S.   1971.   Studies on *Onchocerca gutturosa* (Neumann, 1910) and
    its development in *Simulium ornatum* (Meigen, 1818).   I.—Observations on *O. gutturosa* in
    cattle in South-East England.   *J. Helminth.* **45**: 245–258.

ESTEVEZ, C.   1947.   *Algunas consideraciones sobre la patologia y transmisión de la oncocercosis.*
    Universidad de San Carlos de Guatemala, Guatemala City, 37 pp.   [Numbered 219–255 in
    reprint.]

FAIN, A. & HALLOT, R.   1965.   Répartition d'*Onchocerca volvulus* Leuckart et de ses vecteurs
    dans le bassin du Congo et les régions limitrophes.   *Mém. Acad. r. Sci. Outre-Mer*, n.s.
    17 (1): 1–86.

——, —— 1965.   Nouvelles localités pour *Simulium neavei* Roubaud en République du Congo-
    Léopoldville.   *Revue Zool. Bot. afr.* **71**: 327–335.

FIGUEROA MARROQUIN, H.   1967.   *Enfermedad de Robles (Oncocercosis humana).   Bibliografia
    guatemalense*, ed. Talleres Graficós Galindo, Guatemala City, 158 pp.   [Valuable biblio-
    graphy on onchocerciasis in Guatemala.]

GARMS, R. & POST, A.   1966.   Die Verbreitung von *Simulium damnosum* in Guinea/Westafrika.
    Z. Tropenmed. Parasit. **17**: 443–466.

GARMS, R. & VOELKER, J.   1969.   Unknown filarial larvae and zoophily in *Simulium damnosum*
    in Liberia.   *Trans. R. Soc. trop. Med. Hyg.* **63**: 676–677.

GARMS, R. & WEYER, F.   1968.   Naturliche infektion von *Simulium damnosum* mit *Onchocerca
    volvulus* in savannengebieten Guineas.   *Z. Tropenmed. Parasit.* **19**: 289–296.

GARNHAM, J. C. & WALLIKER, D.   1965.   Is *Simulium amazonicum* the vector of *Mansonella
    ozzardi*?   *Trans. R. Soc. trop. Med. Hyg.* **59**: 672–674.

GIBBINS, E. G.   1933.   Studies on Ethiopian Simuliidae: *Simulium damnosum* Theo.   *Trans.
    R. ent. Soc. Lond.* **81**: 37–51.

—— 1938.   The mouthparts of the female in *Simulium damnosum* Theobald, with special
    reference to the transmission of *Onchocerca volvulus* Leuckart.   *Ann. trop. Med. Parasit.*
    32: 9–20.

GIBSON, C. L.   1955.   The indiscriminate feeding of anthropophilic *Simulium* upon man and
    domestic animals, and its relation to studies on the transmission of human onchocerciasis in
    Guatemala.   *Boln Of. sanit. pan-am.* **38**: 293–295.   [For Spanish version of this paper see
    Gibson, 1955, *Revta ibér. Parasit.* (Tomo Extraordinario): 307–311.]

GIBSON, C. L. & DALMAT, H. T.   1952.   Three new potential intermediate hosts of human
    onchocerciasis in Guatemala.   *Am. J. trop. Med. Hyg.* **1**: 848–851.

GIUDICELLI, J.   1966.   Récoltes de Simulies en Côte-d'Ivoire.   Étude de l'activité diurne des
    femelles de *Simulium damnosum* Theobald.   *Annls Soc. ent. Fr.* (n.s.) **11**: 325–342.

GRENIER, P. & FERAUD, L. 1960. Étude biometrique et morphologique de la croissance larvaire chez *Simulium damnosum* Theobald. *Bull. Soc. Path. exot.* **53**: 563–581.

GRENIER, P. & OVAZZA, M. 1956. Contribution à l'étude des Diptères vulnérants de l'empire d'Éthiopie. *Bull. Soc. Path. exot.* **49**: 182–196.

GRENIER, P., OVAZZA, M. & VALADE, M. 1960. Notes biologiques et faunistiques sur *S. damnosum* et les Simuliidae d'Afrique occidentale (Haute-Volta, Côte d'Ivoire, Dahomey, Soudan). *Bull. Inst. fr. Afr. noire* **22**: 892–918.

HÄUSERMANN, W. 1969. On the biology of *Simulium damnosum* Theobald, 1903, the main vector of onchocerciasis in the Mahenge Mountains, Ulanga, Tanzania. *Acta trop.* **26**: 29–69.

HOFFMANN, C. C. 1930. Nuevas investigaciones acerca de la transmisión de la oncocercosis de Chiapas. *Revta mex. Biol.* **10**: 131–140.

—— 1930. Investigaciones sobre la transmisión de la oncocercosis de Chiapas. *An. Inst. Biol. Univ. Méx.* **1**: 59–62.

—— 1930. Über *Onchocerca* im süden von Mexiko und die Weiterentwicklung ihrer Mikrofilarien in *Eusimulium mooseri*. *Arch. Schiffs-u. Tropenhyg.* **34**: 461–472.

—— 1931. Estudios entomólogicos y parasitológicos acerca de la oncocercosis en Chiapas. *Salubridad* **3**: 669–697.

LE BERRE, R. 1966. *Contribution à l'étude biologique et écologique de* Simulium damnosum *Theobald, 1903 (Diptera, Simuliidae)*. O.R.S.T.O.M., Paris, 204 pp.

LE BERRE, R., BALAY, G., BRENGUES, J. & COZ, J. 1964. Biologie et écologie de la femelle de *Simulium damnosum* Theobald, 1903, en fonction des zones bioclimatiques d'Afrique occidentale. *Bull. Wld Hlth Org.* **31**: 843–855.

LE BERRE, R. & WENK, P. 1966. Beobachtungen über das Schwarmverhalten bei *Simulium damnosum* (Theobald) in Obervolta und Kamerun. *Verh. dt. zool. Ges. Göttingen* (suppl.) **30**: 367–372.

LEWIS, D. J. 1953. *Simulium damnosum* and its relation to onchocerciasis in the Anglo-Egyptian Sudan. *Bull. ent. Res.* **43**: 597–644.

—— 1956. Biting times of parous and nulliparous *Simulium damnosum*. *Nature, Lond.* **178**: 98–99.

—— 1957. Aspects of the structure, biology and study of *Simulium damnosum*. *Ann. trop. Med. Parasit.* **51**: 340–358.

—— 1958. *Simulium damnosum* in the Tonkolili valley, Sierra Leone. *Proc. 10th Int. Congr. Ent.*, Montreal (1956) **3**: 541–550.

—— 1958. Observations on *Simulium damnosum* Theobald at Lokoja in Northern Nigeria. *Ann. trop. Med. Parasit.* **52**: 216–231.

—— 1960. Observations on *Simulium damnosum* in the Southern Cameroons and Liberia. *Ann. trop. Med. Parasit.* **54**: 208–222.

—— 1963. Simuliidae (Diptera) from the human onchocerciasis area of Venezuela. *Proc. R. ent. Soc. Lond.* (B) **32**: 53–62.

—— 1965. Features of the *Simulium damnosum* population of the Kumba area in West Cameroon. *Ann. trop. Med. Parasit.* **59**: 365–374.

LEWIS, D. J. & DUKE, B. O. L. 1966. *Onchocerca-Simulium* complexes. II.— Variation in West African female *Simulium damnosum*. *Ann. trop. Med. Parasit.* **60**: 337–346.

LEWIS, D. J. & IBAÑEZ DE ALDECOA, R. 1962. Simuliidae and their relation to human onchocerciasis in northern Venezuela. *Bull. Wld Hlth Org.* **27**: 449–464.

LEWIS, D. J., LYONS, G. R. L. & MARR, J. D. M. 1961. Observations on *Simulium damnosum* from the Red Volta in Ghana. *Ann. trop. Med. Parasit.* **55**: 202–210.

MARR, J. D. M. 1962. The use of an artificial breeding-site and cage in the study of *Simulium damnosum* Theobald. *Bull. Wld Hlth Org.* **27**: 622–629.

MARR, J. D. M. & LEWIS, D. J. 1964. Observations on the dry-season survival of *Simulium damnosum* Theo. in Ghana. *Bull. ent. Res.* **55**: 547–564.

MCMAHON, J. P. 1940. *Onchocerca volvulus* and its vector in the Southern Kavirondo district of Kenya. *Trans. R. Soc. trop. Med. Hyg.* **34**: 65–83.

L

McMAHON, J. P.   1951.   The discovery of the early stages of *Simulium neavei* in phoretic association with crabs and a description of the pupa and the male.   *Bull. ent. Res.* **42**: 419–426.

—— 1968.   Artificial feeding of *Simulium* vectors of human and bovine onchocerciasis.   *Bull. Wld Hlth Org.* **38**: 957–966.

MEILLON, B. DE.   1957.   Bionomics of the vectors of onchocerciasis in the Ethiopian geographical region.   *Bull. Wld Hlth Org.* **16**: 509–522.

MERIGHI, B., PARRINELLO, A. E. & RIVOSECCHI, L.   1969.   Presenza di *Simulium damnosum* Th. (Diptera-Simuliidae) in Yemen.   *Annali Ist. sup. Sanità* **5**: 197–199.

MUIRHEAD-THOMSON, R. C.   1956.   Communal oviposition in *Simulium damnosum* Theobald (Diptera, Simuliidae).   *Nature, Lond.* **178**: 1297–1299.

—— 1957.   Effect of desiccation on the eggs of *Simulium damnosum*, Theobald.   *Nature, Lond.* **180**: 1432–1433.

NELSON, G. S.   1965.   Filarial infections as zoonoses.   *J. Helminth.* **39**: 229–250.

NELSON, G. S. & PESTER, F. R. N.   1962.   The identification of infective filarial larvae in Simuliidae.   *Bull. Wld Hlth Org.* **27**: 473–481.

OVAZZA, M., COZ, J. & OVAZZA, L.   1965.   Étude des populations de *Simulium damnosum* Theobald, 1903 (Diptera: Simuliidae) en zones de gîtes non permanents.   I.—Observations sur les variations de quelques-uns des caractères utilisés dans l'estimation de l'âge physiologique.   *Bull. Soc. Path. exot.* **58**: 938–950.

OVAZZA, M., OVAZZA, L. & BALAY, G.   1965.   Étude des populations de *Simulium damnosum* Theobald, 1903 (Diptera: Simuliidae) en zones de gîtes non permanents.   II.—Variations saisonnières se produisant dans les populations adultes et préimaginales. Discussion des différentes hypothèses qui peuvent expliquer le maintien de l'espèce dans les régions sèches. *Bull. Soc. Path. exot.* **58**: 1118–1154.

OVAZZA, M., OVAZZA, L. & BALAY, G.   1965.   Etude des populations de *Simulium damnosum* Theobald, 1903 (Diptera: Simuliidae) en zones de gîtes non permanents.   III.—Corrélation possible entre certains phénomènes météorologiques et la réapparition des femelles en début de saison des pluies.   *Bull. Soc. Path. exot.* **60**: 79–95.

QUELENNEC, G.   1962.   Enquêtes préliminaires sur *Simulium damnosum* et l'onchocercose au Dahomey.   *Méd trop.* **22**: 463–470.

QUELENNEC, G., SIMONKOVICH, E. & OVAZZA, M.   1968.   Recherche d'un type de déversoir de barrage défavorable à l'implantation de *Simulium damnosum* (Diptera, Simuliidae).   *Bull. Wld Hlth Org.* **38**: 943–956.

QUELENNEC, G., VALADE, M. & CORDELLIER, R.   1968.   Bilan des recherches sur la répartition en République du Dahomey de *Simulium damnosum* Theo. vecteur de l'onchocercose humaine. Cartes de répartition des autres espèces de Simulies.   *Cah. Off. Rech. Sci. Tech. Outre-Mer* (Ent. Med.) **6**: 21–54.

RAYBOULD, J. N.   1967.   A study of anthropophilic female Simuliidae (Diptera) at Amani, Tanzania: the feeding behaviour of *Simulium woodi* and the transmission of onchocerciasis. *Ann. trop. Med. Parasit.* **61**: 76–88.

—— 1967.   A method of rearing *Simulium damnosum* Theobald (Diptera: Simuliidae) under artificial conditions.   *Bull. Wld Hlth Org.* **37**: 447–453.

—— 1969.   Studies on the immature stages of the *Simulium neavei* Roubaud complex and their associated crabs in the eastern Usambara Mountains in Tanzania.   I.—Investigations in rivers and large streams.   *Ann. trop. Med. Parasit.* **63**: 269–287.

RAYBOULD, J. N. & YAGUNGA, A. S. K.   1969.   Artificial feeding of East African Simuliidae (Diptera), including vectors of human onchocerciasis.   *Bull. Wld Hlth Org.* **40**: 463–466.

——, —— 1969.   Studies on the immature stages of the *Simulium neavei* Roubaud complex and their associated crabs in the eastern Usambara Mountains in Tanzania.   II.—Investigations in small heavily shaded streams.   *Ann. trop. Med. Parasite.* **63**: 289–300.

STEWARD, J. S.   1937.   The occurrence of *Onchocerca gutturosa* Neumann in cattle in England, with an account of its life history and development in *Simulium ornatum* Mg. *Parasitology* **29**: 212–219.

STRONG, R. P., SANDGROUND, J. H., BEQUAERT, J. C. & MUÑOZ OCHOA, M. 1934. Onchocerciasis with special reference to the Central American form of the disease. *Contr. Dep. trop. Med. & Inst. trop. Biol. Med.*, No. 6, 234 pp., Harvard University Press, Cambridge Press, Cambridge, Mass.

VAN DEN BERGE, L., PEEL, E. & CHARDOME, M. 1964. The filarial parasites of the eastern gorilla in the Congo. *J. Helminth.* **38**: 349–368.

WANSON, M. 1950. Contribution à l'étude de l'onchocercose africaine humaine. *Annls Soc. belge Méd. trop.* **30**: 667–863.

WANSON, M. & HENRARD, C. 1945. Habitat et comportement larvaire du *Simulium damnosum* Théobald. *Recl Trav. Sci. méd. Congo belge* **4**: 113–122

WANSON, M. & LEBIED, B. 1948. Note sur le cycle gonotrophique de *Simulium damnosum*. *Revue Zool. Bot. afr.* **41**: 66–82.

WANSON, M. & PEEL, E. 1945. *Onchocerca volvulus* Leuckart. Indices d'infection des simulies agressives pour l'homme. Cycle de développement chez *Simulium damnosum* Théobald. *Recl Trav. Sci. méd. Congo belge* **4**: 122–136.

WEGESA, P. 1967. *Simulium vorax* Pomeroy, a potential vector of *Onchocerca volvulus*. *Ann. trop. Med. Parasit.* **61**: 89–91.

WILLIAMS, T. R. 1968. The taxonomy of the East African river-crabs and their association with the *Simulium neavei* complex. *Trans. R. Soc. trop. Med. Hyg.* **62**: 29–34.

WORLD HEALTH ORGANISATION. 1954. Expert Committee on Onchocerciasis. First Report. *Wld Hlth Org. Tech. Rep. Ser.*, No. 87, 37 pp., W.H.O., Geneva.

—— 1966. WHO Expert Committee on Onchocerciasis. Second Report. *Wld Hlth Org. Tech. Rep. Ser.*, No. 335, 96 pp., W.H.O., Geneva. [Useful introduction to whole field of onchocerciasis and its vectors.]

(c) Simuliids and pathogens other than filariae

ANDERSON, J. R. & DEFOLIART, G. R. 1961. Feeding behaviour and host preferences of some black flies (Diptera: Simuliidae) in Wisconsin. *Ann. ent. Soc. Am.* **54**: 716–729.

ANDERSON, J. R., LEE, V. H., VADLAMUDI, S., HANSON, R. P. & DEFOLIART, G. R. 1961. Isolation of eastern encephalitis virus from Diptera in Wisconsin. *Mosquito News* **21**: 244–248.

ANDERSON, J. R., TRAINER, D. O. & DEFOLIART, G. R. 1962. Natural and experimental transmission of the waterfowl parasite, *Leucocytozoon simondi* M. & L., in Wisconsin. *Zoonoses Res.* **1**: 155–164.

AUSTIN, F. J. 1967. The arbovirus vector potential of a simuliid. *Ann. trop. Med. Parasit.* **61**: 189–199.

BAKER, J. R. 1970. Transmission of *Leucocytozoon sakharoffi* in England by *Simulium angustitarse*. *Parasitology* **60**: 417–423.

BENNETT, G. F. 1961. On the specificity and transmission of some avian trypanosomes. *Can. J. Zool.* **39**: 17–33.

BENNETT, G. F., GARNHAM, P. C. C. & FALLIS, A. M. 1965. On the status of the genera *Leucocytozoon* Ziemann, 1898 and *Haemoproteus* Kruse, 1890 (Haemosporidiida: Leucocytozoidae and Haemoproteidae). *Can. J. Zool.* **43**: 927–932.

DEFOLIART, G. R., ANSLOW, R. O., HANSON, R. P., MORRIS, C. D., PAPADOPOULOS, O. & SATHER, G. E. 1969. Isolation of Jamestown Canyon serotype of California encephalitis virus from naturally infected *Aedes* mosquitoes and tabanids. *Am. J. trop. Med. Hyg.* **18**: 440–447.

DEFOLIART, G. R. & HANSON, R. P. 1968. Non-Culicine vectors of arboviruses. *8th Int. Congr. trop. Med. Malar.*, Teheran, 1968, Abstracts & Reviews: 698–699.

FALLIS, A. M., ANDERSON, R. C. & BENNETT, G. F. 1956. Further observations on the transmission and development of *Leucocytozoon simondi*. *Can. J. Zool.* **34**: 389–404.

FALLIS, A. M. & BENNETT, G. F. 1958. Transmission of *Leucocytozoon bonasae* Clarke to ruffed grouse (*Bonasa umbellus* L.) by the black flies *Simulium latipes* Mg. and *Simulium aureum* Fries. *Can. J. Zool.* **36**: 533–539.

——, —— 1962. Observations on the sporogony of *Leucocytozoon mirandae*, *L. bonasae*, and *L. fringillinarum* (Sporozoa: Leucocytozoidae). *Can. J. Zool.* **40**: 395–400.

——, —— 1966. On the epizootiology of infections caused by *Leucocytozoon simondi* in Algonquin Park, Canada. *Can. J. Zool.* **44**: 101–112.

GEORGEWITCH, J. 1909. Sur un trypansomide nouveau, *Crithidia simuliae*, n. sp. d'une simulie (*Simulium columbacensis*) de la Serbie septentrionale. *C.r. Séanc. Soc. Biol.* **1909** (2): 480–482.

KHAN, R. A. & FALLIS, A. M. 1970. Life cycles of *Leucocytozoon dubreuili* Mathis & Leger, 1911 and *L. fringillinarum* Woodcock, 1910 (Haemosporidia: Leucocytozoidae. *J. Protozool.* **17**: 642–658.

—— 1971. Speciation, transmission, and schizogony of *Leucocytozoon* in corvid birds. *Can. J. Zool.* **49**: 1361–1367.

MYKYTOWYCZ, R. 1957. The transmission of myxomatosis by *Simulium melatum* Wharton (Diptera: Simuliidae). *C.S.I.R.O. Wildl. Res.* **2**: 1–4.

SKIDMORE, L. V. 1932. *Leucocytozoon smithi* infection in turkeys and its transmission by *Simulium occidentale* Townsend. *Zentbl. Bakt. ParasitKde* I (Orig.) **125**: 329–335.

(d) Morphology, biology and miscellaneous

BENNETT, G. F. 1963. Use of P^{32} in the study of a population of *Simulium rugglesi* (Diptera: Simuliidae) in Algonquin Park, Ontario. *Can. J. Zool.* **41**: 831–840.

BROWN, H. & BERNTON, H. S. 1970. A case of asthma caused by *Simulium jenningsi* (Order Diptera) protected by hyposensitization. *J. Allergy* **45**: 103–104.

CHANCE, MARY M. 1970. The functional morphology of the mouthparts of blackfly larvae (Diptera: Simuliidae). *Quaest. ent.* **6**: 245–284.

FALLIS, A. M. 1964. Feeding and related behavior of female Simuliidae (Diptera). *Expl Parasit.* **15**: 439–470. [Contains table, with references, to recorded hosts of world Simuliidae.]

FALLIS, A. M. & SMITH, S. M. 1964. Ether extracts from birds and CO_2 as attractants for some ornithophilic simuliids. *Can. J. Zool.* **42**: 723–730.

GLICK, P. A. 1939. The distribution of insects, spiders and mites in the air. *Tech. Bull. U.S. Dep. Agric.*, No. 673, 150 pp. [Contains altitude data for black-flies.]

GRENIER, P. 1949. Contribution à l'étude biologique des Simuliides de France. *Physiologia comp. Oecol.* **1** (1948): 165–330.

——, 1959. Remarques concernant le fonctionnement des mandibules chez les femelles de diptères hématophages nématocères et brachycères. *Annls Parasit. hum. comp.* **34**: 565–585.

HINTON, H. E. 1958. The pupa of the fly *Simulium* feeds and spins its own cocoon. *Entomologist's mon. Mag.* **94**: 14–16.

HYNES, H. B. N. 1970. *The ecology of running waters.* 555 pp. Liverpool.

KRSTITSCH, A. & ZIVKOVITCH, V. 1968. Dermatitis, verursacht durch das Insekt *Simulium erythrocephalum*. *XIIIth Int. derm. Congr.*: 288–289.

PHELPS, R. J. & DeFOLIART, G. R. 1964. Nematode parasitism of Simuliidae. *Res. Bull. agric. Exp. Stn Univ. Wis.*, No. 245, 78 pp. [Very useful work on mermithid nematodes in black-flies, but does not deal with filariae as title might imply.]

PURI, I. M. 1925. On the life-history and structure of the early stages of Simuliidae (Diptera, Nematocerca). Part I. *Parasitology* **17**: 295–334.

RUBTSOV, I. A. 1956. Nutrition and facultative bloodsucking in black-flies (Diptera, Simuliidae). *Ent. Obozr.* **35**: 731–751. [In Russian.]

—— 1958. The gonotrophic cycle in bloodsucking black-flies. *Parazit. Sb.* **18**: 255–282. [In Russian.]

—— 1960. The gonotrophic cycle in non-bloodsucking [phytophagous] species of black-flies (Diptera, Simuliidae). *Ent. Obozr.* **39**: 556–573. [In Russian: for English translation see *Ent. Rev., Wash.* **39**: 392–405.]

WENK, P. 1962. Anatomie des Kopfes von *Wilhelmia equina* L. ♀ (Simuliidae syn. Melusinidae, Diptera). *Zool. Jb.* (Anat.) **80**: 81–134.

(e) Control

(Some North American literature containing information pertinent to vector control is included in the following list as well as literature dealing directly with control of vector species.)

ARNASON, A. P., BROWN, A. W. A., FREDEEN, F. J. H., HOPEWELL, W. W. & REMPEL, J. G.   1949.   Experiments in the control of *Simulium arcticum* Malloch by means of DDT in the Saskatchewan River.   *Scient. Agric.* **29**: 527–537.

BARNLEY, G. R.   1958.   Control of *Simulium* vectors of onchocerciasis in Uganda.   *Proc. 10th Int. Congr. Ent. Montreal* (1956) **3**: 535–538.

BROWN, A. W. A.   1962.   A survey of *Simulium* control in Africa.   *Bull. Wld Hlth Org.* **27**: 511–527.

BUCKLEY, J. J. C.   1951.   Studies on human onchocerciasis and *Simulium* in Nyanza Province, Kenya.   II.—The disappearance of *S. neavei* from a bush-cleared focus.   *J. Helminth.* **25**: 213–222.

BURTON, G. J.   1964.   An exposure-tube for determining the mortality of *Simulium* larvae in rivers following larvicidal operations.   *Ann. trop. Med. Parasit.* **58**: 339–342.

BURTON, G. J., NOAMESI, G. K., ZEVE, V. H. & McRAE, T. M.   1964.   Quantitative studies on the mortality of *Simulium damnosum* larvae following dosing with DDT under field conditions.   I.—Mortality within the first three hours.   *Ghana med. J.* **3**: 93–96.

COLLINS, D. L. & JAMNBACK, H.   1958.   Ten years of blackfly control in New York State.   *Proc. 10th Int. Congr. Ent. Montreal* (1956) **3**: 813–818.

CROSSKEY, R. W.   1958.   First results in the control of *Simulium damnosum* Theobald (Diptera, Simuliidae) in Northern Nigeria.   *Bull. ent. Res.* **49**: 715–735.

DAVIES, J. B.   1963.   An assessment of the insecticidal control of *Simulium damnosum* Theobald in Abuja Emirate, Northern Nigeria, from 1955 to 1960.   I.—The effect on the prevalence of onchocerciasis in the human population.   *Ann. trop. Med. Parasit.* **57**: 161–181.

—— 1965.   An assessment of the insecticidal control of *Simulium damnosum* Theobald in Abuja Emirate, Northern Nigeria, from 1955 to 1960.   II.—The effect on the incidence of *Onchocerca* larvae in the vector.   *Ann. trop. Med. Parasit.* **59**: 43–46.

—— 1968.   The *Simulium* control scheme at Abuja, Northern Nigeria, and its effect on the prevalence of onchocerciasis in the area.   *Bull. Wld Hlth Org.* **39**: 187–207.

DAVIES, J. B., CROSSKEY, R. W., JOHNSTON, M. R. L. & CROSSKEY, M. E.   1962.   The control of *Simulium damnosum* at Abuja, Northern Nigeria, 1955–1960.   *Bull. Wld Hlth Org.* **27**: 491–510.

FAIRCHILD, G. B. & BARREDA, E. A.   1946.   DDT as a larvicide against *Simulium*.   *J. econ. Ent.* **38**: 694–699.

FREDEEN, F. J. H.   1962.   DDT and Heptachlor as black-fly larvicides in clear and turbid water.   *Can. Ent.* **94**: 875–880.

FREDEEN, F. J. H., ARNASON, A. P. & BERCK, B.   1953.   Adsorption of DDT on suspended solids in river water and its role in black-fly control.   *Nature, Lond.* **171**: 700.

FREDEEN, F. J. H., ARNASON, A. P., BERCK, B. & REMPEL, J. G.   1953.   Further experiments with DDT in the control of *Simulium arcticum* Mall. in the North and South Saskatchewan Rivers.   *Can. J. agric. Sci.* **33**: 379–393.

GARMS, R. & POST, A.   1967.   Freilandversuche zur Wirksamkeit von DDT und Baytex gegen Larven von *Simulium damnosum* in Guinea, Westafrika.   *Anz. Schadlingsk.* **40**: 49–56.

GARNHAM, P. C. C. & McMAHON, J. P.   1947.   The eradication of *Simulium neavei* Roubaud from an onchocerciasis area in Kenya Colony.   *Bull. ent. Res.* **37**: 619–628.

——, —— 1954.   Final results of an experiment on the control of onchocerciasis by eradication of the vector.   *Bull. ent. Res.* **45**: 175–176.

HYNES, H. B. N. & WILLIAMS, T. R.   1962.   The effect of DDT on the fauna of a Central African stream.   *Ann. trop. Med. Parasit.* **56**: 78–91.

JAMNBACK, H.   1962.   An eclectic method of testing the effectiveness of chemicals in killing blackfly larvae (Simuliidae: Diptera).   *Mosquito News* **22**: 384–389.

JAMNBACK, H. & COLLINS, D. L.   1955.   The control of blackflies (Diptera: Simuliidae) in New York.   *Bull. N.Y. St. Mus.*, No. 350, 113 pp.

JAMNBACK, H. A., DUFLO, T. & MARR, D. 1970. Aerial application of larvicides for control of *Simulium damnosum* in Ghana: a preliminary trial. *Bull. Wld Hlth Org.* **42**: 826–828.

JAMNBACK, H. & EABRY, H. S. 1962. Effects of DDT, as used in black fly larval control, on stream arthropods. *J. econ. Ent.* **55**: 636–639.

JAMNBACK, H. & FREMPONG-BOADU, J. 1966. Testing blackfly larvicides in the laboratory and in streams. *Bull. Wld Hlth Org.* **34**: 405–421.

JAMNBACK, H. & MEANS, R. 1966. Length of exposure period as a factor influencing the effectiveness of larvicides for blackflies (Diptera: Simuliidae). *Mosquito News* **26**: 590–591.

KUZOE, F. A. S. & HAGAN, K. B. 1967. The control of *Simulium damnosum* Theobald (Diptera, Simuliidae) in the region of the Volta Dam. *Ann. trop. Med. Parasit.* **61**: 338–348.

LEA, A. O. & DALMAT, H. T. 1954. Screening studies of chemicals for larval control of black-flies in Guatemala. *J. econ. Ent.* **47**: 378–383.

——, —— 1955. Field studies on larval control of black flies in Guatemala. *J. econ. Ent.* **48**: 274–278.

——, —— 1955. A pilot study of area larval control of black flies in Guatemala. *J. econ. Ent.* **48**: 378–383.

LE BERRE, R., OVAZZA, M. & JUGE, E. 1965. Résultats d'une campagne larvicide contre *Simulium damnosum* Theobald (Diptera-Simuliidae) en Afrique de l'ouest. *Proc. 12th Int. Congr. Ent. London* (1964): 811.

McMAHON, J. P. 1957. DDT-treatment of rivers for eradication of Simuliidae. *Bull. Wld Hlth Org.* **16**: 541–551.

McMAHON, J. P., HIGHTON, R. B. & GOINY, H. 1958. The eradication of *Simulium neavei* from Kenya. *Bull. Wld Hlth Org.* **19**: 75–107.

MUIRHEAD-THOMSON, R. C. 1957. Laboratory studies on the reactions of *Simulium* larvae to insecticides. Parts I, II & III. *Am. J. trop. Med. Hyg.* **6**: 920–934.

—— 1970. The potentiating effect of pyrethrins and pyrethroids on the action of organophosphorus larvicides in *Simulium* control. *Trans. R. Soc. trop. Med. Hyg.* **64**: 895–906.

MUIRHEAD-THOMSON, R. C. & MERRYWEATHER, J. 1970. Ovicides in *Simulium* control. *Bull. Wld Hlth Org.* **42**: 174–177.

NOAMESI, G. K. 1964. A progress report on the control of *Simulium damnosum* Theobald (Diptera, Simuliidae) in northern Ghana. *Ghana J. Sci.* **4**: 157–168.

—— 1964. The tube bioassay technique in tests to evaluate entomologically the effects of *Simulium* control operations in northwest Ghana. *Ghana med. J.* **3**: 163–165.

NOEL-BUXTON, M. B. 1956. Field experiments with DDT in association with finely divided inorganic material for the destruction of the immature stages of the genus *Simulium* in the Gold Coast. *J. W. Afr. Sci. Ass.* **2**: 36–40.

OBENG, L. E. 1967. Oviposition and breeding habits of the Simuliidae in relation to control practices. *Proc. Ghana Acad. Sci.* **5**: 45–64.

PETERSON, D. G. & WEST, A. S. 1960. Control of adult black flies (Diptera: Simuliidae) in the forests of eastern Canada by aircraft spraying. *Can. Ent.* **92**: 714–719.

RAYBOULD, J. N. 1966. A simple laboratory method for testing the susceptibility of larvae of the *Simulium neavei* complex to insecticides. *Bull. Wld Hlth Org.* **35**: 887–892.

ROBERTS, J. M. D., NEUMANN, E., GÖCKEL, C. W. & HIGHTON, R. B. 1967. Onchocerciasis in Kenya 9, 11 and 18 years after elimination of the vector. *Bull. Wld Hlth Org.* **37**: 195–212.

RUIZ REYES, F. 1959. Cinco años de lucha antisimulido. *Boln epidem., Méx.* **23**: 44–54.

RUIZ REYES, F. & ANGUIANO LOPEZ, R. 1952. El hexaclorociclohexano en la lucha contra los simulidos. *Boln epidem., Méx.* **16**: 7–12.

TAUFFLIEB, R. 1955. Une campagne de lutte contre *Simulium damnosum* au Mayo Kebbi. *Bull. Soc. Path. exot.* **48**: 564–576.

—— 1956. Rapport sur la campagne antisimulidienne de 1956 au Mayo Kebbi. *Bull. Inst. Étud. centrafr.* (n.s.) **11**: 53–59.

TENDEIRO, J. 1967. La lutte contre *Simulium damnosum* (Diptera, Simuliidae) dans la Guinée Portugaise. *Wiad. parazyt.* **13**: 491–495.

TWINN, C. R. & PETERSON, D. G. 1958. *Control of black flies in Canada.* Canada Dept. of Agriculture, Science Service, Entomology Division, Publ. No. 940, 9 pp.

WADDY, B. B. 1969. Prospects for the control of onchocerciasis in Africa with special reference to the Volta River basin. *Bull. Wld Hlth Org.* **40**: 843–858.

WANSON, M., COURTOIS, L. & LEBIED, B. 1949. L'éradication du *Simulium damnosum* (Théobald) a Léopoldville. *Annls Soc. belge Méd. trop.* **29**: 373–403. [Paper of historical interest for first attempts at *Simulium* control in Africa but care needed in using it: now known that eradication was not achieved and that *damnosum* was probably not the species under attack.]

WEST, A. S. 1958. *Biting Fly Control Manual.* Woodlands Research Index No. 104. Pulp and Paper Research Institute of Canada, Montreal, 142 pp.

WEST, A. S., BROWN, A. W. A. & PETERSON, D. G. 1960. Control of black fly larvae (Diptera: Simuliidae) in the forests of eastern Canada by aircraft spraying. *Can. Ent.* **92**: 745–754.

WORLD HEALTH ORGANISATION. 1963. Recommended methods for vector control. WHO Expert Committee on Insecticides. *Wld Hlth Org. Tech. Rep. Ser.*, No. 265, W.H.O., Geneva.

(f) Collecting and rearing

FALLIS, A. M. & SMITH, S. M. 1964. Ether extracts from birds and CO_2 as attractants for some ornithophilic simuliids. *Can. J. Zool.* **42**: 723–730.

FREDEEN, F. J. H. 1959. Rearing black flies in the laboratory (Diptera: Simuliidae). *Can. Ent.* **91**: 73–83.

HARTLEY, C. F. 1955. Rearing simuliids in the laboratory from eggs to adults. *Proc. helminth. Soc. Wash.* **22**: 93–95.

McMAHON, J. P., HIGHTON, R. B. & GOINY, H. 1958. The eradication of *Simulium neavei* from Kenya. *Bull. Wld Hlth Org.* **19**: 75–107. [Contains on pp. 88–90 descriptions and figures of traps for crabs required for collecting early stages of *Simulium neavei* group.]

RAYBOULD, J. N. 1967. A method of rearing *Simulium damnosum* Theobald (Diptera: Simuliidae) under artificial conditions. *Bull. Wld Hlth Org.* **37**: 447–453.

RUBTSOV, I. A. 1956. *Methods of studying black-flies.* Academy of Sciences of U.S.S.R., Moscow & Leningrad, 55 pp. [In Russian: contains useful figures of collecting and rearing equipment, dissection techniques etc.]

TARSHIS, I. B. 1968. Use of fabrics in streams to collect black fly larvae. *Ann. ent. Soc. Am.* **61**: 960–961.

—— 1968. Collecting and rearing black flies. *Ann. ent. Soc. Am.* **61**: 1072–1083.

WENK, P. & RAYBOULD, J. N. 1972. Mating, blood-feeding and oviposition of the "Kibwezi" form of *Simulium damnosum* Theobald (Diptera: Simuliidae) in the laboratory. *Wld Hlth Org.* Unpublished document WHO/ONCHO/72.88, 13 pp.

WOOD, D. M. & DAVIES, D. M. 1966. Some methods of rearing and collecting black flies. *Proc. ent. Soc. Ont.* **96** (1965): 81–90.

3c. PHLEBOTOMIDAE AND PSYCHODIDAE (Sand-flies and Moth-flies)

by D. J. Lewis

PHLEBOTOMIDAE (SAND-FLIES)

THE Phlebotomidae or sand-flies are very small brownish hairy flies which can generally be distinguished at first sight, by their nearly erect rather narrow wings and slender bodies (fig. 90A, B), from Psychodidae which have less erect rather broad wings and squat bodies (fig. 93E). The few species of the psychodid subfamily Bruchomyiinae look rather like sand-flies but lack piercing mandibles.

In the Phlebotomidae the eye is round with no eye-bridge (a mesad extension of the eye around the antenna), the mandibles are functional (for blood-sucking), the antennal flagellar segments are usually long and of nearly uniform thickness, the wing is about 1.3 to 2.8 mm long (except in *Warileya*), and the radial vein has five branches (there are two veins between the radial and medial forks). Sand-flies often fly in a series of hops.

These flies were treated as a subfamily of the Psychodidae till Rodendorf (1964) raised them to family rank in view of their blood-sucking adaptations and the relatively dry breeding places of many species.

It is sometimes necessary to refer to sand-flies explicitly as phlebotomid sand-flies in distinction from some Ceratopogonidae and Simuliidae which are called sand-flies in the West Indies and Australia respectively, and in some other areas.

Over 530 species of Phlebotomidae have been described, rather more than half of them from the New World. It is instructive to study the sand-flies in terms of the three main genera. This makes it easy to understand the apparent paradox that, although sand-flies (unlike mosquitoes and black-flies) are chiefly insects of warm countries, the main sand-fly-borne disease (leishmaniasis) occurs, in the Old World, almost entirely north of the equator and largely outside the tropics.

The study of sand-flies and their relation to disease is difficult. Adults are often hard to find, and larvae usually impossible. These are the most difficult of vector arthropods to study in the laboratory (Marsden, 1970), and many of the leishmanial parasites which they transmit are morphologically indistinguishable.

Nearly all research on sand-flies is directed primarily towards the recognition, study and possibly control of vectors of leishmaniasis. The main subjects for consideration in this respect are taxonomy, distribution, physiology, host preference and resting habits. Distribution, host preference and resting habits are the main extrinsic factors which influence the vectorial status of a species. Physiology is the intrinsic factor (affecting adaptation of parasite to insect host) but little is known about it. Taxonomic affinity to known vectors, however, sometimes indicates a possible vector.

Morphology

Adults. The structure of sand-flies is described in several of the works listed among the references, and some of the taxonomic and other features are mentioned below. Various names have been used for certain structures, and can be related to those used here by reference to figures in the relevant publications.

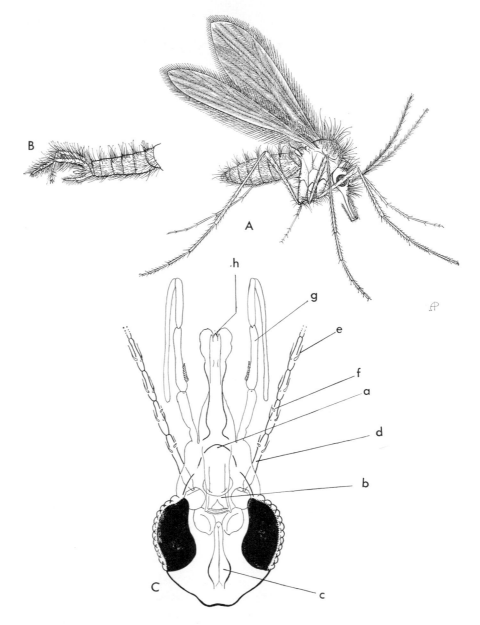

Fig. 90   A, B, ♀ and ♂ abdomen of *Phlebotomus papatasi* (adapted from Whittingham and Rook); C, head of ♀ *Sergentomyia clydei* from below (semi-diagrammatic) showing certain structures.   a, tip of clypeus; b, cibarium; c, pharynx; d, antennal segment 3; e, ascoid; f, papilla; g, palpal segment 3; h, tip of hypopharynx hiding labrum. (A, natural position).

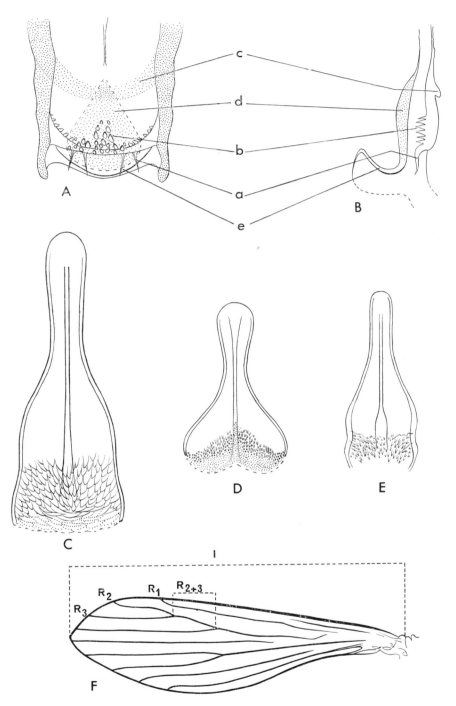

Fig. 91 A, cibarium of ♀ *Lutzomyia panamensis*; B, diagrammatic sagittal section of same; C–E, pharynx of ♀ *Phlebotomus sergenti*, *Sergentomyia antennata* and *S. babu*; F, wing of sand-fly. a, hind teeth; b, fore teeth; c, chitinous arch; d, pigment patch; e, posterior bulge; l, wing length; R₂ etc., branches of radial vein.

The inner ventral surface of the cibarium or buccal cavity (figs 90C, 91A, B) usually bears a transverse row of backwardly pointing teeth which are known as the cibarial teeth, or as hind teeth if fore teeth are present. The fore teeth point upward and look like nodules when seen from below. In front of the teeth is the chitinous arch, a ridge to which muscles of the salivary pump are attached. On the dorsal wall of the cibarium there is often a pigment patch of thick cuticle opposite the teeth. Immediately behind the pigment patch is a bulge which is pronounced and symmetrical in two genera.

The shape of the pharynx (fig. 91C–E) is of some taxonomic value, and so is the nature of posterior teeth or ridges which occur in many species, those on the dorsal plate being usually examined.

The length of the labrum is measured, for taxonomic purposes, from the tip of the clypeus, seen in ventral view, to the tips of the labral papillae.

Each antenna comprises sixteen segments, and the length of segments 3 and of $4 + 5$ are often noted. The relative length of 3 is often expressed by dividing it by the length of the labrum. Segments 3 to 15 bear one or two sensory ascoids.

Each palp comprises five segments, the first of which is sometimes regarded as the palpifer. In measuring the segments the laterad (outer) outline is used to delimit each one. Accuracy is difficult, especially for the first two, but these are the least important. The relative lengths may be expressed in relation to the first one which can be given the value of 10. The palpal formula consists of the serial numbers of the segments arranged in order of increasing length. Some segments overlap each other slightly, so the sum of the segment lengths exceeds that of the palp. Segment 3, and sometimes others, carries small sensilla known as Newstead's sensilla or bulb organs.

Wing lengths (fig. 91F) are often measured from the proximal end of the hairy basal costal node (figured for a psychodid by Quate, 1955, p. 109). The length of vein R_2 is divided by that of R_{2+3} to give a figure which is often useful though variable. 'R_1 apex' indicates either the part of R_1 distad of the tip of R_{2+3}, or, when shown with a minus sign, the distance by which R_1 ends proximad of the tip of R_{2+3}. Upper (post-spiracular) and lower (mesanepisternal) pleural setae are present in some groups.

On the hind margins of abdominal tergites 2 to 6 the presence or absence of erect hairs (usually lost during mounting but indicated by conspicuous round sockets) are important generic characters in the Old World. In the female the furca or genital fork may be used to locate the outlet of the spermathecal ducts which indicate the position of the two spermathecae (fig. 92A–G). These take the form of tubes or of smooth, wrinkled or annulated capsules or other formations. At the tip (the proximal end, farthest from the duct) of each spermatheca are hair-like ductules leading from glands which are invisible in cleared specimens. The accessory glands are relatively large but are of no taxonomic value.

On the abdomen of the male, segment 7 becomes partly, and 8 wholly, rotated dextrally through 180° (Davis, 1967). Segment 9 is not clearly discernible. Behind segment 8 are the aedeagus and parameres (fused together), the claspers (comprising coxite and style) and the surstyles. The inconspicuous segment 10 bears the cerci.

Between the two halves of the aedeagus lie the tips of the genital filaments which lead from the genital pump (fig. 92H). The relative length of a filament is expressed by dividing it by the length of the pump. The coxite bears sensory hairs which may be concentrated on a setiferous lobe (fig. 93A). The style varies in shape and often bears five sensory spines, one or more of which may be reduced or absent (fig. 93A, B). The surstyles (a name discussed by Crampton, 1942, pp. 92 and 111, and used by Quate, 1955, for Psychodidae) are the lateral lobes of tergum 10.

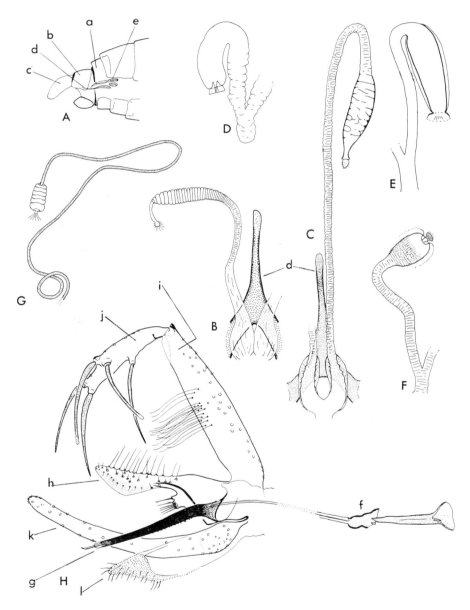

Fig. 92   A, terminalia of ♀ *Phlebotomus papatasi*; B–G, spermathecae of *P. (Larroussius) kandelakii*, *P. (Adlerius) chinensis*, *Sergentomyia (Sergentomyia) punjabensis*, *S. bailyi*, *S. (Grassomyia) squamipleuris* and *S. (Sintonius) christophersi*; H, terminalia of ♂ *P. kandelakii*.   a, tergum 7; b, tergum 9; c, cercus; d, furca; e, spermatheca; f, genital pump and filaments; g, aedeagus; h, paramere; i, coxite; j, style; k, surstyle.

Early stages (fig. 93C, D). The egg is oblong with a sculptured surface. The antennae of the larva are short and leaf-like, the head and body bear numerous pinnate hairs, and the second to fourth stage larva of most species has four long dark setae on segment 9. The pupa is attached to the substratum by the last larval pelt and is usually upright.

Classification

The classification used here is based almost entirely on that of Theodor (1958, 1965). The sand-flies of the world are grouped in a few genera, the three main ones being *Phlebotomus* Rondani and *Sergentomyia* França and Parrot of the Old World and *Lutzomyia* França of the New. The status of one or other of these genera in the northern and southern parts of the Old and New Worlds has an important influence on the epidemiology of leishmanial disease.

Many species of *Sergentomyia* belong to well-defined subgenera but others, especially in the east of the Oriental Region, cannot be grouped. At present it seems advisable not to establish a number of small subgenera for these, but to leave them ungrouped indefinitely, or perhaps until the early stages of many more species can be studied and used for classification as in the Simuliidae.

Forattini (1971) has proposed a new classification of *Lutzomyia*, raising some subgenera to generic status, and Sherlock and Guitton (1970) have suggested some changes in grouping. These alterations may well be necessary but are not incorporated here because the difficulty of identifying many species makes it difficult to evaluate and use the whole of the 1965 system.

A key for genera, and for one sex of subgenera of two genera, is presented later. Keys for the species of certain large areas have been provided by Abonnenc (1972), Perfil'ev (1966, 1968) and Theodor (1958). Works containing keys for some limited areas are included in the list of references. For many areas, including South America as a whole, there are no up-to-date keys because subgeneric classification is difficult, new species are being discovered, and some species are known only from one sex. For some areas it is necessary to study an extensive literature and recognize species without resort to keys.

In addition to identifying species and subspecies, it is sometimes necessary to consider other infraspecific forms. Variation in the appearance of insect species can be due to so many causes (Lewis *et al.*, 1969; Reid, 1970) that it is often advisable to give informal vernacular names to variants, at least until their nature is understood. Such names are useful for certain forms which have become synonyms although they have some biological significance. Some variants must be considered in relation to disease because, as in the case of mosquitoes (Reid, 1970), Simuliidae (Duke *et al.*, 1966) and Mollusca (Wright, 1961), certain pathogens are transmitted only by variants within species.

The early stages of many New World and a few Old World species have been reared from eggs obtained from gravid females, and described (Abonnenc and Larivière, 1957; Barretto, 1941; Guitton and Sherlock, 1969; Hanson, 1968; Mangabeira, 1942; Sherlock, 1964; Vattier-Bernard, 1970 and others), and Hanson has studied the bearing of Panamanian larvae on classification.

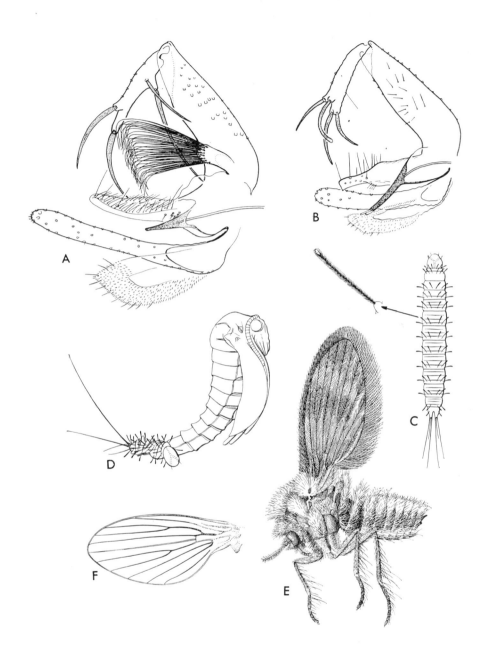

Fig. 93 A, B, terminalia of ♂ *Phlebotomus* (*Paraphlebotomus*) *nuri* and *Sergentomyia babu*; C, last stage of larva of *P. papatasi* (after Grassi); D, pupa of sand-fly; E, ♀ *Pericoma* species pinned specimen with wings artificially raised (Psychodidae); F, wing of *Pericoma* sp. with scales removed.

KEY TO GENERA OF PHLEBOTOMIDAE
AND TO SUBGENERA OF MALE *PHLEBOTOMUS* AND
FEMALE *SERGENTOMYIA*
(subgenera in brackets)

This key is adapted mainly from the works of Theodor (1948, 1958) for the Old World and Fairchild (1955) for the New. Modifications may become necessary in view of proposals by Abonnenc (1972) and Forattini (1971). Use of the subgeneric sections should be supplemented by reference to descriptions of both sexes. At present it is not practicable to make a key for *Lutzomyia*.

1   Radius dichotomously branched, at least R_5 not forking from R_s proximal to R_4. Surstyle of male very short, shorter than cercus. Cibarium without teeth. Palpal formula 1–4–5–3–2 (Neotropical Region)   .   .   .   ***HERTIGIA*** Fairchild

–   Radius pinnately branched, R_5 always arising proximal to R_4. Surstyle of male always longer than cercus. Palpal formula never as above   2

2   Wing broad and rounded, R_5 arising close to R_4. Coxite of male short and rounded, shorter than style. No pleural setae (Neotropical Region)   ***WARILEYA*** Hertig

–   Wing usually long and slender, often pointed, R_5 not arising close to R_4. Coxite of male always considerably longer than wide and at least as long as style   .   .   3

3   Posterior bulge in dorsal wall of cibarium absent or, if present, usually asymmetrical in preparations. Palpal segment 5 the longest. Postspiracular and lower mesanepisternal setae usually absent (Old World)   4

–   Posterior bulge in dorsal wall of cibarium symmetrical in preparations and well defined. Palpal segment 3 often the longest. Postspiracular and lower mesanepisternal setae present (New World)   5

4   Cibarial teeth absent or, if present, usually in the form of spicules, and not arranged as in *Sergentomyia* (below). Pigment patch nearly always absent (a small one in *Parvidens*). Hind end of abdominal tergites 2 to 6 with erect hairs, sockets as large as on 1. Style of male with three to five spines and sometimes (in some species of *Idiophlebotomus*) with two or more strong hairs .   ***PHLEBOTOMUS***   6

–   Cibarial teeth in a posterior transverse row, sometimes with fore teeth which mostly point upward. Pigment patch usually present. Hind ends of abdominal tergites 2 to 6 with all or nearly all hairs recumbent, sockets much smaller than on 1. Style of male with four major spines and an accessory seta   .   ***SERGENTOMYIA***   17

5   Cibarium of female with four longitudinal rows of teeth, the hind tooth of each mesad row being markedly larger than the others. Spermatheca segmented, with enlarged apical segment and very long thin ducts. Style very long, with five spines of which 1 or 2 are terminal and 3 near the middle   .   ***BRUMPTOMYIA*** França & Parrot

–   Cibarium of female with a transverse row of hind teeth, and one or more rows of fore teeth in most species. Spermatheca segmented or not. Style of male with 1–6 large, and 1–3 small, spines, one of the latter being subterminal in many species   ***LUTZOMYIA***

6   (Males of *Phlebotomus*) Antennal segment 3 about three or more times length of labrum and much longer than palp. Rods associated with genital pump (Oriental Region). Females not known to bite man   (***IDIOPHLEBOTOMUS***) Quate & Fairchild

–   Antennal segment 3 subequal to or shorter than labrum, always shorter than palp   .   7

7   Vein M_{1+2} forking at level of r-m; R_2 is 5–6 times length of R_{2+3}. Genital pump with associated rods. One large African cave species, female not known to bite man   (***SPELAEOPHLEBOTOMUS***) Theodor

–   Vein M_{1+2} forking far from level of r-m; R_2 less than five times length of R_{2+3}. Genital pump without associated rods   8

8   Style with three spines, two of them terminal and one near the middle. Aedeagus rudimentary; genital filaments very short and thick (Australia)

                                       (AUSTRALOPHLEBOTOMUS) Theodor

–   Style with four or more spines 9

9   Coxite with hairy process near base. Genital filaments short, 1.3–2.3 times length of pump 10

–   Coxite without such process. Genital filaments 3–11 times length of pump . . 12

10   Coxite long (0.37–0.63 mm); process very small. Style long and cylindrical, with three terminal spatulate spines, and two other spines. Paramere with two long dorsal processes. Surstyle with terminal spines. Pharyngeal teeth resembling network of lines, or scales . . . . **(PHLEBOTOMUS)**

–   Coxite short (0.2–0.33 mm.). Process of coxite large with a tuft of long hairs. End-process of paramere with flat elliptical dorsal surface beset with short hairs. Surstyle without terminal spines 11

11   Style with four long spines, two being near the tip and two near the base

                                       (PARAPHLEBOTOMUS) Theodor

–   Style with five long spines of which two are terminal   **(SYNPHLEBOTOMUS)** Theodor

12   Paramere not divided 13

–   Paramere with two or three lobes (that is, with one or two ventral processes) . . 15

13   Paramere with truncated straight end and ventral subterminal notch. Not yet placed in subgenus . . . . **P. NEWSTEADI** Sinton

–   Paramere with rounded end 14

14   Genital filaments 3–5 times length of pump. Pharynx with punctiform teeth

                                     (LARROUSSIUS) Nitzulescu

–   Genital filaments 6.6–11 times length of pump, except in *P. mascittii* which has scale-like pharyngeal teeth . . . . **(ADLERIUS)** Nitzulescu

15   Paramere with two lobes, the lower one either partly flattened or bearing row of stout spines; aedeagus thick. Pleura with two groups of setae (two species in Africa and one in Iran). Females not known to bite man

                                (PARVIDENS) Theodor & Mesghali

–   Paramere either with two lobes differently formed or with three lobes . . . 16

16   Style with five long spines, two of them terminal and three in the middle; paramere with three lobes (Oriental and Palaearctic Regions)   **(EUPHLEBOTOMUS)** Theodor

–   Style with four long spines, one of them terminal, one subterminal and two nearer the middle; paramere with two or three lobes .   **(ANAPHLEBOTOMUS)** Theodor

17   (Females of *Sergentomyia*) Spermathecal duct having a large bulge near its junction with the spermatheca which is a large sac. Cibarial teeth long and pointed. Legs very long (Ethiopian Region). Cave form . .   **(SPELAEOMYIA)** Theodor

–   Spermathecal duct not like this 18

18   Spermatheca distinctly segmented. Hind ends of abdominal tergites 2–6 with a few erect hairs . . . .   **(SINTONIUS)** Nitzulescu

–   Spermatheca not segmented but sometimes indefinitely striated 19

19   Spermathecae tubular 20

–   Spermathecae capsular 21

20   Spermathecae with smooth walls, of uniform width throughout their length. Wings narrow and lanceolate, R_1 usually 0.3–0.8 length of R_{2+3}. Antennal segment 3 short (0.08–0.19 mm), shorter than 4+5, usually shorter than labrum

                                     (SERGENTOMYIA)

–   Spermathecae transversely striated, more strongly sclerotised distally; ducts usually narrower than distal ends of spermathecae. Wings broad; R_1 usually longer than R_{2+3}. Antennal segment 3 long (0.28–0.4 mm), longer than 4+5, usually longer than labrum . . . .   **(RONDANOMYIA)** Theodor

M

21 Spermathecae are round capsules beset with fine spicules. Antennal segment 3 without ascoid. Cibarial teeth in posteriorly convex row **(*GRASSOMYIA*)** Theodor

 – Spermathecae not shaped thus. Antennal segment 3 with ascoid. Cibarial teeth usually straight or in posteriorly concave row 22

22 Spermathecae are round or elliptical smooth capsules. Cibarial teeth in comb-like row, usually of parallel teeth with short points. Pharynx narrow, bulging near hind end, with many, few or no teeth . . . **(*PARROTOMYIA*)** Theodor

 – With other combinations of characters. Species not placed in subgenera.

Some features of genera and certain subgenera and species

Some taxonomic features mentioned below are by no means diagnostic but give some idea of the appearance of certain taxa.

Hertigia **and** *Warileya* These are small tropical American genera (Fairchild and Hertig, 1951; Fairchild, 1953) which probably evolved earlier than the others. *Hertigia* may be able to bite, and at least one species of *Warileya* can bite man.

Phlebotomus In subgenus *Phlebotomus* the pharynx of the female is armed with scale-like teeth, and the ductules of the spermathecae are associated with a delicate membrane. The most widespread species, *P. papatasi* Scopoli, is easily recognized in many areas but in some must be carefully distinguished from certain others, including *P. duboscqi* Neveu-Lemaire and the Iranian *P. salehi* Mesghali and Rashti (1968) which does not appear in most keys.

In subgenus *Paraphlebotomus* the pharynx of the female has large teeth, and the terminal segment of the spermatheca is differentiated from the others. The widespread species *P. sergenti* Parrot must be distinguished from closely related forms by careful examination of both sexes. Another member is *P. caucasicus* Marzinowski.

Synphlebotomus includes *P. celiae* Minter, *P. martini* Parrot and *P. vansomerenae* Heisch, Guggisberg and Teesdale.

In *Larroussius* the pharynx of the female has punctiform teeth and the spermatheca has a finger-like end-process. Species include *P. ariasi* Tonnoir, *P. kandelakii* Shchurenkova, *P. langeroni* Nitzulescu, *P. longicuspis* Nitzulescu, *P. longipes* Parrot, *P. major* Annandale, *P. perfiliewi* Parrot and *P. perniciosus* Newstead.

In the related subgenus *Adlerius*, to which *P. chinensis* Newstead and *P. simici* Nitzulescu belong, the spermatheca is incompletely segmented.

Euphlebotomus includes *P. argentipes* Annandale and Brunetti.

Idiophlebotomus is a small Oriental subgenus.

Sergentomyia In this large genus is *S. sintoni* Pringle, a member of the *antennata*-group of the subgenus *Sergentomyia*.

Brumptomyia Some of the females of this small Neotropical genus are impossible to identify. They probably feed on armadilloes.

Lutzomyia This large American genus shows great morphological diversity but some groups are ill-defined. Theodor (1965) provisionally placed most of the species in twenty-four groups which comprise eight subgenera and sixteen species-groups more or less equivalent to the subgenera.

In subgenus *Lutzomyia* the spermatheca is segmented, the coxite has a basal tuft, and the paramere bears strong setae. Two of the species are *L. longipalpis* (Lutz and Neiva) and *L. renei* (Martins, Falcão and da Silva).

The *cruciata*-group is one of those with a pear-shaped spermatheca having a rounded end-segment and a segmented base. The few species include *L. cruciata* (Coquillett) and *L. gomezi* (Nitzulescu).

The *migonei*-group is a provisional one and the spermathecae of its species vary from tubes to round or elliptical capsules.

In the *verrucarum*-group, to which *L. colombiana* (Ristorcelli and van Ty) belongs, the spermathecae are striated sacs of various forms.

The *vexator*-group includes a number of divergent species to which *L. sanguinaria* (Fairchild and Hertig) may belong.

In the subgenus *Nyssomyia* Barretto the spermatheca has a large terminal knob, palpal segment 5 of the female is about as long as 3, and the genitalia of the male are simple. Among the species are *L. anduzei* (Rozeboom), *L. flaviscutellata* (Mangabeira), *L. intermedia* (Lutz and Neiva), *L. olmeca* (Vargas and Díaz Nájera), *L. trapidoi* (Fairchild and Hertig), *L. whitmani* (Antunes and Coutinho) and *L. ylephiletor* (Fairchild and Hertig).

In *Psychodopygus* Mangabeira the spermatheca is imbricated, palpal segment 5 is very short, and in most species some of the fore teeth of the females are enlarged. One member is *L. panamensis* (Shannon).

In the *shannoni*-group the spermathecae are usually long smooth capsules with long individual ducts.

In the subgenus *Pintomyia* Costa Lima, which includes *L. fischeri* (Pinto) and *L. pessoai* (Coutinho and Barretto), the spermatheca lacks a terminal knob, and the ductules open through a broad surface.

Biology

Distribution

Sand-flies are widely distributed in the tropics and other warm mainland areas, and extend northwards to latitudes in the region of 50° N, such as 48° in the Soviet Union, Savignies near Beauvais in France, Jersey in the English Channel, and Kamloops in Canada. *Phlebotomus* is the dominant genus in the south of the northern temperate part of the Old World, and *Sergentomyia* and *Lutzomyia* in the tropics of the Old and New Worlds respectively.

The habit of breeding in soil makes sand-flies independent of surface water, their small size enables them to use microclimates, and their nocturnal habits help them to avoid heat. These insects, therefore, despite their delicate structure and rather limited flight range, cover very wide areas. Some species flourish in semi-desert country and thus extend across some of the barriers between zoogeographical regions.

The Old World. *Phlebotomus* is dominant in the Palaearctic Region, and is represented in the other three zoogeographical regions. In the Ethiopian Region there are several species in the north and east but remarkably few in West Africa.

In most of the Old World the majority of sand-fly species flourish in savanna rather than forest.

The New World. In the Nearctic Region there is no genus comparable to *Phlebotomus*, and in the U.S.A. there are few species of sand-flies and none of medical importance.

In the Neotropical Region many species of *Lutzomyia* live in the extensive forests.

The forest environment. In the Old World forests there are few important man-biting sand-flies. In many forests of the New World are vast numbers of potential man-biters which normally have no contact with people. Under such conditions road-building, logging and agricultural settlement can lead to zoonotic leishmaniasis.

The open environment. In the Old World Phlebotomidae are widespread in many areas of savanna and steppe where ease of travel brings many people into contact with man-biting sand-flies. The distribution of sand-fly vectors of leishmaniasis largely determines whether the disease in a particular area is a purely animal disease, a sporadic or a permanent zoonosis, or an urban infection.

Much open country can be so hot and wind-swept by day that rodents and accompanying sand-flies must shelter in microhabitats. These, and therefore dermal leishmaniasis, are often localized, and their study is associated with the concept of landscape epidemiology.

Caves. Many caves shelter sand-flies, some of which are essentially cave species. Caves may have played a part in the evolution of the synanthropic habit of some species.

Early stages

Eggs, larvae and pupae occur in soil, or among leaf litter in forests, and the larvae feed on various kinds of plant and animal material and faeces. Development is slow, and various periods have been quoted in different climates (Perfil'ev, 1968). Foster *et al.* (1970) studied *P. longipes* in Ethiopia and found that at 18° to 20°C. development from oviposition to adult emergence took about 100 days, and at 28° to 29° about 53 days. The respective periods for eggs, larvae and pupae were about 15, 61 and 24; and 6, 37 and 10. Larvae undergo diapause in countries with a cold winter, and in some other areas. Development from egg to adult takes weeks or months, according to latitude, and in some northern areas there is only one generation a year.

Physiological age of adults

Estimation of parous rates is useful in the study of population dynamics and rates of infection with *Leishmania*, but the ovaries of sand-flies are so small that it is difficult to dissect them quickly and to interpret all findings. However, residual secretion in the accessory glands of many Old World and some New World species often shows that a fly is parous. In spite of a small degree of error, the accessory glands could be used for studying general trends in the ecology of many species.

Movement

Sand-flies fly and bite mainly at night. When near their resting places, or when attracted to lights, they often fly in short characteristic hops, but many can travel some distance, especially in calm weather, and flights of about 1 km have been recorded.

Food

Sand-flies of both sexes probably feed on plant sugars, and the females suck vertebrate blood. Some species habitually bite mammals of a particular group and occasionally attack lizards, and some lizard-biters can also feed on man.

Many species of *Phlebotomus* take mammalian blood and can bite man, and in the Old World man-biting sand-flies are particularly numerous in the south of the Palae-arctic Region where this genus is dominant. The main anthropophilic sand-flies of eastern Africa, southern Arabia and eastern India are species of *Phlebotomus* and represent outliers of its main area.

Many, and probably all, species of *Sergentomyia* feed mainly on lizards or other cold-blooded animals, so most sand-flies of the Old World tropics are not anthropo-philic. Some species of *Sergentomyia* and *Lutzomyia* can bite birds but little is known of this.

Many species of certain groups of *Lutzomyia* can bite man, and the South American forests, unlike those of Africa, teem with potential man-biters.

Phlebotomids have a rigid labrum about 0.15 to 0.57 mm. long and are presumably pool feeders.

Reproduction

Hertig (1949) found sand-flies mating in flight. Blood meals are necessary for ovarian development in many species but a few are autogenous.

Resting sites

Natural shelters. Day-time resting sites of sand-flies include tree holes, spaces between buttresses of trees, foliage of forest undergrowth, animal burrows, termite hills, rock crevices in caves and elsewhere, cavities among boulders, and soil cracks.

Buildings. Many sand-flies enter buildings to feed on man, domestic animals or wall-lizards, and many females of some species remain there by day. The domestic habit is particularly prevalent among some species of *Phlebotomus* and therefore common in the south of the Palaearctic Region. In some areas it is a small step from rodent burrows to nearby human dwellings. 'Domestic' is an indefinite term but the habit is so important in relation to disease, sampling and control that it is useful to regard sand-flies as either wild (the vast majority) or domestic forms.

The domestic habit may be defined as 'the habit of remaining within a man-made shelter throughout the whole or a definite part of the gonotrophic cycle'. This is Senior-White's (1954) definition of 'endophilic' for mosquitoes and possibly some other haematophagous arthropods, and the term was used in this sense by Detinova (1962, p. 106). The original usage has tended to change (for mosquitoes), and the following definitions were given by the World Health Organization (1963). Endophagy is the 'tendency . . . to feed indoors'. Endophily is the 'tendency . . . to rest indoors, whether by day or night'. Exophagy is the 'tendency to feed outdoors'. Exophily is the 'tendency . . . to rest outdoors, whether by day or night'. These terms are avoided by some authors, to avoid ambiguity, but can sometimes be useful. They are virtually meaningless in sparsely inhabited country. No species is entirely domestic, and the degree of domesticity depends on several varying factors. The terms refer to behaviour without explaining its underlying causes.

Sand-fly seasons

In the temperate part of the Old World sand-flies occur in summer, their numbers taking some time to build up after the winter larval diapause.   In the tropics some species flourish throughout the year, and others in the wet or the dry season.

Natural enemies

In general sand-flies seem to have relatively few predators or parasites apart from the pathogens mentioned below.   It is curious that no natural filarial infection has been recorded despite a report of experimental infection (Nelson, 1964, p. 85).   Possibly hypostomial or cibarial teeth hinder ingestion of microfilariae.

Relation to Disease

Irritation

P. papatasi and some other species can be a pest owing to irritation from their bites.

Leishmaniasis

Leishmaniasis in general.   This is a group of diseases, often zoonoses, caused by several forms of the protozoan genus *Leishmania*.   Sand-flies are the only known invertebrate vectors, although several other insects were suspected many years ago (Adler and Theodor, 1957).

The Map (fig. 215) based largely on Cahill (1968), Heyneman (1971), Lysenko (1967, 1971), Mackie *et al.* (1954), Manson-Bahr (1966) and Theodor (1964), gives some general information about the distribution of leishmaniasis.   It cannot be concise because a world map will not show local networks, because some infected areas represent a few sporadic cases, not all autochthonous, and because distribution and our knowledge of it are continually changing.   Some large areas without known human cases probably harbour infected animals.   For instance Lysenko (1967) estimated that, in the part of the Soviet Union where the temperature is 20°C. for at least 120 days a year, the actual and potential areas of the wet form of human dermal leishmaniasis were 53 000 and 702 000 km^2 respectively.   In some infected areas the potential threat is not realized because transmission among reservoir animals is maintained in a closed circuit by sand-flies which do not bite man (Shaw, and Lainson, 1968 p. 402).

Leishmania, though a flagellate, has no flagellum when in the vertebrate host.   On entering a sand-fly it develops a flagellum, becoming a promastigote or 'leptomonad', multiplies in the stomach, and, in the case of most forms which affect mammals, spreads forward into the anterior midgut (the narrow part of the midgut behind the cardia, which is often loosely called the cardia), the oesophagus and head.

Human leishmaniases are of two main types, dermal and visceral, of which the wild hosts are largely rodents and Canidae respectively.   For each form the dog may be a liaison carrier to man, or a regular host where the infection has become urban and lost its wild host.   The degree of human involvement in leishmaniases is very variable. Leishmaniasis can be a purely wild-animal sand-fly disease, a zoonosis with few or many human cases, a disease of dogs, sand-flies and man, or one of man and sand-flies.

The epidemiology of the diseases is largely determined by the ecology of their sand-fly vectors. It is not possible to draw up a concise list of these for several reasons. There is a gradual transition from major habitual vectors to minor vectors, occasional vectors and non-vectors. Some species transmit *Leishmania* of man in some areas and not in others. No two areas are epidemiologically similar, and conditions are always changing. Some species are not fully proved vectors although their role is reasonably certain, for instance in some areas where there is only one man-biting species. Although we are concerned with sand-flies which transmit disease to man, it is necessary to take some account of those which are entirely or primarily vectors among animals.

Many subjects have to be considered in assessing the vectorial status of a species of sand-fly. Epidemiological requirements include knowledge of distribution, phenology, the presence, in captured flies, of flagellates, and their position in the gut. These parasites have to be identified by serological or other means, in view of the morphological similarity of the leishmaniae of man and some animals. Flies must be experimentally infected, and the growth and behaviour of flagellates studied. Position of flagellates can vary but is an aid to identification. Transmission experiments are valuable. The paragraphs below on dermal and visceral leishmaniasis include references to some but not all the possible vectors.

Transmission is influenced by infraspecific variation of sand-flies in terms of both man-biting habits and of physiological adaptation of parasites to sand-flies. Lysenko (1967), discussing the three-member parasite system (parasite, vertebrate host and arthropod vector), compared the *Leishmania* sand-fly association with the adaptation (Duke *et al.*, 1966) of local forms of *Onchocerca volvulus* to apparently-infraspecific forms of *Simulium damnosum* Theobald. Comparable situations may well exist among the leishmaniases, particularly in the forests of South America.

In establishing vectors of leishmaniasis it is necessary to treat with caution certain records of the disease until they have been fully investigated. Some may have been imported, and in rare instances transmission may occur without an insect vector (Garnham, 1965).

In the Old World the research for vectors is complicated by the presence of *Leishmania* of lizards, but in the New World this is known only from Martinique and is probably rare or absent on the mainland (Johnson and Hertig, 1970, pp. 297, 298).

Dermal leishmaniasis. Old World dermal leishmaniasis (oriental sore) is due to *Le. tropica* and occurs in Asia as far east as 80° longitude, and in the Mediterranean area and parts of northern Africa. Its distribution corresponds largely with the main area of *Phlebotomus*, the south of the Palaearctic Region.

There is a zoonotic 'wet' form which is associated with the large gerbil, *Rhombomys opimus*, of Central Asia, and may have developed in relation to the microhabitats provided for sand-flies by the gerbil burrows in the harsh environment. Permanent zoonoses are common in this region because many gerbil colonies are close to villages. In some areas *P. caucasicus* transmits the infection among the gerbils and *P. papatasi* conveys it to man. *S. sintoni* bites gerbils to some extent and can be experimentally infected with *Le. tropica* (Saf'janova & Alexeev, 1967) but is probably of no practical significance.

The urban 'dry' form of Old World dermal leishmaniasis is more widespread than the zoonosis and is often associated with dogs. Vectors include the often domestic *P. papatasi* and *P. sergenti*, and probably *P. perfiliewi*. Diffuse dermal leishmaniasis of Ethiopia follows infections probably transmitted by *P. longipes*.

Old World dermal leishmaniasis is especially associated with the subgenera *Phlebotomus* and *Paraphlebotomus*.

American dermal (including muco-cutaneous) leishmaniasis occurs over much of the range of *Lutzomyia*, in various countries from Mexico to Chile, largely in forest country, and is due to several forms of *Leishmania* which normally infect wild rodents. The human infections have various effects. For example, the chiclero's ulcer of Belize affects the ear, the Panamanian disease is like oriental sore, and the espundia of Brazil usually spreads to the mucous membranes. Uta occurs in a dry part of Peru and has diminished after control. *L. olmeca* and the related *L. flaviscutellata* transmit *Leishmania* among rodents in Belize (Williams, 1970) and near Belém, Brazil (Shaw and Lainson, 1968), respectively, and *L. olmeca* evidently transmits to man as well. Various species may transmit the infection to man in different areas. *L. longipalpis*, *L. renei*, *L. cruciata*, *L. gomezi*, *L. migonei*, *L. sanguinaria*, *L. trapidoi*, *L. ylephiletor*, *L. panamensis*, *L. fischeri* and *L. pessoai* have been suspected as vectors, but it now seems likely (Williams, 1970) that most habitual vectors to man belong to the subgenus *Nyssomyia*.

Visceral leishmaniasis. The visceral disease, kala-azar, is due to *Le. donovani* in most areas and to *Le. infantum* in the Mediterranean–Central Asian area. In the Old World kala-azar occurs in various areas north of the equator as far as 45°. The principle ones are the Mediterranean-Central Asian area, north–east China, eastern India, and Africa (Sudan and Kenya). In these four areas there are three main epidemiological types, the main China focus being probably a derivative from Central Asia (Lysenko, 1967). Visceral leishmaniasis thus exists in many parts of the *Phlebotomus* area of the Palaearctic Region, except in some dry parts of northern Africa where the subgenus *Larroussius* does not flourish, and occurs in some distinctly localized tropical areas. The main focus in India is limited partly by climate and topography, and partly by the distribution of the anthropophilic form of the vector.

Wild canids often live far from habitations, so kala-azar is largely a sporadic zoonosis.

Vectors of Old World kala-azar belong mostly to the subgenera *Synphlebotomus*, *Larroussius*, *Adlerius* and *Euphlebotomus* and include *P. martini* and probably *P. celiae* and *P. vansomerenae* (Kenya), *P. langeroni orientalis* Parrot (Sudan), *P. longicuspis* (Algiers), *P. major* and *P. perniciosus* (Mediterranean), *P. simici* (East Mediterranean), *P. chinensis* (China and probably some other foci) and *P. argentipes* (some wetter parts of India).

In the New World, as in the Old, kala-azar exists largely in rather dry country. One of its main areas is Ceará in north-eastern Brazil where the vector is *L. longipalpis*.

Other Protozoa

Sand-flies include the vectors of several trypanosomes of animals (Wallace and Hertig, 1968; Anderson and Ayala, 1968) and probably of an *Endotrypanum* of sloths (Shaw, 1964). A reptilian malaria parasite has been found to develop to the sporozoite stage in two species of *Lutzomyia* by Ayala and Lee (1970).

Bartonellosis

Bartonella bacilliformis, probably a protozoan, is the cause of this disease which takes the form of the dermal verruga peruana or the severe Oroya fever or Carrión's disease. In Peru it occurs in arid country and is transmitted by *L. verrucarum* (Townsend). No animal reservoir is known. The related *L. colombiana* is probably the vector in Colombia.

Viruses

Sand-fly fever occurs in the southern part of the Palaearctic Region and extends up the Nile and into India. It can be transmitted by *P. papatasi*, and the virus can probably pass from the female through the egg to the larva.

Yellow fever virus (Woodall, 1964) and several other viruses (Peralta and Shelokov, 1966, and others) have been isolated from sand-flies but the significance of these findings is not known.

Control

Sand-fly nets and repellents are used in some areas.

Larvae are often impossible to control, but the breeding of domestic species can be reduced by removing refuse and filling cracks in soil and walls. In the Soviet Union development of irrigation has removed gerbils and 'burrow sand-flies' from considerable areas, and in Teheran and some other cities urbanization has destroyed many sand-fly breeding places.

The hopping flight of many sand-flies doubtless ensures repeated contact with residual insecticides, and DDT used in malaria eradication has controlled domestic sand-flies in towns and villages in many countries. New problems may arise as malaria is eradicated unless the malaria eradication service is transformed into a vector-borne disease control organization suggested by Gabaldon (1969).

Destruction of the adults of wild sand-flies is often difficult or impossible, but in the Soviet Union much has been done by blowing insecticides into gerbil colonies.

SOME METHODS OF STUDY

Some methods are summarized below, and these and others may be studied by reference to the extensive literature.

Collecting

Owing to the small size, nocturnal habits and hidden resting sites of sand-flies, special methods are necessary to capture them. Some have been discussed by Muirhead-Thomson (1968).

Larvae. These are occasionally collected in special surveys.

Resting flies. Sand-flies can often be found by day in the more accessible of the resting sites already mentioned. They can be disturbed by dust, smoke or twigs, and caught in a fine-mesh black net. The type with a folding steel-spring frame is useful. The flies can then be removed with a suction catcher and killed with benzine. If the barrel of the catcher is conical (made from a centrifuge tube or an adaptor) flies can easily be ejected.

Sand-flies resting in houses or stables by day can be taken in suction catchers, if necessary fitted with long inlet tubes for high ceilings.

In certain caves and other sites it is advisable to guard against histoplasmosis (Aspin, 1959) or other infections (Rioux and Golvan, 1969, p. 36), either by collecting directly into a net or a vial, or by using a catcher with a mechanical fan or a catcher worked by blowing into it.

Sand-flies in flight. Sheets of paper about 10 by 17 cm or larger may be fixed in cleft sticks 28 cm or more long, smeared with castor oil, and placed about sunset near resting places. In the morning any adhering sand-flies should be removed with a needle. If many traps are used they can be planted in a dish of sand for transport to the laboratory. In wet or very humid weather the oil may run and make such traps ineffective. In rain forests sand-flies tend to be too dispersed for paper traps, flying at various levels and not concentrating in microhabitats.

A Shannon-type trap (Barretto and Coutinho, 1941; Pessoa and Barretto, 1948; Forattini, 1954) is like a large bed net raised above the ground, and can be slung from trees. It often catches males of blood-sucking species, which may be essential for identification. The Damasceno-type trap (Deane, 1956; Sherlock and Pessoa, 1964) is a tent-like truncated cone with a lateral opening, and can be fitted over burrows or against tree trunks.

Flies biting. Sand-flies biting man, horses or other bait animals at and after dusk may be caught with suction catchers or vials. Various baited traps have been devised by Disney (1966), Thatcher (1968) and others. The use of baits is particularly important in areas where man-biting species are seldom caught by other means.

Flies attracted to light. Light traps are useful for collecting certain species, particularly where wet weather or forest conditions reduce the value of unbaited sticky traps. In Africa many sand-flies can sometimes be collected near lights in houses, but most belong to non-anthropophilic species. In France (Rioux *et al.*, 1969) and the USSR some man-biting species come to light traps.

The above-mentioned Shannon-type trap, used in South America, depends partly on attraction to light.

Temporary preservation

Flies caught as dry specimens can be preserved dry. Those taken on sticky traps may be kept in alcohol (about 80 per cent), but after some three years they may become discoloured and difficult to clear. Alternatively, flies can be put in a drop of gum-chloral (Berlese's) mounting medium in a closed plastic or glass vial, the fluid being kept to a minimum to avoid movement and loss of legs. This method avoids risk of shaking and does not harden the delicate spermathecal ducts. Flies may be stored for short or long periods on microscope slides, as described below, many being placed under one cover-glass till they can be processed.

Mounting

Sand-flies are very rarely pinned, and are usually mounted on microscope slides. They may be macerated in caustic potash, stained and mounted in Canada balsam (Sinton, 1932), or mounted unstained in balsam and later examined with phase-contrast illumination. Such mounts should last at least a century and perhaps indefinitely, but do not always show the spermathecae and their ducts clearly. The system of Fairchild and Hertig (1948) has been much used for American species. Many other methods have been published but some are slow and complex. Flies to be mounted unstained in Euparal can be previously treated with hot ten per cent caustic potash for two minutes, and dehydrated in alcohol up to 90 or 95 per cent.

Quate and Steffan (1966) preferred to use Hoyer's medium, a gum-chloral aqueous fluid of the Berlese's type, containing glycerine. Media containing sugar instead of glycerine dry quite well, and a proprietary medium can be very useful because it saves the user from the time-consuming filtration of impurities from gum arabic. For those wishing to make their own medium the following formula, used for aphids by Stroyan (1971), is effective.

Gum arabic, picked lumps	12 g
Chloral hydrate crystals	20 g
Glacial acetic acid	5 ml
50% w/w glucose syrup	5 ml
Distilled water	30–40 ml

The constituents are dissolved at room temperature in the order shown, and filtered over glass wool to remove dust, and the mixture is allowed to evaporate at not more than 30°C. till it reaches the required consistency.

When dry specimens are to be mounted they can be wetted with a one per cent solution of domestic detergent in water, and heated in water at about 80°C. for four minutes to remove bubbles.

Each sand-fly may be mounted under a circular cover-glass one cm. in diameter on a slide not thicker than one mm. (to save weight and space and allow viewing from both sides). The head should be separated from the body and mounted upside down. It is sometimes necessary to mount various parts of a fly under different cover-glasses, and to orientate the abdomen of a female dorso-ventrally. The whole, or the hind end, of the abdomen of males of some species of *Phlebotomus* should be mounted separately so that it can be orientated to show the aedeagus at a particular angle. In order to prevent a cover-glass from compressing a specimen it is often advisable to let the mountant dry for an hour or so before applying the cover-glass, or to run in additional mountant on the following day.

In many temperate or tropical-dry climates the mountant becomes dry at the edge in a few weeks. Slides can be kept over silica gel, or in an oven at about 37°C., either to speed the process or as a routine in wet climates. Then any excess of mountant can be scraped away with a needle, and the cover-glass quickly ringed with Euparal with one stroke of a brush. In a temperate climate damp summer weather may make it necessary to finish the drying in an incubator. Transparent finger-nail varnish is sometimes used for temporary ringing. Three one-mm. beads can be attached with Euparal to each slide to protect cover-glasses if large numbers of slides are to be stored vertically in slotless drawers.

Sample mounts should be checked after a few years, especially in very damp climates without air-conditioning, to see if they have absorbed or lost water. Many specimens are in excellent condition after 15 years in such mounts, and may well last indefinitely without inspection, and be safer than flies on pins or in alcohol. For extra security cover-glasses could be ringed directly with Canada balsam, but the mounts would have to be baked (Stroyan, 1971) and this would change the refractive index of the medium and make some structures difficult to see.

The use of gum-chloral has several advantages over methods involving maceration with caustic potash. It is very quick to use and gives good undistorted preparations of the spermathecae and ducts. The natural colour remains, and the clypeus and pigment patch are well defined. The soft tissues persist, invisible, hold the antennal and palpal segments together during any reorientation, and maintain the natural shape of the body so that a specimen resembles the living insect instead of being a mere skeleton with softened wings. Spirit specimens may be difficult to macerate, but gum-chloral avoids this problem. Nematode parasites, and even some protozoa, have been recovered from gum-chloral mounts after years. If necessary, mounts can be soaked in water and the insects sectioned for the study of soft parts. Very rarely eye pigment escapes on to the cibarium but can be removed by dissection or by putting the head in potash for a minute or less.

The males of many species are useful for confirming the identification of females, but often, especially in some species of *Sergentomyia*, the identification of males takes a long time and furnishes little or no useful information. It is then convenient to store them on slides for possible future reference, without detaching the heads, and ensuring that they are not compressed.

The general appearance of most species is useless for determination but can be used for the preliminary sorting of some species, particularly in the New World.

Breeding

Laboratory rearing requires much time and attention. Methods have been described by Hertig and Johnson (1961), Saf'janova (1964), Vattier-Bernard (1970) and others.

PSYCHODIDAE (MOTH-FLIES)

The papers by Duckhouse (1965), Fairchild (1955), Freeman (1950), Jung (1956, 1958), Quate (1955, 1959, 1960), Quate and Quate (1967) and Tonnoir (1940) are important sources of information and a guide to the other literature of this family. There are three subfamilies.

The Bruchomyiinae were treated by Fairchild (1955) as a tribe of the Phlebotominae (then in Psychodidae), and are among the most primitive psychodids (Quate, 1961). Here they are separated from the sand-flies for the present. There are three genera, *Nemopalpus* (with a few tropical species; Quate, 1962), *Eutonnoiria* (African) and *Bruchomyia* (South American).

The other two subfamilies are known as moth-flies, and their general features were summarized by Quate (1959). The antennae are long, and often each segment has a cupuliform whorl of hairs. The wings are moderately or very broad, the longitudinal veins are equally sclerotized, and cross veins are usually absent. The wings are held roof-like over the body when at rest, and the insect moves in short jerky flights when disturbed. The pupa has a thoracic simple respiratory horn, and a fringe of hairs and spurs on the abdominal segments, and the terminal segment is quadrate in lateral view.

In the Trichomyiinae there is no eye-bridge, the antennal flagellar segments are pyriform or subcylindrical, and the radial vein has four branches. This is a small but widespread group. A few species (*Horaiella* and *Sycorax*) can or may be able to bite cold-blooded vertebrates (Fairchild and Hertig, 1951; Jung, 1958b p. 8; Parrot, 1951, p. 30; Theodor, 1948) but the maxillae of some or all are short and thus differ radically from those of Phlebotomidae.

In the Psychodinae there is an eye-bridge, the antennal flagellar segments are strongly nodiform or barrel-shaped, the palpi have four, or rarely three, visible segments, and the radius has five branches. Quate (1959) summarized some features of the larva. The head and mouth parts are complete, and they and the head are not retractile. The larva has secondary annulations on the body, and tergal plates on the annuli. In *Psychoda alternata* Say and some other species the plates may be absent on the anterior abdominal segments. The abdomen ends in a siphon-tube bearing apically two pairs of conical projections fringed with fan-like groups of hairs.

The Psychodinae are the dominant group of moth-flies. A few species breed in drains or sewage works and may become a minor nuisance, and a very few have been reported to cause myiasis. The tropicopolitan *Telmatoscopus albipunctatus* Williston and the cosmopolitan *Ps. alternata* (both redescribed by Quate (1959, 1960)) have a great range of larval habitats, and their wide distribution with little variation suggests that they have been largely distributed by man (Quate, 1959).

BIBLIOGRAPHY

ABONNENC, E. 1967a. Révision des phlébotomes de l'Afrique au sud du Zambèse . . . *Cah. Off. Rech. sci. Tech. Outre-Mer.* Ent. méd. **5**: 3–19.

—— 1967b. Les phlébotomes de l'Angola (Diptera, Psychodidae). *Publicões cult. Co. Diam. Ang., Lisboa* no. **77**: 57–122.

——, 1972. Les phlébotomes de la région éthiopienne (Diptera, Psychodidae). *Cah. Off. Rech. sci. Tech. Outre-Mer.* **55**: 289 pp.

ABONNENC, E. & LARIVIÈRE, M. 1957. Les formes larvaires de quelques phlébotomes des régions méditerranéenne et éthiopienne. *Archs Inst. Pasteur Algér.* **35**: 391–403.

——, —— 1958. *Phlebotomus duboscqi* Neveu-Lemaire, 1906. Morphologie de l'oeuf et des formes larvaires. *Archs Inst. Pasteur Algér.* **36**: 259–265.

ABONNENC, E. & MINTER, D. M. 1965. Bilingual keys for the identification of the sandflies of the Ethiopian Region [French and English]. *Cah. Off. Rech. sci. Tech. Outre-Mer.* Ent. méd. no. **5**: 1–63.

ADLER, S. & THEODOR, O. 1957. Transmission of disease agents by phlebotomine sandflies. *A. Rev. Ent.* **2**: 205–226.

ANDERSON, J. R. & AYALA, S. C. 1968. Trypanosome transmitted by a *Phlebotomus*: first report from the Americas. *Science* **161**: 1023–1025.

ASPIN, J. 1959. Cave sickness—benign pulmonary histoplasmosis. *Trans. Cave Res. Group* **5**: 105–114.

AYALA, S. C. & LEE, D. 1970. Saurian malaria: development of sporozoites in two species of phlebotomine sandflies. *Science* **167**: 891–892.

BARNETT, H. C. 1962. Sandflies and sandfly-borne diseases. *In* Maramosch, K. (Ed.) *Biological transmission of disease agents.* New York and London. pp. 83–91.

BARRETTO, M. P. 1941. Morfologia dos ovos, larvas e pupas de alguns flebótomos de São Paulo. *Anais Fac. Med. Univ. S. Paulo* **17**: 357–427.

—— 1947. Catalogo dos flebótomos Americanos. *Archos Zool. S. Paulo* **5**: 177–242.

—— 1962. Novos subgêneros de *Lutzomyia* França, 1924. *Revta Inst. Med. trop. S. Paulo* **4**: 91–100.

BARETTO, M. P. & COUTINHO, J. O. 1941. Processo de captura, transporte, dissecão e montagem de flebótomos. *Anais Fac. med. Univ. S. Paulo* **16**: 173–187.

BELOVA, E. M. 1971. Reptiles and their importance in the epidemiology of leishmaniasis. *Bull. Wld Hlth Org.* **44**: 553–560.

BRAY, R. S. 1972. Leishmaniasis in the Old World. *Brit. med. Bull.* **28**: 39–43.

CAHILL, K. M. 1968. Clinical and epidemiological patterns of leishmaniasis in Africa. *Trop. geog. Med.* **20**: 109–118.

CARNEIRO, M. & SHERLOCK, I. A. 1964. Estudo morfológico sôbre as pupas de Phlebotominae. *Revta bras. Malariol.* **16**: 311–327.

CRAMPTON, G. C. 1942. The external morphology of the Diptera. Guide to the insects of Connecticut, part VI. *State geol. nat. Hist. Surv. Bull.* no. **64**: 10–165.

DAVIS, N. T. 1967. Leishmaniasis in the Sudan Republic. 28 . . . *J. med. Ent.* **4**: 50–65.

DEANE, L. de M. 1956. *Leishmaniose visceral no Brasil.* Rio de Janeiro. 162 pp.

DETINOVA, T. S. 1962. *Age-grouping methods in Diptera of medical importance.* World Health Organization monog. ser. 47. 216 pp.

DISNEY, R. H. L. 1966. A trap for phlebotomine sandflies attracted to rats. *Bull. ent. Res.* **56**: 445–451.

DOLMATOVA, A. V. & DEMINA, N. A. (1971. Les phlébotomes (Phlebotominae) et les maladies qu'ils transmettent. *Cah. Off. Rech. Sci. Tech. Outre-Mer.* **18**: 168 pp.

DUCKHOUSE, D. A. 1965. Psychodidae of Southern Australia, subfamilies Bruchomyiinae and Trichomyiinae. *Trans. R. ent. Soc. Lond.* **117**: 329–343.

DUKE, B. O. L., LEWIS, D. J. & MOORE, P. J. 1966. *Onchocerca-Simulium* complexes. I . . . *Ann. trop. Med. Parasit.* **60**: 318–336.

FAIRCHILD, G. B. 1953. A note on *Hertigia hertigi* Fairchild and description of the female. *Proc. ent. Soc. Wash.* **55**: 101–102.

—— 1955. The relationships and classification of the Phlebotominae. *Ann. ent. Soc. Am.* **48**: 182–196.

FAIRCHILD, G. B. & HARWOOD, R. F. 1961. Phlebotomus sandflies from animal burrows in eastern Washington. *Proc. ent. Soc. Am.* **63**: 239–245.

FAIRCHILD, G. B. & HERTIG, M. 1948. An improved method of mounting small insects. *Science* **108**: 20–21.

——, —— 1951. Notes on the Phlebotomus of Panama VIII. Two new species of *Warileya*. *Ann. ent. Soc. Am.* **44**: 422–429.

FLOCH, H. & ABONNENC, E. 1952. Diptères phlébotomes de la Guyane et des Antilles françaises. Paris: Faune Union franc. no. **14**: 5–207.

FORATTINI, O. P. 1954. Algumas observações sôbre biologia de flebotómos em região de Bacia do Rio Paraña (Brasil). *Archos Fac. Hyg. Saúd. públ. S. Paulo* **8**: 15–136.

—— 1971. Sôbre a classificação da subfamilia Phlebotominae nas Américas (Diptera: Psychodidae). *Pap. avulsos Zool. S. Paulo* **24**: 93–111.

FOSTER, W. A., TESFA-YOHANNES, T. M. & TESFAI TECLE. 1970. Studies on leishmaniasis in Ethiopia II . . . *Ann. trop. Med. Parasit.* **64**: 403–409.

FREEMAN, P. 1950. Family Psychodidae. *Handbk Ident. Br. Insects* **9**. Part 2. pp. 77–96. London.

GABALDON, A. 1969. Global eradication of malaria: changes of strategy and future outlook. *Am. J. trop. Med. Hyg.* **18**: 641–656.

GARNHAM, P. C. C. 1965. The leishmanias, with special reference to the role of animal reservoirs. *Am. Zool.* **5**: 141–151.

—— 1971a. The genus *Leishmania*. *Bull. Wld Hlth Org.* **44**: 477–489.

—— 1971b. American leishmaniasis. *Bull. Wld Hlth Org.* **44**: 521–527.

GROVÉ, S. S., DOWNES, R. & ZIELKE, E. 1971. Leishmaniasis in South West Africa: preliminary notes on host reservoir and vector studies. *S. Afr. med. J.* **45**: 293–294.

GUITTON, N. & SHERLOCK, I. A. 1969. Descrição das fases imaturas do *Phlebotomus longipalpis* Lutz & Neiva, 1912 (Diptera, Psychodidae). *Revta bras. Biol.* **29**: 383–389.

HANSON, W. J. 1968. The immature stages of the subfamily Phlebotominae in Panama. *Dissert. Abstr.* (B) **29**: p. 2074.

HARWOOD, R. F. 1965. Observations on the distribution and biology of *Phlebotomus* sandflies from northeastern North America. *Pan-Pacif. Ent.* **41**: 1–4.

HERTIG, M. 1949. The genital filaments of *Phlebotomus* during copulation. *Proc. ent. Soc. Wash.* **51**: 286–288.

HERTIG, M. & JOHNSON, P. T. 1961. The rearing of *Phlebotomus* sandflies I. . . . *Ann. ent. Soc. Am.* **54**: 753–764.

HEYNEMAN, D. 1971. Immunology of leishmaniasis. *Bull. Wld Hlth Org.* **44**: 499–514.

HOOGSTRAAL, H. & HEYNEMAN, D. 1969. Leishmaniasis in the Sudan Republic 30: Final epidemiologic report. *Amer. J. trop. Med. Hyg.* **18**: Suppl.: 1089–1210.

HOUIN, R. 1963. Données épidémiologiques et déductions prophylactiques sur les leishmanioses autochthones en France. *Annls Parasit. hum. comp.* **38**: 379–438.

JOHNSON, P. T. & HERTIG, M. 1970. Behaviour of *Leishmania* in Panamanian phlebotomine sandflies fed on infected animals. *Exper. Parasit.* **27**: 281–300.

JUNG, H. F. 1956. Beiträge zur Biologie, Morphologie und Systematik der europäischen Psychodiden (Diptera). *Dt. ent. Z.* (N.F.) **3**: 97–257.

—— 1958a. Psychodidae. Psychodidae—Bruchomyiinae. In E. Lindner (Ed.) *Fliegen palaearkt. Reg.* **3** (9 & 9a): 10 pp., Stuttgart.

—— 1958b. Psychodidae-Trychomyiinae. *Ibid.* (9b): 16 pp., Stuttgart.

KIRK, R. & LEWIS, D. J. 1951. The Phlebotominae of the Ethiopian Region. *Trans. R. ent. Soc. Lond.* **102**: 383–510.

LAINSON, R. & SHAW, J. J. 1972. Leishmaniasis of the New World: taxonomic problems. *Brit. med. Bull.* **28**: 44–48.

LEE, D. J., REYE, E. J. & DYCE, A. L. 1963. 'Sandflies' as possible vectors of disease in domesticated animals in Australia. *Proc. Linn. Soc. N.S.W.* **87**: 364–376.

LEWIS, D. J. 1967. The phlebotomine sand-flies of West Pakistan. *Bull. Brit. Mus. nat. Hist.* (Ent.) **19**: 1–57.

—— 1971. Phlebotomid sandflies. *Bull. Wld Hlth Org.* **44**: 535–551.

—— 1973. Phlebotomidae. In Catalog of the Diptera of the Oriental Region. In the press.

LEWIS, D. J., BARNLEY, G. R. & MINTER, D. M. 1969. On *Sergentomyia ruttledgei* (Lewis & Kirk) and *S. schwetzi* (Adler, Theodor & Parrot) (Diptera: Psychodidae). *Proc. R. ent. Soc. Lond.* (B) **38**: 53–60.

LEWIS, D. J., LAINSON, R. & SHAW, J. J. 1970. Determination of parous rates in phlebotomine sandflies with special reference to Amazonian species. *Bull. ent. Res.* **60**: 209–219.

LYSENKO, A. Y. 1967. The geographic distribution of leishmaniasis in the Old World. 24 pp. Document prepared for WHO Seminar on Leishmaniasis in Moscow.

—— 1971. Distribution of leishmaniasis in the Old World. *Bull. Wld Hlth Org.* **44**: 515–520.

MACKIE, T. T., HUNTER, G. W. & WORTH, C. B. 1954. *A manual of tropical medicine.* 907 pp. Philadelphia: Saunders.

MANGABEIRA, O. 1942. 13a Contribução ao estudio dos Flebotomus . . . *Mem. Inst. Oswaldo Cruz* **37**: 375–381.

MANSON-BAHR, P. H. 1966. *Manson's tropical diseases.* 16th edn. 1068 pp. London.

MARETT, P. J.   1923.   A note on the capture of a *Phlebotomus permiciosus* ♂ in Jersey, C.I.   *Trans. R. Soc. trop. Med. Hyg.* **17**: p. 267.

MARSDEN, P. D.   1970.   [Abstract.]   *Trop. Dis. Bull.* **67**: 1318.

MARTINS, A. V., MACIEL, C. S. & SILVA, J. E. da.   1968.   Notas sôbre os flebótomos do grupo *squamiventris* do subgenero *Psychodopygus* Mangabeira, 1941.   *Bol. Mus. Hist. nat. Univ. fed. Minas Gerais* (Zool.) **1**: 1–28.

MESGHALI, A. & RASHTI, M. A. S.   1968.   Phlebotomini of Iran IV.   *Bull. Soc. Path. exot.* **61**: 768–772.

MOSHKOVSKIJ, S. D. & DUHANINA, N. N.   1971.   Epidemiology of the leishmaniases: general considerations.   *Bull. Wld Hlth Org.* **44**: 529–534.

MOSHKOVSKIJ, S. D. & SOUTHGATE, B. A.   1971.   Clinical aspects of leishmaniasis with special reference to the USSR.   *Bull. Wld Hlth Org.* **44**: 491–497.

MUIRHEAD-THOMPSON, R. C.   1968.   *Ecology of insect vector populations.*   174 pp.   London

NELSON, G. S.   1964.   Factors influencing the development and behaviour of filarial nematodes in their arthropodan hosts.   pp. 75–119.   Host-parasite relationships in invertebrate hosts.   Oxford: Blackwell.

NERONOV, V. M. & GUNIN, P. D.   1971.   Structure of natural foci of zoonotic cutaneous leishmaniasis and its relationship to regional morphology.   *Bull. Wld Hlth Org.* **44**: 577–584.

PARROT, L.   1951.   Notes sur les phlébotomes LXI.   A propos de classification.   *Archs Inst. Pasteur Algér.* **29**: 28–45.

PERALTA, P. H. & SHELOKOV, A.   1966.   Isolation and characterization of arboviruses from Almirante, Republic of Panama.   *Am. J. trop. Med. Hyg.* **15**: 369–378.

PERFIL'EV, P. P.   1966.   Fauna of the U.S.S.R.   Dipterous insects. **3**, no. 2.   Family Phlebotomidae.   [In Russian.]   Moscow: Inst. Zool. Acad. Sci. U.S.S.R.

—— 1968.   Translation of 1966 book.   Jerusalem: Israel Program of Scientific Translations.

PESSOA, S. B. & BARRETTO, M. P.   1948.   Leishmaniose tegumentar americana.   Rio de Janeiro: Impresa Nacional.

PETRISHCHEVA, P. A.   1965.   Sandflies (Phlebotominae).   pp. 57–85.   *In* P. A. Petrishcheva (Ed.) *Vectors of diseases of natural foci.*   London: Oldbourne Press (distributor of translation of 1962 original).

—— 1971.   The natural focality of cutaneous leishmaniasis in the USSR.   *Bull. Wld Hlth Org.* **44**: 567–576.

QUATE, L. W.   1955.   A revision of the Psychodidae in America north of Mexico.   *Univ. Calif. Publ. Ent.* **10**: 103–273.

—— 1959.   Insects of Micronesia.   Diptera: Psychodidae.   *Insects of Micronesia* **12**: 433–484.

—— 1960.   Psychodidae.   Guide to the insects of Connecticut.   Part VI.   The Diptera or true flies of Connecticut.   Fasc. **7**: 54 pp.

—— 1961.   Zoogeography of the Psychodidae.   *Proc. 11th internat. Congr. Ent.* **1**: 168–173.

—— 1962.   A review of Indo-Chinese Phlebotominae.   *Pacific Insects* **4**: 251–267.

—— 1964.   Phlebotomine sandflies of the Paloich area in the Sudan.   *J. med. Ent.* **1**: 213–268.

—— 1965.   A taxonomic study of Philippine phlebotomines.   *Ibid.* **2**: 17–27.

QUATE, L. W. & QUATE, S. H.   1967.   A monograph of Papuan Psychodidae, including *Phlebotomus.*   *Pacific Insects Monog.* **15**: 216 pp.

QUATE, L. W. & STEFFAN, W. A.   1966.   An alternative method of mounting *Phlebotomus* and other small Diptera.   *J. med. Ent.* **3**: 126.

REID, J. A.   1970.   Systematics of malaria vectors. . . .   *Misc. Publ. ent. Ent. Soc. Am.* **7**: 56–62.

RIOUX, J. A. & GOLVAN, Y. J.   1969.   Epidemiologie des leishmanioses dans le sud de la France.   pp. 223.   Paris: Inst. nat. Santé Rech. méd. Monogr. 37.

RODENDORF, B. B.   1964.   Historical development of Diptera.   *Trudÿ paleont. Inst.* **100**: 1–312.   [In Russian.]

SAF'JANOVA, V. M.   1964.   Laboratory cultivation of sandflies.   *Bull. Wld Hlth Org.* **31**: 573–576.

—— 1971.   Leishmaniasis control.   *Bull. Wld Hlth Org.* **44**: 561–566.

SAF'JANOVA, V. M. & ALEXEEV, A. N.   1967.   Experiments on the susceptibility of sandflies to various strains of leptomonads. *Parasitologia, Leningrad* **1**: 191–200. [In Russian with English Summary.]

SCHMIDT, J. R., SCHMIDT, M. L. & McWILLIAMS, J. G.   1960.   Isolation of phlebotomus fever virus from *Phlebotomus papatasi*. *Am. J. trop. Med. Hyg.* **9**: 450–454.

SENIOR-WHITE, R.   1954.   Adult anopheline behaviour patterns: a suggested classification. *Nature, Lond.* **173**: p. 730.

SHAW, J. J.   1964.   A probable vector of *Endotrypanum schaudini* of the sloth *Choloepus hoffmani*, in Panama. *Nature, Lond.* **201**: 417–418.

SHAW, J. J. & LAINSON, R.   1968.   Leishmaniasis in Brazil: II . . . *Trans. R. Soc. trop. Med. Hyg.* **62**: 396–405.

SHERLOCK, I. A. & GUITTON, N.   1970.   Notas sôbre o subgênero *Trichophoromyia* Barretto, 1961 (Diptera, Psychodidae, Phlebotominae). *Revta brasil. Biol.* **30**: 137–150.

SHERLOCK, I. A. & PESSOA, S.B.   1964.   Métodos practicos para a captura de flebótomos. *Revta brasil. Biol.* **24**: 331–340.

SINTON, J. A.   1932.   Notes on some Indian species of the genus *Phlebotomus*.   Part XXX. Diagnostic table for the females . . . *Indian J. med. Res.* **20**: 55–74.

STROYAN, H. L. G.   1971.   Personal communication.

TAYLOR, R. M.   1967.   Catalogue of arthropod-borne viruses of the world. 898pp. Washington: U.S. Department of Health.

THATCHER, V. E.   1968.   Studies of phlebotomine sandflies using castor oil traps baited with Panamanian animals. *J. med. Ent.* **5**: 293–297.

THEODOR, O.   1948.   Classification of the Old World species of the subfamily Phlebotominae. *Bull. ent. Res.* **39**: 85–111.

—— 1958.   Psychodidae-Phlebotominae.   In E. Lindner. *Fliegen palaearkt. Reg.* **3** (9c): 55 pp. [In German.]

—— 1964.   Leishmaniases.   In J. van der Hoeden (Ed.) *Zoonoses*.   Amsterdam.   pp. 475–493.

—— 1965.   On the classification of American Phlebotominae. *J. med. Ent.* **2**: 171–197.

—— 1973.   Family Psychodidae (Phlebotominae).   A catalogue of the Diptera of the Americas south of the United States.   Fasc. 6a.   Dept. Zool. Agric. S. Paulo.   In press.

TONNOIR, A. L.   1940.   A synopsis of the British Psychodidae with descriptions of new species. *Trans. Soc. Br. Ent.* **7**: 21–64.

VATTIER-BERNARD, G.   1970.   Contribution a l'étude systématique et biologique des phlébotomes cavernicoles en Afrique intertopicale. *Cah. ORSTOM, Ent. méd. Parasit.* **8**: 175–288.

WALLACE, F. G. & HERTIG, M.   1968.   Ultrastructural comparison of promastigote flagellates (leptomonads) of wild-caught Panamanian *Phlebotomus*. *J. Parasitol.* **54**: 606–612.

WILLIAMS, P.   1970.   Phlebotomine sandflies and leishmaniasis in British Honduras (Belize). *Trans. R. Soc. trop. Med. Hyg.* **64**: 317–368.

WOODALL, J. P.   1964.   The viruses isolated from arthropods at the East African Virus Research Institute in the 26 years ending December 1963. *Proc. E. Afr. Acad.* **2**: 141–146.

WORLD HEALTH ORGANIZATION.   1963.   Terminology of malaria and malaria eradication. 127 pp.   Geneva: WHO.

WRIGHT, C. A.   1961.   Taxonomic problems in the molluscan genus *Bulinus*.   *Trans. R. Soc. trop. Med. Hyg.* **55**: 225–231.

N

3d. CERATOPOGONIDAE
(Biting midges, "Sand-flies", "Punkies")
by Paul Freeman

THE small to very small midges forming this family of the Nematocera are equipped with piercing and sucking mouthparts in the female and are properly termed biting midges, to distinguish them from the non-biting midges of the allied family Chironomidae. In various parts of the world they are called by other vernacular names, such as 'punkies' or 'no-see-ums' and are even called 'sand-flies', particularly in America and Australia, though this name is usually reserved for flies of the family Phlebotomidae. They are among the smallest of the blood-sucking flies, being about 1–4 mm. long. Some 50 or more genera occur throughout the world. For a while, until the position was clarified by the International Commission for Zoological Nomenclature, the family was sometimes called the Heleidae.

Ceratopogonids may be separated from other families by the following characters: Antennae generally with 14 visible segments (the proximal segment usually being incorporated in the head capsule), but rarely with 12, 13 or 15, plumed in the male and with plumes somewhat adpressed to the antennal shaft; mouthparts of female formed into a short piercing proboscis; legs short and stout, often with long and unequal claws in the female; wings, at rest, superimposed over the back (fig. 94A), venation with the media forked (except in the Leptoconopinae), frequently with the radial area condensed (see fig. 94), wings often speckled in species of *Culicoides* and other genera.

The larvae (fig. 140G) may be as much as 5–6 mm. long; they are apneustic and are generally either aquatic or semi-aquatic, being found in damp places or in mud. They tend to be long, narrow and smooth, with certain notable exceptions, and to progress through water with a quick eel-like motion. Breeding places of the important genus *Culicoides* include swamps, river banks, wet soil with much decaying plant material, rotting tree stumps and sand. The larval stage may last three weeks in the tropics or up to seven months in temperate regions where most species over-winter as larvae.

Adult midges may feed partly on plant juices but the females of most species, where their habits are known, take animal food. Some of the larger species of no medical importance feed carnivorously on other insects, especially Chironomidae; others, generally the smaller species, feed on the blood of larger insects such as dragonflies, phasmids, lacewings, meloid beetles, butterflies and larvae of Lepidoptera and sawflies.

The medically important species are those that feed on blood of vertebrates, especially mammals, and belong to four genera; *Leptoconops* (subf. Leptoconopinae), *Forcipomyia* subgenus *Lasiohelea* (subf. Forcipomyiinae), *Culicoides* (subf. Ceratopogoninae) and *Austroconops* (? subf. Ceratopogoninae). In addition, species of *Dasyhelea* have sometimes been accused of sucking human blood, but there is no definite proof of this habit. An interesting Malayan species, *Culicoides anophelis* Edwards, attacks engorged mosquitoes for the purpose of obtaining vertebrate blood from their crops.

There seems little doubt that in the future Ceratopogonidae will be regarded as of much more medical importance than they are at present, particularly in their role as vectors of virus diseases.

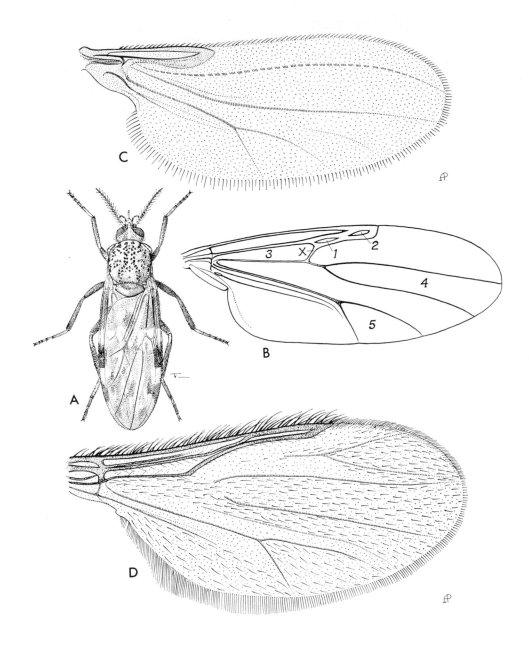

Fig. 94 A, *Culicoides nubeculosus*, resting attitude of living adult ×16; B, wing of *Culicoides*; 1 and 2, first and second radial cells; 3, basal cell; 4, median fork; 5, cubital fork; x, cross-vein; C, wing of *Leptoconops rhodesiensis* (after Carter); D, wing of *Forcipomyia lefanui*.

KEY TO SUBFAMILIES OF CERATOPOGONIDAE

1   Crossvein r-m absent, media unforked, wings without macrotrichia (fig. 94C); female
    antenna apparently 11–13 segmented; larva without sclerotised head capsule and
    prolegs and mouthparts reduced; pupa free from larval skin   ***LEPTOCONOPINAE***

–   Crossvein r-m present, media forked (fig. 94B, D), wings usually with at least some
    indication of macrotrichia; female antenna apparently 14-segmented; larva with
    head capsule present and well-developed mouthparts   2

2   Empodium well developed, at least in female, claws markedly curved; wings usually
    with numerous macrotrichia; larva with both anterior and posterior prolegs, body
    usually with short spines or long processes; larval skin usually attached to posterior
    segments of pupa .   .   .   .   ***FORCIPOMYIINAE***

–   Empodium small or vestigial, claws gently curved, larval skin free from pupa   .   .   3

3   Antennal segments sculptured; first radial cell nearly or completely obliterated, second
    obliterated or square-ended, ending at or before middle of wing; female claws small
    and equal; eyes with very short pubescence; larva with anterior prolegs absent,
    posterior pair retractile and with hooks   .   .   .   .   ***DASYHELEINAE***

–   Antennal segments not sculptured; one or both radial cells present, the second not
    markedly square-ended, ending beyond middle of wing; eyes usually bare; larva
    vermiform, no prolegs, in aquatic species swimming with eel-like motion
                                                            CERATOPOGONINAE

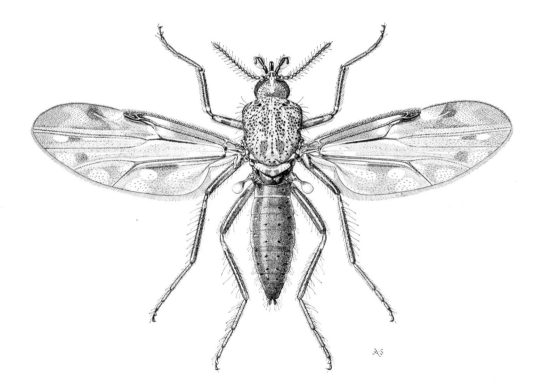

Fig. 95   *Culicoides nubeculosus* female ×20 (from a 'set' specimen).

LEPTOCONOPS **Skuse** includes a number of serious man-biting pest species in the warmer parts of the Old and New World, one of which, *L. kerteszi* Kieffer, seems more or less cosmotropical.   Smith and Lowe (1948) describe the biology of *L. kerteszi* (Kieffer) and *L. torrens* (Townsend).   Some species have milky wings, short spines on the frons and a whitish abdomen.   They bite during the day and one Australian species that bites around the eyes causes a swollen condition lasting for about 3 days.

LASIOHELEA **Keiffer** now considered a subgenus of *Forcipomyia*, contains about 50 species, including a few day-time biting species, also from the warmer parts of the world. The wings are hairy and unmarked, and there is a large empodium between the tarsal claws.   Species are found in both the Old and New world.   Some species are vicious biters of man and other warm-blooded animals in the tropics.

AUSTROCONOPS **Wirth & Lee** contains only one, rather aberrant species, *A. macmillani* Wirth & Lee, placed for the moment in the subfamily Ceratopogoninae. It is only known from Western Australia, where it bites man during the day.

CULICOIDES **Latreille** with about 800 species, is practically a world-wide genus, many species being severe pests of man and other vertebrates.   The genus can usually be recognized by well-defined humeral pits on the thorax and the speckled wings of most species.   Many species feed on birds and at least some of these are peculiar in having many sensory pits on the antennae (Jamnback, 1965).   Most of the man-biting *Culicoides* are crepuscular and their flight range is generally rather small.   They rarely enter houses to bite, though the important African species, *C. milnei* Austen (=*C. austeni* Carter) and *C. grahamii* Austen will do so.

Man-biting species can cause much irritation when they are numerous, but in addition some species of *Culicoides* transmit parasites of both domestic mammals and man.   For example, the virus of blue tongue of sheep, and probably of African horse sickness, are transmitted by these midges.   Arboviruses of medical importance have been isolated from *Culicoides* in Australia (Murray, 1970).   Other viruses of potential medical importance have been isolated and some important recent references to this work are included in the bibliography.

Three species of filarial worms, *Dipetalonema perstans* (mainly African), *D. streptocerca* (Africa) and *Mansonella ozzardi* (tropical America) are transmitted by *Culicoides*, but do not appear to be of much clinical significance.   *Onchocerca cervicalis* of horses and *O. gibsoni* of cattle are transmitted by *Culicoides* in England and Malaya respectively.

CONTROL.   Destruction of larvae in their extensive breeding places is often impracticable, and much use is made of repellents.   When certain holiday resorts are affected, funds are available for control and much has been done by controlling water levels and applying larvicides.

BIBLIOGRAPHY

AKIBA, K.   1960.   Studies on the Leucocytozoon found in the chicken in Japan.   II.   On the transmission of L. caulleryi by *Culicoides arakawae*. *Japan J. Vet. Sci.* **22**: 309–319.

ARNAUD, P. H.   1956.   The heleid genus *Culicoides* in Japan, Korea, and Ryukyu Islands (Insecta, Diptera).   *Microentomology* **21**: 84–207.

ARNAUD, P. H. & WIRTH, W. W.   1964.   A name list of world *Culicoides*, 1956–1962.   *Proc. ent. Soc. Wash.* **66**: 19–32.

BENNETT, G. F. 1961. On the specificity and transmission of some avian trypanosomes. *Can. J. Zool.* **39**: 17–33.

BENNETT, G. F., GARNHAM, P. C. C. & FALLIS, A. M. 1965. On the status of the genera *Leucocytozoon* Zeimann, 1898 and *Haemoproteus* Kruse, 1890 (Haemosporidiida: Leucocytozoidae and Haemoproteidae). *Can. J. Zool.* **43**: 927–932.

BUCKLEY, J. J. C. 1934. On the development, in *Culicoides furens* Poey, of *Filaria* (= *Mansonella) ozzardi* Manson, 1897. *J. Helminth.* **12**: 99–118.

CAMPBELL, J. A. & PELHAM-CLINTON, E. C. 1960. A taxonomic review of the British species of *Culicoides* Latreille. *Proc. R. Soc. Edinb.* (B) **67**: 181–302.

CARTER, H. F. 1921. A revision of the genus *Leptoconops* Skuse. *Bull. ent. Res.* **12**: 1–28.

CHARDOME, M. & PEEL, E. 1949. La répartition des filaires dans la région de Coquilhatville et la transmission de *Dipetalonema streptocerca* par *Culicoides grahami*. *Annls Soc. belge Med. trop.* **29**: 99.

CLASTRIER, J. & WIRTH, W. W. 1961. Notes sur les cératopogonidés. XIV. Cératopogonidés de la région éthiopienne (2). *Arch. Inst. Pasteur Algér.* **39**: 302–337.

DAVIES, J. B. 1967. The distribution of sandflies (*Culicoides* spp.) breeding in a tidal mangrove swamp in Jamaica and the effect of tides on the emergence of *C. furens* Poey and *C. barbosai* (Wirth & Blanton). *W. Ind. med. J.* **16**: 39–50.

DOWNES, J. A. 1955. Observations on the swarming flight and mating of *Culicoides* (Diptera: Ceratopogonidae). *Trans. R. ent. Soc. Lond.* **106**: 213–236.

—— 1958. The feeding habits of biting flies and their significance in classification. *Ann. Rev. Ent.* **3**: 249–266.

DUKE, B. O. L. 1954. The uptake of the microfilariae of *Acanthocheilonema streptocerca* by *Culicoides grahamii*, and their subsequent development. *Ann. trop. Med. Parasit.* **48**: 416–420.

—— 1956. The intake of the microfilariae of *Acanthocheilonema perstans* by *Culicoides grahamii* and *C. inornatipennis*, and their subsequent development. *Ann. trop. Med. Parasit.* **50**: 32–38.

—— 1958. The intake of the microfilariae of *Acanthocheilonema streptocerca* by *Culicoides milnei*, with some observations on the potentialities of the fly as a vector. *Ann. trop. Med. Parasit.* **52**: 123–128.

DU TOIT, R. M. 1944. The transmission of bluetongue and horse sickness by *Culicoides*. *Onderstepoort J. vet. Sci. Anim. Ind.* **19**: 7.

FALLIS, A. M. & BENNETT, G. F. 1961. Sporogony of *Leucocytozoon* and *Haemoproteus* in Simuliids and Ceratopogonids and a revised classification of Haemosporidiida. *Can. J. Zool.* **39**: 215–228.

——, —— 1961. Ceratopogonidae as intermediate hosts for *Haemoproteus* and other parasites. *Mosquito News* **21**: 21–18.

FOOTE, R. H. & PRATT, H. D. 1954. The *Culicoides* of the eastern United States. Washington: U.S. Public Health Monog. No. 18. 53 pp.

FORATTINI, O. P. 1957. *Culicoides* da regiao Neotropical. *Arcos Fac. Hig. Saúd. Públ.* **11**: 161–526.

FORATTINI, O. P., RABELLO, E. X. & PATTOLI, D. 1958. *Culicoides* da regiao Neotropical. II—Observacoes sôbre biologia em condicoes naturais. *Ibid.* **12**: 1–52.

FOX, I. 1955. A catalogue of the bloodsucking midges of the Americas (*Culicoides, Leptoconops* and *Lasiohelea*) with keys to the subgenera and Nearctic species, a geographical index and bibliography. *Univ. Puerto Rico J. Agric.* **39**: 214–285.

GAD, A. M. 1951. The head-capsule and mouthparts in the Ceratopogonidae. *Bull. Soc. Fouad I Ent.* **35**: 17–75.

GARNHAM, P. C. C., HEISCH, R. B. & MINTER, D. M. 1961. The vector of *Hepatocystis* (=*Plasmodium) kochi*; the successful conclusion of observations in many parts of tropical Africa. *Trans. R. Soc. Trop. Med. Hyg.* **55**: 497–502.

GUTSEVICH, A. V. 1960. Blood-sucking Ceratopogonidae of the U.S.S.R. fauna. 131 pp. Moscow: *Zool. Inst. Akad. Nauk.* S.S.S.R. [In Russian.]

HARDY, J. L., SCRIVANI, R. P., LYNESS, R. N., NELSON, R. L. & ROBERTS, D. R. 1970. Ecological studies on Buttonwillow virus in Kern county, California, 1961–1968. *Am. J. trop. Med. Hyg.* **19** (3): 552–563.

HOPKINS, C. A. 1952. Notes on the biology of certain *Culicoides* studied in the British Cameroons, West Africa, together with observations on their possible role as vectors of *Acanthocheilonema perstans*. *Ann. trop. Med. Parasit.* **46**: 165–172.

HOPKINS, C. A. & NICHOLAS, W. L. 1952. *Culicoides austeni*, the vector of *Acanthocheilonema perstans*. *Ann. trop. Med. Parasit.* **46**: 276–283.

JAMNBACK, H. 1965. The *Culicoides* of New York State. *Mus. Bull. N.Y. St. Mus. Sci. Serv. Bull.* no. 399. 154 pp. Albany.

JORGENSEN, N. M. 1969. The systematics, occurrence and host preferences of *Culicoides* (Diptera: Ceratopogonidae) in south-eastern Washington. *Melanderia* **3**: 1–47.

KARSTAND, L., FLETCHER, O., SPALATIN, J., ROBERTS, R. & HANSON, W. J. 1957. Eastern equine encephalitis virus from three species of Diptera from Georgia. *Science, N.Y.* **125**: 395–396.

KETTLE, D. S. 1961. A study of the association between moorland vegetation and breeding sites of *Culicoides*. *Bull. ent. Res.* **52**: 381–411.

—— 1962. The bionomics and control of *Culicoides* and *Leptoconops*. *A. Rev. Ent.* **7**: 401–418.

—— 1965. Biting Ceratopogonids as vectors of human and animal diseases. *Acta Trop.* **22**: 356–362.

—— 1969. The ecology and control of blood-sucking Ceratopogonids. *Acta trop.* **26**: 235–248.

KETTLE, D. S. & LAWSON, J. W. H. 1952. The early stages of British biting midges, *Culicoides* Latreille and allied genera. *Bull. ent. Res.* **43**: 421–467.

KHAMALA, C. P. M. & KETTLE, D. S. 1971. The *Culicoides* Latreille (Diptera: Ceratopogonidae) of East Africa. *Trans. R. ent. Soc. Lond.* **123**: 1–95.

KREMER, M. 1965. Contribution à l'étude du genre *Culicoides* Latreille particulièrement en France. *Encycl. ent.* (A) **39**: 300 pp.

LAURENCE, B. R. & MATHIAS, P. L. 1972. The biology of *Leptoconops* (*Styloconops*) *spinosifrons* (Carter) (Diptera, Ceratopogonidae) in the Seychelles Islands, with descriptions of the immature stages. *J. Med. Ent.* **9**: 51–59.

LAWSON, J. W. H. 1951. The anatomy and morphology of the early stages of *Culicoides nubeculosus* Meigen (Diptera: Ceratopogonidae = Heleidae). *Trans. R. ent. Soc. Lond.* **102**: 511–574.

LEE, D. J. 1948–1955. Australasian Ceratopogonidae, Pts. I–X. *Proc. Linn. Soc. N.S.W.* **72–87**.

LEE, D. J., REYE, E. J. & DYCE, A. L. 1962. 'Sandflies' as possible vectors of disease in domesticated animals in Australia. *Ibid.* **87**: 364–376.

MURRAY, M. D. 1970. The identification of blood-meals in biting midges (*Culicoides*: Ceratopogonidae). *Ann. trop. Med. Parasit.* **64**: 115–122.

NICHOLAS, W. L. 1953. The dispersal of *Culicoides grahamii* and *C. austeni* from their breeding sites prior to their taking a blood-meal. *Ann. trop. Med. Parasit.* **47**: 309–323.

—— 1953. The bionomics of *Culicoides austeni*, vector of *Acanthocheilonema perstans* in the rain-forest of the British Cameroons, together with notes on *C. grahamii* and other species which may be vectors in the same area. *Ann. trop. Med. Parasit.* **47**: 187–206.

NICHOLAS, W. L., KERSHAW, W. E. *et al.* 1953 & 1956. Studies on the epidemiology of filariasis in West Africa, with special reference to the British Cameroons and the Niger Delta: III. The distribution of *Culicoides* spp. biting man in the rain-forest, the forest fringe and the mountain grasslands of the British Cameroons; VII. Further records of the distribution of *Culicoides* spp., with a note on the taxonomic status of *C. austeni*. *Ann. trop. Med. Parasit.* **47**: 95–111; **49**: 455–460.

NICHOLAS, W. L. & KERSHAW, W. E. 1954. Studies on the intake of microfilariae by their insect vectors, their survival, and their effect on the survival of their vectors III.—The intake of the microfilariae of *Acanthocheilonema perstans* by *Culicoides austeni* and *C. grahamii*. *Ann. trop. Med. Parasit.* **48**: 201–206.

PETRISHCHEVA, P. A. 1965. Heleidae pp. 108–130. *In* P. A. Petrishcheva (Ed.). Vectors of diseases of natural foci. London: Oldbourne Press (distributor of translation of 1962 original).

REEVES, W. C., SCRIVANI, R. P., HARDY, J. L., ROBERTS, D. R. & NELSON, R. L. 1970. Button-willow virus, a new Arbovirus isolated from mammals and *Culicoides* midges in Kern county, California. *Am. J. trop. Med. Hyg.* **19** (3): 544–551.

SHARP, N. A. & DYCE, A. L. 1928. *Filaria perstans*; its development in *Culicoides austeni*. *Trans. R. Soc. trop. Med. Hyg.* **21**: 371.

SMITH, L. M. & LOWE, H. 1948. The black gnats of California. *Hilgardia* **18**: 157–183.

TOKUNAGA, M. 1937. Sand flies from Japan. *Tenthredo* **1**: 234–338.

—— 1959. New Guinea biting midges. *Pacif. Insects* **1**: 177–313.

WIRTH, W. W. 1952. The Heleidae of California. *Univ. Calif. Publ. Ent.* **9**: 95–266.

WIRTH, W. W. & Blanton, F. S. 1959. Biting midges of the genus *Culicoides* from Panama. *Proc. U.S. nat. Mus.* **109**: 237–482.

WU, CHIAO-JU & WU, SHU-YIN. 1957. Isolation of virus B type Encephalitis from *Lasiohelea taiwana* Shiraki, a blood-sucking midge. *Acta microbiol. sin.* **5**:(1): 22–26.

3e. CHIRONOMIDAE
("Non-biting midges")

by Paul Freeman

MIDGES of this family form the familiar swarms of dancing flies seen over or near freshwater, especially in the evening. Until recently, when the nomenclature was stabilized, they were sometimes referred to as the family Tendipedidae. They are abundant and usually harmless insects and are not even mentioned in some older books on medical entomology. Under certain circumstances, however, in various parts of the world, some species become important pests, chiefly because they gather in immense numbers near houses and around lights, and thus cause intense annoyance. In some places people appear to have become allergic to them, and asthma and other conditions cause much distress in the chironomid season. The midges are sometimes accompanied by numbers of Mayflies (Ephemeroptera), Caddisflies (Trichoptera) and plume midges (Diptera, Chaoboridae).

Like many other Nematocera, the larger species bear a superficial resemblance to mosquitoes and are frequently mistaken for them, but a closer inspection shows the complete absence of a piercing proboscis, the scaleless wings and the quite different venation (fig. 96A). In many species the front legs are long and held in a slightly raised position. Chironomids hold their wings, at rest, either apart or roof-like along the sides of the abdomen; rarely are they folded flat over the abdomen. The males have plumose antennae (fig. 96B) and they dance in swarms towards evening, generally over fixed objects, near water. The females enter the swarms and are seized by the males, mating taking place on nearby surfaces. Eggs are usually laid in water and embedded in a gelatinous string.

Chironomid larvae are mostly aquatic and are frequently coloured red ('blood-worms'), green or yellow. They do not come to the surface for air, but obtain what they need from the water; some live in tubes, others in the mud, amongst detritus, or they may burrow into water plant leaves. Breeding places are known where there may be as many as 100,000 larvae per square metre of bottom and outbreaks of midges are often caused by natural increases. This happens in Lake Nicaragua and in Lake Victoria where chironomids are common and where they contribute to the plagues of 'lake flies', though these in fact consist mainly of Chaoboridae. It is not uncommon for the larvae (fig. 141C) to be found in drinking water systems, either on account of a deficient filter or because the adult females can gain access to storage tanks, for instance, through broken covers. The larvae are of no known medical significance.

On the Nile the effect of dams has evidently increased the breeding and local emergence of *Tanytarsus* species which occur in well-aerated water. Contributory causes of the pest were riverine urban development, which brought people in contact with the midges, and provided gardens which sheltered these delicate insects.

Water pollution, which favours some species and destroys others, is responsible for some outbreaks. These are likely to increase as a result of urban development.

Chironomid pests are made worse by the effect of lights which concentrate the insects near people. In such cases even minute species may accumulate in heaps measurable by the kilogramme.

In various areas different degrees of control have been achieved by water management, larvicides, introduction of fish and by insecticidal fogging. At Khartoum adequate control may prove impracticable, at least for the present, owing to the vast breeding area of moving water and the need for more biological research. Palliative measures include the use of strong decoy lights to attract midges away from houses which are lit with weak reddish lamps.

There are large numbers of species, but their identification is really a matter for the expert, especially as there is often a succession of species throughout the season. Sometimes the nuisance is caused mainly by single species as, for instance, *Tanytarsus lewisi* at Khartoum, *Glyptotendipes paripes* in Florida, and *Chironomus decorus* in Long Island, New York. There are many papers on the numerous genera of Chironomidae in various parts of the world. Some of the more recent or more important ones are mentioned in the references. Grodhaus (1963) gives a good account of Chironomidae as a nuisance.

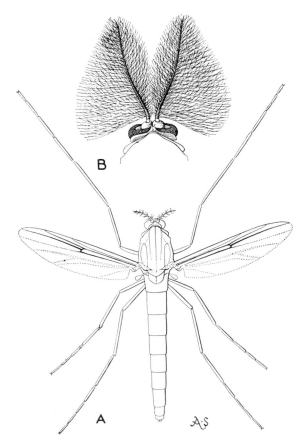

Fig. 96   *Chironomus*, non-biting midge: A, female ×6; B, head of male at higher magnification, showing plumed antennae.

BIBLIOGRAPHY

ANDERSON, L. D., BAY, E. C. & INGRAM, A. A.   1964.   Studies of Chironomid midge control in water-spreading basins near Montebello, California.   *Calif. Vector Views* **11**: 13–22.

BAY, E. C.   1964.   An analysis of the 'sayule' nuisance at San Carlos, Nicaragua, and recommendations for its alleviation. [Unpublished World Health Organization Document (WHO/EBL/20).   18 pp.]

BAY, E. C. & ANDERSON, L. D.   1965.   Chironomid control by carp and goldfish.   *Mosquito News* **25**: 310–316.

BAY, E. C., ANDERSON, L. D. & SUGARMAN, J.   1965.   The abatement of a chironomid nuisance on highways at Lancaster, California.   *Cal. Vector Views* **12**: 29–34.

BRUNDIN, L.   1956.   Zur systematik der Orthocladiinae (Dipt., Chironomidae).   *Rep. Inst. Freshwat. Res. Drottningholm.* **37**: 5–185.

—— 1967.   Transantarctic relationships and their significance, as evidenced by Chironomid midges with a monograph of subfamilies Podonominae and Aphroteniinae and the Austral Heptagyinae.   *K. svensk. Vetensk.-Akad. Handl.* **11**: 1–472.

BRYCE, D. & HOBART, A.   1972.   Biology and Identification of the larvae of the Chironomidae (Diptera).   *Entomologist's Gaz.* **23**: 175–217.

COE, R. L.   1950.   Family Chironomidae.   pp. 121–206.   *Handbk. Ident. Br. Insects.*   Diptera. **2**.   Nematocera.   216 pp.   London.

EDWARDS, F. W.   1931.   Chironomidae.   *Diptera of Patagonia and South Chile.* **2**, fasc. 5: 233–316.

EDWARDS, R. W., EGAN, H., LEARNER, M. A. & MARIS, P. J.   1964.   The control of chironomid larvae in ponds, using TDE (DDD).   *J. appl. Ecol.* **1**: 97–117.

FITTKAU, E. J.   1962.   *Die Tanypodinae . . .* pp. x + 453.   Berlin: Akad.-Verlag.

FREEMAN, P.   1955–1958.   A study of the Chironomidae of Africa south of the Sahara.   *Bull. Brit. Mus. nat. Hist.* (Ent.).   Pt. I, **4**: 1–67; Pt. II, **4**: 287–368; Pt. III, **5**: 323–426; Pt. IV, **6**: 263–363.

—— 1959.   A study of the New Zealand Chironomidae (Diptera, Nematocera).   *Bull. Br. Mus. nat. Hist.* (Ent.) **7**: 395–437.

—— 1961.   The Chironomidae of Australia.   *Austral. J. Zool.* **9**: 611–737.

FREY, D. G.   1963.   *Limnology in North America.*   734 pp.   Wisconsin: University of Wisconsin Press.

GRODHAUS, G.   1963.   Chironomid midges as a nuisance.   *Calif. Vector Views* **10**: 19–24, 27–37.

HILSENHOFF, W. L.   1967.   Ecology and population dynamics of *Chironomus plumosus* (Diptera: Chironomidae) in Lake Winnebago, Wisconsin.   *Ann. ent. Soc. Amer.* **60**: 1183–1194.

LEWIS, D. J.   1957.   Observations on Chironomidae at Khartoum.   *Bull. ent. Res.* **48**: 155–184.

MACAN, T. T.   1959.   *A guide to freshwater invertebrate animals.*   London: Longmans.

MUNDIE, J. H.   1957.   The ecology of Chironomidae in storage reservoirs.   *Trans. R. ent. Soc. Lond.* **109**: 149–232.

OLIVER, D. R.   1971.   Life history of the Chironomidae.   *Ann. Rev. Ent.* **16**: 211–230 [useful bibliography].

PROCEEDINGS OF THE 4TH INTERNATIONAL SYMPOSIUM ON CHIRONOMIDAE.   1971.   *Can. Ent.* **103**: 289–486.   [25 papers.]

TOWNES, H. K.   1945.   The Nearctic species of Tendipedini.   *Am. Midl. Nat.* **34**: 1–206.

WÜLKER, W.   1963.   Prospects for biological control of pest Chironomidae in the Sudan. [Unpublished World Health Organization Document (WHO/EBL/11).   23 pp.]

3f. NEMATOCERA OF MINOR MEDICAL IMPORTANCE

by Kenneth G. V. Smith

ANISOPODIDAE (WINDOW GNATS)

FLIES of this family bear a superficial resemblance to mosquitoes with which they are sometimes confused. However, the lack of scales, short proboscis and venation of the wing (fig. 4F) distinguish the Anisopodidae from the Culicidae. The family is widely distributed, but apparently rare in the Ethiopian region. The adults have no known medical importance but are occasionally reported as a nuisance in the vicinity of sewage works where they breed in the percolating filters. In such areas the flies may occur in large numbers and invade nearby houses. The larvae (fig. 139F) have been recorded from a variety of habitats other than sewage, such as decaying leaves in tree rot holes, slime flux exuding from tree wounds, rotting potatoes, mouldy decaying cardboard and even liver which had been preserved in formalin for 7 years. The adults will oviposit on almost any moist surface.

Silvicola (=*Anisopus*) *fenestralis* (Scopoli) has been reported in the literature as causing intestinal and urino-genital myiasis, the larvae being passed in the urine and stools. The recorded cases are reviewed by Smith and Taylor (1966) since when, only one further case has been reported (Morris, 1968).

TIPULIDAE (CRANE-FLIES, DADDY-LONG-LEGS)

THESE slender fragile flies are of no known medical importance in the adult stage. Some of the smaller species may bear a superficial resemblance to mosquitoes but like the previous family can be distinguished by the lack of scales and proboscis. Species of *Geranomyia*, *Toxorrhina*, and a few others have an elongated proboscis giving them a mosquito-like appearance. However, the construction of the proboscis is quite different from that of mosquitoes and they are unable to pierce the skin and suck blood of vertebrates due to the absence of piercing stylets. They have from time to time been reported as attempting to bite humans (McCrae 1967). It is possible that they were absorbing moisture from the skin surface. The family has a cosmopolitan distribution.

The larvae (fig. 141D, E) of *Tipula* are the well-known leather-jackets which are normally found in the soil where they feed on plant roots and do considerable damage to cereal crops. They are dirty brownish-grey in colour with a black head capsule which is capable of considerable retraction into the anterior part of the larva.

There are a few recorded cases of intestinal myiasis supposedly due to Tipulid larvae, but these are almost certainly due to accidental ingestion of the larvae with foodstuffs or dirty drinking water.

BIBLIOGRAPHY

ALEXANDER, C. P.   1919–20.   The Crane-flies of New York, pts 1, 2.   *Mem. Cornell Univ. agric. Exp. Stn.* **25**: 765–993; **38**: 691–1133.

EDWARDS, F. W.   1928.   Anisopodidae In *Genera Insect.* **190**.   Bruxelles.

GUEGAN, J.   1970.   Myiase intestinale humaine provoquée par des larves de *Tipula paludosa* Meigen (Diptera, Tipulidae) en Vendée.   *Annls. Parasit. hum. comp.* **45**: 243–246.

KEILIN, D. & TATE, P.   1940.   The early stages of the families Trichoceridae and Anisopodidae (Rhyphidae) (Diptera: Nematocera).   *Trans. R. ent. Soc. Lond.* **90**: 39–62.

McCRAE, A. W. R.   1967.   Unique record of crane flies biting man.   *Uganda J.* **31**: 127.

MORRIS, R. F.   1968.   A case of urinogenital myiasis caused by larvae of *Anisopus fenestralis* (Diptera: Anisopodidae).   *Can. Ent.* **100**: 557.

SMITH, K. G. V. & TAYLOR, E.   1966.   *Anisopus* larvae (Diptera) in cases of intestinal and urinogenital myiasis.   *Nature, Lond.* **210**: 852.

4a. TABANIDAE
(Horse-flies, clegs, deer-flies, etc.)
by Harold Oldroyd

STOUTLY built flies, ranging in size from 5–25 mm., with large heads; eyes in life often brilliantly coloured, with green, red and purple spots, bands or zig-zag patterns. The wing-venation shown in fig. 97 is characteristic of the family: note the complete venation, with a discal cell, and 5 posterior cells; note particularly the big fork at the wing-tip formed by the veins R_4 and R_5.

There are over three thousand species in the world, but the ones that attract attention belong mostly to three genera: *Chrysops* with banded wings and spotted eyes, called *deer-flies* because they often appear in the broken woodland that is characteristically deer-country; *Haematopota*, with speckled wings, and eyes marked with zig-zag bands, called *clegs* or *stouts*; and *Tabanus*, a very large and complex genus, usually with clear wings, and with eyes either uniformly coloured or with one or more horizontal bands, called *horse-flies*, and sometimes *hippo-flies*, *buffalo-flies* (not buffalo *gnats*, which are Simuliidae) or *elephant-flies*.

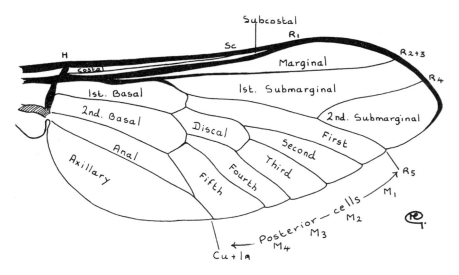

Fig. 97   Nomenclature of the wing veins and cells of the Tabanidae.

Although Tabanidae belong to the Suborder Brachycera they have the same biting equipment as the blood-sucking families of Nematocera, mosquitoes, biting midges and so on. The mandibles and maxillae are elongated into piercing organs. Only females can pierce and suck blood: males of all species, and the females, too, of a number, have lost the mandibles. Both sexes feed on the nectar and pollen of flowers, especially immediately after emergence from the pupa.

O

Female Tabanidae suck the blood of mammals, reptiles and amphibia, but few have been reported to attack birds. In a number of genera of the tribe Pangoniini the labium has become elongated, in association with feeding from deep flowers, the elongation varying considerably in different genera, and even between species of the same genus. The Indian species *Philoliche longirostris* Hardwicke has the labium more than twice as long as the insect's body. The piercing stylets are not correspondingly elongate, and when the female is sucking blood the stylets are applied to the skin of the victim by pushing aside or folding back the labium.

KEY TO THE TRIBES OF TABANIDAE

1 Hind tibiae with spurs (fig. 98A). Vertex of head often, though not always, with three ocelli. Chitinised spermathecal ducts (after dissection) without enlargement into 'trumpets' 2

– Hind tibiae without spurs. Vertex of head without functional ocelli, though an ocellar tubercle may be present. Chitinised spermathecal ducts (after dissection) enlarged into 'trumpets' (fig. 98B) . . . . . Subfamily **TABANINAE** 4

2 Mouthparts vestigial. Grey flies, living on the coasts of Africa and of S. America Subfamily **SCEPSIDINAE**; Tribe **SCEPSIDINI** (Genera *Adersia, Braunsiomyia* in Africa: *Scepsis* in S. America.)

– Mouthparts as a whole well-developed, though mandibles may be reduced or lacking 3

3 Antennae with three large segments, plus six or seven terminal segments, making nine or ten in all (figs 98C, 100A) . . . . Subfamily **PANGONIINAE** (Three Tribes: PANGONIINI, N. and S. American, Australian, Palaearctic; SCIONIINI, N. and S. American, Australian; PHILOLICHINI, African, Oriental.)

– Antennae with three large segments, plus four or fewer terminal segments, making a total of seven (figs 98D, 100B) . . . . . Subfamily **CHRYSOPSINAE** (Three Tribes: BOUVIEROMYIINI, S. American, Australian, S. African; RHINOMYZINI, African, Oriental, S. American (1 species); CHRYSOPSINI, predominantly Holarctic and Oriental, penetrating S. of Equator in all continents except Australia).

4 Terminal segments of antennae reduced to three. Eyes (♀) widely separated, frons usually broader than high (fig. 99B) . . . . Tribe **HAEMATOPOTINI** (*Haematopota* is the only important genus: Palaearctic, Oriental and African, almost extinct in N. America.)

– Terminal segments of antennae almost uniformly four in number. Eyes (♀) nearly always close together, mostly frons higher than broad 5

5 Basicosta (subepaulet) of wing without macrotrichiae . . Tribe **DIACHLORINI** (Primitive genera, resembling BOUVIEROMYIINI, mostly Australian and N. and S. American.)

– Basicosta with microtrichiae (figs 98E, 99A) . . . . Tribe **TABANINI** (A complex of indistinct genera, '*Tabanus sensu lat.*' Well-defined segregates are *Atylotus*, Holarctic, African, Oriental; and *Hybomitra*, Holarctic. In N. and S. America a number of genera—*Chlorotabanus, Leucotabanus, Hamatabanus,* are fairly well defined, see Stone, 1965; Barretto, 1960.)

WORLD DISTRIBUTION

The family Tabanidae appears to have originated either in S. America or in an earlier subantarctic continent, and the primitive elements of the family have dispersed into other continents by a southerly route. Entomologically the real core of the family lies

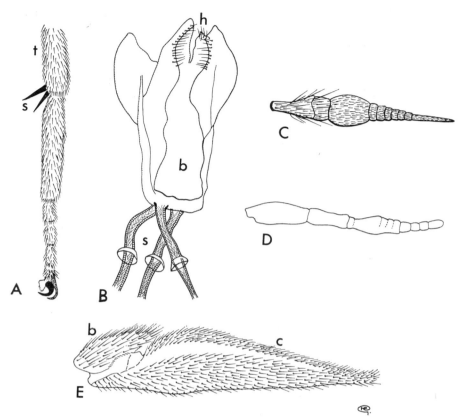

Fig. 98   A, hind leg showing tip of tibia (t) with spurs (s); B, bursa copulatrix of *Tabanus secedens*; s, chitinised bases of spermatheca; b, cavity of bursa; h, hair tufts; C, antenna of ♀ *Philoloche* (*Philoloche*) *aethiopica*; D, antenna of ♀ *Chrysops perpensa*; E, *Tabanus xanthomelas*, basicosta (subepaulet) with microtrichiae (b) and costa (c) with longitudinal groove.

in these members of the tribes PANGONIINI, SCIONIINI, BOUVIEROMYIINI, DIACHLORINI.   The tribe RHINOMYZINI, with larvae living in rot-holes in trees, is an African and Oriental offshoot from the southerly radiation.   Many of these primitive elements have lost the ability to suck blood, even in the female, and none is of medical importance, except possibly the Rhinomyzine genus *Tabanocella* in Africa and Madagascar.

Of more specialized groups S. America has the greatest concentration of closely-related genera, which are little known either taxonomically or biologically.   *Lepiselaga crassipes F.*, a small black horse-fly of the Amazon Basin, is often abundant and troublesome, but no S. American Tabanid is of any known medical importance.

The three well-known genera, *Chrysops*, *Haematopota* and *Tabanus*, which are the flies that come to mind when Tabanidae are mentioned, are offshoots from the S. American complex which have spread by a northerly route.   They appear to have arisen in association with the great evolution of ungulate mammals of the late Tertiary Period, and to have had their maximum success in the past.   Now, with the rapid decline of wild ungulates, and increasing aridity through climatic changes and through the effects of land drainage, cultivation and soil erosion, they are likely to decrease still further.

Haematopota must have flourished formerly in N. America, as did the tsetse-fly, *Glossina*, but *Glossina* has long vanished from there, and *Haematopota* is almost extinct there too.

LARVAE.   Tabanid larvae live in wet places, and are usually described as being aquatic.   They cannot, however, live indefinitely submerged, and need to reach the atmosphere periodically.   They breathe air by means of a posterior siphon, which varies in length and shape in different genera.   Their usual habitat is in the wet mud at the margins of streams, ponds, lakes and rivers; a few live in sand on the seashore, even between tidemarks.   Some of the most interesting biologically, though of little or no medical importance, breed in the wet debris that accumulates in rot-holes in trees.

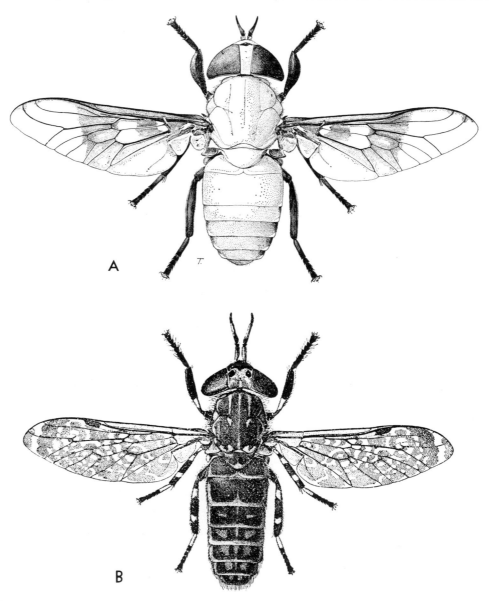

Fig. 99   A, *Ancala africana* ♀ × 4 (from Castellani & Chalmers); B, *Haematopota maculosifacies* ♀ × 9 (after Austen).

The larvae of *Haematopota* occur scattered through areas of wet soil, and some species of *Tabanus* are present in relatively dry areas such as chalk uplands, making use of dew ponds, or of pockets of damp soil, for larval development.

The larvae of a great many Tabanidae, including those of *Chrysops*, feed on vegetable debris, whereas the larvae of *Haematopota*, *Tabanus*, and a number of other genera, are carnivorous, feeding on other insects, crustacea, worms, and even their own species. Hence *Chrysops* larvae can be reared to the adult merely by keeping them collectively in the debris in which they were found; to do this with *Tabanus* or *Haematopota* results in only one large larva surviving in each container.

Tabanidae breed, therefore, in any area that is not entirely arid. The larvae are very long-lived, and in areas which have a prolonged dry season the adult flies emerge only at

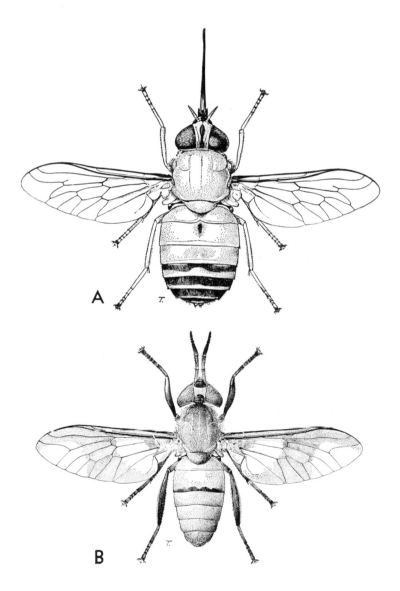

Fig. 100   A, *Pangonia magrettii* ♀, ×3; B, *Chrysops fixissimus* ♀, ×4.

a suitable season for subsequent egg-laying.   If the larva misses the opportunity to pupate in one year, by being insufficiently fed, for example, it may defer pupation for another year, or longer.   One or two species may survive desiccation by constructing cylinders of mud, and this habit may be more widespread than is at present known.

HABITAT

Tabanidae are most particularly flies of open woodland, and savannah country.   In open country they are most numerous near streams, rivers, lakes, or boggy areas suitable for breeding.   There both sexes may be found in abundance after a mass emergence, when they sit on the vegetation, hardening off and feeding on nectar and pollen.   Females leave the breeding sites to seek a blood-meal, which is usually necessary before being able to lay viable eggs, but not invariably so.   The flies follow moving objects, such as grazing animals, men on foot or on horseback, cars and even trains, up to about 40 km/h (25 mph).   Males are not normally attracted in this way, but often come to light at night.

In high equatorial forest, with a closed canopy, there are very many suitable breeding places, but relatively few Tabanidae make use of them.   The flies are not concentrated near breeding places as they are in open country, partly because the breeding sites themselves are more uniformly distributed, and partly because few species look for blood-meals on the forest floor.   Emergent flies of either sex usually fly up to the canopy, where mating swarms of male Tabanidae form above the treetops at dusk and dawn, and sometimes in between.   The females find a blood-meal among the monkeys of the canopy, but some readily descend to ground level to feed if they see movement below.   There is thus relatively little biting by horse-flies in the deep shade of the forest, but more risk of this in clearings, around villages, along forest roads, at rivers—especially at ferries where activity is concentrated—and at the forest fringe.   Female horse-flies will follow movement over water and attack people in boats and even on ships at sea within a mile or so of land.   A few species breed on the seashore, though some of these (the tribe Scepsidini) are incapable of biting.

MEDICAL AND VETERINARY IMPORTANCE

Tabanidae in large numbers draw an appreciable amount of blood, and also greatly disturb people or grazing animals.   The milk yield or the weight increase of grazing stock, and the output of workers, is materially reduced by heavy attacks from horse-flies. A horse-fly driven off by the tail of one animal often resumes its feeding from a different beast, and so is liable to transmit blood-parasites by mechanical contamination.   Although this risk exists with other biting flies—e.g. *Stomoxys*—Tabanidae are particularly dangerous because the spongy labella of the proboscis hold a greater amount of uncongealed blood.   Tabanidae are among the many mechanical vectors of such diseases as anthrax and anaplasmosis which may sometimes spread to man.   Only three pathogens, however, are especially associated with Tabanid vectors.

The trypanosome *T. evansi* is a blood-parasite of camels, which also affects horses, dogs, cattle and domestic buffalo, and which causes the disease *surra* (*gufar: el-debab*).

The disease is often fatal, and may cause serious economic losses in these domestic animals all over the area from the Sahara to the Philippines. Though any mechanical agent is a potential vector, Tabanidae are the principal vectors. No cyclical development takes place in the insect.

Tularaemia is a febrile disease, sometimes fatal, caused by the bacterium *Pasteurella tularensis*, and widespread in the Holarctic Region, both in N. America and in Eurasia. It is a disease of rodents, especially hares, rabbits and voles, and is transmitted mechanically. Fur trappers may become infected through handling and skinning these animals, or by eating insufficiently cooked meat from them. The common arthropod vectors are ticks and horse-flies, especially the deer-fly *Chrysops discalis* Will. in the U.S.A.

The only parasite that is known to be cyclically transmitted by Tabanidae is the nematode worm *Loa loa* Guyot in the African rain-forest. The adult worms live subcutaneously in monkeys and in man, and may survive for ten or more years. They move in the subcutaneous connective tissue and may become visible if they invade the conjunctiva of the eye, where they set up temporary irritation, but do not usually cause permanent damage. The principal symptom of loiasis is the occurrence of *Calabar Swellings*, hot, painful swellings of the joints, especially at the wrists and ankles. They last for a few days and then disappear, but may reappear after quite long intervals.

The adult *Loa*, if both sexes are present, may give rise to larval forms known as *microfilaria*, which circulate in the blood. They never develop beyond this stage in the blood of the original mammalian host, but need to be picked up by a biting *Chrysops*. In the insect host they develop into a second larval form, the *infective form*. This erupts from the proboscis of the *Chrysops* during a subsequent blood-meal, and enters a new monkey or man, where the infective larvae develop into adult worms, and live subcutaneously.

Infection therefore occurs only from the bite of a *Chrysops* that has previously fed on infected blood. It was at one time thought that the same species of *Loa* occurred in monkeys and in man, but there is evidence that the two may be different, and have different habits and different insect vectors. The number of adult *Loa* is limited to the number of larvae injected, and is less in practice because of the death of some larvae or adults. The symptom of Calabar swellings is thought to be linked with the death of a worm.

The vectors of *Loa* are species of *Chrysops* living in the canopy, and four species are concerned: *C. silacea* Austen; *C. dimidiata* Wulp; *C. centurionis* Austen and *C. langi* Bequaert. The biology and habits of these flies were worked out by a team from the Liverpool School of Tropical Medicine, operating at Kumba in the Cameroons. Kumba was a notorious focus of loiasis and this was found to arise from the situation of houses on a hill at the same level as the canopy of the lowland forest. Work by Duke (1955–1960) showed that *C. silacea* and *C. dimidiata* are active by day, and are attracted down to ground level by activity, and by the smoke of fires; they transmit human loiasis, the microfilariae of which appear in the peripheral skin by day. *C. centurionis* and *C. langi* are mainly crepuscular and nocturnal, feeding upon sleeping monkeys in the treetops, whose peripheral blood carries microfilariae at night.

This epidemiology compares with that of jungle yellow fever, first recognized as a distinct form of the disease in S. America, and later discovered in Uganda. The use of high towers in Uganda to study the mosquitoes of the treetops led to many new discoveries about the habits of Tabanidae (see Haddow *et. al.*, 1961).

BIBLIOGRAPHY

ANTHONY, D. W. 1962. Tabanidae as disease vectors. IV. *In* Maramosch K. *Biological transmission of disease agents* New York and London. pp. 93–107.

BARRETTO, M. P. 1960. Chave para os generos neotropicais de Tabanidas. *Papélis avuls. Zool. S. Paulo* **14**: 57–69.

CHVALA, M., LYNEBORG, L. & MOUCHA, J. 1972. *The Horse Flies of Europe.* Ent. Soc. Copenhagen. 500 pp. Copenhagen.

DUKE, B. O. L. 1955–1960. Studies of the biting habits of *Chrysops* I–VII. *Ann. trop. Med. Parasit.* **49**: 193–202, 260–272, 362–367, 369–375; **52**: 24–35; **53**: 203–214; **54**: 147–155.

FAIRCHILD, G. B. 1950. The generic names for Tabanidae (Diptera) proposed by Adolfo Lutz. *Pysche Camb.* **57**: 117–127.

—— 1956. Synonymical notes on Neotropical flies of the family Tabanidae. *Smithson. Misc. Collns.* **131**: no. 3, 33 pp.

—— 1961. The Adolfo Lutz collection of Tabanidae. *Mem Inst. Oswaldo Cruz* **59**: 185–249, 279–295.

FAIRCHILD, G. B. & PHILIP, C. B. 1960. A revision of the Neotropical genus *Dichelacera*. *Studia Ent.* **3** (1–4): 96 pp.

HADDOW, A. J. & others. 1961. Entomological studies from a high tower in Mpanga Forest, Uganda. *Trans. R. ent. Soc. Lond.* **113**: 249–368.

KRÖBER, O. 1934. Catalogo dos Tabanidae da America do Sul e Central, incluindo o Mexico e as Antilhas. *Revta. Ent., Rio de J.* **4**: 222–276, 291–333.

LECLERCQ, M. 1952. Introduction à l'étude des Tabanides et revision des espèces de Belgique. *Mém. Inst. roy. Sci. nat. Belg.* **123**: 1–80.

LUTZ, A. 1909. Sobre a systematica dos Tabanideos, subfamilia Tabaninae. *Brasil. Med.* **27**: 486–487.

MACKERRAS, I. M. 1954–1955. The classification and distribution of Tabanidae. *Aust. J. Zool.* **2**: 431–454; **3**: 439–511, 583–633.

—— 1957. Tabanidae (Diptera) of New Zealand. *Trans. R. Soc. N.Z.* **84**: 581–610.

—— 1964. The Tabanidae (Diptera) of New Guinea. *Pacif. Insects* **6**: 69–210.

MARCHAND, W. 1920. The early stages of Tabanidae (horse-flies). *Rockefeller Inst. Med. Res. Monogr.* **13**: 203 pp.

OLDROYD, H. 1954–1957. *Horseflies of the Ethiopian Region* I–III. 211+341+483 pp. London, B.M.(N.H.).

OLSOUFIEV, N. G. 1937. *Faune de l'U.R.S.S.* Insectes Diptères **7** (2). Tabanidae 433 pp. [In Russian, German keys.]

OTSURU, M. & OGAWA, S. 1959. Observations on the bite of tabanid larvae in paddy-fields (Diptera, Tabanidae). *Acta med. et Biologica* **7**: 37–50.

PHILIP, C. B. 1931. The Tabanidae (horseflies) of Minnesota, with special reference to their biologies and taxonomy. *Tech. Bull. Minn. Agric. Exp. Stn* **80**: 132 pp. (And many other papers on Tabanidae listed in STONE, A. *et al.* 1965.)

RICARDO, G. 1911. A revision of the species of *Tabanus* from the Oriental Region. *Rec. Ind. Mus.* **4**: 111–258.

—— 1911. A revision of the Oriental species of the genera of the family Tabanidae other than *Tabanus*. *Rec. Indian Mus.* **4**: 321–397.

SCHUURMANS-STEKHOVEN, J. H. 1926. The Tabanids of the Dutch East Indian Archipelago. *Treubia* **6** (suppl.): 551 pp.

STONE, A., SABROSKY, C. W., WIRTH, W. W., FOOTE, R. H. & COULSON, J. R. 1965. *A Catalog of the Diptera of America North of Mexico.* U.S. Dept. Agric. Handb. **276**: 319–342.

WRIGHT, R. E. 1970. Isolations of LaCrosse virus of the California group from Tabanidae in Wisconsin. *Mosquito News* **30**: 600–603.

4b. BRACHYCERA AND CYCLORRHAPHA ASCHIZA OF MINOR MEDICAL IMPORTANCE

by Kenneth G. V. Smith

BRACHYCERA

RHAGIONIDAE (SNIPE-FLIES)

THE family Rhagionidae (=Leptidae) contains several genera recorded as biting man.

Species of *Symphoromyia* are vicious biters occurring in the Palaearctic and Nearctic Regions, alighting quite silently and inflicting a painful bite before their presence is known. *Atherix* is another blood-sucking genus in the Nearctic and Neotropical Regions and in Australia *Spaniopsis* (fig. 101) and *Austroleptis* are troublesome.

The immature stages are little known, but the larvae are predaceous and habitats include moist soil with a high oxygen content and rotting wood, while *Atherix* is aquatic and *Vermileo*, a genus of non-medical importance, is the 'worm lion'.

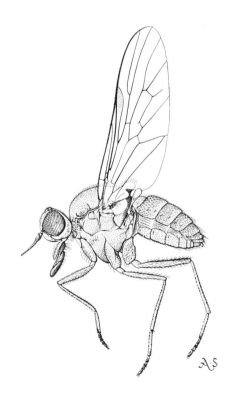

Fig. 101   *Spaniopsis longicornis,* ×8.

STRATIOMYIDAE (SOLDIER-FLIES)

This family is of no known medical importance in the adult stage, but occasionally the larvae of *Hermetia illucens* (L.) have been involved in cases of enteric pseudo-myiasis. The fly breeds in a variety of substances including animal and human cadavers (Malloch, 1917) and oviposits on overripe or decaying fruit and may thus be accidentally ingested.   On the credit side, the larvae of *Hermetia* have been shown to render faeces in privies too liquid for the optimal development of house-flies (Furman *et al.*, 1959). The larva is a typical terrestrial Stratiomyid, broad and rather flattened with a distinct narrow head, the surface is clothed with short hairs and some transverse rows of bristles (fig. 138E).

Although primarily an American species *Hermetia illucens* has been transported via ships to parts of the Australian region, Europe, Africa and Asia and is likely to establish itself wherever climatic conditions are favourable (Leclercq, 1969).

THEREVIDAE

The adults of Therevidae are of no known medical importance, but there are cases of the larvae being coughed up (Austen, 1912) or vomited (James, 1947).   The larva (fig. 139A, B) resembles that of *Scenopinus* (see below and Chapter 6) and has been recorded (Wilson, 1924) from cabbage plants and potato tubers, though normally they feed upon humus and leaf-mould in the soil.

SCENOPINIDAE (WINDOW-FLIES)

This family shares with the Anisopodidae the popular name of window-flies due to their frequent occurrence on windows indoors.   However, Scenopinids are blackish in colour and have their wings folded tightly over their backs.   The biology of the family is little known, but several species are known to be predaceous in the larval stage on the larvae of Dermestidae (Coleoptera) in food storehouses and museums.   They may also occur in carpets indoors, where they are predaceous upon the larvae of carpet moths. Dumitresco & Ionesco (1946) record accidental myiasis in the maxillary sinus involving a *Scenopinus* larva and Thompson *et al.* (1970) report a case of urinogenital myiasis.

The larvae (fig. 139C) are yellowish-white, slender, active creatures about 20 mm long when fully grown.   Many of the abdominal segments are subdivided so that they appear to be 17 abdominal segments and this should distinguish them from larvae of other insects found in similar situations.   Larvae of Therevidae (fig. 139A, B) are superficially similar, but may be distinguished by the differences in the structure of the internal head skeleton as given in the key in Chapter 6 (compare figs 139A, B and 139D, E).

CYCLORRHAPHA

PHORIDAE

Phoridae are small active flies, easily recognized by their rather hump-backed appearance and their characteristic venation (fig. 102A), though some genera have wingless species. Phoridae are not easily identified to species and specimens should be submitted to a specialist. The adults are not commonly encountered, but may sometimes occur in swarms in buildings. The 'coffin-fly', *Conicera tibialis* Schmitz, is associated with the human corpse after burial and may thus be encountered in medico-legal cases (see Chapter 17).

Larvae (fig. 143B) are more frequently seen than adults since they are found in accumulations of decaying organic matter and may frequent faeces. The larvae are dirty white in colour, about 5 mm. long and the integument has small fleshy tubercles on it. The posterior spiracles are situated on the top of two small processes towards the hinder end. The puparia (fig. 102B) are angular in shape, and have two horns protruding from the dorsal surface towards the anterior end and are usually found in situations near the material in which the larvae feed. The larvae of *Paraspinophora bergenstammi* Mik feed in the dried up curds in the bottom of unwashed milk bottles and the

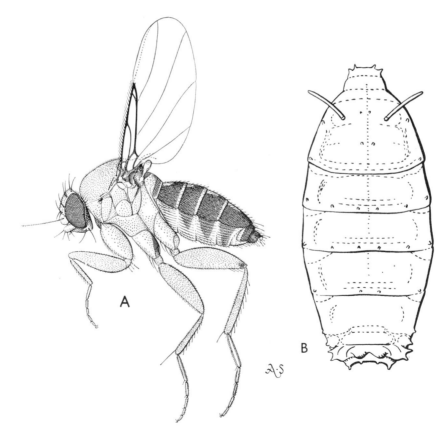

Fig. 102   *Megaselia scalaris*; A, adult ×25; B, puparium, ×25.

pupae may be found attached to the sides of the bottle. These pupae are firmly cemented to the glass and may survive normal washing processes.

Larvae may be ingested in food and then be passed in the faeces. *Megaselia scalaris* (Loew) is the species usually reported in such cases. Phoridae are almost world-wide in distribution. Robinson (1971) summarises the known biology of *Megaselia*.

SYRPHIDAE (HOVER-FLIES, DRONE-FLIES)

Flies of this family vary very much in size, shape and external form. Often they are brightly coloured, some black and yellow species strongly resemble wasps, while others are hairy and resemble bees. The wing venation is fairly characteristic having a 'vena spuria', or false vein (fig. 4D) and often the outer cells are closed, the veins forming a sort of false margin to the hind edge of the wing. Conspicuous as they are, the adults will seldom come before the medical entomologist, but some of the larvae may be found to cause intestinal myiasis.

The larvae of *Eristalis* are the well-known 'rat-tailed maggots' (fig. 142C) that are to be found in stagnant water heavily contaminated with organic matter. They may also occur in gutters or drains where dead leaves and water accumulate, and in sewage. Sometimes specimens occur in drinking water or in vegetable salads and may thus be ingested by human beings. There are several cases in which larvae have been passed in the stools (Mumford, 1926; Chagnon & Leclercq, 1949; Meissner, 1950; etc.).

The main characteristic of the larva is, as the name implies, the 'tail' which is in fact a long respiratory siphon at the tip of which are situated the posterior spiracles. This long telescopic siphon enables the larva to breathe at various depths when submerged in the liquid or semi-liquid media in which it lives. *Eristalis* species occur throughout the world. The larvae measure up to 2.5 cm. in length, exclusive of the 'tail'.

Fig. 103   *Eristalis tenax* female ×3.

E. tenax (L.) (fig. 103), the common drone-fly, is the commonest species, and is cosmopolitan.

Larvae of *Helophilus* species have also been recorded in cases of intestinal myiasis. They closely resemble *Eristalis*, but are distinguished by the undulating tracheal trunks which are straight in *Eristalis*.

Larvae of the subfamily Syrphinae particularly the genus *Syrphus* frequently occur on vegetation, where they prey on aphids and may be swallowed along with salads. In most terrestrial Syrphid larvae the posterior spiracles are usually situated together on short projecting tubercles.

BIBLIOGRAPHY

ALDRICH, J. M. 1915. The Dipterous genus *Symphoromyia* in North America. *Proc. U.S. natn. Mus.* **49**: 113–142.

AUSTEN, E. E. 1912. British flies which cause myiasis in man. *Rep. loc. Govt. Bd publ. Hlth* **66**: 5–15.

BORGMEIER, TH. 1963–1964. Revision of the North American Phorid Flies. *Studia ent.* **6**: 1–256; **7**: 257–416; **8**: 1–160.

—— 1968. A Catalogue of the Phoridae of the World. *Studia Ent.* **11**: 1–367. [Includes bibliography of all taxonomic works.] [Supplement, 1971, *Studia ent.* **14**: 177–224].

CHAGNON, G. & LECLERCQ, M. 1949. Myiase intestinale à *Eristalis tenax* L. (Dipt., Syrphidae). *Rev. méd. Liège* **4**: 634–635.

COLYER, C. N. 1954. The 'Coffin Fly', *Conicera tibialis* Schmitz (Dipt., Phoridae). *J. Soc. Br. Ent.* **4**: 203–206.

COOKSON, H. A. & OLDROYD, H. 1937. Intestinal infestation by larvae of a drone fly. *Lancet.* Oct. 2nd. **1934**: 804.

DIXON, T. J. 1960. Key to and descriptions of the third instar larvae of some species of Syrphidae (Diptera) occurring in Britain. *Trans. R. ent. Soc. Lond.* **112**: 345–376.

DUMITRESCO, M. & IONESCO, V. 1946. Un cas de penetration accidentelle de la larvae d'*Omphrale* sp. dans le sinus maillaire de l'homme. *Bull. Acad. Sci. Bucharest* **26**: 475–478.

FERGUSON, E. W. 1915. Descriptions of new Australian bloodsucking flies belonging to the family Leptidae. *J. Proc. R. Soc. N.S.W.* **49**: 233–243.

FURMAN, D. P., YOUNG, R. D. & CATTS, E. P. 1959. *Hermetia illucens* as a factor in the natural control of *Musca domestica* Linnaeus. *J. econ. Ent.* **52**: 917–921.

HALL, M. C. & MUIR, J. T. 1913. A critical study of a case of myiasis due to *Eristalis*. *Archs intern. Med.* **11**: 193–203.

HARTLEY, J. C. 1961. A taxonomic account of the larvae of some British Syrphidae. *Proc. zool. Soc. Lond.* **136**: 505–573.

HEISS, E. M. 1938. A classification of the larvae and puparia of the Syrphidae of Illinois, exclusive of the aquatic forms. *Illinois biol. Monogr.* **16**: 1–142.

HOY, J. B. & ANDERSON, J. R. 1966. Snipe flies (*Symphoromyia*) attacking man and other animals in California. *Proc. Pap. 33rd Ann. Conf. Calif. Mosquito Control Assoc.* **1966**: 61–64.

HULL, F. M. 1949. The Morphology and Inter-relationships of the genera of Syrphid Flies, Recent and Fossil. *Trans. zool. Soc. Lond.* **26**: 257–408.

JAMES, M. T. 1947. The flies that cause myiasis in man. *Misc. Publs U.S. Dep. Agric.* **631**: 1–175.

KELSEY, L. P. 1969. A revision of the Scenopinidae of the World. *Bull. U.S. natn. Mus.* **277**, pp. 336.

—— 1970. The Scenopinidae of Australia; including the description of one new genus and six new species. *J. Aust. ent. Soc.* **9** (2): 103–148.

LECLERCQ, M.   1969.   *Entomological parasitology.*   158 pp. London, New York, etc.

LEONARD, M. D.   1930.   A revision of the Dipterous family Rhagionidae in the United States and Canada.   *Mem. Am. ent. Soc.* **7**: 1–181.

MALLOCH, J. R.   1917.   A preliminary classification of Diptera exclusive of Pupipara, based upon larval and pupal characters, with keys to imagines in certain families.   Part I. *Bull. Ill. St. Lab. nat. Hist.* **12**: 161–407.

MEISSNER, E.   1950.   Die 'Rattenschwanzmade' der Schwirrfliege (*Eristalis tenax*) als Schmarotzer im menschlichen Darm.   *Med. Klin.* **45**: 1474.

MUMFORD, E. P.   1926.   Three new cases of myiasis in man in the north of England, with a survey of earlier observations by other authors.   *Parasitology* **18**: 375–383.

PARAMONOV, S. J.   1962.   A review of the Australian Leptidae.   *Aust. J. Zool.* **10**: 113–169.

ROBINSON, W. H.   1971.   Old and new biologies of *Megaselia* species (Diptera, Phoridae).   *Studia ent.* **14**: 321–348.

SCHMITZ, H.   1929.   *Revision der Phoridae.*   211 pp.   Dummler, Berlin and Bonn.

—— 1938–1965.   Phoridae in Lindner, E. *Die Fliegen der Palaearktischen Region* **7** (33). Stuttgart.

STUCKENBERG, B. R.   1960.   Diptera (Brachycera) Rhagionidae.   *S. Afr. anim. Life* **7**: 216–308.

THOMPSON, Jr., J. H., KNUTSON, L. V. & CUPP, O. S.   1970.   Larva of *Scenopinus* sp. (Diptera, Scenopinidae) causing human urinogenital myiasis.   *Mayo Clin. Proc.* **45**: 597–601.

VOCKEROTH, J. R.   1969.   A revision of the genera of the Syrphini (Diptera: Syrphidae).   *Mem. ent. Soc. Can.* **62**: 3–176.

WILSON, G. FOX   1924.   Contributions from the Wisley Laboratory. xlii. *Thereva plebeia* as a pest of economic importance.   *Jl. R.hort. Soc.* **49**: 197–202.

5a. GLOSSINIDAE
(Tsetse-flies)

by W. H. Potts

TSETSE-FLIES are two-winged flies (Diptera) of various shades of brown ranging from yellowish or greyish to dark or almost blackish-brown, sometimes with transverse black bands on the abdomen. They all belong to the genus *Glossina*, the sole constituent genus of the family Glossinidae (see key to families, Chapter 2). This genus has generally been included in the family Muscidae, being placed either in the subfamily Stomoxyinae or in a separate subfamily, the Glossininae; the whole question has recently been thoroughly examined and discussed by Pollock (1971).

The length of individual flies, exclusive of their proboscis, varies from about 6–8 mm. (about the size of a common house-fly, *Musca*), in the smaller species, to about 10–14 mm. (about the size of a bluebottle, *Calliphora*, though more slender and less robustly built), in the larger species.

They may be recognized in life by their habit of sitting with the wings folded scissor-like over the back, the tips of the wings then extending slightly beyond the end of the abdomen (fig. 104A), and by their long biting mouth-parts projecting forwards from underneath the head (figs 104A, 106A). Two other characters absolutely diagnostic for *Glossina* are the hatchet-shaped discal cell in the wing (fig. 105D) and the secondary branching of the hairs of the antennal arista, on the upper surface only of which the hairs are borne (fig. 120). Neither of these characters is found in any other known genera of Diptera. This peculiar shape of the discal cell, which lies between veins IV and V, is given by the upward curve of the former vein to meet the anterior cross vein that links vein IV with vein III; the cutting edge of the hatchet blade is formed by the portion of vein IV that runs between the anterior and posterior cross-veins. It was the shape of this cell, and the presence of the long biting mouth-parts, that provided the basis for the assignment of certain fossil insects from the Oligocene shales of Florissant, Colorado, to the genus *Glossina*.

FEEDING AND HOSTS

The mouth-parts consist of a proboscis or haustellum, sheathed by two maxillary palps equal to it in length, which lie along each side of it. The proboscis is needle-like and straight, arising from a swollen bulb-like base lying directly under the head (fig. 106A), an arrangement quite different from that found in *Stomoxys* (fig. 129A) or *Haematobia* (fig. 106B), in which the base of the proboscis is angled and not bulb-like. The proboscis of *Glossina* consists of three parts, the labrum-epipharynx, the labium and the hypopharynx (fig. 106C). These structures, which are described in detail by Newstead *et al.* (1924) and Buxton (1955), form a tube, the food channel, up which the blood of the host is sucked, and the hypopharynx, which is hollow and continuous with the main duct

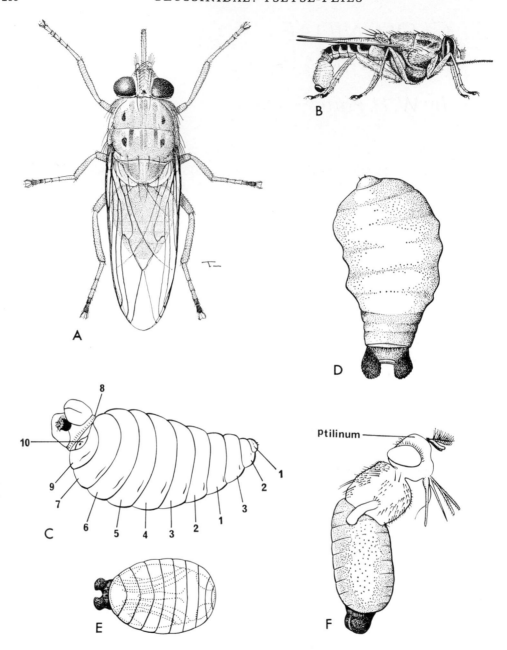

Fig. 104

A   Tsetse-fly (*G. longipennis*), resting attitude of the living fly, showing proboscis, (×4) (after Austen).

B   Female of *G. morsitans* expelling larva (about ×3) (after Newstead *et al.*, 1924);

C   Third-instar larva of *G. swynnertoni*, ventro-lateral aspect, diagrammatic, showing segmentation (about ×11) (after Burtt and Jackson, 1951);

D   Third-instar larva of *G. swynnertoni* at 3/8 sec. after beginning of peristaltic movement (about ×10) (after Burtt and Jackson, 1951);

E   Puparium of *G. morsitans* with pupa (dotted lines) inside (about ×7) (after Buxton);

F   Adult of *G. morsitans* emerging from puparium, showing frontal sac (ptilinum), by alternate expansion and contraction of which the fly breaks off the anterior end of the puparium and forces its way through the soil (about ×7) (after Buxton).

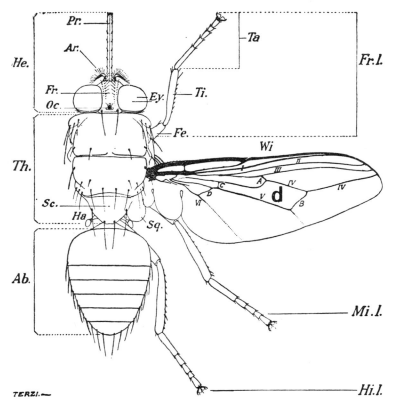

Fig. 105   Diagram of a tsetse-fly.   H*e*. head; *Th*, thorax; *Ab*, abdomen; *Fr*, l, fore-leg;
Mi. l, mid-leg; H*i. l*, hind-leg; *Pr*, proboscis with ensheathing palpi; *Ar*, arista;
Fr, frons; *Oc*, occiput; *Ey*, eye; *Sc*, scutellum; H*a*, haltere; *Sq*, squama; *Wi*, wing;
Fe, femur; *Ti*, tibia; *Ta*, tarsus.   Wing veins, etc: *I, II, III, IV, V & VI*, 1st, 2nd,
3rd, 4th, 5th and 6th longitudinal veins respectively; A, anterior cross-veins; B,
posterior cross-vein; C & D indicate other cross-veins, d, discal cell.   (From Castellani
& Chalmers, after Austen.)

of the salivary glands, forms another tube down which the salivary secretion is pumped
(fig. 106D).   This secretion contains an anti-coagulant which keeps the blood liquid
during its passage into the digestive portion of the alimentary canal.   The tip of the
labium, the labellum, is armed with rasps and teeth, which pierce the skin of the host
and which sever the capillaries in the tissues underneath, producing a pool of blood from
which the fluid is sucked up by the action of the muscular pharyngeal pump.   The
process is described in detail in Buxton (1955).   This method of feeding is known as
pool feeding, as opposed to feeding directly from the capillaries into which the proboscis
is inserted, as in mosquitoes, which, however, also use the pool-feeding method (Gordon
& Crewe, 1948).   These authorities regarded pool-feeding as the usual, if not the sole,
method used by *Glossina*, but in a later paper, Willett & Gordon (1957) record tsetse-flies
as also feeding directly from the capillary.   The blood of vertebrate animals is the sole
source of food for *Glossina*, both sexes feeding with equal voracity.   Serological tests
have shown that the different species have different host preferences.   Although most
species, if not all, will feed on a wide range of animals, including man, some are normally
more discriminative in their choice than others.   Thus one group feeds largely on

P

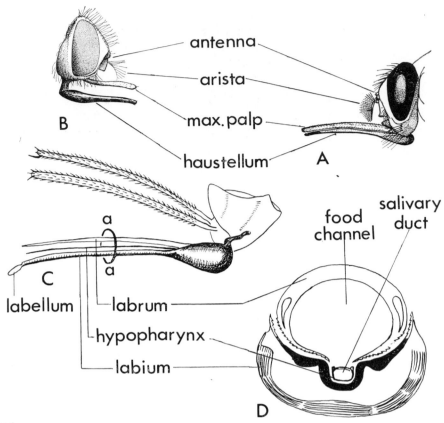

Fig. 106

A   Side view of head of *Glossina*, showing proboscis with bulb-like base, beneath paired palps (slightly raised), only the near member of which is shown (modified from Smart, 1965);

B   Side view of head of *Haemotobia irritans* (after Smart, 1965);

C   Details of proboscis and palps of *Glossina*, with palps separated from the haustellum (after Newstead *et al.*, 1924);

D   Cross section of haustellum of *Glossina* at a . . . a (in C), diagrammatic (after Newstead *et al.*, 1924).

antelopes and another on Suids (warthog, bushpig, etc.) and one species shows a predilection for the bushbuck and another for the rhinoceros, but another group is much more indiscriminate in its feeding, including birds and reptiles in its diet and feeding more readily on man, who is not much favoured by many species (Weitz, 1963 and in Chapter 13 of Mulligan, 1970; Ford, 1971, pp. 50–53, 297–299, 454–455).

The eyes are large, brown or sometimes reddish, and widely separated in both sexes (which are nevertheless readily distinguished by the presence of a knob-like hypopygium underneath the end of the abdomen of the male and by the absence of any such knob in the female (fig. 107A, B). Sight plays a large part in the finding of its host and resting places by the tsetse, which are largely diurnal in their activities, although individuals may occasionally be encountered during the night; there appear to be no signs of any adaptation of the eyes to nocturnal vision.

The olfactory sense probably also plays a part in host-finding, more so in some species than in others; there is some evidence that this sense may be associated with the antennae.

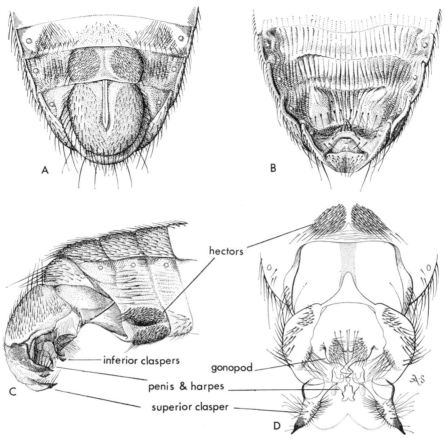

Fig. 107

A   Hind end of abdomen of male *Glossina* viewed from beneath, showing knob-like appearance of hypopygium drawn up into the abdomen;

B   Hind end of abdomen of female *Glossina* viewed from beneath, showing absence of any knob-like hypopygium;

C   Hind end of abdomen of male *Glossina* with hypopygium extended, viewed from a ventro-lateral position;

D   Hypopygium of *Glossina* after maceration and flattening under a cover slip.

REPRODUCTION

In its method of reproduction, *Glossina* is somewhat unusual amongst Diptera in being viviparous, a condition shared by few other families. The particular form this takes in *Glossina* is known as adenotrophic viviparity, because the egg contains enough yolk for the embryo to complete its development, and the larva is nourished in the uterus by material derived from the mother. The paired ovaries each contain only two ovarioles and these discharge their fully developed eggs alternately into a common oviduct, which enlarges to form a uterus. This method of discharge, alternately from each ovary, provides regularly succeeding cycles of ovarian development that afford a means of determining the age of the female for up to about 80 days. (For a full account of this see Saunders in Mulligan, 1970, Ch. 14.) The sperms are transferred to the female in a spermatophore (Pollock, 1970) and are stored in two spermathecae, from which they pass down ducts that unite to discharge a common opening into the anterior end of the uterus.   One act of mating is sufficient to render a female fertile for life, and though there is evidence that more than one such act may take place, this possibly does not occur very often in nature.

Fertilization is controlled by regulation of the sperm supply from the spermathecae by means of a sphincter-like valve at the opening of the spermathecal duct, which is opened only at ovulation, remaining closed during the rest of the ovarian cycle.   It is also controlled by the fact that the micropyle of the egg comes to rest opposite the opening of the spermathecal duct only when ovulation is complete, its correct position being determined by the choriothete (Roberts, 1971a).

The formation of the male genital organs and spermatogenesis in *Glossina tachinoides*, *G. morsitans morsitans*, *G. austeni* and *G. fuscipes fuscipes* are fully described for the first time by Itard (1970).

The egg hatches in the uterus, the shell or chorion being ruptured by an egg-tooth (Jackson, 1948; Roberts, 1971b) at the anterior end of the first-instar larva.   The broken remains of the chorion and subsequently of the first- and sometimes of the second-instar exuvia are often to be found associated with the choriothete, an organ situated towards the anterior end of the uterus on the inside of its ventral wall.   This organ was described by Jackson (1948) and Bursell & the late Jackson (1957), who thought its function was to remove the chorion and later the exuvium of the first-instar larva from the emerging larvae.   Recently, Roberts (1971) has shown that the function of this organ is to form a platform holding the embryo and the first instar larva, which are considerably smaller than the uterine chamber, firmly against the walls of the uterus.   Roberts suggests that the secretory cells of this organ produce a cementing fluid, rather than the chitinase suggested by Hoffman (1954), as no evidence of lysis of the chorion and exuvia has been found.   Subsequently, the second- and third-instar larvae are of sufficient size to be held in position firmly by several radially circular ridges on the inside of the uterine wall.

The larvae develop one at a time in the uterus, being nourished by a fatty fluid discharged into the uterus from specially developed accessory glands (the 'milk' glands) situated in the abdomen.   The anterior of the ridges in the uterine wall and the securing of the embryonic and early larval stages firmly against the uterine wall serve to hold the secretion of the milk glands in a pool round the mouth of the developing larvae; see Roberts (1971b).   Larval development is completed in the uterus, and the fully developed third-instar larvae are deposited at intervals of 8–25 days, the length of the interval depending on the temperature; the general mean is about 10–12 days at the temperatures at which the flies usually breed.   The larvae quickly burrow into the soil, becoming quiescent within a few inches of the surface.   The third-instar skin is transformed into an almost black sclerotized shell, the puparium, in which pupation and metamorphosis to the adult stage take place.   These processes may take a further 20–30 days at the temperatures usually experienced in the breeding sites of the fly in the field, though they may be prolonged to 80 or even 100 days at low temperatures.

The puparium (see fig. 104E, pl. 3B) is characterized by two prominent rounded black lobes that form its posterior end.   These are commonly referred to as respiratory lobes, and are present in the third-instar larva, in an un-melanized form, and only greyish in colour; it is, however, only during the development of the larva in the uterus that they have any respiratory function, and possibly particularly during its short active extra-uterine life; during the development of the pupa and the metamorphosis of the imago inside the puparium they perform no respiratory function (Bursell, 1955; Roberts, 1971c). Their structure is described in detail in Newstead *et al.* (1924) and their transformation from the respiratory lobes of the third-larval instar by Bursell (1955). These are the very much modified stigmatic plates of the third-instar Dipterous larva, adapted to serve the respiratory needs of the larva during its intra-uterine life.

BREEDING SITES AND HABITAT

The puparia are to be found in numbers in a variety of sites, generally where the soil is dry, loose and friable. Such 'traditional' sites are commonly to be found under fallen trees or decumbent tree trunks, at the bases of thicket stems, under overhanging rocks and in tree holes (either hollows at the bases of the trunks or in rot holes, even up to a height of 4·5 m (15 ft) or more). Nevertheless, it is likely that large numbers of larvae are deposited in shaded sites scattered throughout the floor of the woodlands through which the females range, and it is likely that such scattered sites may make a larger contribution to the general tsetse population than do the restricted sites in which they can profitably be collected. Recent investigations in Northern Botswana (Atkinson, 1971) have given ample evidence confirming this.

Although it is difficult to find the puparia in any quantities during the rainy season, there is no clear indication of any particular breeding season. Though reports have been made from one area that breeding is less in the rains, in another an increase has been claimed at that season; such discrepancies may well arise through one area lying at the wet extreme of the range of a species and another at its dry extreme.

The general habitat of the tsetse is woodland or forest, the shade from trees interspersed with thicket patches being essential for the survival of the flies. It is generally believed that different portions of their woodland environment are put to different use by the tsetse, certain areas, more open, being used as feeding grounds, and others, with closer tree and thicket cover, as resting areas or breeding grounds. There is, however, no clear indication that definite breeding grounds exist, in the sense of areas of any appreciable extent specially used by tsetse-flies for breeding in, though certain vegetation associations may well contain greater concentrations of the restricted 'traditional' types of breeding site enumerated above than do other associations. Much has been written on this subject of the relation of tsetse-flies to vegetation (see Buxton, 1955), but it is difficult to make any satisfactory general statement about this. The most one can say is that although flies may range far into, or even completely through, areas of uniform vegetation, it would seem that they do need a variety of diverse vegetation associations for the full satisfaction of their requirements, and they do seem to be especially associated with lines of contact between one type of vegetation and another. However, recent investigations (Pilson & Pilson, 1967; Bursell, 1966; also Bursell in Mulligan, 1970, Ch. 16) have thrown some doubt on the reality of the 'concentrations' associated with certain types of vegetation that have been suggested by some methods of sampling tsetse populations. Nevertheless, alterations of the vegetative cover of such 'concentrations', or restriction of the application of insecticides to them, have resulted in great reductions in the numbers of tsetse over wide stretches of country, even if not always eliminating them. However, different species of *Glossina* can certainly be associated with different types of vegetation, and an attempt has been made in Table 5 to indicate knowledge of this in summary fashion.

In the headings defining the types of vegetation in this table, 'forest' is used to cover associations of trees, mainly evergreen and with a continuous canopy, and with an under-storey of bushes and shrubs without any grass growth, except where there are gaps in the top canopy; there may be one or more storeys of canopy, afforded by trees of different heights, between the shrub layer and the canopy given by the tallest trees; lianas, strag-gling up the trees, and epiphytes are a frequent feature. By 'savanna woodlands' are indicated the grassed woodlands with trees generally so spaced as to allow easy passage in any direction; the canopy is seldom quite continuous and there are no definite storeys of canopy, except those provided by an intermittent under-storey afforded by scattered groups of bushes and shrubs and by the taller trees. The five habitat categories of this table must not be regarded as being adhered to too strictly by the species of tsetse, as individuals and populations of a species may often be found in atypical habitats, especially in those adjacent to characteristic ones.

The following notes amplify the information given in the table and particularly draw attention to instances of the occurrence of various species in atypical habitats.

1. *G. caliginea* has only been found near surface water, as are generally the species in category III and in parts of its range it shows a special association with mangrove swamps.

2. *G. pallicera* is often associated with the presence of surface water, but is sometimes found at some distance from it.

3. *G. brevipalpis* is also found throughout the savanna woodlands of East Africa, but always based on riverine and stream thickets where these contain a fair proportion of evergreen elements, though it is not dependent on the presence of surface water; it may be found either with *G. fuscipleuris* (category II), on the one hand, and on the other with *G. longi-pennis* (category V) as well as with *G. pallidipes* (category IV).

4. *G. palpalis* has, exceptionally, been found well away from surface water, in semi-artificial sites (e.g. in Southern Nigeria).

5. Established populations of *G. fuscipes* have also been found in situations quite far from surface water, generally in semi-artificial habitats and even under peridomestic conditions (e.g. in the Nyanza region, Kenya).

6. *G. tachinoides* has recently been found in natural situations well removed from surface water, and in semi-artificial habitats (e.g. in cultivation and mango groves) and peri-domestic situations (e.g. in villages, feeding on the domestic pigs).

7. *G. longipalpis* and *G. pallidipes* are both also found based on forests fringing permanent surface water and lakes as well as in the vegetation of the banks of seasonal pools and rivers and streams that are dry for most of the year, but they are not, as the species of category III generally are, dependent on the proximity of permanent water.

8. *G. morsitans* and *G. swynnertoni* are essentially insects of the dry savanna woodlands of East Africa, and so closely associated with the wild ungulates of those woodlands that they are often referred to as the 'game tsetse'; they still, however, have seemed generally to be ultimately dependent, for their survival through the dry season, on the shelter afforded by the heavier vegetation of the water courses that are dry for most of the year and in which water only flows intermittently during the rainy periods; lately, investigations in Rhodesia have thrown some doubt as to the reality of this dependence, at least in the area under investigation.

9. *G. longipennis* is an anomalous species in that, although a member of the group of tsetse-flies often termed 'the forest tsetse-flies' that form the major constituents of categories I and II, it is found in more arid conditions than any other species of *Glossina* (except possibly *G. pallidipes*); it is found both in dry grasslands with patches of bushes and trees, and associated with thicket patches on the banks of seasonally dry streams and rivers, or it may even be found together with *G. brevipalpis* in the drier parts of the range of that fly.

Table 5 Habitats of *Glossina* species

I	II	III	IV	V
Species of the rain-forest, including swamp forest & mangrove forests	Species of the forests surrounding the rain-forest & of relict forest patches in savanna woodlands	Waterside species of the West and Central Forests & of the riverine & lake-side fringing forests of rivers & lakes of the surrounding savanna woodlands	Species based on dry thickets (always with some evergreen elements) and secondary scrub	Species of the savanna woodland
G. tabaniformis	G. brevipalpis[3]	G. palpalis (s.l.)[4]	G. longipalpis[7]	G. morsitans (s.l.)[8]
G. haningtoni	G. fuscipleuris	G. fuscipes (s.l.)[5]	G. pallidipes[7]	G. swynnertoni[8]
G. nashi	G. medicorum	G. tachinoides[6]		G. longipennis[9]
G. vanhoofi	G. severini			
G. caliginea[1]	G. schwetzi			
G. pallicera[2]	G. austeni[10]			

G. fusca
G. nigrofusca

10. *G. austeni* is another anomalous species in that morphologically it is a member of the *morsitans* group, though a somewhat aberrant one, whereas it is an inhabitant of forests. It has been suggested that it is a tsetse that, having adapted itself to living under arid conditions, it has made a secondary return to moister ones (Bursell, 1958).

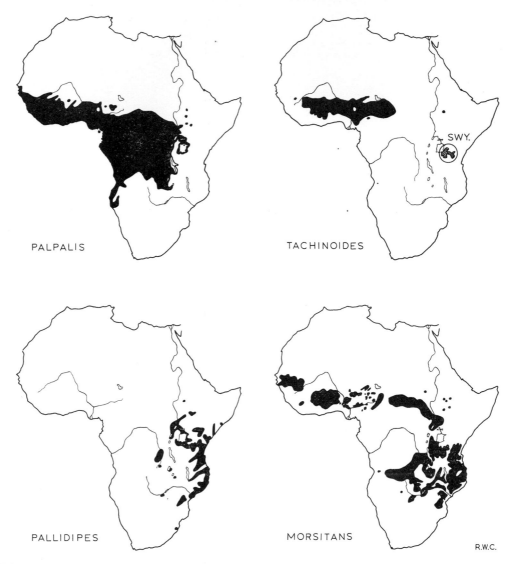

Fig. 108 Approximate distribution of the tsetse vectors of human sleeping sickness. N.B. the disease distribution is *not* co-extensive with the vector distribution (see fig. 109). (N.B. *palpalis* is not distinguished from *fuscipes* in this map.)

GEOGRAPHICAL DISTRIBUTION

The genus *Glossina* is now confined to subtropical and tropical regions of Africa, from latitude 5°N. to 20°S., except for the record of one species (*G. tachinoides*) having been found in the south-west corner of Arabia in 1910, from which area it has never been reported again. Indications of the political territories comprising the range of each species are given in the key for the identification of species.   Fig. 108 shows the distribution of the tsetse vectors of human sleeping sickness.   In Table 6 an attempt is made to show in summary fashion the distribution of the 22 species currently recognized by setting out the taxa (species, sub-species and other infra-specific forms) that occur characteristically in the West African, in the Central African and in the East African regions in three columns respectively, with the northern and southern limits of their ranges.   Some taxa are considered to belong essentially to two of the regions instead of being limited to one only, and others extend to a greater or lesser degree into adjacent

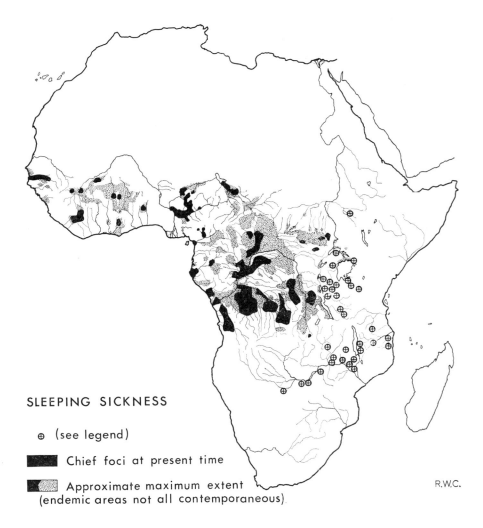

SLEEPING SICKNESS

⊕ (see legend)

■ Chief foci at present time

▨ Approximate maximum extent
(endemic areas not all contemporaneous)

R.W.C.

Fig. 109   The distribution of human sleeping sickness. Solid black areas represent the chief foci according to recent evidence, and the black and stippled areas taken together approximate to the maximum known historical extent of the endemic areas of *T. gambiense* sleeping sickness (but note that not all foci shown were in existence at the same time). Circle-symbols represent diffuse areas of virulent infection with *T. rhodesiense* sleeping sickness that developed during the 20th century (maximum extent). (Map compiled from maps 2 and 3 of Duggan, in Mulligan, ed., 1971, with additional Ethiopian locality—northernmost circle-symbol—from McConnell *et al.*, 1970.)

regions.   The extent to which the latter occurs is shown by placing the figures showing the northern and southern limits of the taxa in parentheses.   The political territories comprising the three regions are as follows, the allocation being essentially that used by Buxton (1955), which follows the definitions accepted by the International Conference on Tsetse and Trypanosomiasis held at Brazzaville in February, 1948.

I. *West Africa:* the former French West Africa [comprising the territories now known as Mauritania, Senegal, Republic of Guinea (formerly known as French Guinea), Mali (French Sudan), Ivory Coast, Upper Volta, Niger, Togo, Dahomey]; the former British West African Territories [comprising the Gambia, Sierra Leone, Ghana (formerly the Gold Coast), Nigeria (as formerly, but with the northern part of the former British Cameroons added as the Sardauna Province)]; Portuguese Guinea; Liberia.

II. *Central Africa:* the former French Equatorial Africa [comprising the territories now known as Gabon (formerly Gaboon), Republic of Congo (Brazzaville) (formerly the Middle Congo), Central African Republic (formerly Oubangui-Shari) and Chad]; the Cameroon Republic [comprising the former French Cameroons and the southern part of the former British Cameroons]; the Equatorial Republic of Guinea (formerly Spanish Guinea); the Congo (Kinshasa) (formerly the Belgian Congo); and Angola.

III. *East Africa:* Ethiopia (formerly Abyssinia); the Somalilands [comprising the Somali Republic (composed of the former British and Italian Somalilands) and the French Territory of the Afars and Issas (formerly French Somaliland)]; the Sudan Republic (formerly the Anglo-Egyptian Sudan); Uganda; Kenya; Tanganyika (now with Zanzibar known as Tanzania); Zambia and Rhodesia (formerly known as Northern and Southern Rhodesia, respectively); Malawi (formerly Nyasaland); Mozambique, sometimes referred to as Portuguese East Africa; and the former High Commission Territories [comprised of Lesotho (formerly Basotuland), Botswana (formerly Bechualand), and Swaziland].

The infra-specific forms of *Glossina morsitans* included in the table are those now recognised by Machado (1970).   The western occurrences of *G.m. submorsitans* are considered to be a form of this subspecies (20b) corresponding to the *G.m. ugandensis* of Vanderplank (1949) and designated as *G.m. submorsitans* form *ugandensis*.   The subspecies *G.m. morsitans* of Vanderplank (1949) cannot be regarded as belonging to this form which constituted the type as originally described by Westwood, which has been examined by Machado, who pronounced it to be a specimen belonging to the form *G.m. orientalis* as described by Vanderplank (1949); for the latter form, which Machado recognises as a subspecies the name *G.m. centralis* is proposed, and the *G.m. orientalis* of Vanderplank becomes a synonym of *G.m. morsitans* Westwood.

From Table 6 it would appear that, of the 30 taxa shown therein, 7 are essentially West African (1a, 7[@], 13a, 15a[@], 16[@], 17[@], and 18[@]), 5 of them (marked with [@]) extending to a greater or less degree into the Central region and one of them (17) even as far as the Eastern region as well; 5 can be regarded as both Central and Western taxa (1b, 2, 5a, 13b and 20a[@]), the last extending into the East African region also, and one only as a Central and East African form (20b); 11 are characteristically Central African (5b[@], 6[@], 8, 9, 10[@], 11, 12[@], 14a[@], 14b, 14c[@], and 15b), 6 of them also extending into one or other of the adjacent regions; the remaining 6 are East African (3, 4[@], 19[@], 20c, 21 and 22), two only entering into the Central region.

Table 6 Distribution of *Glossina* species, Northern and Southern limits of the taxa in:

		West Africa	Central Africa	East Africa
1a	G. fusca fusca	10°N–4°S		
1b	G. fusca congolensis	7°N–3°S	5°N–9°S	
2	G. tabaniformis	7°N–2°S	5°N–7°S	
3	G. longipennis			7°N–4°S
4	G. brevipalpis		((3°–10°S))	3°N–27°S
5a	G. nigrofusca nigrofusca	7°–2°N	5°–1°N	
5b	G. nigrofusca hopkinsi		5°N–10°S	(((1°N–0°)))
6	G. fuscipleuris		7°N–8°S	((1°N–2°S))
7	G. medicorum	10°–3°N	(((3°–5°S)))	
8	G. severini		0°–8°S	
9	G. schwetzi		3°–12°S	
10	G. haningtoni	(((5°N–3°S)))	5°N–6°S	
11	G. vanhoofi		2°N–2°S	
12	G. nashi	(((4°N)))	4°N–13°S	
13a	G. palpalis gambiensis	15°–4°N		
13b	G. palpalis palpalis	9°–2°N	5°N–13°S	
14a	G. fuscipes fuscipes		8°N–8°S	((8°N–3°S))
14b	G. fuscipes quanzensis		2°N–10°S	
14c	G. fuscipes martenii		3°–12°S	(((9°S)))
15a	G. pallicera pallicera	9°–2°N	(((4°–2°N)))	
15b	G. pallicera newsteadi		4°N–9°S	
16	G. caliginea	7°–2°N	((4°N–2°S))	
17	G. tachinoides	13°–4°N	(11°–4°N)	(((12°–7°N)))
18	G. longipalpis	13°–4°N	((5°–1°N))	
19	G. pallidipes		(((3°N–9°S)))	7°N–20°S
20a	G. morsitans submorsitans	14°–5°N	10°–2°N	((10°N–0°))
20b	G. morsitans 'centralis'		5°–17°S	2°–19°S
20c	G. morsitans 'orientalis'			5°–21°S
21	G. swynnertoni			2°–5°S
22	G. austeni			10°N–26°S

(Adapted from the tables given by Smart, 1965, and Buxton, 1955 and amended in accordance with the details given in Key B.)

The parentheses indicate spread from the regions considered characteristic for the species or infra-specific form concerned into an adjacent region.

( ) indicates occurrence in a large portion or considerable portions of the adjacent area, but never to the same degree as in the region or regions regarded as characteristic of the taxon concerned.

(( )) indicates occurrences in a number of localities or over a moderately large portion of the adjacent area.

((( ))) indicates occurrence in only a few localities or in only a small portion of the adjacent area.

SYSTEMATIC POSITION AND IDENTIFICATION OF SPECIES

The genus *Glossina* belongs to the family Glossinidae (see Key to the Families, Chapter 2). This family is regarded by some as a sub-family of the Muscidae. It consists of some 20 species, the exact number depending on the status accorded to certain forms which are regarded by some as full species and by others as taxa of lower rank. In this work, the number of species recognized is 22, six of which are regarded as being constituted by 2, 2, 2, 3, 2, and 3 subspecies respectively, as shown in Table 6. These species fall into three well-defined groups, which have been regarded as having the status of subgenera, the names of which are given in parentheses in the Key A (couplet 1). These three groups correspond very broadly with the grouping of the species according to the habitats in which they are found as shown in Table 5 (p. 217) those of the *fusca* group constituting the species found associated with forest habitats (columns I & II in the table), those of the *palpalis* group the species associated with water-side habitats (some of columns I & all of III) and those of the *morsitans* group with the savanna woodlands and associated thicket formations (columns IV & V).

The keys below are based on the key given by Potts in Mulligan (1970, Ch. 9). This has been modified and divided into three parts, so that the reader may identify most of the 22 species of *Glossina* by the external characters alone (key A), the males of all the species by examination of the male terminalia (key B) and the females of the *fusca* and *palpalis* groups by examination of the signa or genital plates (key C). Specimens that prove difficult to identify by means of key A can be identified by means of keys B and C (except for females of the *morsitans* group, which are, however, generally easily identifiable in key A). Identifications are often difficult to arrive at without authoritatively named specimens for comparison, and doubtful specimens should be submitted to taxonomists experienced in the group for confirmation.

The terms used in the key for various structures will generally be readily understood from the figures, but definitions of some that may still prove difficult are given in the following list:

aedeagus: a sclerotic tube forming principal part of male intromittent organ.

external genital armature (of females): chitinized plates surrounding the anal and genital openings.

genital fossa: a thick gelatinous structure in the dorsal wall of the uterus towards its anterior end (see fig. 113).

harpes: chitinized plates and appendages in the male terminalia functioning as clasping organs. The structure described and figured by Newstead *et al.* (1924) as the harpes of *G. severini* (his fig. 29, p. 81) is regarded by Machado (1959, pp. 37 & 38 & fig. 27) as an unpaired sclerite underneath the gonopore; this latter worker considers the true harpes of this species to be an unidentified part of the structure labelled by Newstead *et al.* as the 'juxta'; the harpes of fig. 117A is *sensu* Machado.

hectors: the two lobes of a median thickened plate, covered with short stiff hairs, on the sternum of the 5th segment, probably helping to hold the female during coupling see fig. 107C, D.

inferior claspers (of males): appendages of male genital armature, see fig. 107C; these appendages have been referred to by other authorities variously as parameres and gonopods.

infra-alar bulla: a roughly oval protuberance situated on the side of the mesothorax, anterior of top upper angle of the pteropleuron, just in front of the base of the wing (see figs 11A, 122).

pleuron (-a, plural; **-al,** adj.): sides of body, between **tergum** (dorsal surface of the body) and **sternum** (ventral surface of body).

pteropleuron: the sclerite on the pleuron below the base of the wing (fig. 122).

signum (-a, plural): a chitinized plate on the inner side of the genital fossa (fig. 113).

sternopleuron: lower part of the pleuron (fig. 122).

superior claspers (of males): terminal appendages of the males (see fig. 107C, D).

tergite: a dorsal chitinized plate or sclerite.

thoracic squama (-ae, plural): lobe attached along one side to pleuron, above the haltere; called by some authorities the calypter (fig. 122).

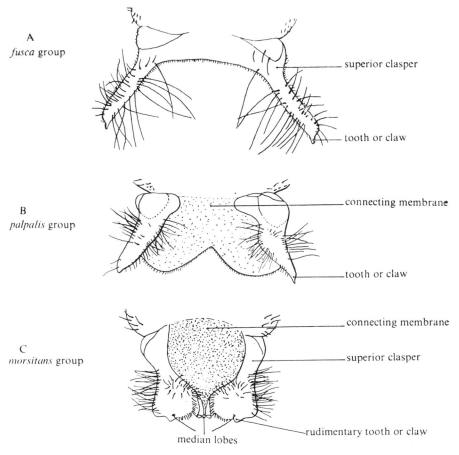

Fig. 110   Diagrammatic representation of typical forms of superior claspers of *Glossina* illustrating the three groups into which the genus is divided (from Newstead, Evans and Potts, 1924).

Preparations of the male terminalia and female genital armatures and signa suitable for examination microscopically are generally necessary for sure identification of the species. These may be made by macerating the abdomen, or the dissected hypopygium of the male, in 10% caustic potash, followed by thorough washing in distilled water. Permanent mounts can then be made by dehydration in successively greater strengths of ethyl alcohol, clearance in xylol and finally mounting on a slide under a coverslip in Canada balsam. A detailed description of the processes involved has been given by Potts in Mulligan (1970, Ch. 19).

Specific identifications of *Glossina* are sometimes difficult to arrive at without authoritatively named specimens for comparison, particularly with members of the *fusca* group, and doubtful specimens should always be submitted to taxonomists experienced in the group for confirmation of their identity.

The four infra-specific forms mentioned on p. 220 can, according to Machado (1970), be distinguished from one another by various characters, principally in the male terminalia, but the recognition of these is a matter for the expert, and the forms have therefore all been included under *G. morsitans* in the keys and elsewhere.

KEY FOR THE IDENTIFICATION OF ADULTS OF THE SPECIES OF *GLOSSINA*

A. By external characters (visible with naked eye or pocket lens).

1 Pteropleuron (figs 11A, 122) bearing a few strong bristles of size equal to those on the sternopleuron, these bristles being clearly distinct from the general vestiture of the shorter setulose hairs on the pleura; hairs fringing the thoracic squamae (figs 11A, 122) curly and numerous, giving a woolly appearance; large to medium flies (9½–14 mm.)   .   .   .   .   .   **fusca** group (subgenus *Austenina*)   2

− Pteropleuron bearing only setulose hairs, of which some may be longer than others, but none equal in size to the clearly differentiated bristles that project from amongst the setulose hairs of the sternopleuron; hairs fringing the thoracic squamae not curly but giving a neat fringe-like appearance; medium to small flies (6½–11 mm.)
palpalis and *morsitans* groups (subgenera *Nemorhina* and *Glossina s.s.*)   13

2 Palps shorter than width of head, or not exceeding it by more than a ninth of their length   3

− Palps longer than width of head by a sixth to a third of their length   .   .   .   6

3 Ground tint of wings dusky; antennal fringe a quarter to a third of greatest width of antenna   .   .   .   .   **tabaniformis** Westwood

− Ground tint of wings pale; antennal fringe a fifth of greatest width of antenna or less   4

4 Dorsum of thorax with a conspicuous dark brown spot towards each corner; a pale species (generally light yellowish-brown); under side of bulb of proboscis with dark apex   .   .   .   .   **longipennis** Corti

− Dorsum of thorax without any conspicuous dark brown spots; general colour greyish to dark brown; under side of bulb of proboscis uniformly coloured   .   .   .   5

5 Anterior cross-vein of both sexes with thickened portion strongly chitinized and darkened, forming a dark spot on the wing (fig. 121A)   .   **brevipalpis** Newstead

− Anterior cross-vein of female only showing above dark spot   .   **schwetzi** Newstead & Evans

− Anterior cross-vein of neither sex showing dark spot (as in fig. 121B)   **medicorum** Austen

6　All segments of hind tarsi uniformly dark dorsally　.　.　.　.　.　.　.　7
–　Only last two segments of hind tarsi dark dorsally, contrasting with paler coloration of
　　remaining segments　.　.　.　.　.　.　.　.　.　.　.　.　8

7　Last two segments of fore and middle tarsi pale, or at most showing some darkening
　　at distal margins; pleura and hind coxae fuscous grey; third segment of antenna
　　strongly and gradually recurved at tip, as in *G. nigrofusca* (fig. 120B)
　　　　　　　　　　　　　　　　　　　　　　　　severini* Newstead

–　Last two segments of fore and middle tarsi dark, penultimate at least with dark band
　　at distal extremity and last segment entirely dark dorsally, forming a marked
　　contrast to the remaining tarsal segments; third segment of antenna with a blunt
　　tip, only slightly and abruptly recurved, as in *G. fusca* (fig. 120A)　　**nashi*** Potts

8　Antennal fringe half to three-quarters of greatest width of third antennal segment;
　　hind tibia with broad dark suffusions in middle and much less distinct one at
　　apex　.　.　.　.　.　.　.　.　**nigrofusca nigrofusca** Newstead
–　Antennal fringe less than a quarter of greatest antennal width; hind tibiae with or
　　without dark suffusions　.　.　.　.　.　.　.　.　.　.　9

9　Infra-alar bulla (figs 11A, 122) dark brown to fuscous, without any pale vertical streak
　　in centre　.　.　.　.　.　.　.　.　.　.　**fuscipleuris** Austen
–　Infra-alar bulla testaceous, often with pale vertical streak in centre　.　.　.　10

10　Antennal fringe about a fifth of greatest width of third antennal segment
　　　　　　　　　　　　　　　　　　　　haningtoni Newstead & Evans
–　Antennal fringe less than a sixth of greatest width of third antennal segment (fig. 120A)　11

11　Hind tibiae with dark diffusions in middle and a much less distinct one at apex
　　　　　　　　　　　　　　　　　　　　nigrofusca hopkinsi van Emden

–　Hind tibia with or without dark suffusions, and if former, with a scarcely less distinct
　　infuscation at the apex than at the base　.　.　.　.　.　.　.　12

12　Palps grey-black; first three segments of hind tarsi brown　.　**vanhoofi*** Henrard
–　Palps buff or dusky to grey-brown; first three segments of hind tarsi yellowish or
　　ochraceous-buff (but may sometimes tend to brownish)　.　.　**fusca*** Walker

13　All segments of hind tarsi dark brown or blackish when viewed from above; dorsum
　　of abdomen usually uniformly brown, generally dark brown, not showing distinct
　　transverse dark bands on a paler background
　　　　　　　palpalis group (subgenus *Nemorhina*) and some forms of **austeni**　14
–　Only distal segments of hind tarsi dark brown or blackish, generally contrasting
　　strongly with paler proximal segments; dorsum of abdomen generally with distinct
　　dark bands showing against a paler background
　　　　　　　　　　　　　　morsitans group (subgenus *Glossina* s.s.)　22

14　Dorsal surface of abdominal segments with interrupted dark bands on a pale yellowish
　　background　.　.　.　.　.　.　.　.　**tachinoides** Westwood
–　Dorsal surface of abdominal segments without distinct banding on a pale yellowish
　　background　.　.　.　.　.　.　.　.　.　.　.　15

15　Dorsal surface reddish-ochraceous; small flies (7½–8½ mm.)　.　.　**austeni** (*Glossina* s.s.)
–　Dorsal surface of abdomen brown to dark brown (sepia or clove); large flies (8½–11mm)　16

16　Fringe of hairs on anterior edge of third antennal segment a quarter of the antennal
　　width or longer　.　.　.　.　.　.　.　.　.　.　(**pallicera** *s.l.*)　17
–　Fringe of hairs on anterior edge of third antennal segment a sixth of antennal width
　　or shorter　.　.　.　.　.　.　.　.　.　.　.　.　18

17　Third antennal segment usually narrow relatively to its length and strongly curved
　　at apex (fig. 120E); antennal fringe about three-fifths of antennal width
　　　　　　　　　　　　　　　　　　　　pallicera pallicera Bigot
–　Third antennal segment not as preceding, more like a pea-pod in shape (cf. fig. 120F);
　　antennal fringe a quarter of antennal width　.　**pallicera newsteadi** (Austen)

　　　　　* These taxa are better separated by examination of the genitalia.

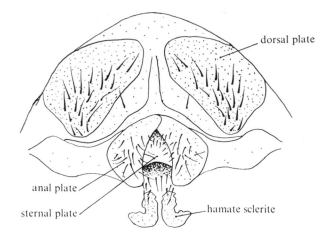

Fig. 111 External genital armature of *Glossina brevipalpis* (♀) (from Newstead, Evans and Potts, 1924).

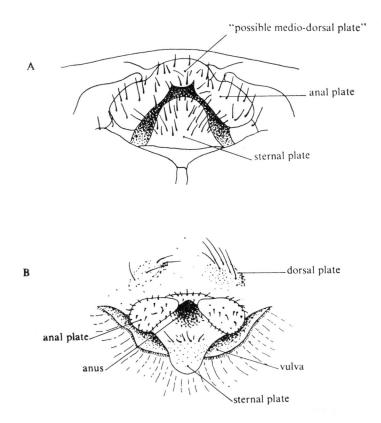

Fig. 112 Postero-lateral views of external genital armatures: A, *Glossina morsitans* (♀), mounted with slight pressure after maceration, showing anal plates united by a slender bridge, possibly representing the medio-dorsal plate of *G. palpalis* (from Newstead, Evans and Potts, 1924); B, *Glossina austeni* (♀), with the median sternal plate reflexed to show basal hinged area and shape of anal plates, which are united by a slender bridge, as in A (after Nash and Kernaghan, 1965).

18 Dorsal surface of abdomen dark to sepia-brown, hind margins of segments not narrowly paler than rest; a wide, more or less square median pale area on second tergite . . . . **caliginea** Austen

–   Hind margins of dorsal surfaces of abdominal segments narrowly paler, and median pale area on second tergite narrow and elongated 19

19 Colour of dorsal surface of abdomen variable but general tendency is to be very dark; posterior margin of hectors (see fig. 107C, D) in form of a shallowly concave curve, or straight . . . . **(fuscipes s.l.)*** 20

–   Colour of dorsal surface of abdomen variable but general tendency is to be less dark; posterior margin of hectors deeply cleft by a forwardly pointed triangle **palpalis** *s.l.**

20 Posterior margin of hectors straight, and with uninterrupted covering of hairs
                                                              fuscipes quanzensis* Pires

–   Posterior margin of hectors shallowly concave, and with median interruption of the covering of hairs 21

21 General coloration pale; glabrous interruption on hind margin of hectors a narrow line . . . . **fuscipes martinii*** Zumpt

–   General coloration darker; glabrous interruption on hind margin of hectors triangular, with apex directed forwards . . . . . **fuscipes fuscipes*** Newstead

22 Antennal fringe a fifth to a third of antennal width 23
–   Antennal fringe not more than a sixth of antennal width 24

23 Front and middle tarsi uniformly yellowish-brown; length of third antennal segment about five times its width, tip strongly recurved and tapering; antennal fringe a third of antennal width (fig. 120I) . . . . . **pallidipes** Austen

–   Front and middle tarsi with strongly defined dark tips; length of third antennal segment about three and a half times its width, tip only slightly and abruptly, recurved and not tapering; antennal fringe a fifth to a quarter of antennal width (fig. 120H) . . . . **longipalpis** Wiedemann

24 Dorsal surface of abdomen reddish-ochraceous to yellowish-buff, with only rather indistinct darker transverse bands; last two segments of hind tarsi not very much darker than the brownish proximal segments . . . . **austeni** Newstead

–   Dorsal surface of abdomen yellowish, or greyish-yellow, with distinct medially interrupted dark transverse bands, dark brown to black; last two segments of hind tarsi dark, strongly contrasted with the pale yellowish-brown proximal segments. 25

25 Hind margins of abdominal dark bands generally not very clearly defined and inner corners rounded (only occasionally somewhat truncate, e.g. some forms of subspecies *submosritans*), so that median pale line is not very sharply defined **morsitans** (*s.l.*)*

–   Hind margins of dark bands clearly defined and inner corners squarely truncate, so that narrow median pale line is very distinct . . . . **swynnertoni*** Austen

B. By the characters of the male terminalia.

1   Superior claspers narrowing to distal extremity, which terminates in a tooth or claw; claspers free, not joined by a membrane (fig. 110A) *fusca* group (subgenus *Austenina*) 15

–   Superior claspers joined by a membrane; may terminate in a tooth or claw as in preceding (fig. 110B) or distal extremity may be dilated (fig. 110C) 2

2   Superior claspers terminating in a tooth or claw *palpalis* group (subgenus *Nemorhina*) 3
–   Superior claspers dilated and distal extremity club-like
                                  morsitans group (subgenus *Glossina* s.s.) 11

3   Superior claspers with free tooth or claw very long, almost a third of the length of the clasper; inferior claspers (see fig. 107C) with head swollen, not foot-like, and notched, giving bi-lobed appearance, but not bifurcated (fig. 114H). West Africa, from Ghana to Cameroon, Gaboon and Oubangui . . . . . **caliginea**

–   Superior claspers with free tooth or claw short, very much less than a third of the length of the clasper; inferior clasper with head foot-like (fig. 114AGI) . . 4

        * These taxa are better separated by examination of the genitalia.

Q

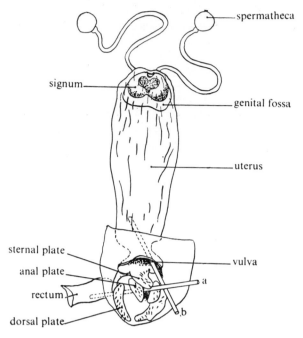

Fig. 113 Diagrammatic representation of external armature of *Glossina fusca* (♀), with extended uterus after maceration in caustic potash, ventral aspect, showing genital fossa and signum on the antero-dorsal wall of the uterus, with bristles (a and b) inserted into anus and vulva, respectively (after Newstead, Evans and Potts, 1924).

4 Inferior clasper with head bifurcated (fig. 114F, G) 5

– Inferior clasper with head not bifurcated (fig. 114A–E, I) . . . 6

5 Internal lobe of inferior clasper (see fig. 114C) with flattened outline (fig. 114F). West Africa, Sierra Leone to Cameroon . . . . . **pallicera pallicera**

– Internal lobe of inferior clasper with pointed outline (fig. 114 G). Gaboon, Ubangi-Shari, Belgian Congo, north-western Angola . . . **pallicera newsteadi**

6 Neck of inferior clasper (see fig. 114C) short, about as broad as long (fig. 114 I). Hinterland of West Africa to Sudan and Abyssinia; S. Arabia (?) . **tachinoides**

– Neck of inferior clasper long, plainly longer than broad (fig. 114A–E) . . 7

7 Inferior claspers with external lobe (see fig. 114C) prominent and projecting at least slightly upwards; internal lobe present and generally prominent (fig. 114C–E)
                      (*fuscipes* s.l.) 8

– Inferior claspers with external lobe not prominent and not projecting, even slightly upward; no internal lobe (fig. 114A, B) . . . . (*palpalis* s.l.) 10

8 Terminal dilatation of inferior claspers ('head') in form of a curved pointed hook, the curve prolonging that of the neck; internal lobe of 'body' of inferior clasper not projecting strongly (fig. 114C). Central and Eastern Africa generally (in region of the great forest and the central African lakes) . . . **fuscipes fuscipes**

– Terminal dilatation of inferior claspers more or less foot-like; internal lobe of inferior claspers may or may not project strongly (fig. 114D, E) 9

9 Terminal dilatation of inferior claspers markedly foot-like, with pronounced 'head'; 'sole' markedly concave; internal lobe of inferior claspers not projecting strongly (fig. 114D). Congo, Tanganyika, Zambia (from Upper Lualaba to Luapula rivers and Lake Tanganyika) . . . . . **fuscipes martinii**

– Terminal dilatation of inferior claspers not so markedly foot-like, heel not very pronounced; 'sole' more or less flat; internal lobe of inferior claspers projecting strongly (fig. 114E). French Equatorial Africa, former Belgian Congo, Angola
                      fuscipes quanzensis

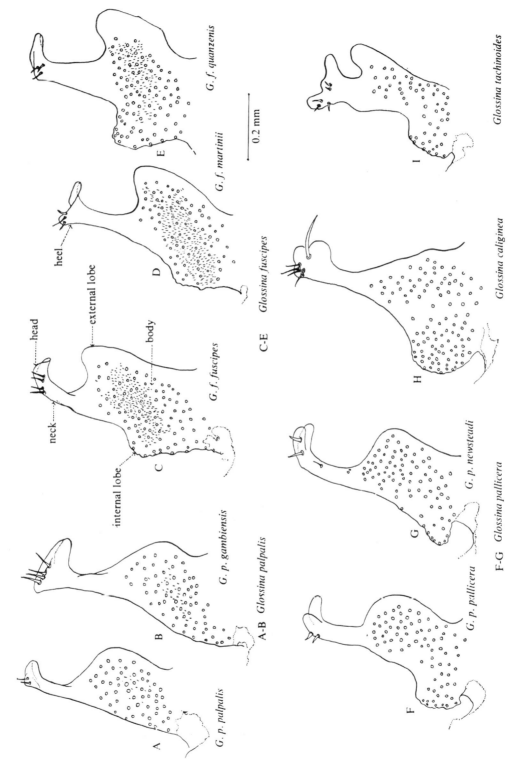

Fig. 114 Inferior claspers of males of *palpalis* group: A, B, *Glossina palpalis* ssp.; C, D, E, *Glossina fuscipes* ssp.; F, G, *Glossina pallicera* sp.; H, *Glossina caliginea*: I, *Glossina tachinoides* (after Machado, 1954).

10   Terminal dilatation ('head') of inferior clasper relatively small, its width plainly less than length of 'neck', which emerges abruptly from the 'body' of the clasper (fig. 114A). West Africa, from Nigeria to southern Angola   .   **palpalis palpalis**

\-   Terminal dilatation of inferior claspers relatively large, its width markedly exceeding length of "neck", which merges gradually into the "body" (fig. 114B). West Africa, from Senegal to Ivory Coast (with form transitional to *G. palpalis palpalis* between Liberia and Ghana)   .   .   .   .   . **palpalis gambiensis**

11   Outer lateral angle of superior claspers forming a blunt tooth (fig. 116A, B) .   .   **12**

\-   Outer lateral angle of superior claspers either rounded or strongly produced, not forming a tapering tooth (fig. 116C–E) **13**

12   Tooth of superior claspers subterminal; length of line of junction between the two inner flange-like extensions of the claspers about equal to greatest width of the claspers (fig. 116B). Central and Eastern Africa, Somalia and Abyssinia to Zululand .   .   .   . **pallidipes**

\-   Tooth of superior claspers terminal; length of line of junction between the inner extensions of the claspers plainly less than the greatest width of the claspers (fig. 116A). West Africa, from Zambia to Cameroon .   .   .   . **longipalpis**

13   Outer lateral angles of superior claspers strongly produced and narrowly rounded (fig. 116E). East African coastal regions, Somalia to Zululand, extending inwards as far as 33° and 38°E in Tanganyika and Mozambique, respectively .   . **austeni**

\-   Outer lateral angles of superior claspers rounded and not strongly produced (fig. 116C, D) **14**

14   Median lobes of superior claspers with broad tips, turned outwards and generally ending level with or projecting slightly beyond the swollen distal portion of the claspers (fig. 116C). Western Central and Eastern Africa .   .   **morsitans** *s.l.***

\-   Median lobes of superior claspers with small pointed tips, not generally reaching the level of the distal edge of the claspers (fig. 116D). Kenya and Tanganyika **swynnertoni**

15   A very prominent median process projecting between the inferior claspers for twice their length or more (fig. 117C). East and central Africa, from south-western Abyssinia and southern Somali through Kenya to Zululand and to south-eastern Congo (Kinshasa) .   .   .   . **brevipalpis**

\-   Median processes not prominent and generally projecting only slightly if at all between the inferior claspers (fig. 117A, D.) and never projecting for more than their length (fig. 117B, 118C) **16**

16   Harpes poorly developed and not properly differentiated from the general chitinisation of the aedeagus, without projecting processes (fig. 118A). Eastern part of former Belgian Congo .   .   .   . **savereni**

\-   Harpes well developed, with one or more freely projecting processes (fig. 117A, B etc.)   **17**

---

* Machado (1970) recognizes four subspecies.

a   Superior claspers with external distal angle rounded; rudimentary tooth of the claspers distinctly projecting beyond the level of the external angle of the clasper (as in fig. 116C) .   .   .   .   b

\-   Superior claspers with distal angle bluntly pointed; rudimentary tooth not projecting beyond the level of the external angle of the claspers .   .   . **(m. submorsitans)** c

b   Median lobes relatively feeble and narrow, with tips only slightly divergent   **m. morsitans**

\-   Median lobes robust and relatively wide, with tips markedly divergent (as in fig. 116C) **m. centralis**

c   Median lobes relatively feeble and narrow, with tips only slightly divergent **m. submorsitans** (typ)

\-   Median lobes robust and relatively wide, with tips markedly divergent (as in fig. 116C) **m. submorsitans ugandensis**

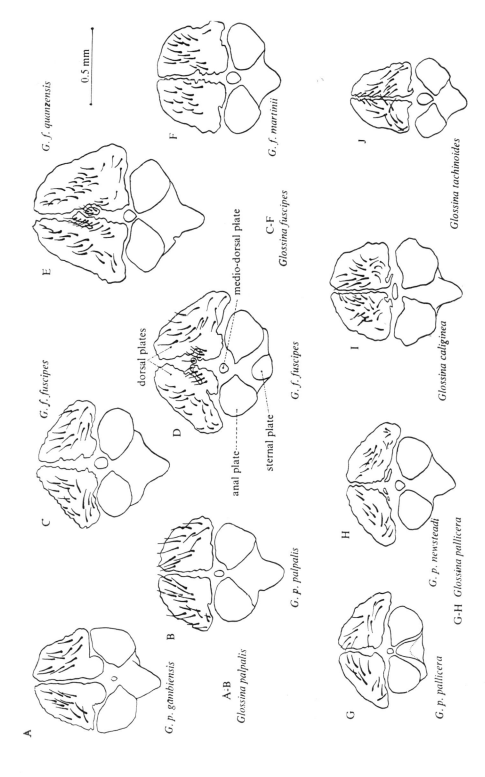

Fig. 115 Plates of genital armature of females of *palpalis* group: A, B, *Glossina palpalis* ssp.; C, D, E, F, *Glossina fuscipes* ssp.; G, H, *Glossina pallicera*; I, *Glossina caliginea*; J, *Glossina tachinoides* (after Machado, 1954).

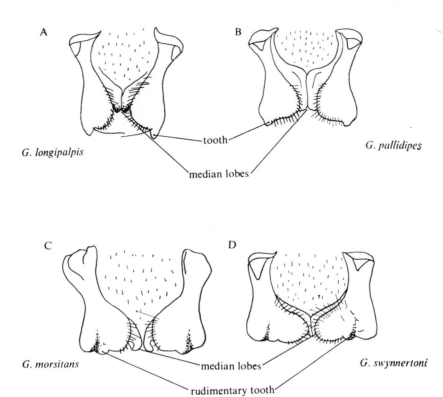

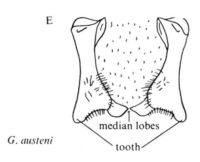

Fig. 116 Superior claspers of males of *morsitans* group: A, *Glossina longipalpis*; B, *Glossina pallidipes*; C, *Glossina morsitans*; D, *Glossina swynnertoni*; E, *Glossina austeni*; (after Newstead, Evans and Potts, 1924).

17 Harpes with one pair of freely projecting processes 18
– Harpes with three such pairs 20

18 Processes of harpes bifid; (fig. 117E).   Coast of Gulf of Guinea and Liberia to Nigeria
medicorum
– Processes of harpes simple, not bifid 19

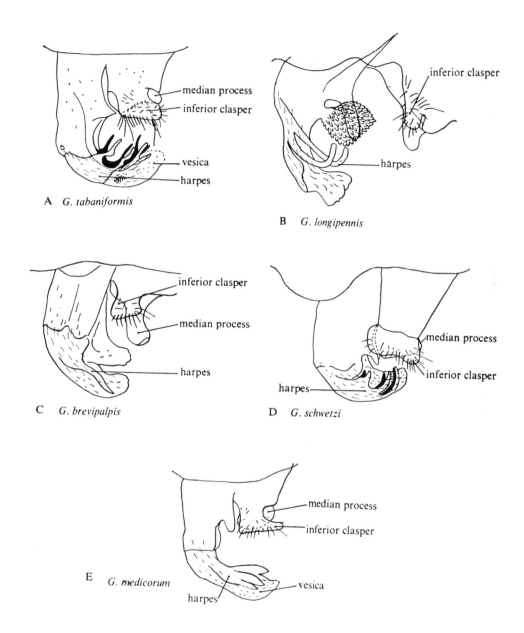

Fig. 117 Lateral views of male terminalia of species of *fusca* group: A, *Glossina tabaniformis*; B, *Glossina longipennis*; C, *Glossina brevipalpis*; D, *Glossina schwetzi*; E, *Glossina medicorum* (after Newstead, Evans and Potts, 1924) with modifications based on Nash and Jordan, 1959, and Machado, 1959).

19 Harpes consist of a pair of long slender processes curving upwards and tapering (fig.
    117B).   Northern East Africa including the southern border of the Sudan, Abys-
    sinia, Somalia, Kenya, northern Uganda and northern border of Tanganyika
                                                                        longipennis

–   Harpes consist of a basal triangular portion, the bottom corner of which is drawn out
    into a bluntly ending process, strongly chitinized; covered by a membrane thickly
    studded with short squamiform (scale-like) spines (fig. 118C).   [The male termi-
    nalia do not differ in the two subspecies of *G. nigrofusca* but these are easily differen-
    tiated by other characters (see key A, couplets 8–11, p. 225).]   West Africa,
    Liberia, Ivory Coast, Ghana, Nigeria and along northern border of former Belgian
    Congo and adjacent portions of territories to north and west of these, as **n. nigro-
    fusca**; extreme east of former Belgian Congo and extreme west of Uganda, as
    G. n.hopkinsi .   .   .   .   **nigrofusca***

20 None of the three pairs of processes with bifid members (fig. 117D, 118F)   .   .   21
–   Distal pair of processes with bifid members (117A, 118B)   23

21 Harpes with the processes of the proximal and middle pairs dilated towards their tips,
    only the distal ones tapering to a point; all three pairs of approximately the same
    length (fig. 118F).   Eastern edge of equatorial forest in former Belgian Congo
    from its boundary with the Central African Republic to the Kivu area   **vanhoofi**

–   Harpes with none of the processes dilated distally; pairs of processes not all of ap-
    proximately the same length (figs 117D, 118D)   22

22 Harpes with proximal pairs of processes the shortest of the three pairs (fig. 117D).
    Central Africa, Congo (Brazzaville), Western Congo (Kinshasa) and Angola (on the
    Congo River system) .   .   .   .   **schwetzi**

–   Harpes with the middle one of the three pairs of processes the shortest; the processes
    of the last pair characteristically with a dark base, the rest of it being clear and
    somewhat transparent (fig. 118D).   Central Africa (former French Cameroons,
    former Belgian Congo), southern Sudan, Uganda, western Kenya and (?) Botswana*
                                                                        fuscipleuris

23 Harpes with process of all three pairs tapering to a point   24
–   Harpes with processes of one or other of pairs dilated distally or in form of blunt
    protuberances (fig. 118E)   25

24 Harpes with proximal pair of processes markedly longer than the other two pairs,
    characteristic form and disposition of the processes as shown in fig. 118B.   Southern
    Sardauna province of Nigeria (former British Cameroons), Central African Repub-
    lic, the Congo Republic (Brazzaville), Gaboon and Angola (Belize, in Kabinda
    area) .   .   .   .   **nashi**

–   Harpes with proximal pair of processes not markedly longer than the other two pairs;
    characteristic form and disposition of the processes as shown in fig. 117A.   West
    coast of Africa—Ivory Coast and Ghana, and from Nigeria to former Belgian
    Congo, where stretches inland nearly to the eastern boundary of that territory
                                                                        tabaniformis

25 Proximal and middle pairs of processes of harpes peg-like; processes of distal (bifid)
    pair in form of blunt protuberances (fig. 118E).   West and Central Africa—south-
    western corner of Nigeria to Republic of Congo, the Congo (Kinshasa) and Angola
    (in the Kabinda area) .   .   .   .   **haningtoni**

–   Middle pair of harpes with processes dilated distally, the proximal and distal (bifid)
    pairs with pointed processes (fig. 118G)   26

        * Supported by the discovery of a puparial case (Dias, 1961, 1962) in riverine forest
    on a tributary of the lower Limpopo river in Mozambique (see also Ford, 1971).

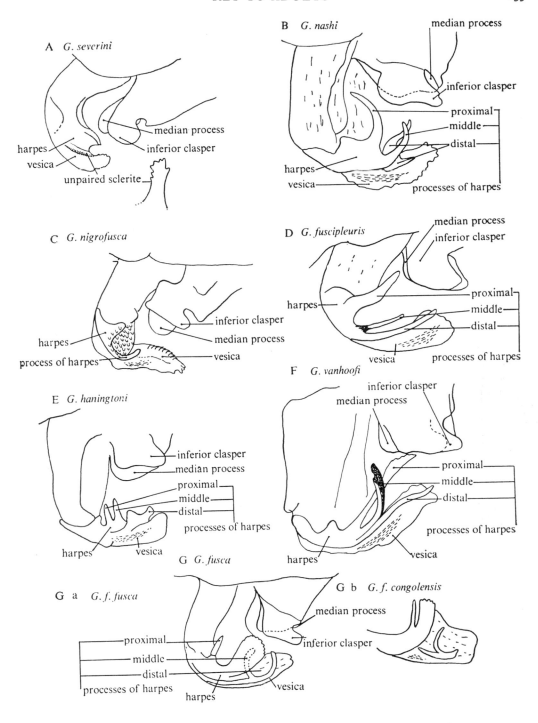

Fig. 118 Lateral views of male terminalia of species of *fusca* group: A, *Glossina severini*; B, *Glossina nashi*; C, *Glossina nigrofusca*; D, *Glossina fuscipleuris*; E, *Glossina haningtoni*; F, *Glossina vanhoofi*; G, *Glossina fusca*; G(a) *G. f. fusca*; G(b) *G. f. congolensis* (A, C, D, E and G, after Newstead, Evans and Potts, 1924, modified and interpreted in light of Nash and Jordan, 1959, and Machado, 1959; B and F after Potts, 1955) (for the sake of simplicity, only one of each of the pairs of processes of the harpes has been shown).

26 A macrophallic form; proximal pair of processes stout, short and peg-like reaching to about middle of the second pair, members of which are broadly dilated distally, with shallowly convex serrated distal margin, and relatively broad shafts; distal arm of the bifid sickle-shaped distal process as long as the proximal part (fig. 118Ga).   West Africa—Republic of Guinea to central Southern Ghana   **fusca fusca**

 - A microphallic form; proximal pair of processes straight, more slender, longer and tapering gradually to a point, about as long as the second pair, members of which are less broadly dilated and truncated distally, with straight serrated distal margin; distal arm of bifid sickle-shaped distal process about half the length of the proximal one or less (fig. 118Gb).   West and Central Africa—western south Ghana and Dahomey to western Uganda and former Belgian Congo, as far south as the southeastern corner near the border of Angola   .   .   .   .   **fusca congolensis**

C.   By the female genital armatures and signa.

1   External genital armature consists of five or six well-defined chitinous plates (figs 111 and 115D); signum (a chitinized plate on the anterior end of the uterus (see fig. 113)) may or may not be present   2

 - External genital armature very much reduced, well-defined chitinous plates being generally absent and not exceeding three if present (fig. 112A); signum never present
                                                    morsitans group (subgenus *Glossina* s.s.)†

2   External genital armature consists of five well-defined chitinous plates (fig. 111); signum present in all species except one   .   .   ***fusca*** group (subgenus *Austenina*) **3**

 - External genital armature consists of six well-defined chitinous plates (fig. 115D); signum never present   .   .   .   .   ***palpalis*** group (subgenus *Nemorhina*) **15**

3   No signum; in addition to the five well-defined chitinous plates there are two 'hamate' (hook- or comma-shaped) sclerites at the base of the anal plates (found only in this species) (fig. 111)   .   .   .   .   **brevipalpis**

 - Signum always present, though it may be only very weakly chitinized in freshly emerged specimens, particularly in some species (e.g. *G. longipennis*)   .   .   .   4

4   Signum consists of separated paired chitinous plates   5
 - Signum consists of single unpaired chitinous plates   6

5   Signum much reduced, the 'plates' being only two widely separated sub-medial strips of pale chitin running upwards from the bottom of the genital fossa to about half-way up it (fig. 119G)   .   .   .   .   **nigrofusca***

 - Signum not reduced, the plates being well-defined, and of characteristic shape, expanded at the tip and approximated to one another, occupying most of the genital fossa (fig. 119E)   .   .   .   .   **severini**

6   Signa elongated vertically, somewhat lyriform in shape (fig. 119A, H)   .   .   .   7
 - Signa of various shapes, but never lyriform   8

7   Signum strongly flexed or even bent double in middle of its length, sides roughly parallel, with two transverse constrictions towards bottom and top (fig. 119H)
                                                                                   fuscipleuris

 - Signum not flexed and sides not roughly parallel, divided into two unequal portions by a transverse constriction, the bottom margin with divergent horns laterally and the top one more or less deeply bifurcate (fig. 119A)   .   .   .   **tabaniformis**

8   Signum cordiform in outline (fig. 119D)   .   .   .   .   .   **medicorum**
 - Signum not cordiform   9

† The females of this group must be identified by use of key A.

* So far as is known at present, there is no difference between the signa or the external genital armatures of the two sub-species of *G. nigrofusca*, which however can be readily differentiated from each other by external characters (key A, couplets 8–11).

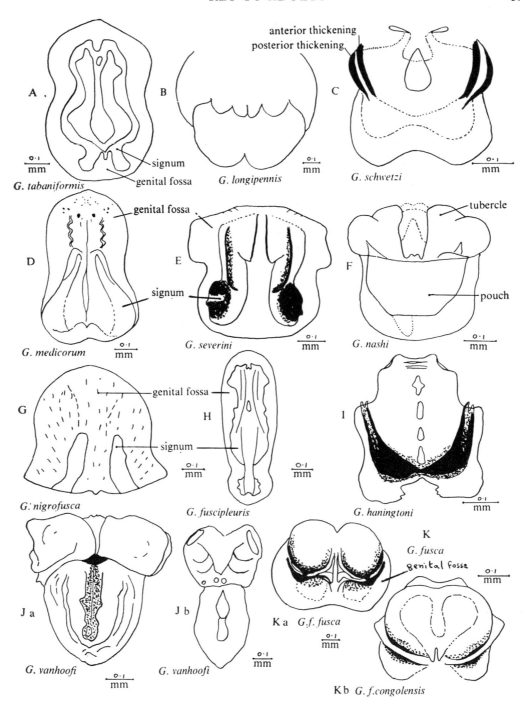

Fig. 119 Signa of females of species of the *fusca* group: A, *Glossina tabaniformis*; B, *Glossina longipennis*; C, *Glossina schwetzi*; D, *Glossina medicorum*; E, *Glossina severini*; F, *Glossina nashi*; G, *Glossina nigrofusca*; H, *Glossina fuscipleuris*; I, *Glossina haningtoni*; J, *Glossina vanhoofi* (a) well chitinised specimen, (b) weakly chitinised specimen; K, *Glossina fusca* (a) *G. f. fusca*, (b) *G.f. congolensis* (A,C,D,E,G,H,I, and K after Newstead, Evans and Potts, 1924; B after Patton, 1934; F from Jordan, 1962; J(a) from Henrard, 1952), (b) from Potts, 1955). (The genital fossa is not shown in B, C, E and K(b)).

9 Signa with conspicuous paired dark curved chitinous thickenings (as in fig. 119C) .   10

– Signa without such thickenings   13

10 Signa sub-rotund   11

– Signa not sub-rotund   12

11 Signum mainly composed of two lobes separated anteriorly by a deep V-shaped depression (fig. 119Ka) . . . . **fusca fusca**

– Signum mainly composed of two lobes separated anteriorly by only a shallow depression (fig. 119Kb) . . . . **fusca congolensis**

12 Shape of signum as in fig. 119C, particularly characteristic being the small processes directed outwards from the anterior corners; the paired chitinous thickenings situated anteriorly and not continuous medially, sometimes a separated second pair behind the first . . . . **schwetzi**

– Shape of signum as in fig. 119 I, no outwardly directed anterior processes; the paired chitinous thickenings situated posteriorly, and continuous medially, so forming a crescent . . . . **haningtoni**

13 Signum in form of the bottom part of a circle, the curved portion directed posteriorly and cut into medially by a V-shaped notch, separating the half circle into two lobes; tends to be very weakly chitinized, so much so in freshly emerged specimens that it was originally missed when the female genital armatures were described; occupying only a small portion of the genital fossa, posteriorly . . . **longipennis**

– Signum not so shaped and occupying most of genital fossa   14

14 Signum of uterus with two anterior bilobed tubercles tapering posteriorly into hollow stalks that lead into a sporran-shaped pouch (fig. 119F) . . . **nashi**

– Signum of uterus consisting of two parts, the upper a hollow lobe capping the lower one, a truncated cone in which there is a median chitinous plate in the form of a spear head with the point directed anteriorly (fig. 119J) . . . **vanhoofi**

15 Median plate of external genitalia broader than tall (fig. 115 I) . . **caliginea**

– Median plate as tall as broad or taller   16

16 Dorsal plates taller than broad   17

– Dorsal plates as tall as broad or nearly so   20

17 Dorsal plates nearly twice as tall as broad (fig. 115J) . . . . **tachinoides**

– Dorsal plates never nearly twice as tall as broad   18

18 Dorsal plates extending laterally well beyond width of anal plates (fig. 115E)

                                                          fuscipes quanzensis

– Dorsal plates not extending laterally beyond width of anal plates   19

19 Inner angles of dorsal plates projecting markedly downwards below base of plates; median plate very small (fig. 115A) . . . . . **palpalis gambiensis**

– Inner angles of dorsal plates not so projecting; median plate large (fig. 115F)

                                                         fuscipes martinii

20 Dorsal plates markedly broader than tall (fig. 115G, H) . . **pallicera s.l.***

– Dorsal plates about as broad as tall or only slightly broader   21

21 Dorsal plates very close together; hairs more robust and absent from median space and internal angles (fig. 115B) . . . . . . **palpalis palpalis**

– Dorsal plates comparatively widely separated; hairs less robust and almost always present on internal angles and often on median space as well (fig. 115C, D)

                                                         fuscipes fuscipes

---

* The females of the subspecies *G. pallicera pallicera* and *G. p. newsteadi* cannot be separated by their external genitalia, but are easily distinguished from each other by external characters (key A).

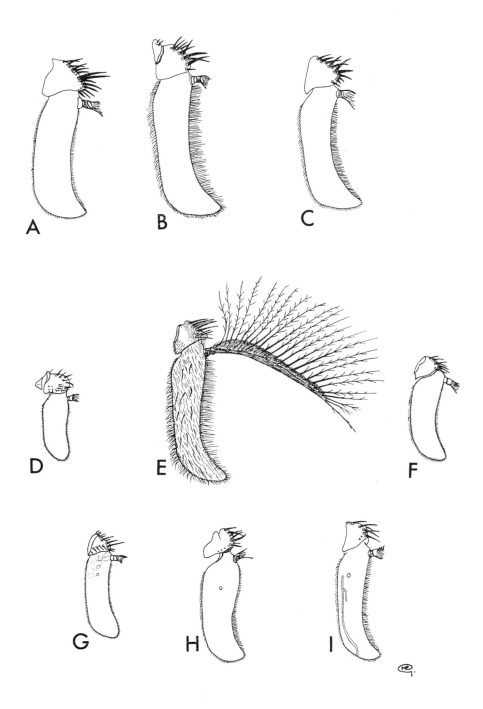

Fig. 120   Antennae of *Glossina* spp., showing antennal fringe (semi-diagrammatic, based on Pl. VII of Newstead *et al.*, 1924): A, *G. fusca*; B, *G. nigrofusca*; C, *G. tabaniformis*; D, *G. tachinoides*; E, *G. pallicera*; F, *G. palpalis*; G, *G. morsitans*; H, *G. longipalpis*; I, *G. pallidipes*.

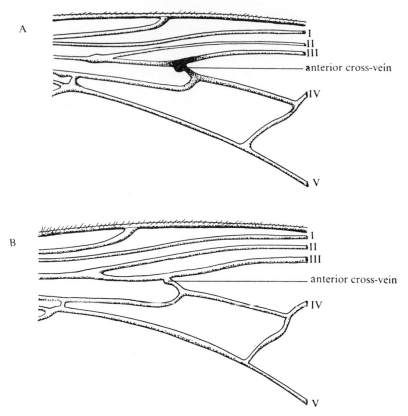

Fig. 121   Portions of wing showing anterior cross-vein: A, *Glossina brevipalpis*; B, *Glossina schwetzi* (from Buxton, 1955).

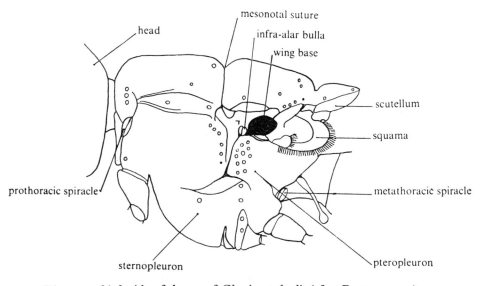

Fig. 122   Diagram of left side of thorax of *Glossina palpalis* (after Buxton, 1955).

IDENTIFICATION OF THE PUPARIA OF *GLOSSINA*

The puparia of 14 species of *Glossina* have been described and figured. A peculiar feature of these puparia is the pair of rounded prominent black posterior lobes (fig. 104E, pl. 3B) to which reference has already been made. Characteristically, the two lobes are separated by a cavity the shape of which varies from species to species. In one species, however (*G. longipennis*, fig. 123D), these two lobes are fused, and thus resemble those of the Conopid fly illustrated in fig. 123C. Puparia of two other Dipterous flies that have prominent posterior lobes and might be mistaken for those of *Glossina* are also shown in this figure (A & B), but the differences are obvious (Smith & Baldry, 1969.).

The differences in the shapes of the posterior lobes of the different species *Glossina*, and of the cavities between them, are not easy to see, and great care must be exercised in aligning the puparia correctly in order that these shapes may be observed properly. The puparium must be viewed from above, the flat (dorsal) surface being uppermost, and must be very carefully aligned so that it is not tilted from the horizontal from side to side, and is so tilted from back to front that the cavity between the lobes is seen as a single-line profile. The following key and figures are based on such profiles. The size of the puparia of any one species of tsetse, which can vary considerably in laboratory cultures, does not seem to vary very much in nature (though there may be a slight seasonal variation, the puparia produced in the hottest times of the years tending to be somewhat smaller than those produced during the cooler periods); nevertheless, the size of the puparium can generally be regarded as a useful criterion, and is used in the key. This key must, however, be applied with some caution, and is generally likely to be most useful in a particular area, for distinguishing the puparia collected in that area from one another. After all, one generally knows from the adult flies present what species the puparia collected are likely to belong to, and it is often necessary to be able to assign the individual puparia to their particular species.

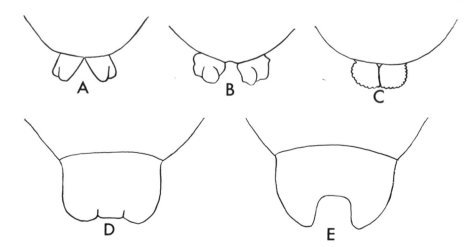

Fig. 123 Posterior lobes of puparia (semi-diagrammatic): A, *Plesiocyptera* sp. (Tachinidae); B, *Argyrophylax aureiventrum* (Tachinidae); C, Conopid, (?) *Physocephala* sp. (all based on Smith and Baldry, 1969); D, *Glossina longipennis*; E, *Glossina fuscipleuris*.

KEY TO PUPARIA OF *GLOSSINA*

1 No cavity separating the posterior lobes, which appear fused into a single mass, presenting only a slight, barely noticeable concavity posteriorly (fig. 123D)   **longipennis**

– A well-marked cavity between the posterior lobes **2**

2 Cavity V-shaped; a large puparium (7–8 mm) (fig. 124A)   .   .   **brevipalpis**

– Cavity not V-shaped **3**

3 Cavity with constriction giving it the appearance of a key-hole; small to medium-sized puparium (3–6 mm) **4**

– Cavity without any constriction; puparium large (7–8.5 mm)   .   .   .   **11**

4 Constriction of cavity between puparial lobes very marked, giving cavity a very marked key-hole shape (cf. fig. 124B) **5**

– Constriction of cavity not so pronounced, but key-hole shape still quite obvious (cf. fig. 124E) **7**

5 Puparium very small (3–5.4 mm); lobes closely approximated, gap between them at narrowest point about one-fifth width of each lobe or less; inner cavity of key-hole more rounded, its depth nearly equalling its width (fig. 124B)   **tachinoides**

– Puparium small- to medium-sized (5–5.6 mm); inner cavity of key-hole shallower, its depth plainly less than its width **6**

6 Puparial lobes closely approximated, narrowest distance between them about a quarter of width of each lobe (fig. 124C) .   .   .   .   **palpalis**

– Puparial lobes comparatively widely separated, narrowest distance between them not much less than half the width of each lobe (fig. 124D)   .   .   **fuscipes**

7 Small puparium (about 5.5 mm); lobes appear large in comparison with size of puparium, particularly in relation to their breadth, which sometimes nearly approaches that of puparium (fig. 124E) .   .   .   .   .   **austeni**

– Small- to medium-sized puparium (5.6–6 mm); lobes not markedly large in comparison with the size of the puparium (cf. fig. 124F) **8**

8 Bottom of cavity between lobes about equidistant from their proximal margins and their distal ends **9**

– Bottom of cavity between lobes nearer to their proximal margins than to their distal ends **10**

9 Constrictions of cavity between lobes very slight and narrowest part of key-hole very little below their distal ends (fig. 124G) .   .   .   **pallidipes**

– Constriction of cavity between lobes more pronounced and narrowest part of key-hole about half-way between bottom of cavity and distal ends of lobes (fig. 124I)   **swynnertoni**

10 Length of 'waist' of key-hole about half total depth of cavity; maximum width of inner cavity of key-hole at least half maximum width of each lobe (fig. 124H)   **longipalpis**

– Length of 'waist' of key-hole about two-thirds total depth of cavity; width of inner cavity of key-hole markedly less than half maximum width of each lobe (fig. 124F)   **morsitans**

11 Tips of posterior lobes slightly incurved, so that mouth of cavity between them is slightly narrower than widest part of cavity **12**

– Tips of posterior lobes not incurved, so that cavity is in form of a straight-sided 'U' .   **13**

12 Tips of lobes only very slightly incurved; distance between them only very slightly less than total width of cavity (fig. 125A)   .   .   .   **tabaniformis**

– Tips of lobes very noticeably incurved; distance between them markedly less than maximum width of cavity (fig. 125B)   .   .   .   .   **fusca**

13 Tips of lobes definitely tapering, cavity between the lobes comparatively deep, depth nearly equal to distance between bottom of cavity and junction of lobes with body of pupa (fig. 123E) .   .   .   .   **fuscipleuris**

– Tips of lobes much more rounded and barely tapering; cavity comparatively shallow, depth obviously less than distance between bottom of cavity and junction of lobes with body of pupa (fig. 125C)   .   .   .   .   .   **schwetzi**

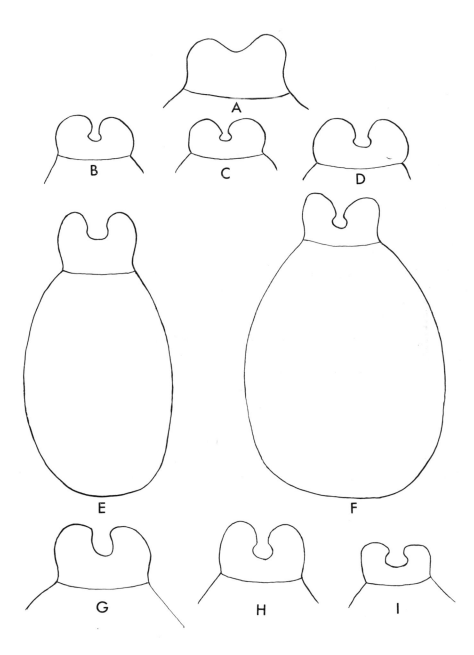

Fig. 124   Posterior lobes of puparia of *Glossina* spp. (semi-diagrammatic): A, *G. brevipalpis*;
B, *G. tachinoides*; C, *G. palpalis*; D, *G. fuscipes*; E, *G. austeni* (whole puparium);
F, *G. morsitans* (whole puparium); G, *G. pallidipes*; H, *G. longipalpis*; I, *G. swynnertoni*.

R

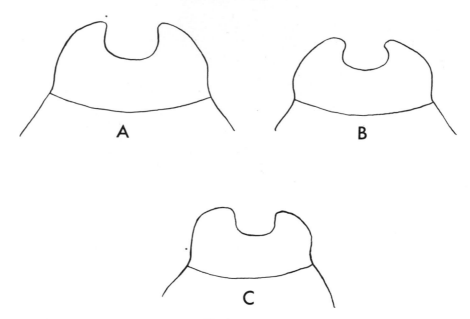

Fig. 125   Posterior lobes of puparia of *Glossina* spp. (semi-diagrammatic): A, *G. tabaniformis*; B, *G. fusca*; C, *G. schwetzi*.

TSETSE SPECIES AND TRYPANOSOMIASIS

Individuals of all the 22 species of *Glossina*, with the exception of three (*G. severini*, *G. schwetzi* and *G. nashi*), have been shown experimentally to transmit trypanosomes pathogenic either to man or to his domestic animals, or have been found infected with them naturally.   It is therefore safe to assume that any species of *Glossina* can act as a vector of trypanosomiasis.

Flies become infected by imbibing blood containing trypanosomes from hosts already infected with these organisms, and they transmit the infection again by feeding on uninfected host vertebrates, the parasites being injected through the mouth-parts with the salivary secretion (see p. 211).   The transmission may be direct (when it is often termed mechanical), the insect merely acting as a flying hypodermic syringe and only remaining infective for a matter of hours after the infecting feed, or it may be indirect, when it is frequently referred to as cyclical.   In this, the trypanosomes ingested by the fly undergo a cycle of development in its proboscis or alimentary canal (and in some instances, its salivary glands), the fly remaining incapable of infecting another host for a definite period after the infecting feed.   The developmental cycle of the fly may take place entirely within the channels of the proboscis (e.g. *Trypanosoma vivax*, a member of the subgenus of *Trypanosoma* recently designated as *Duttonella*), or initially in the mid-gut of the tsetse and finishing in the proboscis (e.g. *T. congolense*, belonging to the recently designated subgenus *Nannomonas*); these two groups of trypanosomes are pathogenic to one or more of the species of domestic animals.   In a third type of developmental cycle, the trypanosomes develop first in the mid-gut, then pass *via* the proboscis and hypopharyngeal tube into the salivary glands, in which they complete their development, becoming infective forms.   This group of trypanosomes, often unfortunately called the polymorphic trypanosomes, is now designated as the subgenus *Trypanozoon*, and contains the species *T. brucei*, pathogenic to domestic animals, and *T. gambiense* and *T. rhodesiense*, pathogenic to man; there is, however, a general tendency to regard these three species rather as subspecies of *T. brucei*, though specific status is often retained to avoid the cumbersome trinomials otherwise required. (Hoare, 1964; Ormerod, 1967; & in Ford, 1971, pp. 59–62.)

The species of tsetse that have been mainly associated with outbreaks of human trypanosomiasis (sleeping sickness) are *G. palpalis*, *G. fuscipes*, *G. tachinoides*, *G. morsitans* and *G. pallidipes*. Their importance, and the nature of the disease they give rise to in man (whether the acute form known as Rhodesian sleeping sickness, caused by *T. rhodesiense* or the milder and chronic but equally lethal Gambian form, due to *T. gambiense*) depends largely on the degree of contact between the species of *Glossina* concerned and man, and this varies from time to time and from place to place, according to the relations existing between the tsetse-flies, their natural hosts and man. Thus in West and Central Africa, human trypanosomiasis is generally of the Gambian type, and is carried by one of the waterside species of tsetse (Category III, Table 5); it is with these species that contact between man and tsetses is most frequent and close in these regions, man there being frequently the favoured host of the flies. In East Africa, however, sleeping sickness is mainly of the Rhodesian type, carried by one or other of the species of game tsetse (Categories IV and V, Table 5). In this region, man is only rarely, and in exceptional circumstances, brought into sufficiently close contact with the game tsetses, which anyhow do not usually favour him as a host, for outbreaks of human sleeping sickness to arise. When they do, they therefore tend to be of a violent nature. The whole subject of the epidemiology of human sleeping sickness is a very complex one, and one still the subject of much discussion, controversy and research. The reader who is interested should refer for guidance to recent views and literature bearing on this subject to Mulligan (1971); publications of particular interest in this respect are: Ashcroft (1959), Ormerod (1961), Apted (1962), Van den Bergh *et al.* (1963), Apted *et al.* (1963), Onyango (1969) and Goodwin (1970, pp. 797–800). A very recent review of the subject with special reference to *G. fuscipes fuscipes*, but also with some original observations on the biology of this fly, is to be found in Van Vegten (1971).

The distribution of the disease is shown in fig. 109 (p. 219).

CONTROL OF TSETSE AND TRYPANOSOMIASIS

Much has been written on this subject, and it has been fully discussed and reviewed by Buxton (1955, Ch. 14), who had earlier, in his 1954 presidential address to the Royal Entomological Society of London (Buxton, 1955a), described how ideas on tsetse control had developed from ecological investigations. The present position has more recently been reviewed in Mulligan (1970, Chs 21–30 & 43).

Control of trypanosomiasis can be sought by two broadly different avenues of approach. The first consists in the control of the disease along medical and veterinary lines, mainly by chemotherapy and the use of prophylactic drugs; these may be combined with the manipulation of human and cattle populations in such a way as to break completely or to minimize contact between man and his domestic animals on the one hand, and the tsetse on the other.   The second line of approach is the elimination of the insect vectors or their reduction to such low numbers that the chain of contact between them and man and his animals essential for the maintenance of the disease cycle is broken.   Both these lines are dealt with in Mulligan (1970) whereas Buxton (1955) is concerned solely with the vectors.

Methods used for vector control can be divided into direct and indirect ones.   In the first, destruction of the flies by direct attack on the insects themselves is attempted. Such methods consist of the destruction of the flies by handcatching on a large scale, by use of traps or by the application of insecticides, which may be directed against the insects themselves or as deposits on their resting and perching sites.   Slightly different methods of direct attack are of a biological nature, either by the use of parasites and pre-dators of the tsetse or by attempts to sterilize the insects.   The last may be attempted by the irradiation with gamma rays of the adults or the pupae; the latter then produce only females that are sterile and males that mate readily with the wild females impregnating them with sperm that do not result in the production of viable offspring, and so rendering the wild females sterile.   Similar effects can be obtained by the use of chemosterilants applied to the pupae or to the adults, again as with the insecticides, applied either directly against the insects or to their perching places.   The last two methods, attempting sterilization of the vectors, are at present only in the exploratory stage.

The indirect methods of control are those in which attempts are made to alter the environment of the fly in such a way as to make it impossible for the flies to continue to exist therein.   These methods consist in the clearance, complete or partial, of the woody vegetation of the tsetse habitat, or in the interference with the fly's food supply, either by the killing of its mammalian hosts or by driving these away or reducing their numbers to such an extent that the flies cannot obtain sufficient food either to prevent starvation or to allow reproduction.

Of the various methods of control outlined above, only the last has ever been completely successful on a large scale.   In the first attempts all the wild ungulate hosts of the tsetse were shot.   Now, however, thanks to the work of Weitz (1963) and his collaborators in various parts of Africa, it would seem probable that elimination of the fly may be achieved by attacking only those ungulates that have been shown to be the most favoured hosts of the flies (Cockbill, 1967).   The other principal methods of control now in use are the application of insecticides and the clearance of woody vegetation.   In both of these, the ecological investigations of the past 50 years have shown how these applications may be limited to certain portions of the habitat of the fly only and still give some considerable measure of success which often appears to amount to complete elimination. Failures in this respect may be attributed to the doubts as to the reality of those 'con-centrations' of fly that have been associated with certain parts of the habitat to which reference has been made on p. 215, or they may be attributed to a more fundamental cause as has been suggested by Bursell (1968) in an interesting discussion of the relation of problems of control to our knowledge of the ecology of the fly.   In this, Bursell has pointed out the great independence of its physical environment achieved by *Glossina* by virtue of its peculiar method of its reproduction and by the various delicate adaptations of the adult stage to life in an arid climate.   These make it more difficult to achieve success in control by alteration of the physical environment than by interference with its food supplies, from which *Glossina* has by no means succeeded in achieving the same degree of independence.

The subject of control cannot be dismissed without mention of a recently published book by Ford (1971), in which the epidemiology of both human and animal trypanosomiasis is discussed, in great detail and with a wealth of historical background, on a broad ecological basis.   The author stresses the necessity for an integrated approach to the problems of trypanosomiasis and its control which has so far been appreciated only somewhat vaguely by workers in this field.   He points out that merely to aim at the elimination of the disease or the vector without paying attention to the broad ecological situation can at best give only temporary success, and is all too often followed by the return of trypanosomiasis and the vector and a situation possibly worse than the original one.   He concludes that the problem now really facing those concerned is how best to use the knowledge available to control the wildlife ecosystems of Africa in such a way as to reconcile them with man's development and best use of the resources of that continent.

BIBLIOGRAPHY

APTED, F. I. C.   1962.   Sleeping sickness in Tanganyika, past, present and future.   *Trans. R. Soc. trop. Med. Hyg.* **56**: 15–23.

APTED, F. I. C., ORMEROD, W. E., SMYLY, D. P., STRONACH, B. W. & SZLAMP, E. L.   1963.   A comparative study of the epidemiology of endemic Rhodesian sleeping sickness in different parts of Africa.   *J. trop. Med. Hyg.* **66**: 1–16.

ASHCROFT, M. T.   1959.   A critical review of the epidemiology of human trypanosomiasis in Africa.   *Trop. Dis. Bull.* **56**: 1073–1092.

ATKINSON, P. R.   1971.   A study of the breeding distribution of *Glossina morsitans* Westwood in northern Botswana.   *Bull. ent. Res.* **60**: 415–426.

AZEVADO, J. F. DE (Ed.).   1970.   Criacão da mosca tsé-tsé em laboratório e sua aplicão prática. [Tsetse-fly breeding under laboratory conditions and its practical application, 1st International Symposium.]   524 pp.   Lisbon.   [Various languages.]

BURCHARD, R. P. & BALDRY, D. A. T.   1970.   Polytene chromosomes of *Glossina palpalis* Robineau-Desvoidy (Diptera: Muscidae). I.—The preliminary demonstration.   *Proc. R. ent. Soc. Lond.* (A) **45**: 182–183.

BURSELL, E.   1955.   The polypneustic lobes of the tsetse larva (*Glossina*, Diptera).   *Proc. R. Soc.* (B) **144**: 275–286.

—— 1958.   The water balance of tsetse pupae.   *Phil. Trans. R. Soc.* (B). **241**: 179–210.

—— 1966.   The nutritional state of tsetse flies from different vegetation types in Rhodesia. *Bull. ent. Res.* **57**: 171–180.

—— 1968.   '*A Prospect of Tsetse Flies*'.   An inaugural lecture given in the University College of Rhodesia.   Salisbury.

BURSELL, E. & the late JACKSON, C. H. N.   1957.   Notes on the choriothete and milk gland of *Glossina* and *Hippobosca*.   *Proc. R. ent. Soc. Lond.* (A), **32**: 30–34.

BUXTON, P. A.   1955a.   Tsetse and climate; a consideration of the growth of knowledge. *Proc. R. ent. Soc. Lond.* (6).   **19**: 71.

—— 1955.   The Natural history of tsetse flies.   *Mem. Lond. Sch. trop. Med. Hyg.*, No. 10, 816 pp.   London.

COCKBILL, G.   1967.   Recent developments in tsetse and trypanosomiasis control.   The history and significance of trypanosomiasis problems in Rhodesia.   *Proc. Trans. Rhod. scient. Res.* **52**: 7.

DIAS, J. A. TRAVASSOS DOS SANTOS.   1961.   Resultado de um reconhecimento glossinico em algumas circumscricoas do Distrito de Cabo Delgado.   *Anais Servs Vet. Ind. anim. Moçamb.* **7**: 167.

—— 1962.   The status of the tsetse fly in Mozambique before 1896.   *S. Afr. J. Sci.* **58**: 243–247.

FORD, J. 1971. *The role of the Trypanosomiases in African ecology.* 568 pp. Oxford.

GLASGOW, J. P. 1963. *The distribution and abundance of tsetse flies.* 241 pp. Oxford.

—— 1967. Recent fundamental work on tsetse flies. *A. Rev. Ent.* **12**: 421–438.

GOODWIN, L. G. 1970. The pathology of the African trypanosomiasis. *Trans. R. Soc. trop. Med. Hyg.* **64**: 797–817.

GORDON, R. M. & CREWE, W. 1948. The mechanism by which mosquitoes and tsetse-flies obtain their blood-meal, etc. *Ann. trop. Med. Parasit.* **42**: 334–356.

HENRARD, C. 1952. Une tsetse nouvelle du groupe *fusca, Glossina vanhoofi* sp. nov. *Rev. Zool. Bot. afr.* **45**: 193–197.

HOARE, C. A. 1964. Morphological and taxonomic studies on mammalian tryanosomes X.—Revision of the systematics. *J. Protozool.*, **11**: 200–207.

HOCKING, K. S., LAMERTON, J. F. & LEWIS, E. A. 1963. Tsetse-fly control and eradication. *Bull. Wld Hlth Org.* **28**: 811–823.

HOFFMANN, R. 1954. Zur Fortptflanzungensbiologie und zur intrauterinen Entwicklung von *Glossina palpalis. Acta trop.* **11**: 1.

HULLEY, P. E. 1968. Mitotic chromosomes of *Glossina pallidipes* Aust. *Nature, Lond.* **217**: 977.

ITARD, J. 1966. Chromosomes de Glossines (Diptera: Muscidae) (Note présentée par M. Clément Bressou, Séance du 24 Octobre, 1966) *C. R. Acad. Sc. Paris.* **263**: 1395 (Ser. D).

—— 1969. Les caryotypes de six espèces de Glossines. First Symposium on the breeding of tsetse fly in the laboratory and its practical applications, April 22–23, 1969. Lisbon.

—— 1970. L'appareil reproducteur mâle des Glossines (Diptera–Muscidae). Les étapes de sa formation chez la pupe. La spermatogénèse. *Revue Élev. Méd. vét. Pay strop.* **23**: 57–81.

—— 1971. Chromosomes de *Glossina fusca congolensis* Newstead et Evans, 1921 (Diptera–Muscidae). (Note de M. Jacques Itard, présentée par M. Clement Bressou, Séance du 3 Mai 1971) *C. R. Acad. Sc. Paris* **272**: 2561–2564 (Sér. D).

JACKSON, C. H. N. 1948. The eclosion of tsetse (*Glossina*) larvae (Diptera). *Proc. R. ent. Soc. Lond.* (A) **23**: 36–38.

JORDAN, A. M. 1962. A re-description of the signum of *Glossina nashi* Potts. *Ann. trop. Med. Parasit.* **56**: 70–72.

—— 1965. The status of *Glossina fusca* Walker (Diptera: Muscidae) in West Africa. *Ann. trop. Med. Parasit.* **59**: 219–225.

McCONNELL, E., HUTCHINSON, M. P. & BAKER, J. R. 1970. Human trypanosomiasis in Ethiopia: the Gils river area. *Trans. R. Soc. trop. Med. Hyg.* **64**: 683–691.

MACHADO, A. DE BARROS. 1954. Révision systématique des Glossines du groupe *palpalis* (Diptera). *Publços cult. Co. Diam. Angola* **22**: 1–89.

—— 1959. Nouvelles contributions à l'étude systématique et biogéographique des Glossines. *Publços cult. Co. Diam. Angola* **46**: 17–90.

—— 1970. 'Les races géographiques de *Glossina* morsitans'. In J. Fraza de Avedo, Ed. Criacão da mosca tsé-tsé em laboratório e sua aplicacão prática. [1st International Symposium on Tsetse Fly Breeding under Laboratory conditions and its practical application. 22nd and 23rd April, 1969.] pp. 471–486. Lisbon. [In French.]

MAUDLIN, I. 1970. Preliminary studies on the karyotypes of five species of *Glossina. Parasitology* **61**: 71–74.

MULLIGAN, H. W. (Ed.). 1970. *The African trypanosomiases.* 950 pp. London.

NASH, T. A. M. 1969. *Africa's Bane—the tsetse fly.* 224 pp. London.

NASH, T. A. M. & JORDAN, A. M. 1959. A guide to the identification of the West African species of the *fusca* group of tsetse flies by dissection of the genitalia. *Ann. trop. Med. Parasit.* **53**: 72–88.

NEWSTEAD, R., EVANS, A. M. & POTTS, W. H. 1924. Guide to the study of tsetse-flies. *Mem. Lpool Sch. trop. Med.* (new series) **1**: pp. 382. Liverpool.

ONYANGO, R. J. 1969. New concepts in the epidemiology of Rhodesian sleeping Sickness. *Bull. Wld Hlth Org.* **41**: 815–823.

ORMEROD, W. E. 1961. The epidemic spread of Rhodesian Sleeping sickness 1908–1960. *Trans. R. Soc. trop. Med. Hyg.* **55**: 525–538.

—— 1967. Taxonomy of the sleeping sickness trypanosomes. *J. Parasit.* **53**: 824–830.

PILSON, R. D. & PILSON, B. M. 1967. Behaviour studies of *Glossina morsitans* Westw. in the field. *Bull. ent. Res.* **57**: 227–257.

POLLOCK, J. N. 1970. Sperm transfer by spermatophores in *Glossina austeni* Newstead. *Nature, Lond.* **225**: 1063–1064.

—— 1971. The origin of tsetse-flies. *J. ent.* (B) **40**: 101–109.

POTTS, W. H. 1955. A new tsetse-fly from the British Cameroons. *Ann. trop. Med. Parasit.* **49**: 218–226.

RIORDAN, K. 1968. Chromosomes of the tsetse fly, *Glossina palpalis. Parasitology* **58**: 835–838.

—— 1970. Polytene chromosomes of *Glossina palpalis* Robineau-Desvoidy (Diptera: Muscidae). II.—Their improved demonstration. *Proc. R. ent. Soc. Lond.* (A) **45**: 184–186.

ROBERTS, M. J. 1971. 'The role of the choriothete in tsetse flies'. *Parasitology* (in press).

—— 1971a. 'The control of fertilisation in tsetse flies'. *Ann. trop. Med. Parasit* (in press).

—— 1971b. 'The functional anatomy of the head in the larva of the tsetse fly *Glossina austeni* Newst. (Diptera, Glossinidae)'. *The Entomologist* **104**: 190–203.

—— 1971c. Thesis, Salford University.

SMITH, K. G. V. & BALDRY, D. A. T. 1969 [1968]. Some Dipterous puparia resembling and found among those of tsetse flies. *Bull. ent. Res.* **59**: 367–370.

SMART, J. 1965. *Insects of Medical Importance*, 5th ed. xi+303 pp. London.

VAN DEN BERGHE, L. & LAMBRECHT, F. L. 1963. The epidemiology and control of human trypanosomiasis in *Glossina morsitans* fly belts. *Amer. J. trop. Med. Hyg.* **12**: 129–164.

VANDERPLANK, F. L. 1948. Experiments in cross-breeding tsetse-flies (*Glossina* species). *Ann. trop. Med. Parasit.* **42**: 131–152.

—— 1949. The classification of *Glossina morsitans* Westwood, Diptera, Muscidae, including a description of a new subspecies, varieties and hybrids. *Proc. R. ent. Soc. Lond.* (B) **18**: 56–64.

VAN VEGTEN, J. A. 1971. The tsetse fly *Glossina fuscipes* Newstead, 1911, in East Africa; some aspects of its biology and its role in the epidemiology of human and animal trypanosomiasis. Academisch Proefschrift. pp. 132.

WEITZ, B. 1963. The feeding habits of *Glossina. Bull. Wld Hlth Org.* **28**: 711.

WILLETT, K. C. & GORDON, R. M. 1957. Studies on the deposition, migration and development of the blood forms of trypanosomes belonging to the *T. brucei* group. Pt II. *Ann. trop. Med. Parasit.* **51**: 471–492.

5b. MUSCIDAE
(House-flies, Stable-flies, etc.)
by Adrian C. Pont

THE family Muscidae of older authors is now generally divided into three families, Fanniidae, Anthomyiidae, and Muscidae s. str., but for convenience these are considered together in this chapter, and the Anthomyiidae and Fanniidae are treated as sub-families of the Muscidae.

The Muscidae may be defined as Calyptrate Diptera which lack a row of strong setae on the hypopleuron. From the Glossinidae, which also possess these characters, they may be distinguished as indicated in the key to families (Chapter 2).

A few genera are, in the adult or larval stage or both, of considerable medical importance to man, and figure prominently in textbooks devoted to medical entomology. But there are also many species which may be of some hygienic importance to man because of their adult habits or biology: species that breed in or settle on excrement, garbage, sewage or compost that also enter human dwellings or settle persistently on human beings are all potential mechanical vectors of pathogenic organisms though in most cases this is at present suspected, but not proved. Such species are included in the key to genera given below, and are not discussed further here, but are treated in Chapter 18.

KEY TO THE GENERA OF MUSCIDAE OF MEDICAL IMPORTANCE

1 Vein 6 complete, reaching wing-margin even if only faintly or as a fold (fig. 126A); or if not reaching the wing-margin then proclinate upper orbital setae present *and* lower sternopleural seta equidistant from the upper two. Scutellum in most species with a number of soft pale hairs at apex on ventral surface. Hind metatarsus usually with a short ventral basal seta. (See discussion below)
ANTHOMYIINAE

– Vein 6 incomplete, never reaching wing-margin. When lower sternopleural seta equidistant from the two upper setae, upper orbital setae never proclinate. Scutellum very seldom with soft pale hairs below. Hind metatarsus without a basal ventral seta 2

2 Vein 6 very short, not extending half the distance from its base to wing-margin, vein 7 strongly curved so that an imaginary extension would intersect an imaginary extension of vein 6 before the wing-margin (fig. 126B). Hind tibia with a submedian seta in an exact dorsal position. Male mid legs always modified on ventral surface (fig. 127). Female parafrontalia broad, the inner margins convex (fig. 128A). When three sternopleural setae present, lower one much closer to posterior one than to anterior one. (See discussion below) . . . **FANNIINAE**

– Vein 6 longer, extending over half the distance from its base to wing-margin, or if short then vein 7 more shallowly curved so that an imaginary extension would not intersect an imaginary extension of vein 6 before wing-margin (fig. 126C) (*Azelia*) or three strong sternopleural setae present with the lower one equidistant from the upper two (Coenosiinae). Hind tibia without a true dorsal submedian seta, though most species have a seta slightly posterior of dorsal (a 'calcar'), between *d* and *pd*, or a posterodorsal seta. Male mid legs rarely modified. Female parafrontalia slender, the inner margins concave . . . . **(MUSCINAE)** 3

3   Proboscis of the piercing type (fig. 129A), strongly sclerotized, mentum at least as long as head, slender, strongly tapering to apex, non-retractile, without distinct labella. Sternopleural setae 0+1 or 1+1 4

–   Proboscis of the licking type (fig. 129C), rather short and stout, mentum shorter than head and only moderately tapering, retractile, labella well developed; but if adapted for piercing (*Musca crassirostris* Stein, fig. 129B) then three sternopleural setae present 5

4   Sternopleurals 1+1. Palpi about as long as mentum of proboscis. (See discussion below) . . . . *HAEMATOBIA* Le Peletier and Serville

–   Sternopleurals 0+1. Palpi not half as long as mentum of proboscis. (See discussion below). . . . . *STOMOXYS* Geoffroy

5   Pteropleuron setulose 6
–   Pteropleuron bare 15

6   Lower squama of the *Phaonia*-type (fig. 128B) 7
–   Lower squama of the *Musca*-type (fig. 128C) 10

7   Old World species (*Neomuscina transporta* Snyder from Ghana is believed to be a chance importation from South America) . . . .
    New World species . . . .

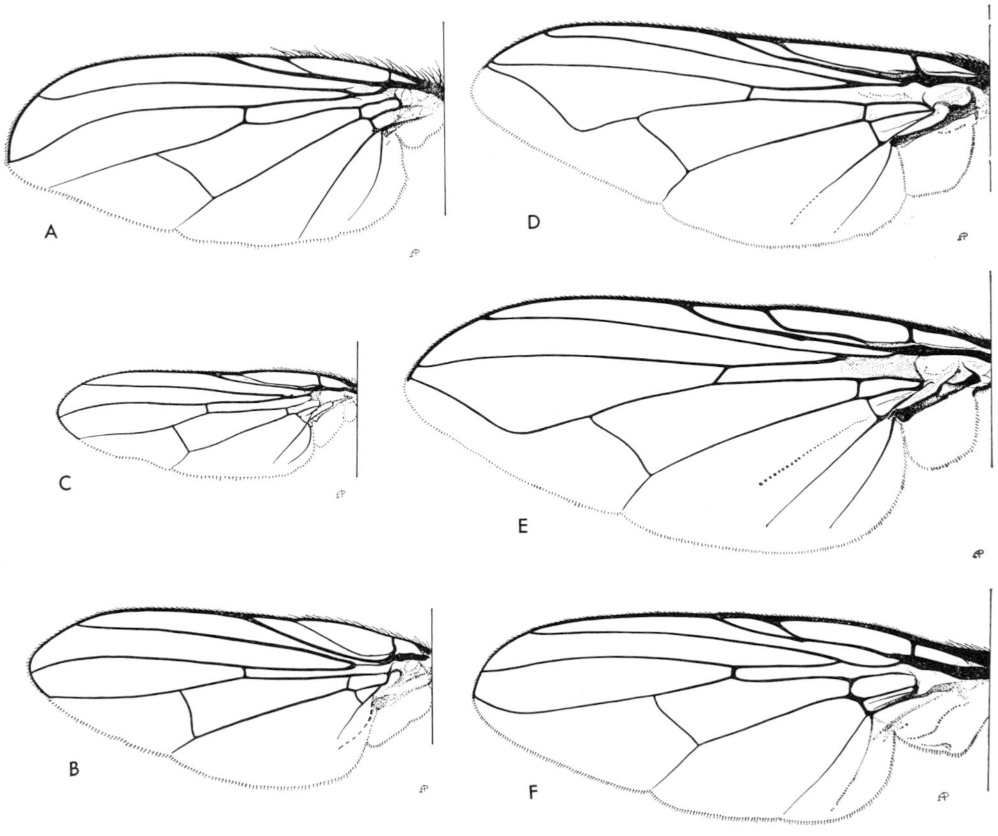

Fig. 126   Wings of Muscidae: A, *Anthomyia spinigera*; B, *Fannia canicularis*; C, *Azelia triquetra*; D, *Musca domestica*; E, *Orthellia timorensis*; F, *Muscina stabulans*.

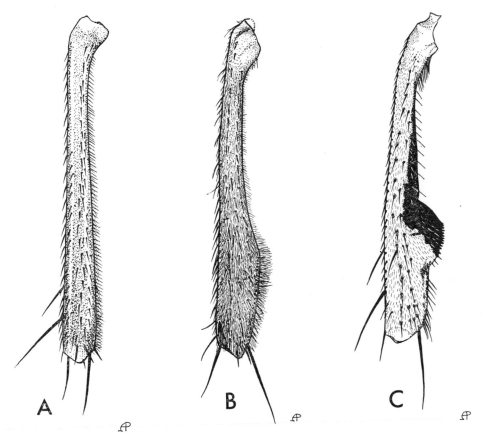

Fig. 127 Mid tibiae of male *Fannia*: A, *F. canicularis*; B, *F. manicata*; C, *F. scalaris*.

8 Metathoracic spiracle with a row of black setulae along lower margin and overlying
operculum of spiracle. Proclinate upper orbital setae absent in both sexes. Cell
R5 wide at apex, where it is not or only slightly narrower than at middle. Dorsal
surface of all veins bare, except costa and sometimes vein 3 at base. Body never
metallic blue or green, except in Malagasy subregion. Distributed throughout
the Old World tropics. Keys to Palaearctic (Pont, 1966), Ethiopian (Emden, 1942),
Oriental (Emden, 1965), Australian (Pont, 1969) and Micronesian (Snyder, 1965)
species. Adults of certain species frequent human and animal faeces and rotting
fruit, and also enter houses. Some larvae live in dung   **DICHAETOMYIA** Malloch

– Metathoracic spiracle bare along lower margin, at most with a few setulae on posterior
margin. Proclinate upper orbital setae present in both sexes. Cell R5 very
strongly narrowed at apex, where it is not much wider than length of small cross-
vein. At least vein 3 setulose on dorsal surface from base half-way to small cross-
vein, and vein 1 and often other veins setulose. Thorax metallic green or blue,
abdomen similar to partly or wholly orange. Ethiopean Region, keys in Zielke
(1971). Adults on human faeces, carcasses and sap, and sometimes entering
houses. Larvae in rotting fruit . . . . .   *PYRELLINA* Malloch

9 Apical part of stem-vein always setulose on ventral surface and usually with 1–2
setulae on dorsal surface; basal part of stem-vein always bare on both surfaces.
Prosternum always bare. Vein 4 strongly curved forward and ending at or close
to wing-tip; vein 3 ending well before wing-tip. Neotropical Region, keys in
Snyder (1954). Adults on excrement, rotting vegetable matter and fresh fruit,
and often entering houses, tents, caves; they have been observed ovipositing on
human faeces . . . .   *NEOMUSCINA* Townsend

– Apical part of stem-vein always bare on dorsal and ventral surfaces; basal part setulose or bare.   Prosternum setulose or bare.   Vein 4 ending well behind wing-tip, and vein 3 ending at or near wing-tip except in *continens* Snyder.   Neotropical Region, keys in Snyder (1954).   Habits and hygienic significance as for *Neomuscina*

CYRTONEURINA Giglio-Tos

10   Vein 4 sharply bent forward towards vein 3 (fig. 126D, E), often with a slight dip just beyond the bend   11

–   Vein 4 with a rounded forward curvature towards vein 3 (fig. 126F)   .   .   .   12

11   Shining metallic green species.   Mid tibia with a strong posteroventral seta.   Infra-alar bulla and subcostal sclerite setulose.   Supra-squamal ridge setulose.   All regions, keys to New World (Huckett, 1965), Palaearctic (Hennig, 1963a), Ethiopian (Zielke, 1971), Oriental (Emden, 1965) and Australian (Pont, in press) species.   Adults very common on dung, in which the larvae live; rarely in dwellings or attracted to man, of little hygienic importance.

ORTHELLIA Robineau-Desvoidy, p.p.

–   Non-metallic species.   Mid tibia without a posteroventral seta.   Infra-alar bulla and subcostal sclerite bare.   Supra-squamal ridge bare or setulose.   (See discussion below)   .   .   .   .   **MUSCA** Linnaeus

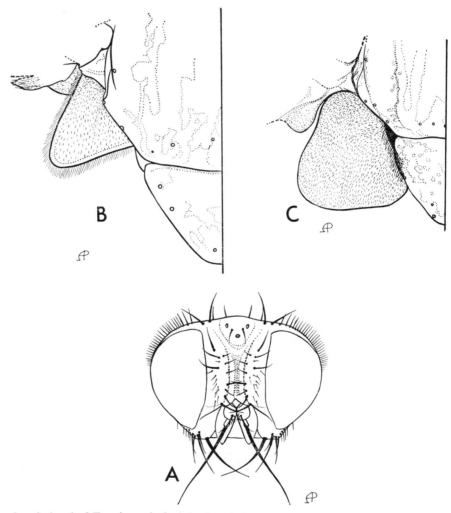

Fig. 128   A, head of *Fannia canicularis* in dorsal view; B, thoracic squama of *Phaonia variegata*; C, thoracic squama of *Musca domestica*.

12 Non-metallic species.   Mid tibia usually without a posteroventral seta.   All regions. Keys to Neotropical (Albuquerque, 1956), Nearctic (Huckett, 1965), Palaearctic (Hennig, 1964), Ethiopian (Zielke, 1971) and Oriental (Emden, 1965) species. Adults on flowers and vegetation, on excrement ,and also attracted as 'sweat-flies' to sweat and mucus of man, cattle and horses, extremely irritating in late summer in Europe; occasionally in houses.   Larvae in dung
MORELLIA Robineau-Desvoidy, p.p.

- Metallic green species.   Mid tibia with or without a posteroventral seta   .   .   13

13 Supra-squamal ridge setulose.   Mid tibia with a posteroventral seta.   (See above)
ORTHELLIA Robineau-Desvoidy, p.p.

- Supra-squamal ridge bare   14

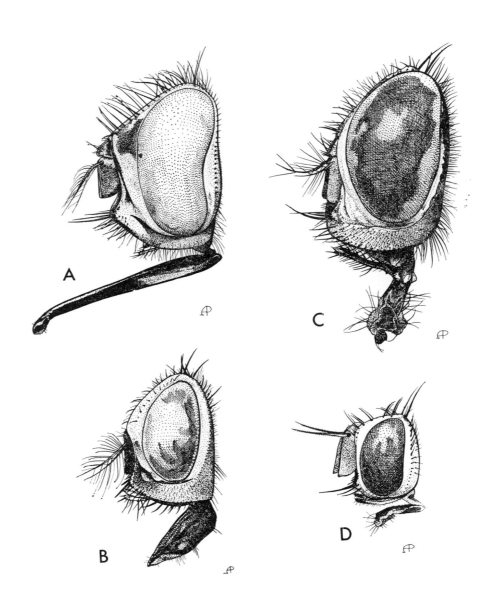

Fig. 129   Heads in lateral view: A, *Stomoxys calcitrans*; B, *Musca crassirostris*; C, *Musca domestica*; D, *Atherigona orientalis*.

14 Mid tibia with a strong posteroventral seta. Prosternum slender and bare, or if broad and setulose (two Australian species) then subcostal sclerite setulose. All regions, except Neotropical. Keys to Nearctic (Huckett, 1965), Palaearctic (Hennig, 1963a), Ethiopian (Zielke, 1971), Oriental (Emden, 1965) and Australian (Pont, in press) species. Biology and significance as for *Orthellia*.

    PYRELLIA Robineau-Desvoidy

– Mid tibia without a posteroventral seta, or if with one (some Neotropical species) then prosternum setulose and subcostal sclerite bare. Prosternum always broad. (See above) . . . . . . **MORELLIA** Robineau-Desvoidy, p.p.

15 Lower squama of the *Musca*-type (fig. 128C). Vein 4 always conspicuously curved forward towards vein 3 in apical part (fig. 126F). Scutellum often yellow at apex   16

– Lower squama of the *Phaonia*-type (fig. 128B) 17

16 Arista bare. Prosternum setulose. Hypopleuron setulose on metepisternum and below spiracle. (See discussion below) **SYNTHESIOMYIA** Brauer & Bergenstamm

– Arista plumose. Prosternum bare. Hypopleuron setulose only on metepisternum. (See discussion below) . . . . . **MUSCINA** Robineau-Desvoidy, p.p.

17 Hind tibia with a calcar, i.e. a strong seta in apical half placed slightly posterior of dorsal 18

– Hind tibia without a calcar 22

18 Vein 4 conspicuously curved forward towards vein 3 in apical part (fig. 126F). Hypopleuron setulose on metepisternum. Scutellum yellow at apex. (See discussion below) . . . . **MUSCINA** Robineau-Desvoidy, p.p.

– Vein 4 running straight to wing-margin or, in a few *Phaonia* and *Hydrotaea*, very weakly inclined forward towards vein 3 in which case either hypopleuron bare on metepisternum or scutellum wholly dark 19

19 Hind tibia with a row of 4–6 setae above and below the calcar, i.e. with many calcar setae. Female with proclinate upper orbital setae and crossed interfrontal setae. Four postsutural dorsocentral setae. Holarctic Region. Key to Nearctic species (Huckett, 1965), discussion of Palaearctic species (Hennig, 1963b). Adult males retiring in habit. Adult females are sweat-flies; often abundant and irritating in late summer, attracted to sweat and mucus of man and other animals, particularly around the head. Larvae in dung . . . . **TRICHOPTICOIDES** Ringdahl

– Hind tibia with a single calcar, at most with 1–2 short setae above it in basal half . 20

20 Sternopleural setae 1+2 or 2+2. If vein 4 is weakly inclined forward towards vein 3, then hypopleuron setulose on beret. Males without modified fore legs. Females without proclinate upper orbital setae, rarely with crossed interfrontal setae. All regions, keys to Nearctic (Malloch, 1923; Huckett, 1965), Palaearctic (Hennig, 1963a), Ethiopian (Emden, 1943) and Oriental (Emden, 1965) species. Adults on vegetation, flowers, trees, faeces, occasionally also in houses so perhaps of some hygienic significance; larvae in decaying vegetable and animal matter

    PHAONIA Robineau-Desvoidy

– Sternopleural setae 1+1, or if 1+2 then vein 4 weakly inclined forward towards vein 3 and hypopleuron bare on beret. Males often with modified fore-legs. Females with proclinate upper orbital setae and crossed interfrontal setae . . . 21

21 Males with modified fore-legs, fore femur with teeth or tubercles and fore tibia with corresponding indentations. Females usually with the frontal triangle dusted (females with a glossy frontal triangle are separable only with difficulty from *Ophyra*). (See discussion below) . . **HYDROTAEA** Robineau-Desvoidy, p.p.

– Males with simple fore-legs. Females with the frontal triangle undusted glossy black (females of Australian species with dusted frontal triangle are separable only with difficulty from *Hydrotaea*). (See discussion below) **OPHYRA** Robineau-Desvoidy

22 Vein 3 with setulae on the node at base on lower wing surface 23

– Vein 3 bare above and below 24

23 Hind tibia with the anterodorsal preapical seta hardly as long as tibial depth. Pre-alar seta absent. Holarctic, Ethiopian and Oriental Regions. Keys to Nearctic (Huckett, 1965), Palaearctic (Hennig, 1956), Ethiopian (Emden, 1951) and Oriental (Emden, 1965) species. Larvae in dung. Adults often active as sweat-flies or around wounds caused by other biting flies, also on dung; possible mechanical vectors of pathogens . . . . . . ***HEBECNEMA*** Schnabl, p.p.

– Hind tibia with the anterodorsal preapical seta longer than tibial depth. Pre-alar seta present, even if only short. All regions. Keys to Nearctic (Snyder, 1949; Huckett, 1965), Palaearctic (Hennig, 1959), Ethiopian (Emden, 1951), Oriental (Emden, 1965) and Australian (Malloch, 1925b) species. Adults generally retiring and innocuous in habits, occasionally on faeces or entering houses. Larvae in rotting matter, usually vegetables or wood . ***HELINA*** Robineau-Desvoidy, p.p.

24 Prosternum setulose. (See above) . . . ***HELINA*** Robineau-Desvoidy, p.p.
– Prosternum bare 25

25 Sternopleural setae 1+2, the lower seta equidistant from the upper two. Head of striking shape, subquadrate (fig. 129D). (See discussion below) ***ATHERIGONA*** Rondani
– Sternopleural setae 1+1, 1+2, or 2+2, when 1+2 the lower seta much closer to posterior one than to anterior one. Head of normal shape, never subquadrate . 26

26 Males with modified fore-legs, fore femur with teeth or tubercles and fore tibia with corresponding indentations. Females with proclinate upper orbital setae and crossed interfrontal setae. (See discussion below)
HYDROTAEA Robineau-Desvoidy, p.p.
– Males with simple fore-legs. Females with neither proclinate upper orbital setae nor crossed interfrontal setae 27

27 Hind tibia with the anterodorsal preapical seta longer than tibial depth. Pre-alar seta usually present. (See above) . . ***HELINA*** Robineau-Desvoidy, p.p.
Hind tibia with the anterodorsal preapical seta weak or absent. Pre-alar seta absent. 28

28 Arista bare or short-pubescent. (See discussion below)
GYMNODIA Robineau-Desvoidy
Arista plumose. (See above) . . . . . . ***HEBECNEMA*** Schnabl, p.p.

Subfamily ANTHOMYIINAE

Adult Anthomyiinae may be recognized by the complete 6th (anal) vein which in nearly all cases reaches the wing-margin even if only faintly (fig. 126A). The scutellum usually has a number of soft pale hairs on the ventral surface at apex, and the hind metatarsus has a short ventral basal seta. Females have proclinate upper orbital setae and, in most genera, both sexes possess interfrontal setae.

The adults of this subfamily are of no known medical or hygienic importance. *Anthomyia* Meigen, which can be distinguished by the presence of fine setulae on the propleural depression, may be of some hygienic importance as adults are sometimes found on faeces, carrion and garbage in residential areas and also occur indoors.

Most larvae so far as is known are phytophagous, and the larvae of certain genera may therefore cause accidental intestinal myiasis (see Chapter 6). The genera in question are *Paregle* Schnabl, *Erioischia* Lioy, *Pegomya* Robineau-Desvoidy, *Delia* Robineau-Desvoidy, *Pegohylemyia* Schnabl, and *Leptohylemyia* Schnabl.

Adults and larvae of this difficult subfamily should always be submitted to a specialist for identification.

Subfamily FANNIINAE

Adult Fanniinae may be recognized by the extremely short 6th (anal) vein, which does not extend half the distance from its base to the wing-margin, and by the curvature of the 7th vein, an imaginary extension of which would intersect an imaginary extension of the 6th vein well before the wing-margin (fig. 126B); and by the presence of a true dorsal submedian seta on hind tibia, in exact alignment with the preapical dorsal seta. Males are characteristic slender flies (pl. 5B), with a triangular or trimaculate abdominal pattern, and the mid legs are always modified ventrally, with either tubercles (fig. 127C) or a mat of fine erect hairs (fig. 127A, B). Females have unusually broad parafrontalia, the inner margins of which are convex (fig. 128A).

Fannia Robineau-Desvoidy is the only genus of medical importance, and all the species concerned in this volume have the hind coxa setulose behind.

Adult *Fannia* are generally encountered out-of-doors and the males soar singly or in swarms under trees, whilst females are more retiring in habit and move amongst the ground vegetation. Females of some species are attracted to carcasses or to mammalian sweat and mucus, and can act as vectors of nematodes. Certain species, such as *canicularis* (Linnaeus) and *scalaris* (Fabricius), occur commonly indoors. They rarely settle on humans or on food, but their presence and their potential as carriers of bacteria and nematodes render them a continual source of danger.

Larval *Fannia* feed in a wide variety of material, such as fungi, rotting plant material, leaf mould, decaying animal matter, decaying foodstuffs, excrement, and nests of vertebrates and insects. Several species cause intestinal, urino-genital, vesicular, aural and dermal myiasis in man and other mammals (see Chapter 6).

'*Fannia*' *desjardinsii* (Macquart) is recorded in cases of intestinal myiasis by Castellani and Chalmers (1913: 734): the larvae in question may well have been *Fannia*, but *desjardinsii* is a species of *Gymnodia*.

The genus contains some 220 described species. There are revisions of Palaearctic (Hennig, 1955), Nearctic (Chillcott, 1961) and Ethiopian (Emden, 1941) species. Adults and larvae of this difficult subfamily should always be submitted to a specialist for confirmation or identification, but the adults most likely to be encountered in the medical or hygienic context may be distinguished as follows:

KEY TO ADULT *FANNIA* OF MEDICAL IMPORTANCE

1 Palpi and basal antennal segments yellow . . . . *benjamini* Malloch
– Palpi and antennae black 2

2 Males only: tergites 4 and 5 trimaculate, each with a pair of lateral black spots separated by pale dust from the black median vitta 3
– Males without trimaculate pattern on abdomen, and all females 4

3 Hind tibia with numerous anteroventral and posteroventral setae   *pusio* (Wiedemann), ♂
– Hind tibia with one anteroventral and no posteroventral setae   *leucosticta* (Meigen), ♂

4 Hind tibia with 1–2 anteroventral setae 5
– Hind tibia with three or more anteroventral setae 8

5   Post-ocular setulae with a regular second row of setulae almost from vertex.   Knob of halteres dark.   Lower squama very long, projecting far beyond upper one. Hind femur with 1–2 preapical anteroventral setae and no posteroventral setae

                                                                                           australis Malloch

–   Post-ocular setulae strictly uniserial. Knob of halteres yellow. Lower squama not so long, not projecting far beyond upper.   Hind femur rarely as above .   .   6

6   Hind tibia with two anteroventral setae.   Mesonotum yellowish- to brownish-grey dusted, usually with three brown vittae.   Abdomen usually yellow at base, never with trimaculate pattern.   Female with the dusted frontal triangle confined to extreme upper part of frons, hardly reaching beyond upper reclinate orbital seta

                                                                       canicularis (Linnaeus)

–   Hind tibia with one anteroventral seta.   Mesonotum differently marked.   Abdomen never yellow at base.   Female with the dusted frontal triangle extending to level of lower reclinate orbital seta   7

7   Abdomen subshining black, without dusted pattern.   Mesonotum subshining black, without dusted vittae, with some light grey dust on humeri and before scutellum. Parafrontalia thinly grey dusted, partly subshining especially on inner margin

                                                                     pusio (Wiedemann), ♀

–   Abdomen with some grey dusting on at least tergites 4 and 5, often more extensively dusted, these tergites often with a trimaculate pattern.   Mesonotum dull grey dusted, often tinged with bluish, with three more or less conspicious dark brown vittae.   Parafrontalia entirely grey dusted   .   .   .   **leucosticta** (Meigen), ♀

8   Male: Fore tibia with an apical tuft of dense adpressed posteroventral setae; fore coxa with a thorn behind.   Female: Fore tibia without an anterodorsal seta at apical fifth; mesonotum, in dorso-lateral view, uniformly black or greyish-black without vittae   .   .   .   .   **manicata** (Meigen)

–   Male: Fore tibia without a preapical tuft of setae; fore coxa without a thorn behind. Female: Fore tibia with a short anterodorsal seta at apical fifth; mesonotum, in dorso-lateral view, thinly grey dusted with a pair of brownish vittae between acrostichal and dorsocentral rows   9

9   Male: Mid coxa with two strong coarse spines; mid tibia with a conspicuous triangular projection at about apical third of ventral surface.   Female: Upper post-ocular setulae uniserial, at most with an occasional setula below this row; parafacialia more slender, at level of apex of third antennal segment not as wide as fore tibia at base

                                                                     scalaris (Fabricius)

–   Male: Mid coxa without spines; mid tibia without a projection, ventral surface covered with a mat of uniform dense short fine hairs.   Female: Upper post-ocular setulae with a second irregular row of setulae below this row; parafacialia broader, at level of apex of third antennal segment at least as wide as fore tibia at base

                                                                  incisurata (Zetterstedt)

Fannia australis Malloch.   An Australian species.   The larva, like that of *canicularis*, can breed in a large range of media.   There are few published records of adult habits, but it is unlikely to be of more than hygienic importance.   It is a tertiary sheep-fly (Mackerras and Fuller, 1937).

Fannia benjamini Malloch.   A North American species.   Adult females pester man and animals by flying around the head, attracted by sweat and mucus.   It is an intermediate host for the nematode *Thelazia californiensis* Price, a parasite of man and domestic animals, and it seems probable that other species could fulfil a similar rôle as vectors of other nematodes in other parts of the world.

Fannia canicularis (Linnaeus).   Cosmopolitan.   This is the most important species of the genus medically, and is popularly known as the Lesser or Little House-fly.   The adult is most commonly found in houses, though it is rarely attracted to man or his food.

S

Females also visit excrement and carcasses. Although not a sweat-fly, it is an intermediate host for the nematode *Thelazia californiensis* Price. The larvae are found in any of the media given in the introductory remarks under this genus, and they may also be found in domestic situations such as rotting vegetables, rodents' and birds' nests, excrement, and urine-soaked napkins. They are a pest of deep litter chicken houses. It is the species most frequently involved in cases of urino-genital myiasis: flies may be attracted to oviposit by discharges from the male and female genital organs, and the young larvae feed on these discharges and progress through the urino-genital tract. It has also been reported in cases of intestinal myiasis, but it is equally possible that larvae have been ingested with food as that oviposition in the perianal region has taken place. The life-cycle, in summer in temperate regions, is completed in under a month, 18–22 days at 27°C. (see Chapter 6; also James (1947) and Zumpt (1965) for discussion of pathogenesis and cases).

Fannia incisurata (Zetterstedt). Holarctic Region and South America. Although the larvae have been found breeding in cesspools, the adult fly is of little known hygienic importance. The larva has also been recorded in cases of intestinal and aural myiasis.

Fannia leucosticta (Meigen). North America, Mediterranean, Ethiopian and Oriental Regions. The larva has been found in nests, carcasses and latrines, and the adult is likely to be of some hygienic importance in the tropics.

Fannia manicata (Meigen). A Holarctic species. The adult is of no known medical or hygienic importance, but the female is attracted to carcasses. The larva has been found in a case of human intestinal myiasis.

Fannia pusio (Wiedemann). Cosmotropical. The larva has been reared from a very large range of decaying animal and vegetable matter. The adult occasionally oviposits on fresh meat, so there is always a possibility of accidental intestinal myiasis. The medical status and potential of this fly on the Pacific island of Guam has been studied by Bohart & Gressitt (1951), and their conclusions are of wide relevance.

Fannia scalaris (Fabricius). Cosmopolitan. After *canicularis*, this is the most important member of the genus from the medical point-of-view. It is known popularly as the Latrine-fly. Adults are found indoors where standards of sanitation are not high, and are abundant around earth and chemical closets, latrine pits, etc. The larvae too breed in privies, and also in fungi, nests, carcasses and excrement. The larva has been involved in cases of urino-genital and intestinal myiasis, and the adult *modus operandi* is the same as that of *canicularis;* some cases of aural and vesicular myiasis have also been reported. The life-cycle, in summer in temperate regions, is completed in a month.

Subfamily MUSCINAE

Genus *Ophyra* Robineau-Desvoidy

Distributed in all regions. Neotropical species are covered by Albuquerque (1958), and species from other regions can be identified with the keys of Sabrosky (1949), Hennig (1962) and Emden (1965). Several species are widespread and abundant in the tropics.

Adults are usually found in the open, where they frequently swarm, but they sometimes enter houses and are abundant around poultry farms. In the tropics they are common on fresh faeces and carrion, and enter houses where they congregate in kitchens and walk over food, utensils, etc. They are also found around feeding troughs of domestic animals.

The larvae feed in carcasses, nests, excrement and manure, and rotting vegetation. The Australian *rostrata* (Robineau-Desvoidy) is a tertiary sheep-fly, but does not cause myiasis in man nor does it attack healthy tissue.

The adults, particularly those of *chalcogaster* (Wiedemann), *nigra* (Wiedemann) and *aenescens* (Wiedemann), are of great hygienic importance in the tropics and rank after species of *Musca* and *Atherigona* as the principal vectors of faecal pathogens. In temperate regions the genus is of minor significance.

Genus *Hydrotaea* Robineau-Desvoidy

Distributed in all regions, mainly Holarctic. Keys are available to Neotropical (Albuquerque, 1957), Nearctic (Huckett, 1954), Palaearctic (Hennig, 1962), Ethiopian (Emden, 1943) and Indo-Australasian (Emden, 1965) species.

Adults, like those of *Ophyra*, are found in the open, on flowers, foliage, decaying matter, etc. Males frequently swarm. Females of some species (*irritans* (Fallén), *meteorica* (Linnaeus), *occulta* (Meigen), *albipuncta* (Zetterstedt)) are the most persistent and irritating of the sweat-flies in late summer in temperate regions and *irritans* has been the subject of a study by Nielsen *et al.* (1971). Some females (*militaris* (Meigen)) are attracted to trampled grass and thus swarm around man and cattle. *H. dentipes* (Fabricius) is commonly found on dung, human faeces, and carcasses, and also enters dwellings. They are also attracted to the blood oozing from wounds, mucus, and other secretions from the nose, eyes and mouth.

Because of these habits, and quite apart from the discomfort they cause, the adults may act as vectors of various pathogens and eye-worms.

The larvae have been reared from dung, human faeces, carrion, and various decomposing plant and animal materials. Some species live in the nests of birds and burrows of mammals. The larvae of *meteorica* have been reported in cases of intestinal myiasis.

Genus *Musca* Linnaeus

This is the most important genus of the family from the medical point-of-view. There are about 60 species, all of which are confined to the Old World except for two, the cosmopolitan *domestica* Linnaeus and the Holarctic *autumnalis* Degeer which has recently spread to the New World.

The species fall into three groups according to the adult habits. Firstly, there are the synanthropic species that are intimately connected with man's person and dwellings. *M. domestica* with its subspecies, and members of the *sorbens*-complex, are the only species here. They can and do transmit pathogens to human food and the human body, both biologically and mechanically. This group has been well studied, and its relationship to disease, genetics, biology and control are now partially understood. Secondly, there are the true blood-sucking species: *crassirostris* Stein, *seniorwhitei* Patton, *fletcheri* Patton & Senior-White, *planiceps* Wiedmann, and *inferior* Stein. These are equipped with a piercing proboscis (*crassirostris*) or with strong prestomal teeth with which they can break the surface of skin and draw blood. They are potential mechanical transmitters of pathogens. Thirdly, there are the haematophagous species, and the bulk of the genus belongs here, including a number of rare or little-known species. These cannot scratch or pierce, but they disturb other biting flies and suck the blood and serum oozing from wounds. They are also attracted to ocular, nasal and buccal secretions, and, although only found in the wild and usually attacking animals other than man, they are also potential transmitters of pathogens and may affect man. The species *conducens* Walker and *mesopotamiensis* Patton are intermediate between the second and third groups, for they cannot pierce skin but can scratch off scabs and enlarge existing wounds with their prestomal teeth.

In general only the members of the *domestica*- and *sorbens*-complexes are of medical importance. There are many observations on the adult habits and biology of the other species, though little laboratory or pathological work, and the majority of them are associated with cattle or domestic animals in the field, with the larvae living in cow dung. Some species, such as *vitripennis* Meigen, are equally attracted to human sweat and can be extremely irritating. But they are primarily vectors of veterinary rather than medical importance, and although it is clear that their habits and biology could lead to contamination of man's person, they require no further discussion here.

Keys are available to the Palaearctic (Hennig, 1964), Oriental (Emden, 1965), Ethiopian (Zielke, 1971), and Australian (Pont, in press) species. In additional, regional keys to all synanthropic species have been published by Pont & Paterson (in Greenberg, 1971).

The following couplets will facilitate identification of the species of prime medical importance:

PARTIAL KEY TO THE ADULTS OF *MUSCA* LINNAEUS

1 Propleural depression setulose . . . . . . *domestica* Linnaeus, s. lat.
- Propleural depression bare 2

2 Proboscis enormously dilated, boat-shaped (fig. 129B) and adapted for piercing. Mid tibia with a submedian anteroventral seta. Palpi yellow, abdomen nowhere yellow, thorax and abdomen densely grey to yellowish-grey dusted *crassirostis* Stein
- Proboscis normal (fig. 129C), not adapted for piercing. Mid tibia without a submedian anteroventral seta 3

3 Hypopleuron with setulae present below spiracle as well as on metepisternum. Sternite 1 bare. Prothoracic spiracle white. Fore tibia without a posterior seta. Mesonotum with the vittae on each side fused behind suture, thus with two broad vittae . . . . *sorbens* Wiedemann, s. lat.
- Without the above combination of characters . . . . . Other species

Musca domestica Linnaeus: Cosmopolitan, known as the House-fly. There are a number of closely related forms that have in the past been accorded specific, subspecific or lower status, *vicina* Macquart, *nebulo* Fabricius, *calleva* Walker (=*cuthbertsoni* Patton), *curviforceps* Saccà and Rivosecchi, but these all possess setulae on the propleural depression which character will distinguish the *domestica*-complex from the rest of the genus. The elucidation and status of these forms is extremely complex, particularly in the Ethiopian Region, and such problems can only be solved by the geneticist. These problems are discussed by Paterson (1956, 1963). The species has been the subject of a monograph by West (1951), and the results of more recent research are summarized by Saccà (1964). A revised edition of West (1951) is in preparation.

The adult is a true synanthrope, and has followed man around the world. It actively seeks to enter houses, where it will alight on garbage, food, or man himself. The adult will feed on anything with a moist surface, especially milk, sweets, meat, excrement and liquefying garbage. Females are particularly attracted to substances containing the protein necessary for the maturation of the ovaries. Its rôle in the contamination of milk is most important as it often vomits or defaecates on its food. It is not a persistent seeker of sweat, mucus or other secretions. It defaecates at random, and regurgitates its food at frequent intervals to aid digestion ('fly spots').

The adult is oviparous. Up to 1000 eggs may be laid, singly or in small to large batches, at the rate of 100–150 a day. In the country they are deposited principally upon fresh horse-dung, but the larvae also feed readily on human, cow and poultry dung, bedding and other rubbish contaminated with excrement, decaying vegetable matter and garbage, decomposing foodstuffs, meat and carcasses.

The life-cycle, from egg to adult, takes 8 days under optimum conditions at 33–35°C. At 30°C. it takes 9–11 days, at 25°C. 15–18 days, at 20°C. 19–22 days, at 18°C. 23–30 days, and at 10–15°C. 40–50 days. Adults generally avoid direct sunlight, preferring the shade of buildings, and are most active during cooler times of the day. Below 10°C. adult activity ceases. Adults can fly up to 3.5 km (2 miles) from their site of emergence with ease, and have been found up to 8 km (5 miles) away.

The enemies of the adult are fungi, of which there are many species, most notably *Empusa muscae* Cohn, and bacteria. Dead adults are often found stuck to windows in autumn surrounded by fungal spores. Immature stages are parasitized by Hymenoptera, and preyed upon by ants, beetles, other fly maggots, and mites.

Because of their habits, the adults are capable of transmitting a large number of diseases to man but they are rarely true biological vectors of pathogens. The transmission takes place mechanically: pathogens may be ingested by the fly and deposited with the faecal spots, ingested and deposited with the vomit, or spread around by the labella and legs. Larvae feeding on infected material can produce infected adults. The house-fly is thereby able to transmit virus diseases (poliomyelitis and Coxsackie virus, both *via* human faeces, and Q fever); bacterial diseases (many diarrhoeal and enteric fevers, infantile summer dysentery, typhoid and paratyphoid fevers, bacillary dysentery, all *via* human faeces; conjunctivitis, tuberculosis, leprosy, plague; streptococci and staphylococci); protozoan parasites (cysts and trophozoites, trypanosomes, amoebic dysentery); tapeworms and nematodes (*Thelazia*, etc.): and other arthropods (*Cordylobia* eggs, young *Pediculus*). Ordinary methods of hygiene, particularly hygienic disposal of faeces and covering of food, are sufficient to avoid danger in temperate climates.

The larva has also been recorded in cases of intestinal, urino-genital, traumatic, aural and nasopharyngeal myiasis, but it is not of great significance.

Musca sorbens Wiedemann.   Abundant in subtropical and tropical regions of the Old World, where it complements *domestica* in biology, ecology and pathology.   The Australian Bush-fly, and certain Oriental and African populations, have in the past been treated as a distinct species, *vetustissima* Walker, and recent work on the genetics of the *sorbens*-complex has shown that at least three species are involved (Paterson and Norris, 1970): two of these are African, whilst the third, for which the correct name is *vetustissima*, is Australian.   Genetics of Oriental and Pacific populations remain to be investigated.   The complex may be identified by its characteristic mesonotal pattern and the other characters given in the key.

Adults are rarely found indoors and are not true domestic flies.   They avoid shade and seek light and sun, being most active during the hottest parts of the day.   They are better able to resist high temperature and desiccation than *domestica*.   They have a special tropism to the human body and are attracted to sores, lesions, abscesses and particularly to the face and around the eyes whether healthy or diseased (pl. 4). They also feed on foodstuffs in the open and have been called Bazaar-flies.

The adult is oviparous.   Up to 80 eggs are laid, in 1–5 batches, usually in human excrement unlike most other species of *Musca*.   Pig, dog and cow dung, refuse, and carcasses also provide food for the larvae.   The life-cycle from egg to adult takes 9 days at 25–28°C.

M. sorbens (pl. 4) is more dangerous than *domestica* in its attraction to eyes and sores. Adults are frequently attracted in swarms to such sites, and up to 50 flies have been counted on one open sore of yaws (Bohart & Gressitt, 1951).   Continual brushing or waving with the hand is not sufficient to discourage them.   They are principal vectors of certain eye infections (ophthalmia, blepharitis, corneal ulcers) which may result in permanent damage to the eyes.   They can also transmit viral, bacterial and parasitic diseases (tuberculosis, leprosy, yaws; streptococci and staphylococci).   The larvae have also been reported in cases of traumatic myiasis.

Musca crassirostris Stein.   Mediterranean, Ethiopian and Oriental Regions.   The adult is a true blood-sucker, but rarely attacks man.   The saliva contains a powerful anti-coagulant.   The larva breeds in cow dung, but has been reported in some cases of intestinal myiasis.   This is almost certainly accidental, and may be due in part to oriental religious rites connected with cows that involve coprophagy.

Genus *Muscina* Robineau-Desvoidy

One cosmopolitan species, *stabulans* (Fallén), and seven other Holarctic species.   Keys are available to Palaearctic (Hennig, 1962) and Nearctic (Snyder, 1955) species.

Muscina stabulans can be distinguished from the other species by the following combination of characters: all tibiae, and apical part of mid and hind femora, yellow; palpi, epaulet and basicosta yellow; male parafrontalia separated throughout by the slender black interfrontalia.   The adults are frequently found in houses, but are commoner in the open around stables, byres, poultry houses, and on decaying organic matter.   They feed on decaying meat, fruit and vegetables.   Their hygienic significance is probably slight.   The larvae live in rotting fungi and fruits and in carrion, and prey on dipterous maggots.   They are facultative parasites of other insects, and attack nestling birds, causing their death. They can also cause intestinal myiasis in man, and traumatic and intestinal myiasis in other mammals. Infection in man takes place by ingestion of eggs deposited on food that is slightly tainted or spoiled. The life-cycle, in summer in temperate regions, takes 4–5 weeks; at 27°C. it takes 20–25 days.

Muscina assimilis (Fallén) and *M. pabulorum* (Fallén) have similar adult habits and larval habitats, but the adults rarely enter houses: they are thus of no hygienic importance, and the chance of their causing accidental myiasis in man is very remote. There are a few records of the larvae causing myiasis in other vertebrates, viz. *assimilis* in nesting birds and *pabulorum* in sheep. These and other species of the genus have been caught in meat-bait traps.

Genus *Synthesiomyia* Brauer and Bergenstamm

There is only one species, *nudiseta* (Wulp), which is cosmopolitan. The adult is seldom common, and rarely enters houses. It has been found resting on wood, and is also attracted to carcasses and dead molluscs. The larva usually breeds in carrion and also, more rarely, in human faeces and refuse. The adult is probably of little hygienic significance, and the larva has been involved in cases of secondary traumatic myiasis but is similarly of minor importance.

Genus *Atherigona* Rondani

The genus is common in tropical and subtropical regions of the Old World, and one species has also spread to the New World. There are keys to the Palaearctic (Hennig, 1961), Ethiopian (Deeming, 1971), Indian (Malloch, 1925a) and Micronesian (Snyder, 1965) species. The species *orientalis* (Schiner) (=*excisa* (Thomson)) is cosmotropical, and its occurrence in the New World is dealt with by Aldrich (1921).

Only the species *orientalis* is of medical and hygienic importance. All other species are retiring and innocuous in the adult stage, and the larvae are phytophagous: these may perhaps, like the Anthomyiinae, occasionally cause accidental intestinal myiasis in man.

Atherigona orientalis is extraordinarily abundant in the tropics. The larvae are scavengers and have been recorded from decaying fruit and vegetables, kitchen refuse, carrion, cow dung and human faeces. They are occasionally predacious. Adults in all parts of the world are attracted to carrion, decaying food, table food, and human excrement. They are equally at home in human dwellings and in the wild.

The adult ranks with *Musca* and *Ophyra* as one of the most important vectors of faecal and other filth-borne pathogens. Its great abundance and its readiness to enter human dwellings and to crawl on food are a continual source of danger.

Genus *Gymnodia* Robineau-Desvoidy

Distributed in all regions. Keys are available to the Nearctic (Huckett, 1932), Palae-arctic (Hennig, 1959), Ethiopian (Emden, 1951), Oriental (Emden, 1965) and Micro-nesian (Snyder, 1965) species.

The larvae live in cow dung. The adults are also found on dung, and sometimes enter houses to hibernate: they may therefore act as mechanical vectors of pathogens. One species, *desjardinsii* (Macquart), is recorded by Castellani & Chalmers (1913: 734) in cases of intestinal myiasis in Angola.

Members of the following two genera are true blood-suckers, equipped with a sharp horny non-retractile proboscis with which to pierce the skin and strong prestomal teeth for scratching and tearing. Victims are generally animals other than man, especially cattle, buffalo and horses. They are primarily of veterinary importance, because they can act as vectors of diseases affecting these animals and also because their continued attentions often cause loss of condition and general debilitation, lower milk yield, poor quality meat, etc. They are mechanical but not biological vectors of pathogens, but do not transmit faecal pathogens because they are not attracted to human excrement or food.

Genus *Stomoxys* Geoffroy

One species, *calcitrans* (Linnaeus), is cosmopolitan; otherwise Ethiopian and Oriental Regions, with some species impinging upon the Palaearctic region. Keys are available to the World species (Zumpt, in press), and older regional keys to Palaearctic (Hennig, 1964), Ethiopian (Zumpt, 1950), and Oriental (Emden, 1965) species.

The cosmopolitan *calcitrans* is known popularly as the Stable-fly or Biting House-fly. The adult has been shown to be capable of transmitting poliomyelitis. It is also suspected of transmitting trypanosomes to man and other animals, and to be occasionally implicated in the spread of nagana, surra, anthrax, tularemia, and other diseases. It is the invertebrate host of *Habronema microstoma*, a parasitic nematode in horses' stomachs. Apart from this it is of little importance to man: it worries animals and can cause serious loss of condition, but seldom bites man.

Adults are abundant around stables and farmyards, sunning themselves on walls, implements, vehicles, etc. When it bites man, it prefers to stab through socks or stockings and rarely attacks bare skin. It is often most persistent after rain. The blood-meal is essential for maturation of the ovaries, and the female is oviparous. 150–450 eggs are laid, in many batches. The larvae live in dung- and urine-soaked straw, wet straw sacks, sand and earth rich in organic matter, and decaying vegetable refuse. They are more selective than, for example, *Musca* or *Ophyra* species, and require a high moisture content. The life-cycle from egg to adult can take as little as 12 days under optimum conditions; at 21°C. it takes 33–36 days.

In rare cases, the larva of *calcitrans* can cause traumatic or intestinal myiasis in man.

Other species of *Stomoxys* are not well studied in respect of adult habits, biology and medical potential. Adults attack cattle, oxen, mules, horses, dogs and occasionally man. They may transmit trypanosomes. The larvae have been reared from cow dung, horse dung, and heaps of fermenting grass.

Genus *Haematobia* Le Peletier and Serville

The nomenclature within this group and the status of the taxonomic segregates are both problems of much complexity and controversy. Application has been made to the International Commission on Zoological Nomenclature to fix the type-species of the controversial generic names, and the results anticipated from this application have been adopted by Zumpt (in press) in his monograph on the Stomoxyinae.

The Neotropical "*Neivamyia*" includes the only South American species of the genus: four species are keyed by Souza Lopes & Mangabeira (1938), and one species has subsequently been described (Souza Lopes, 1955). The three Nearctic species are keyed by Huckett (1965, as *Haematobia* and *Lyperosiops*). There are keys for all other regions: Palaearctic (Hennig, 1964, as *Siphona*), Ethiopian (Zumpt, 1950, as *Haematobia*), Oriental (Emden, 1965, as *Haematobia* and *Lyperosia*), and Australasian (Snyder, 1965, as *Haematobia* and *Bdellolarynx*); and a key to all World species (Zumpt, in press).

So far as is known, the species attack animals other than man (horses, cattle, buffalo, mules, dogs) and, apart from causing general loss of condition, can transmit bovine and equine diseases; adults of "*Neivamyia*" carry eggs of the Oestroid *Dermatobia hominis* (Linnaeus, jr.). The larvae live in dung. *H. irritans* (Linnaeus) and *H. exigua* de Meijere are known respectively as the Horn-fly and the Buffalo-fly and are serious pests where they occur. They have been accidentally introduced into Australia and various Pacific islands where, in the absence of natural enemies, they multiply unchecked and can have disastrous effects on native domestic animals and economy.

BIBLIOGRAPHY

ALBUQUERQUE, D. DE O. 1956. Fauna do Distrito Federal. XII. Sôbre o gênero *Morellia* R.-D. (Diptera: Muscidae). *Bolm Mus. nac. Rio de J., Zool.*, No. **131**: 45 pp.

—— 1957. Contribuição ao conhecimento de *Hydrotaea* R.D., 1830 na América do Sul (Diptera: Muscidae). *Bolm Mus. nac. Rio de J., Zool.*, No. **160**: 18 pp.

—— 1958. Sôbre *Ophyra* R.-D., 1830 na América do Sul, com descrição de uma espécie nova (Diptera: Muscidae). *Bolm Mus. nac. Rio de J., Zool.*, No. **181**: 13 pp.

ALDRICH, J. M. 1921. The Anthomyiid genus *Atherigona* in America. *Insecutor Inscit. menstr.* **9**: 93–98.

BOHART, G. E., and GRESSITT, J. L. 1951. Filth-inhabiting flies of Guam. *Bull. Bernice P. Bishop Mus.*, No. **204**: 152 pp.

CASTELLANI, A., and CHALMERS, A. J. 1913. *Manual of Tropical Medicine.* 2nd edition. 1747 pp. London.

CHILLCOTT, J. G. 1961. A revision of the Nearctic species of Fanniinae (Diptera: Muscidae). *Can. Ent.* **92**, Suppl. 14: 295 pp.

DEEMING, J. C. 1971. Some species of *Atherigona* Rondani (Diptera: Muscidae) from northern Nigeria, with special reference to those injurious to cereal crops. *Bull. ent. Res.* **61**: 133–190.

EMDEN, F. I. VAN. 1941. Keys to the Muscidae of the Ethiopian Region: Scatophaginae, Anthomyiinae, Lispinae, Fanniinae. *Bull. ent. Res.* **32**: 251–275.

—— 1942. Keys to the Muscidae of the Ethiopian Region: *Dichaetomyia*-group. *Ann. Mag. nat. Hist.* (11) **9**: 673–701 and 721–736.

—— 1943. Keys to the Muscidae of the Ethiopian Region: *Phaonia*-group. *Ann. Mag. nat. Hist.* (11) **10**: 73–101.

—— 1951. Muscidae: C.–Scatophaginae, Anthomyiinae, Lispinae, Fanniinae and Phaoniinae, in *Ruwenzori Expedition 1934–5* 2 (6): 325–710. London.

—— 1965. *The Fauna of India and the adjacent countries.* Diptera **7**, Muscidae, part 1. 647 pp. Delhi.

GREENBERG, B. 1965. Flies and disease. *Scient. Am.* **213**: 92–99.

—— 1971. *Flies and Disease.* I.—*Ecology, Classification and Biotic associations.* 856 pp. Princeton.

HENNIG, W. 1955–1964. Muscidae, in Lindner, E., *Fliegen palaearkt. Reg.* **7** (63b): 1110 pp. Stuttgart.

—— 1963b. Eine neue Art der Gattung *Trichopticoides* aus Zentral-Asien (Diptera: Muscidae). *Mitt. dt. ent. Ges.* **22**: 55–57.

HUCKETT, H. C. 1932. The North American species of the genus *Limnophora* Robineau-Desvoidy, with descriptions of new species (Muscidae, Diptera). *Jl N.Y. ent. Soc.* **40**: 25–76, 105–158, and 279–339.

—— 1954. A review of the North American species belonging to the genus *Hydrotaea* Robineau-Desvoidy (Diptera: Muscidae). *Ann. ent. Soc. Am.* **47**: 316–342.

—— 1965. The Muscidae of Northern Canada, Alaska and Greenland (Diptera). *Mem. ent. Soc. Can.* **42**: 369 pp.

JAMES, M. T. 1947. The Flies that cause Myiasis in Man. *Misc. Publs U.S. Dep. Agric.*, No. **631**: 175 pp.

LINDSAY, D. R. & SCUDDER, H. I. 1956. Nonbiting Flies and disease. *A. Rev. Ent.* **1**: 323–346.

MACKERRAS, I. M., and FULLER, M. E. 1937. A survey of the Australian Sheep Blowflies. *J. Coun. scient. ind. Res. Aust.* **10**: 261–270.

MALLOCH, J. R. 1923. Flies of the Anthomyiid genus *Phaonia* Robineau-Desvoidy and related genera, known to occur in North America. *Trans. Am. ent. Soc.* **48**: 227–282.

—— 1925a. Some Indian species of the Dipterous genus *Atherigona* Rondani. *Mem. Dep. Agric. India ent. Ser.* **8**: 111–125.

—— 1925b. Notes on Australian Diptera, No. V. *Proc. Linn. Soc. N.S.W.* **50**: 35–46.

NIELSEN, B. OVERGAARD, NIELSEN, B. MØLLER, and CHRISTENSEN, O. 1971. Bidrag til plantage-fluens, *Hydrotaea irritans* Fall., biologi (Diptera: Muscidae). *Ent. Meddr* **39**: 30–44, 7 figs. [With English summary.]

NORRIS, K. R. 1966. Notes on the ecology of the Bushfly, *Musca vetustissima* Walk. (Diptera: Muscidae), in the Canberra district. *Aust. J. Zool.* **14**: 1139–1156.

PATERSON, H. E. 1956. Status of the Two Forms of Housefly occurring in South Africa. *Nature, Lond.* **178**: 928–930.

—— 1963. On the naming of the indigenous houseflies of the Ethiopian Region. *J. ent. Soc. sth. Afr.* **26**: 226–227.

PATERSON, H. E., and NORRIS, K. R. 1970. *The Musca sorbens* complex: the relative status of the Australian and two African populations. *Aust. J. Zool.* **18**: 231–245, 7 figs, 3 tables.

PONT, A. C. 1966. A second palaearctic species of the genus *Dichaetomyia* Malloch (Diptera: Muscidae). *Stuttg. Beitr. Naturk.*, No. **160**: 4 pp.

—— 1969. Studies on Australian Muscidae (Diptera). II.—A revision of the tribe Dichaeto-myiini Emden. *Bull. Br. Mus. nat. Hist.* (Ent.) **23**: 191–286, 105 figs.

—— In press. Studies on Australian Muscidae (Diptera). IV.—A revision of the subfamilies Muscinae and Stomoxyinae. *Aust. J. Zool.*

SABROSKY, C. W. 1949. The Muscid genus *Ophyra* in the Pacific Region (Diptera). *Proc. Hawaii. ent. Soc.* **13**: 423–432.

SACCÀ, G. 1964. Comparative Bionomics in the genus *Musca*. *A. Rev. Ent.* **9**: 341–358.

SCOTT, H. G. & LITTIG, K. S. 1962. Flies of public health importance and their control. *Publ. Hlth Serv. Publs Wash.* **799**: 1–40.

SMITH, T. A., LINSDALE, D. D. & BURDICK, D. J. 1966. An annotated bibliography of the face fly, *Musca autumnalis* Degeer in North America. *Calif. Vector Views* **13**: 43–54; supplements, **14**: 74–76; **15**: 119–121.

SNYDER, F. M. 1949. Nearctic *Helina* Robineau-Desvoidy (Diptera: Muscidae). *Bull. Am. Mus. nat. Hist.* **94**: 107–160.

—— 1954. A revision of *Cyrtoneurina* Giglio-Tos, with notes on related genera (Diptera: Muscidae). *Bull. Am. Mus. nat. Hist.* **103**: 417–464.

—— 1955. Notes and descriptions of *Muscina* and *Dendrophaonia* (Diptera: Muscidae). *Ann. ent. Soc. Am.* **48**: 445–452.

—— 1965. Muscidae, in *Insects of Micronesia* **13** (6): 191–327. Honolulu.

SOUZA LOPES, H. DE, and MANGABEIRA, O. 1938. Contribuiçao ao conhecimento do genera *Neivamyia* Pinto et Fonseca, 1930 e descripção de uma nova especie (Diptera: Muscidae). *Livro jubilar Prof. Travassos*: 287–290. Rio de Janeiro.

SOUZA LOPES, H. DE. 1955. Sôbre uma nova espécie de mosca hematófaga do gênero 'Neivamyia' P. & F., da Côlombia (Diptera: Muscidae). *Revta bras. Biol.* **15**: 415–418.

WEST, L. S. 1951. *The Housefly.* 584 pp. New York.

WEST, L. S. & PETERS, O. B. [1972]. Annotated bibliography of *Musca domestica* [*circa* 775 pp.] In press.

ZIELKE, E. 1971. Revision der Muscinae der äthiopischer Region. *Series Ent.* **7**: 199 pp.

ZUMPT, F. 1950. 5th preliminary study to a monograph of the Stomoxydinae. Key to the Stomoxydinae of the Ethiopian Region, with description of a new *Haematobia* and a new *Rhinomusca* species from Zululand. *Anais Inst. Med. trop., Lisb.* **7**: 397–426.

—— 1965. *Myiasis in man and animals in the old world.* 267 pp. London.

—— 1973. *The Stomoxyine biting flies of the world (Diptera: Muscidae). Taxonomy, biology, economic importance and control measures.* 175 pp. Stuttgart.

5c. CALLIPHORIDAE & SARCOPHAGIDAE
(Blow-flies and Flesh-flies)

by Kenneth G. V. Smith

THE Calliphoridae and Sarcophagidae are of medical importance mostly in the larval stage and the family is dealt with fully in the section on eggs and larvae (Chapters 6 and 17). However, adults may often be encountered, especially if the larvae are reared through and the resulting adults submitted for identification.

The adult flies are the well-known bluebottles and greenbottles, also known as blow-flies, and the grey flesh-flies. The latter are now regarded as a separate family, the Sarcophagidae, but the two families are treated together here. Although bearing a superficial resemblance to flies of other families (e.g. Muscidae, Tachinidae) the Calliphoridae are distinguished by having well-developed thoracic squamae, postscutellum absent or poorly developed and a row of bristles on the hypopleuron (fig. 11). The Calliphoridae occur throughout the world and mostly breed in decaying animal matter. Some choose specialized habitats, but many will lay their eggs on cooked or uncooked flesh, fish and offal. The resulting larvae soon render the meat 'flyblown' and quite unfit for use as food.

Some Calliphorids lay their eggs in open wounds on man and animals. This habit may be accidental, but in some species is obligatory. Thus infested with actively feeding maggots the wound may be considerably enlarged with extensive damage to neighbouring tissues. Other parts of the body may be invaded, such as the nasal passages, but these cases are included in the section on myiasis (Chapter 6). Other species breed in animal or human excrement and may thus, by subsequently settling on foodstuffs, transmit pathogenic organisms.

The adults of genera of medical importance may be separated by the following key and adults of some of the more important species are illustrated. The larvae are keyed in Chapter 6.

KEY TO THE PRINCIPAL GENERA OF ADULT CALLIPHORIDAE AND SARCOPHAGIDAE OF MEDICAL IMPORTANCE

1   Base of stem-vein (radius) (fig. 4G) with a row of bristly hairs above   .   .   .   2

\-   Base of stem-vein bare above   6

2   Hind coxae hairy behind, green to violet-green species, with three prominent black longitudinal vittae on the mesonotum; palpi short and slender

            COCHLIOMYIA Townsend (=*Callitroga* Brauer)

\-   Hind coxae bare behind, green to bluish-black species, at most with two narrow longitudinal thoracic stripes   3

3   Lower (thoracic) squamae with fine hairs above   ***CHRYSOMYA*** Robineau-Desvoidy
–   Lower squamae bare   4

4   Prothoracic spiracle with bright orange hair   .   ***PHORMIA*** Robineau-Desvoidy
–   Prothoracic spiracle dark-haired   5

5   Acrostichal bristles weak, alar (upper) squamae with long erect black hairs above
           PROTOPHORMIA Townsend
–   Acrostichal bristles strong, alar (upper) squamae bare   ***PROTOCALLIPHORA*** Hough

6   Greyish flies with three broad black longitudinal thoracic stripes (Sarcophagidae)   .   7
–   Metallic blue or green, yellow or brown flies   8

7   Abdomen with black spots (fig. 130B); antennal arista bare
           WOHLFAHRTIA Brauer & Bergenstamm
–   Abdomen chequered (fig. 130C); arista with long hairs above and below
           SARCOPHAGA Meigen

8   Flies of wholly metallic blue-green or coppery coloration (fig. 132B)
           LUCILIA Robineau-Desvoidy
–   Flies of yellow or brownish coloration, at most only partly metallic   .   .   .   9

9   Lower squamae with long hairs above; propleuron hairy
           CALLIPHORA Robineau-Desvoidy
–   Lower squamae bare; propleuron bare   10

10   Second abdominal segment especially long; vein R_2 with hairs almost to cross vein
    r-m; a small but distinct costal spine present; eyes of both sexes broadly separated
    (fig. 131A)   .   .   .   .   .   ***AUCHMEROMYIA*** Brauer & Bergenstamm*
–   Second abdominal segment of ordinary length; vein R_2 with hairs less than half-way
    to cross vein r-m; costal spine absent; eyes of male only narrowly separated (fig.
    131B)   .   .   .   .   ***CORDYLOBIA*** Grünberg*

* The genus *Bengalia* has been erroneously recorded in cases of myiasis probably being confused with *Auchmeromyia* and *Cordylobia*. The eyes are widely separated, as in the former genus, but the abdominal segments are normal with a narrow dark band along the hind margins of their dorsal surfaces (fig. 130A) and the proboscis is swollen and shiny.

CALLIPHORIDAE (BLOW-FLIES)

Genus *Auchmeromyia* Brauer and Bergenstamm

The larva of *Auchmeromyia luteola* (Fab.) is the well-known Congo floor maggot distributed throughout Africa south of the Sahara. The adult (fig. 131A) is yellowish-brown with dark thoracic stripes and is some 1–12 mm. long with slightly brownish wings and the posterior part of the abdomen blackish. Viewed from above the second segment of the abdomen, especially in the female appears lengthened to occupy about half the length of the abdomen. This distinguishes it at once from the Tumbu-fly (fig. 131B).

The larva (fig. 146C, pl. 3A) lives in cracks in the earth floors of native huts and similar places, feeding at night by piercing the skin of the sleeping occupants. Sleeping on beds raised from the ground helps prevent attacks. The larvae can survive for long periods without feeding.

Zumpt (1965) gives a key to adults of the five known species, but only *A. luteola* attacks man. The larva is treated in Chapter 6.

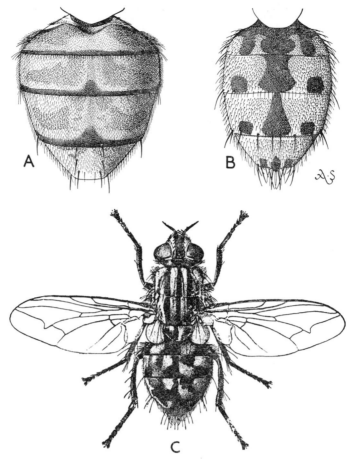

Fig. 130   A, abdomen of *Bengalia depressa* in dorsal view; B, abdomen of *Wohlfahrtia magnifica* in dorsal view; C, *Sarcophaga carnaria* a flesh-fly ×3½ (from Castellani & Chalmers).

Genus *Calliphora* Robineau-Desvoidy

These flies are the common bluebottles or blow-flies (fig. 6F), mostly of bluish or bluish-black coloration and some 8–14 mm. long.   They occur commonly around houses in search of breeding material and food.   They are attracted to meat and carrion upon which females lay their eggs for preference, although other human foods such as cheese may be chosen.

Adults are also attracted to faeces for feeding and the possibility of their flying from faeces to fresh meat gives them some medical importance.

The eggs are white and elongate and are laid in batches in crevices of the meat or dead animal, especially around the natural orifices.

The maggots commence feeding on the flesh which soon putrifies.   They then leave their feeding site and wander in search of a place in which to pupate.   Normally pupation occurs in the soil but in domestic situations suitable sites may be difficult to locate and fully grown maggots may be found wandering in quite unlikely situations some distance from their larval food.   Occasionally the larvae of *Calliphora* are ingested with foodstuffs and are sometimes involved in wound myiasis (see Chapter 6).

Genus *Cochliomyia* (=*Callitroga*) Brauer

These are the New World 'screw worm'. The adults are dull green to bright green or greenish-blue flies of medium size with a usually largely orange-yellow head. There are three longitudinal black thoracic stripes. The genus is restricted to North and South America.

C. hominovorax (Coquerel) is an obligatory parasite on healthy tissues of man or animals. *C. macellaria* (Fabricius) is not an obligatory parasite and feeds in wounds and dead tissues.

James (1947) and Garcia (1952) key the adults and the larvae are treated in Chapter 6.

Genus *Chrysomya* Robineau-Desvoidy

These are large metallic green blow-flies (fig. 132A) which largely replace *Calliphora* and *Lucilia* in the Old World tropics. The adults are a bluer-green than *Lucilia* and less bristly. About ten species are involved in cases of myiasis of which *Chrysomya bezziana* Villeneuve is found in Africa, India and other oriental regions, including the Philippines, Celebes and New Guinea, and is an obligatory parasite in wounds and abcesses of man and animals. This species will also oviposit in the nose, mouth, eye, ear, urinogenital passages and cuts or sores. Other species normally breed in decomposing organic matter, but may become facultative parasites in wounds. Zumpt (1965) keys the adults of species associated with wound myiasis, the larvae are keyed in Chapter 6.

Genus *Cordylobia* Grunberg (including *Stasisia*)

The Tumbu-fly *Cordylobia anthropophaga* (Blanchard) is a yellowish-brown fly just under 10 mm. in length (fig. 131B). There are darker longitudinal stripes on the dorsal surface of the thorax and the posterior part of the abdomen is blackened; the wings are brownish tinged.

Eggs are laid in the sand or dust especially if it is contaminated with urine or faeces. The larvae hatch and penetrate the skin of the human host forming boil-like swellings.

C. rodhaini Gedoelst closely resembles *C. anthropophaga* both as adult and larva. Its main hosts are antelopes and the giant rat (*Crocetomys*) and it is less frequently found in man (see Chapter 6).

Genus *Lucilia* Robineau-Desvoidy

These flies are commonly known as greenbottles and are mostly of metallic, shining green flies usually somewhat smaller and less bristly than *Calliphora*. The general habits and biology are similar to *Calliphora* but some species are the well-known sheep maggot-flies. *Lucilia sericata* Meigen (fig. 132B) is the commonest attacker of sheep and has a cosmopolitan distribution.

The larvae of several species of *Lucilia* will infect wounds and body cavities in man (see Chapters 6 & 17). Larvae of *Lucilia sericata* have been used in the treatment of osteomyelitis. In Britain and probably elsewhere the larvae are sometimes present in the soil around houses and may be forced up in huge numbers after a rain-storm has caused waterlogging of the soil.

Nuortova (1959) has suggested that carrion species such as *Lucilia sericata* and *Phormia regina* transmit Poliomyelitis.

Genus *Phormia* Robineau-Desvoidy

These flies are somewhat smaller than *Calliphora* and of a deeper more purplish-blue colour. The genus is confined to the North Temperate Regions.

The larvae of *Phormia regina* (Meigen) have been involved in cases of dermal and enteric myiasis (see Chapter 6).

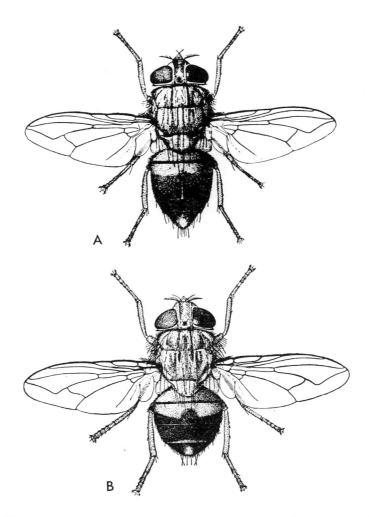

Fig. 131 A, *Auchmeromyia luteola*, female, adult of the Congo floor maggot; B, *Cordylobia anthropophaga*, female, the tumbu-fly (both Castellani & Chalmers).

T

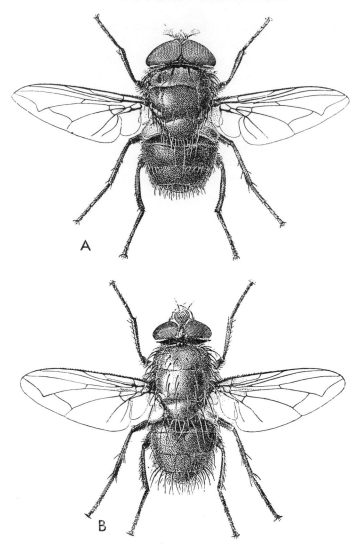

Fig. 132   A, *Chrysomyia megacephala*, male ×*c*4½; B, *Lucilia sericata*, male, green bottle, ×*c*6 (both from Shtakelberg).

SARCOPHAGIDAE (FLESH-FLIES)

Genus *Sarcophaga* Meigen

These are large greyish flies with a greyish-black chequered abdomen and black-striped thorax.   A few species are brownish or brownish-yellow in colour but the thoracic stripes are always present.   The genus is world wide.

The adults (fig. 130C) are attracted to carrion and faeces in which they normally breed.   However, the larvae are found in a great variety of decaying organic matter and sometimes infest wounds or sores and are involved in intestinal myiasis.   As the adults are larviparous relatively few maggots are usually involved in such infestations (see Chapter 6).

Genus *Wohlfahrtia* Brauer and Bergenstamm

The adults of this genus resemble *Sarcophaga*, although the abdomen is not chequered but marked with well-defined black spots (fig. 130B).

Only two species are involved in myiasis. *Wohlfahrtia magnifica* (Schiner) is restricted to the Old World (S. Europe, Middle East, Egypt, S. Russia) and is an obligatory parasite in the wounds and natural cavities of warm-blooded animals, including man. The New World species *Wohlfahrtia vigil* (Walker) is also an obligatory parasite but is capable of penetrating healthy tissue. Like *Sarcophaga* the genus is larviparous and there are only a few larvae are involved in each infestation (see Chapter 6).

BIBLIOGRAPHY

DAVIES, W. T. 1928. *Lucilia* flies anticipating death. *Bull. Brooklyn ent. Soc.* **23**: 118.

FULLER, M. E. 1932. The larvae of the Australian sheep blowflies. *Proc. Linn. Soc. N.S.W.* **57**: 77–91.

GARCIA, M. 1952. Consideraciones generales sobre el genero *Cochliomyia* Townsend, 1916 y description de *C. fontanai* n.sp (Diptera, Calliphoridae). *Publ. Inst. Regional Entomol. Sanitaria.* 1–3 (1948–1950: 68–80.

HALL, D. G. 1948. *Blowflies of North America.* 477 pp. Baltimore.

JAMES, M. T. 1948[7]. The flies that cause myiasis in Man. *U.S.D.A. Misc. Pub.* **631.** 175 pp. Washington.

—— 1970. Calliphoridae in *A Catalogue of the Diptera of the Americas South of the United States,* 28 pp. São Paulo [useful bibliography to the Neotropical fauna].

KAMAL, A. S. 1958. Comparative studies of thirteen species of sarcosaprophagous Calliphoridae and Sarcophagidae (Diptera). 1.—Bionomics. *Ann. Ent. Soc. Am.* **51**: 261–271.

KANO, R. & SHINONAGA, S. 1968. Calliphoridae (Insecta: Diptera). *Fauna Japonica.* 181 pp. Tokyo.

KURAHASHI, H. 1970. Tribe Calliphorini from Australian and Oriental Regions. 1—*Melinda*-group (Diptera: Calliphoridae). *Pacif Insects* **12** (3): 519–542.

LOPES, H. SOUZA. 1969. Sarcophagidae in *A Catalogue of the Diptera of the Americas South of the United States.* 88 pp. São Paulo [useful bibliography to the Neotropical fauna].

NORRIS, K. R. 1965. The bionomics of Blowflies. *A. Rev. Ent.* **10**: 47–68.

NUORTOVA, P. 1959. Studies on the significance of flies in the transmission of poliomyelitis. *Ann. Ent. Fen.* **25**: 1–14.

SCHUMANN, H. 1971. Die Gattung *Lucilia* (Goldfliegen). *Merkbl. augen. Parasitenk,* **18**: 1–20.

SENIOR-WHITE, R., AUBERTIN, D. & SMART, J. 1940. Diptera, VI.—Calliphoridae, *Fauna Br. India.* 288 pp. London.

SHTAKELBERGH, A. A. 1956. *Diptera associated with man from the Russian fauna.* 164 pp. Moscow. [In Russian].

ZUMPT, F. 1956a. Calliphorinae In *LINDNER, E. Die Fliegen der Palaearktischen Region* **8**: 140 pp. Stuttgart.

—— 1956b. Calliphoridae (Diptera: Cyclorrhapha), Part 1.—Calliphorini and Chrysomyiini. *Explor. Parc natn. Albert Miss. G. F. de Witte.* **8**: 200 pp. Brussels.

—— 1965. *Myiasis in man and animals in the Old World.* 267 pp. London.

5d. CYCLORRHAPHA OF MINOR MEDICAL IMPORTANCE

by B. H. Cogan

DROSOPHILIDAE (LESSER FRUIT-FLIES, VINEGAR, POMACE OR WINE-FLIES)

COMMONLY called Fruit-flies, but as this term is more generally applied to the members of the family Tephritidae, it is preferable that they should be referred to as Lesser or Small fruit-flies. The family contains large numbers of species, in relatively few genera, and in recent years a number of species have become cosmopolitan in distribution, usually through the agency of man.

Drosophilids are small flies of characteristic appearance (fig. 133A), which usually have a branched antennal arista. In a majority of the common species branches arise on both sides of the arista, and the last lateral branch and the arista form a characteristic apical fork.

Under natural conditions they breed in decaying and fermenting organic matter, usually of vegetable origin. The adults, however, are attracted to ripe and rotting fruit and fungi, kitchen waste, and in the case of *D. repleta* Wollaston, and *D. ananassae* Doleschall, to human faeces.

Occasionally members of this family are found in sufficient numbers in restaurants, kitchens, public houses and hospitals, to cause a nuisance. The larvae (fig. 145F) if ingested may cause a transitory or false myiasis due to irritation of the bowel. As with so many species of flies of catholic taste that frequent both general refuse and human food, passive transmission of disease organisms is always a possibility and a danger (see also Chapters 6, 17 & 18).

EPHYDRIDAE (SHORE-FLIES)

Commonly called Shore-flies, but in fact the majority of species are associated with fresh or brackish water, or wet grassland. Members of the family are easily recognized by the lack of an anal wing-vein, anal and second basal cell. The mouth opening is usually large with a strongly convex facial region. The face usually has at least a few strong hairs or bristles in longitudinal series and the arista when plumose, bears only a few dorsal hairs.

With the exception of *Teichomyza fusca* Macquart, whose larvae (fig. 142D) abound in water containing a high concentration of organic matter, and have been recorded as causing intestinal myiasis in man, very few Ephydrids are of medical importance. However, a number of species are troublesome as sweat feeders or eye-flies (Bohart & Gressitt, 1951). When abundant enough to be accidentally ingested, dung frequenting

Ephydridae, in fact the majority of small Diptera of similar habits, are of potential medical importance (see also Chapter 18).

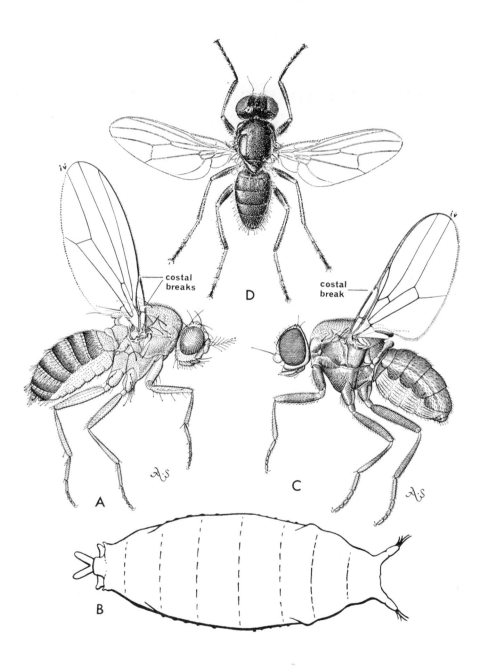

Fig. 133   A, *Drosophila funebris*, a lesser fruit-fly ×15; B, *Drosophila melanogaster* puparium ×31; C, *Siphunculina funicola* the Oriental eye-fly ×25; D, *Piophila casei* male ×15.

SPHAEROCERIDAE (=BORBORIDAE) (LESSER DUNG-FLIES)

Flies of the family Sphaeroceridae are small, usually of sombre hue, and with, typically, incrassate hind metatarsal segments. The mouth opening is very large and the mouth edge usually projects from a concave face; vibrissae are typically present. The third antennal segment is usually rounded with a relatively long arista. The costal vein is interrupted twice, once near the humeral cross-vein and again close to where the first longitudinal vein joins the costa. The large cosmopolitan genus *Leptocera* (s.l.) is easily recognized by the absence or reduction in pigmentation, of the apical sections of the 4th and 5th longitudinal veins.

Under natural conditions Sphaerocerids are associated with decaying organic matter of vegetable or animal origin. The fact that large numbers of species are attracted to, and develop in, animal excrement makes potential disease vectors of those species, found in numbers in houses, dairies and factories. *Leptocera caenosa* Rondani, is commonly taken in houses, and may often be traced to small accumulations of sewage arising from faulty or damaged drainage systems (see Fredeen & Taylor, 1964).

CHLOROPIDAE (FRIT-FLIES, EYE-GNATS OR EYE-FLIES)

A large family of predominantly small flies usually associated with plant material, but also including the Eye-gnats or Eye-flies.

Flies of this family may be recognized by means of the large and usually distinct frontal triangle, scarcity of strong head bristles, the single costal break and reduced venation of the wing. *Hippelates*, the most medically important genus is, unfortunately, atypical in possessing numerous head bristles, in addition to the characteristically curved hind tibial spurs.

It appears that the genera *Hippelates* (New World) and *Siphunculina* (Old World) (fig. 133C) possess species acting as passive transmitters of conjunctivitis, yaws and tropical ulcers (Kumm & Turner, 1936). Apart from the importance of these species from a medical standpoint, the constant attention of these flies around the eyes causes considerable irritation to man. The larvae feed on grass stems and roots, or live amongst decaying vegetation in the soil.

Some species of Chloropidae cause a nuisance by aggregating in numbers indoors (see Chapter 18).

SEPSIDAE

Members of the family Sepsidae are similar in general form and colouration to Piophilids, but on closer inspection are easily distinguished by means of the strongly waisted abdomen, less robust build, and the absence or reduction of the maxillary palpi. The wings of many Sepsids have a dark apical spot. The adult flies have a rather ant-like appearance and a curious habit of standing and twisting their wings about slowly in the air.

Larvae have been recorded in cases of both intestinal and urino-genital myiasis. Under natural conditions the larvae are found in excrement and decaying organic matter.

PIOPHILIDAE (CHEESE SKIPPER OR BACON-FLIES)

The adult flies are small, predominantly dark blue or black, shiny flies with rounded heads and bare antennal aristae. *Piophila casei* (L.) (fig. 133D) the only species of any known medical importance, is distinguished from other synanthropic species in the genus by means of the three distinct rows of mesonotal setulae.

The larvae (fig. 145B) are commonly called Cheese Skippers, a reference to their habit of progressing rapidly by a series of springing movements. They may be found on any dried or cured food of high protein content, e.g. cheese, bacon, dried fish and meat. The resistance of the larvae to the action of mammalian digestive enzymes is such that when they are swallowed with cheese, a deliberate practice of some gourmets in the past, they may cause serious intestinal scarification and be voided in the living state.

OESTRIDAE (s.l.) (WARBLE-FLIES)

Adult Oestrid flies are squat, broad-headed flies, usually with a warty or tuberculose frons and thorax (fig. 135A). The mouth-parts and oral aperture are considerably reduced.

The natural larval habitats of Oestrids are the nasal and cranial sinuses of sheep and goats of both domesticated and wild varieties. The viviparous females lay the young larvae in the nostrils of the host, from where the larvae migrate to the nasal cavities or frontal sinuses in which they develop.

Cases of infection of man by Oestrids are not common and usually occur in areas where there are low density populations of the natural hosts. In cases of casual infection of man the flies drop their larvae into the orbit of the substitute host, in preference to the mouth, nostrils or outer ear. Infection of the eye results in inflammation which invariably lasts for only a short period as the larvae are unable to develop further.

The genus *Hypoderma* may be included here, although it is often given the status of a separate family. The adult flies (fig. 135B) are characteristically bumble-bee-like, of heavy build and densely pilose. The larvae of the various species of *Hypoderma* are well known as the causative agents of cattle-warbles and live in swellings under the hide of cattle, goats and game animals. The mature larva is a fat maggot, slightly curved along the long axis and rounded at both ends.

Eggs are laid attached to hairs on the lower parts of the legs of cattle. On hatching the 1st instar larva enters the skin at the base of the hairs and migrates around the hosts body finally taking up a position under the skin, usually on the back. A swelling is produced, due to the reaction of the host's tissue to the presence of the larva, and the larva perforates the skin to provide a breathing hole. In such situations the larva develops to maturity, eventually emerging and pupating on the ground. The presence of warbles does not usually render the carcasses unfit for human consumption unless secondary infection has extensively affected the flesh.

Numerous cases of infection of man by *Hypoderma* are recorded and the records are summarized by Zumpt (1965). First instar larvae cause a creeping eruption in the skin, while later instars may cause abscesses on various parts of the body, most frequently the back, head and legs.

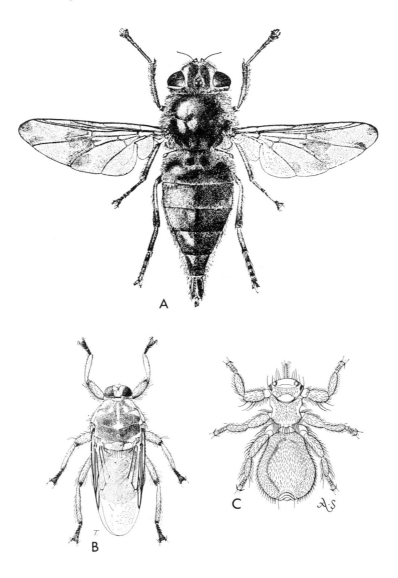

Fig. 134   A, *Gasterophilus intestinalis*, the horsebot-fly ×3½, female; B. *Hippobosca rufipes* ×3; C, *Melophagus ovinus*, the sheep ked ×3.

In central and tropical South America, *Dermatobia hominis* Linnaeus Jr., the human warble-fly is frequently found infecting the skin of man. This species, usually placed in the family Cuterebridae, lays its eggs on biting-flies and sweat-sucking-flies which in turn infect the primary host. Development is similar to that described for *Hypoderma* (see also Chapter 6).

GASTEROPHILIDAE (HORSE BOT-FLIES)

As adult flies, species of *Gasterophilus* (fig. 134A), the major genus in the family, resemble hive- and bumble-bees, both in form and colouring. The mouth-parts are greatly reduced and the mouth opening, in the adult, is reduced to a narrow aperture.

Larvae of species of *Gasterophilus* are found only in the alimentary tract of Equidae, e.g. horses, mules and zebras, where they may often occur in very large numbers. Mature larvae, which are large spinose maggots with strong mouth hooks (fig. 145G) are voided with the faeces and pupate on the ground.

In man the larvae gain entrance through the skin of the hands or feet, either through close contact with horses, or with wet grass upon which some species, e.g. *G. pecorum* F., lay their eggs.

Infection of the skin of man by 1st instar larvae results in a creeping myiasis and intense irritation; the larvae never develop beyond the 1st instar (Zumpt, 1965).

HIPPOBOSCIDAE (LOUSE-FLIES, FLAT-FLIES, TICK-FLIES, KEDS)

A small family of about 120 species of structurally highly modified flies, members of which are external blood-sucking parasites upon a variety of mammals and birds. In form they are dorso-ventrally flattened, with powerful legs bearing strong claws. The wings are usually fully developed as in *Hippobosca rufipes* Macquart (fig. 134B) and *H. equina* L., on horses, and *H. camelina* Dufour, the camel-fly. Other species may have the wings reduced, caducous or even absent as in *Melophagus ovinus* L., the sheep ked (fig. 134C).

The female fly lays fully developed larvae either in the nest of the host or in the host's surroundings. The larvae pupate almost immediately.

In the absence of their natural hosts or, in the case of the fully winged species, as a matter of general behaviour, louse-flies will leave their normal habitats and may attack man. Accidental biting of man is common with certain species (see Bequaert for a critical review of the subject). In the United States the pigeon-fly, *Pseudolynchia canariensis* Macquart, frequently bites man (Soroker, 1958), while other species, *Crataerina pallida* Latreille, a parasite of swifts, and the camel-fly, *H. camelina* are also reported to do so.

The bite is painful, but to the average person far less disturbing than the physical presence of these tenacious insects, which cling tightly to the skin and hair.

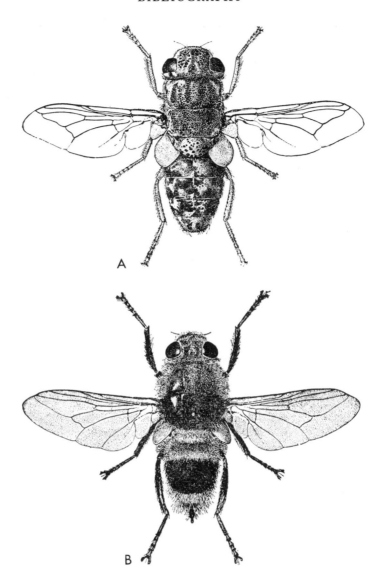

Fig. 135   A, *Oestrus ovis*, sheep-nostril-fly ×4; B, *Hypoderma bovis*, the ox warble-fly ×2½ (both from Castellani & Chalmers, drawn by Terzi).

BIBLIOGRAPHY

BASDEN, E. B.   1954.   The distribution and biology of Drosophilidae (Diptera) in Scotland, including a new species of 'Drosophila'.   *Trans. R. Soc. Edinb.* **72**: 603–654.

BECKER, T.   1926.   Ephydridae.   *In*, Lindner, E., *Fliegen palaearkt. Reg.* **6** (**56a**): 1–104.

BEQUAERT, J.   1942.   A monograph of the Melophaginae, or ked-flies, of sheep, goats, deer and antelopes (Diptera: Hippoboscidae).   *Entomologica am.* **22**: 1–210.

——— 1953.   The Hippoboscidae or louse-flies (Diptera) of mammals and birds.   Part I.—Structure, physiology and natural history.   *Entomologica am.* [1952] **32**: 1–209; (1953) **33**: 211–442.

——— 1954–1957.   Part II. Taxonomy, evolution and revision of American genera and species. *Entomologica am.* (1954) **34**: 1–232; (1955) **35**: 233–416; 1957 [1956] **36**: 417–611.

BOHART, G. E., & GRESSITT, J. L. 1951. Filth inhabiting flies of Guam. *Bull. Bernice P. Bishop Mus.* **204**: 1–152.

BURLA, H. 1954. Zur Kenntnis der Drosophiliden der Elfenbeinkuste (Französisch West-Afrika). *Revue suisse Zool.* **61** (Suppl.): 1–218.

CRESSON, E. T. 1945. A systematic annotated arrangement of the genera and species of the Indo-Australian Ephydridae (Diptera). I.—the subfamily Psilopinae. *Trans. Am. ent. Soc.* **71**: 47–75.

—— 1945. A systematic annotated arrangement of the genera and species of the Neotropical Ephydridae (Diptera). 1.—the subfamily Psilopinae. *Trans. Am. ent. Soc.* **71**: 129–163.

—— 1946. A systematic annotated arrangement of the genera and species of the Ethiopian Ephydridae (Diptera). 1.—the subfamily Psilopinae *Trans. Am. ent. Soc.* **72**: 241–264.

—— 1947. A systematic annotated arrangement of the genera and species of the Ethiopian Ephydridae (Diptera). 2.—the subfamily Notiphilinae. *Trans. Am. ent. Soc.* **73**: 105–124.

—— 1947. A systematic annotated arrangement of the genera and species of the Neotropical Ephydridae (Diptera). 2.—the subfamily Notiphilinae. *Trans. Am. ent. Soc.* **73**: 35–61.

—— 1948. A systematic annotated arrangement of the genera and species of the Indo-Australian Ephydridae (Diptera). 2.—the subfamily Notiphilinae and supplement to part 1 on the subfamily Psilopinae. *Trans. Am. ent. Soc.* **74**: 1–28.

DUDA, O. 1924. Beitrag zur Systematik der Drosophiliden unter besonderer Berucksichtigung der paläarktischen u. orientalischen Arten (Dipteren). *Arch. Naturgesch.* **90** A(3): 172–234.

—— 1926. Monographie der Sepsiden (Dipt.). *Annln naturh. Mus. Wien* [1925] 1, **39**: 1–153; 2, **40**: 1–110.

—— 1930. Die neotropischen Chloropiden (Dipt.). *Folia zool. hydrobiol.* **2**: 46–128.

—— 1931. Die neotropischen Chloropiden (Dipt.). 1.—Fortsetzung: Nachtrag, Erganzungen, Berichtigungen und Index. *Folia zool. hydrobiol.* **3**: 159–172.

—— 1932–1933. Chloropidae. *In*, Lindner, E., *Fliegen palaearkt. Reg.* **6** (61): 1–48 (1932); 49–248 (1933).

—— 1935. Drosophilidae. *In*, Lindner, E., *Fliegen palaearkt. Reg.* **6** (58g): 1–118.

—— 1938. Sphaeroceridae. *In*, Lindner, E., *Fliegen palaearkt. Reg.* **6** (57): 1–182.

—— 1939. Revision der afrikanischen Drosophiliden (Diptera). 1.—*Annls hist.-nat. Mus. natn. hung.* **32**: 1–57.

—— 1940. Revision der afrikanischen Drosophiliden (Diptera). 2.—*Annls hist.-nat. Mus. natn. hung.* **33**: 19–53.

FREDEEN, F. J. H. & TAYLOR, M. E. 1964. Borborids (Diptera: Sphaeroceridae) infesting sewage disposal tanks, with notes on the life cycle, behaviour, and control of *Leptocera* (*Leptocera*) *caenosa* (Rondani). *Can. Ent.* **96**: 801–808.

GRUNIN, K. J. 1964–1969. Hypodermatidae. *In*, Lindner, E., *Fliegen palaearkt. Reg.* **8** (64b): 1–40 (1964); 41–154 (1965); 155–160 (1969).

—— 1966. Oestridae. *In*, Lindner, E., *Fliegen palaearkt. Reg.* **8** (64a): 1–96.

—— 1969. Gasterophilidae. *In*, Lindner, E., *Fliegen palaearkt. Reg.* **8** (64a): 1–61.

GUIMARAES, J. H. & PAPAVERO, N. 1966. A tentative annotated bibliography of *Dermatobia hominis* (Linnaeus Jr., 1781) (Diptera: Cuterebridae). *Archos Zool. Est. S. Paulo* **14**: 223–294.

HAMMER, O. 1941. Biological and ecological investigations on flies associated with pasturing cattle and their excrement. *Vidensk. Meddr dansk naturh. Foren.* **105**: 141–393.

HARRISON, R. A. 1952. New Zealand Drosophilidae (Diptera). 1. Introduction and descriptions of domestic species of the genus *Drosophila* Fallén. *Trans. R. Soc. N.Z.* **79**: 505–517.

—— 1959. Acalypterate Diptera of New Zealand. *Bull. N.Z. Dep. scient. ind. Res.* **128**: 1–382.

HENNIG, W. 1943. Piophilidae. *In*, Lindner, E., *Fliegen palaearkt. Reg.* **5** (40): 1–52.

—— 1949. Sepsidae. *In*, Lindner, E., *Fliegen palaearkt. Reg.* **5** (39a): 1–91.

JAMES, M. T. 1948. The flies that cause myiasis in man. *Misc. Publs U.S. Dep. Agric.* [1947] **631**: 1–175.

KUMM, H. W. & TURNER, T. B. 1936. The transmission of yaws from man to rabbits by an insect vector, *Hippelates pallipes* Loew. *Am. J. trop. Med.* **16**: 245–267.

LAURENCE, B. R. 1955. The ecology of some British Sphaeroceridae (Borboridae: Diptera). *J. Anim. Ecol.* **24**: 187–199.

MAA, T. C.   1963.   Genera and species of Hippoboscidae (Diptera): types, synonymy, habitats and natural groupings.   *Pacif. Insects Monogr.* **6**: 1–186.

MELANDER, A. L.   1924.   Review of the dipterous family Piophilidae.   *Psyche, Camb.* **31**: 78–86.

MELANDER, A. L. & SPULER, A.   1917.   The dipterous families Sepsidae and Piophilidae.   *Bull. Wash. agric. Exp. Stn* **143**: 1–103.

NEIVA, A. & GOMES, J. F.   1917.   Biologia da mosca do Berne (*Dermatobia hominis*) observada em todas as suas fases.   *Anais paul. Med. Cirurg.* **8**: 197–209.

OKADA, T.   1956.   *Systematic study of Drosophilidae and allied families of Japan.*   183 pp.   Tokyo.

—— 1966.   Diptera from Nepal.   Cryptochaetidae, Diastatidae and Drosophilidae.   *Bull. Br. Mus. nat. Hist.* (Ent.) Supplement **6**: 1–129.

—— 1968.   *Systematic study of the early stages of Drosophilidae.*   188 pp.   Tokyo.

RICHARDS, O. W.   1930.   The British Sphaeroceridae (Borboridae: Diptera).   *Proc. zool. Soc. Lond.* (2) **18**: 261–345.

SABROSKY, C. W.   1935.   The Chloropidae of Kansas (Diptera).   *Trans. Am. ent. Soc.* **61**: 207–268.

—— 1941.   The Hippelates flies or eye gnats: Preliminary notes.   *Can. Ent.* **73**: 23–27.

—— 1951.   Chloropidae.   *Ruwenzori Expedition 1934–5.*   **2**: 711–828.

SIMMONS, P.   1927.   The Cheese Skipper as a pest in cured meats.   *Bull. U.S. Dep. Agric.* **1453**: 1–55.

SOROKER, R. H.   1958.   Pigeon fly problem in Southern California.   *Calif. Vector Views* **5**: 46.

SPULER, A.   1923.   North American genera and subgenera of the dipterous family Borboridae.   *Proc. Acad. nat. Sci. Philad.* **75**: 369–378.

STEYSKAL, G. C.   1943.   Old-World Sepsidae in North America, with a key to the North American genera (Diptera).   *Pan-Pacif. Ent.* **19**: 93–95.

STURTEVANT, A. H.   1942.   The classification of the genus *Drosophila*, with description of nine new species.   *Univ. Tex. Publs* **4213**: 5–51.

STURTEVANT, A. H. & WHEELER, M. R.   1954.   Synopses of Nearctic Ephydridae (Diptera).   *Trans. Am. ent. Soc.* **79**: 151–261.

THEODOR, O. & OLDROYD, H.   1964.   Hippoboscidae.   *In*, Lindner, E., *Fliegen palaearkt. Reg.* **8** (65): 1–70.

WETZAL, H.   1970.   Die Fliegen der unterfamilie Oestrinae (Diptera: Oestridae) in der Aethiopischen Region under deren vetinarmedizinische Bedeutung.   *Z. angew. Ent.* **66**: 322–336.

WHEELER, M. R.   1949.   Taxonomic studies on the Drosophilidae.   157–195, 2 figs.   *In*, Patterson, J. T., Studies in the genetics of *Drosophila*.   VI.—Articles on genetics, cytology and taxonomy.   *Univ. Tex. Publs* **4920**: 1–223.

—— 1952.   The Drosophilidae of the Nearctic Region exclusive of the genus *Drosophila*.   162–218, 1 fig.   *In*, Patterson, J. T., Studies in the genetics of *Drosophila*.   VII.—Further articles on genetics, cytology and taxonomy.   *Univ. Tex. Publs* **5204**: 1–251.

—— 1957.   Taxonomic and distributional studies of Nearctic and Neotropical Drosophilidae.   *In*, Patterson, J. T., Studies in the genetics, taxonomy, cytology and radiation.   *Univ. Tex. Publs* **5721**: 1–316.

WHEELER, M. R. & TAKADA, H.   1964.   Diptera, Drosophilidae.   *Insects Micronesia* **14**: 163–242.

WIRTH, W. W. & STONE, A.   1956.   Aquatic Diptera.   372–482, 64 figs.   *In*, Usinger, R. L., *Aquatic insects of California, with keys to North American genera and Californian species.*   508 pp.   Berkeley.

ZUMPT, F.   1965.   *Myiasis in man and animals in the Old World.*   267 pp.   London.

ZUSKA, J. & LASTOVKA, P.   1965.   A review of the Czechoslovak species of the family Piophilidae with special reference to their importance to Food Industry (Diptera: Acalyptrata).   *Acta ent. Bohemoslavaca.* **62**: 141–157.

——, —— 1965.   Species-composition of the Dipterous fauna in various types of food processing plants in Czechoslovakia.   *Acta ent. Bohemoslavaca* **66**: 201–221.

6. EGGS AND LARVAE OF FLIES
by Harold Oldroyd and Kenneth G. V. Smith

DIPTERA are *holometabolous* insects, undergoing a complete metamorphosis, with a life-cycle divided into four stages: egg, larva, pupa and adult.

A few flies do not actually lay their eggs. The families Streblidae, Nycteribiidae and Hippoboscidae are collectively known as 'Pupipara' because not only the egg, but the larva as well is retained inside the body of the female until the larva is fully grown and ready to pupate. The same habit is universal in the family Glossinidae.

Some blow-flies hatch the egg before laying it, and so give birth to larvae. In the family Sarcophagidae there are species in which this is normal, others in which it occurs occasionally, and others which always lay eggs. Calliphoridae usually lay eggs, but if oviposition is delayed through failure to find suitable material for larval food, then the eggs may hatch in the oviduct.

Other variations of the normal life-cycle occur. Since the germ-plasm is segregated from the beginning of larval life, it is possible for development of the ova to begin without waiting for the parent to become adult. This occurs in certain Cecidomyiidae e.g. *Miastor*) where under certain conditions 'daughter larvae' appear within the body of a 'parent larva' and destroy it before it reaches the pupal stage. In some Cecidomyiidae and Chironomidae, larvae may emerge from the adult fly before it has broken out of the pupal skin, and so give rise to so-called 'pupal paedogenesis'.

These devices are adaptations which shorten the normal life-history by omitting one or more free stages. With these few exceptions all flies lay eggs.

EGGS

In most families of flies the egg-stage is inconspicuous, and eggs of Diptera have been little studied. Exceptions are Chironomidae, Culicidae, Syrphidae and some of the muscoid flies, notably house-flies and blow-flies. The aquatic Chironomidae lay their eggs in gelatinous strings in the water, where they are an important food for fish, and where they may be attacked with insecticides if the adult flies are a pest. The eggs of Culicidae, with their rafts and other flotation devices, have contributed to knowledge of both the taxonomy and the biology of mosquitoes. The eggs of house-flies and blow-flies, and even more their egg-laying habits, are of importance in hygiene, and in forensic entomology (Chapters 17 & 18); Hinton (numerous papers); Chandler (1968).

LARVAE

The larvae of flies should be regarded as being biologically independent of the adults. It is as if the fly lived two completely different lives, with different structure, physiology, senses and different powers of movement. Whereas the young (nymphs) of hemimetabolous insects such as aphids live in the same situations, and upon the same food as their adults, and have much of their biology in common, in contrast, all flies live in an en-

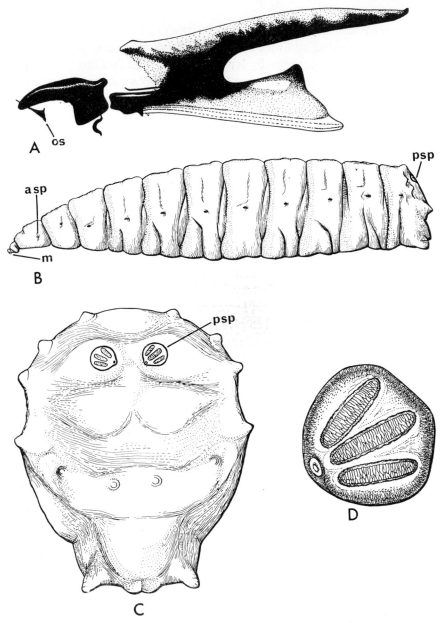

vironment totally different from that of their larvae. There is thus no link between the medical importance, if any, of the adult fly and that of the larva. Only the House-fly, the Lesser House-fly (*Fannia canicularis*), and the Bluebottle have a place in this book both as adults and as larvae.

The present chapter, therefore, has two functions: the identification of dipterous larvae that are found in circumstances that give them a medical importance, and some account of those families in which the larva is the medically important stage.

Fig. 136   *Calliphora vicina* (= *erythrocephala*): A, cephalopharyngeal skeleton, os = accessory oral sclerite (after Hall); B, third instar larva, asp = anterior spiracles, psp = posterior spiracles, m = mouth-hooks; C, ditto, anal view, psp = posterior spiracles; D, ditto, posterior spiracle (after Hall).

Identification of Larvae

Dipterous larvae are soft-bodied, generally with no clear distinction between thorax and abdomen (except in Culicidae and one or two related families). The primitive condition is to have a well-sclerotized head-capsule, complete with mandibulate mouth-parts, and distinct antennae and palpi, but no ocelli. The head-capsule may be capable of being retracted into the thorax, or it may be permanently visible.

A distinct head-capsule, retractable or not, is present in Nematocera (reducing to vanishing point in Cecidomyiidae) and in most Brachycera. In Empididae and Dolichopodidae the head-capsule is reduced to an internal skeleton. In Cyclorrhapha the head-capsule is hardly distinguishable, except for the small antennae and palpi, and the primitive mandibulate mouth-parts are represented by a set of *mouth-hooks* which, with their associated sclerites, are known as a *cephalo-pharyngeal skeleton* (fig. 136A). Such larvae are known as *maggots* if they have a pointed head and truncate posterior end, and *grubs* if they are oval in outline and generally swollen-looking.

No dipterous larva ever has true segmental appendages, either on the thorax or the abdomen, and it is thus technically a legless larva (*apodous*). Most dipterous larvae, however, have some kind of secondary organs of locomotion (*pseudopods: prolegs*) on at least some of the segments. These take the form of swellings, armed with spines or with curved crochets, are are used to push against a firm surface, the spines and crochets increasing adhesion. The term *proleg* is usually applied to more or less foot-like organs, with crochets, on the ventral surface only (figs 140H, 141B, C); *pseudopods* are more generalized swellings, and may occur in a ring all round the segment, as in many maggots and grubs (fig. 136B, C), where the pseudopods help to squeeze the body through small openings.

Respiration of larvae is usually by *spiracles* which can be seen externally as small pores, sometimes enlarged into spiracular discs, or into digitate external structures. The maximum number is ten pairs in *Bibio*, which has a pair on every segment except the second and eleventh. Even here the first and last spiracles are larger than the others, and in Diptera generally these—the *prothoracic* and the *caudal* pairs—are the important ones functionally, and also the ones that provide the most useful taxonomic characters. Often the mouth-parts and these two pairs of spiracles are the only practical means of identifying larvae, especially among the smooth maggots of the Cyclorrhapha (figs 136C, D, 137, 149, 150).

The general evolution of dipterous larvae has been one of simplification, especially of the head. In opposition to this has been adaptation by the acquisition of secondary characters. This is particularly true of locomotory adaptations to living in water, which have produced the characteristic features of the larvae of certain families. Mosquito larvae (fig. 138A) are the outstanding example of aquatic adaptation among Nematocera, closely followed by Simuliidae (fig. 138B) and Chironomidae (fig. 141C) among the larger families, and by Blepharoceridae, Deuterophlebiidae, Thaumaleidae, Chaoboridae and Dixidae. Of these only the first two are of medical importance.

Among Brachycera, the Tabanoidea (Rhagionidae, Tabanidae, and possibly Stratiomyidae) have a strong larval association with water, which is possibly ancestral; whereas the aquatic larvae of some Empididae and Dolichopodidae, as well as those of Syrphidae,

U

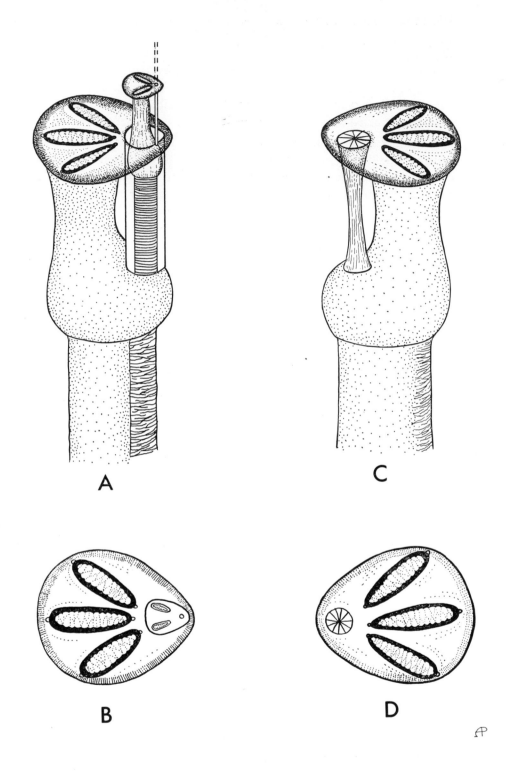

Fig. 137   Schematic figures of the moulting process of spiracles from second to third instar in Cyclorrhaphous larvae: A, B, side views; C, D, end views (after Keilin).

Ephydridae, some Muscidae and other families of Cyclorrhapha appear to be secondarily adapted to living in water. The adaptive structures of both groups of larvae show remarkable similarities.

Aquatic life also involves problems of respiration. Open spiracles are unsuitable, except for a limited period, and the larva must either extend a siphon to the open air—as in mosquitoes, Tabanidae, some Stratiomyidae, and the aquatic Syrphidae—or close the spiracles altogether and breathe through the skin, as in Simuliidae, Chironomidae, Ceratopogonidae and Psychodidae. Some of these receive enough oxygen by transpiration through the skin, but those which have a more active existence have extended the closed tracheal system by developing external gills (figs 140D, 141C).

Parasites encounter some of the problems of aquatic life, especially those of respiration. While parasites do not generally move about much, they need to hang on against forces tending to expel them—e.g. peristalsis in the gut—and themselves need to be able to escape when they are fully fed—e.g. a warble larva from a cavity beneath the skin of the host. So parasitic larvae tend to have strong mouth-hooks, large and complex caudal spiracles, and extensive spinulation of the body (figs 144, 145G).

These facts, both the absence of some features and the presence of others, are additional taxonomic characters and aids to recognition.

On the whole, therefore, dipterous larvae have few external characters that can be used in classification, and those of medical importance in themselves belong mainly to the maggots and grubs of the Cyclorrhapha. The larvae of Nematocera and Brachycera can be identified down to the family, and often to the genus and species; certain families of Cyclorrhapha—Syrphidae, Ephydridae, Tephritidae, Agromyzidae, Oestridae—can be recognized by peculiarities of the spiracles and sometimes can be identified more precisely. Otherwise, the identification of larvae of Cyclorrhapha is a matter of inference from a combination of pointers: structure, biology, and the circumstances of capture. It may be necessary to be able to rear the larva before a precise identification can be made.

KEY TO THE LARVAE OF DIPTERA

Only those families are mentioned that have actual or potential medical importance.

1 Larvae with an obvious head, though this may be withdrawn into the thorax, and become visible only under moderate pressure 2

- Larvae without an obvious head; mouth-hooks may be prominent, but they are not enclosed in a definite head-capsule 22

2 Body obviously divided into different regions by the fusion of the thoracic segments, and perhaps also of some abdominal segments. All aquatic 3

- Body not obviously divided into different regions, though individual segments may have appendages of various kinds. Aquatic, terrestrial or parasitic . . . 5

3 Head and thorax large, disc-like, thorax much broader than abdomen. A posterior siphon, anal gills, mouthbrushes, and an elaborate pattern of long, fine hairs (fig. 138A). Mosquito larvae . . . . **CULICIDAE**

- Thorax sometimes flattened, but not disc-like 4

4 Head with mouth-brushes, and thorax with one proleg. Last three segments of abdomen partly fused and swollen. Larvae attached to objects in flowing water (fig. 138B) . . . . **SIMULIIDAE**

  – Head with antennal spines for grasping prey, but no mouth-brushes.   Thorax a
     little swollen, but no abdominal swelling.   Often two silvery air-sacs visible through
     transparent cuticle (fig. 138C).   Larva moves through water by active twitching
                                          CHAOBORIDAE

  5 Head capsule distinct, not retracted into thorax 6
  – Head capsule retracted into thorax 21

  6 Head with mandibles moving vertically, parallel to each other, and often visible as a
     pair of hooks 7
  – Head with mandibles moving horizontally, so that their tips can be brought together
     (some CERATOPOGONIDAE have parallel mouth-hooks) 9

  7 Body shagreened, i.e. roughened like shark-skin, terrestrial species broad and flattened
     (fig. 138E); aquatic species often have a tapering abdomen, ending in a respiratory
     siphon (fig. 138D) . . . . **STRATIOMYIDAE**
  – Body not shagreened.   Smooth, white, elongate larvae, deceptively like worms, but
     with a distinct small head, and, behind it and visible through the cuticle, an internal
     rod (fig. 139B, D).   Segments subdivided so that they appear to number about
     twenty 8

  8 Internal rod ending in a spatulate tip (fig. 139A, B).   Larvae normally found in soil,
     or among plants . . . . **THEREVIDAE**
  – Internal rod pointed, not at all spatulate (fig. 139C, D).   Larvae occur naturally in
     soil, litter, birds' nests and so on, but are most often found indoors, particularly
     in carpets . . . . **SCENOPINIDAE**

  9 Spiracles at tip of body surrounded by four or five fleshy lobes (fig. 139E, F) . . 10
  – Spiracles at tip of body not surrounded by lobes 11

10 Four lobes at tip of body.   Abdominal segments often subdivided into three indis-
     tinct rings (fig. 139E) . . . . **TRICHOCERIDAE**
  – Five lobes, or terminal processes, at tip of body.   Abdominal segments with only
     two divisions, one narrower than other (fig. 139F) . . **ANISOPODIDAE**

11 Spiracles visible on most or all abdominal segments.   Head usually complete, and
     very distinct; if it is reduced, then mouthparts are also reduced.   Larvae of midges,
     mostly in vegetable matter or in soil 12
  – Spiracles present only on thorax and tip of abdomen.   Head complete, or incomplete
     posteriorly, but with well-developed mouthparts.   Larvae of gnats and midges,
     mostly in water, a few in wet media 15

12 Posterior spiracles each carried at the end of a horn-like process.   No eye-spot on
     head (fig. 139G) . . . . **SCATOPSIDAE**
  – Posterior spiracles not raised on a process 13

13 First thoracic segment, and eighth abdominal segment, each divided into two.   A
     spiracle on third thoracic segment.   Head with eye-spot (fig. 139H) **BIBIONIDAE**
  – Neither subdivision nor metathoracic spiracle usually present.   No eye-spot . . 14

14 Prolegs well-developed.   Epicranial plates underneath head meeting only at one
     point (fig. 140A) . . . . **MYCETOPHILIDAE**
  – Prolegs absent or weakly developed.   Epicranial plates meeting at two points (fig.
     140B, C) . . . . **SCIARIDAE**

15 Tip of abdomen drawn out into a long siphon, with two finger-like gills at tip (fig.
     140D).   In water . . . . **PTYCHOPTERIDAE**
  – Siphon, if present, much shorter than this 16

16 Abdomen without prolegs 17
  – Abdomen with some prolegs 18

17 Usually some, or all segments have a narrow, transverse band as in fig. 140E.   Some-
     times with sucker-discs ventrally . . . . . . **PSYCHODIDAE**

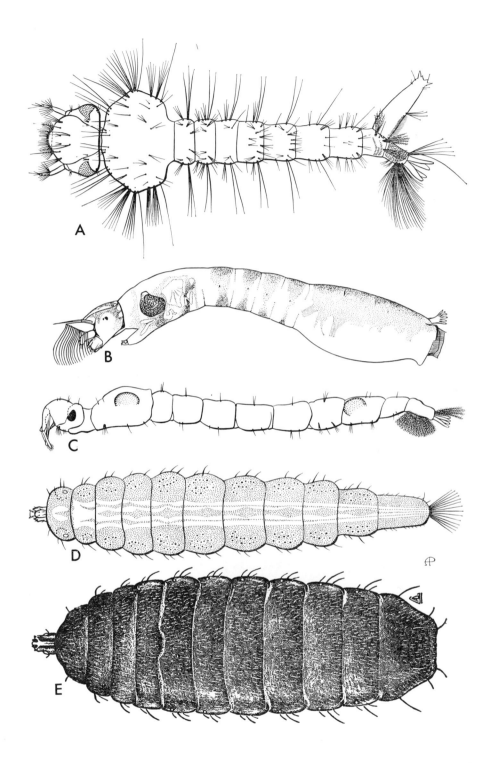

Fig. 138 Third instar larvae: A, *Aedes stimulans* (Culicidae) dorsal; B, *Simulium vittatum* (Simuliidae), lateral; C, *Chaoborus* (Chaoboridae), lateral; D, *Odontomyia* (Stratiomyidae), dorsal; E, *Hermetia* (Stratiomyidae), (after James).

–   Without such bands, but often with retractable respiratory filaments at tip of abdomen
    (fig. 140G) .   .   .   .   **CERATOPOGONIDAE**

18 Prolegs only on first and second abdominal segments.   Tip of abdomen with a fairly
    short median tube, flanked by two lobes fringed with hairs.   In water (fig. 140H)
                                                                          DIXIDAE

–   Prolegs on first and last abdominal segments   19

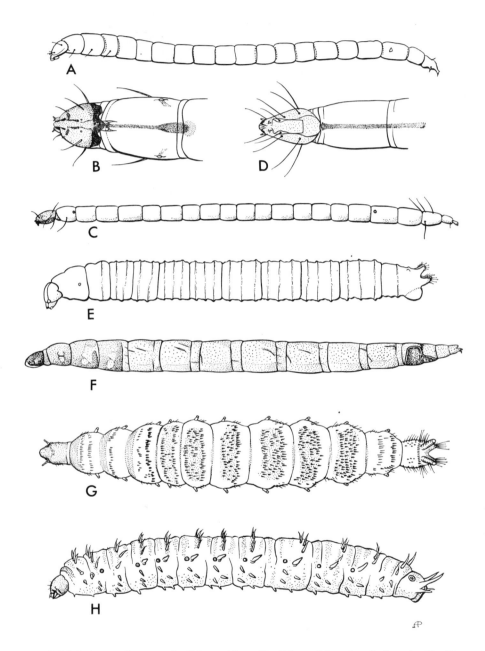

Fig. 139   Third instar larvae: A, Therevidae, B; Therevidae, head dorsal; C, *Scenopinus*
(Scenopinidae); D, *Scenopinus* head, dorsal; E, *Trichocera* (Trichoceridae) lateral;
F, *Sylvicola* (Anisopodidae) (after Keilin & Tate); G, *Scatopse* (Scatopsidae) dorsal;
H, *Bibio hortulanus* (Bibionidae) lateral.

19 First thoracic segment with a pair of short respiratory tubes; eighth abdominal segment with a single spiracle opening between the two finger-like processes. In shallow, running water (fig. 141A) . . . . **THAUMALEIDAE**

– Without respiratory tubes, and without the single spiracle on the eighth abdominal segment 20

20 Body slender, often adorned with long bristles or scales, arranged in groups on each segment. Terrestrial (fig. 141B) . . . some **CERATOPOGONIDAE**

– Body more plump and smooth, never adorned with processes, though it may be hairy. Last segment usually with anal gills. Aquatic (fig. 141C) . . **CHIRONOMIDAE**

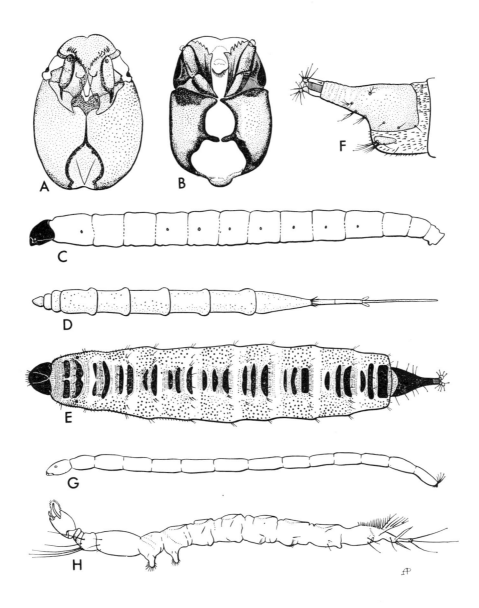

Fig. 140 Third instar larvae: A, *Leia bimaculata* (Mycetophilidae) head, ventral; B, *Bradysia* (Sciaridae) head, ventral; C, *Bradysia* lateral; *Ptychoptera* (Ptychopteridae) dorsal (after Satchell); E, *Psychoda* (Psychodidae), dorsal; F, *Psychoda*, posterior siphon, lateral; G, Ceratopogonidae, lateral; H, *Dixa* (Dixidae) lateral (after Peterson).

21   Mandibles working in a horizontal plane, so that their tips can be brought together.
      The common 'leather-jackets' have two large posterior spiracles in a hollow, sur-
      rounded by a crown of finger-like processes (fig. 141D, E)   .   .   .   **TIPULIDAE**

 -    Mandibles working vertically, parallel to each other, so that their tips cannot be
      brought together. Body scored with longitudinal striations, and each segment
      with a ring of pseudopods (fig. 141F)   .   .   .   .   **TABANIDAE**

22   Small, or tiny larvae, without mouthparts or any other external features, but often
      with a sclerotized sternum of 'breast-bone' of varying shape, visible underneath
      the thorax (fig. 142A).   Often in galls in living plants, but sometimes free-living
                                                                        CECIDOMYIIDAE

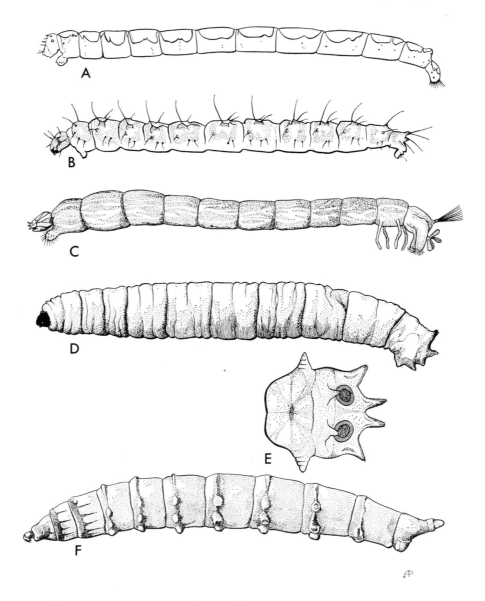

Fig. 141   Third instar larvae: A, Thaumaliidae lateral; B, *Forcipomyia* (Ceratopogonidae)
lateral; C, *Chironomus dorsalis* (Chironomidae) lateral (after Miall & Hammond);
D, *Tipula* (Tipulidae) lateral; E, *Tipula*, spiracular disc in posterior view; F, *Tabanus*
(Tabanidae) lateral.

– Always showing some external structures; at least mouth-hooks and a pair of posterior spiracles; often with more elaborate structures, but never with a sclerotized, subcutaneous sternum . . . . **CYCLORRHAPHA** 23

N.B. The great majority of families of Cyclorrhapha are of no medical importance, and since the separation of larvae of many acalyptrate families is difficult, or impossible, the following key is confined to those families known to be medically significant.

23 A flattened, slug-like larva, with long processes on thorax and tip of abdomen; other abdominal segments distinctly separated, and with a lateral fringe running round the body. Mouthparts small, and an isolated sclerite of head visible above them. In dead leaves and vegetable matter (fig. 142B)
     LONCHOPTERIDAE (MUSIDORIDAE)

– Not so. If a flattened, slug-like larva, then hind spiracles stand together on a papilla (Syrphidae) 24

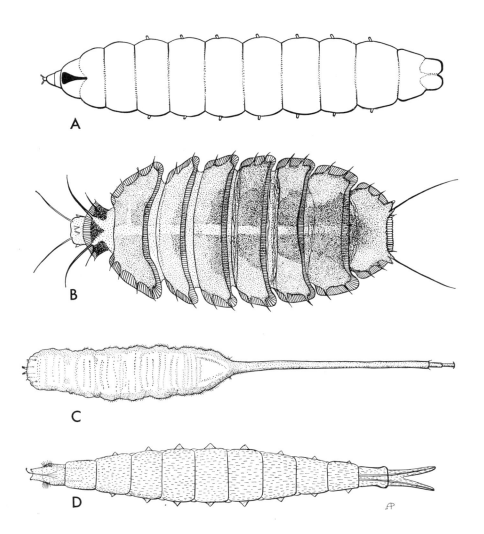

Fig. 142   Third instar larvae in dorsal view: A, Cecidomyiidae (after Peterson); B, *Lonchoptera* (Lonchopteridae); D, *Eristalis* (Syrphidae); C, *Teichomyza* (Ephydridae).

24 Hind spiracles close together, mounted on a single breathing-tube, not forked at tip.
    Tube may be brief in terrestrial larvae, very long and telescopic in aquatic larvae
    many **SYRPHIDAE** (fig. 142C); **EPHYDRIDAE**, GENUS *TEICHOMYZA* (fig. 142D)

 – Hind spiracles separated; or if close together on a tube, then this is forked, with one
    spiracle at each tip (*N.B.* 'one spiracle' means one spiracular plate, which may have
    three slits, or a multitude of small pores)   25

25 Larvae with tuberculate, fleshy or spinous processes, dorsally and laterally   .   . 26

 – Larvae without such processes, though the integument itself may have strong spines as
    in Oestridae (fig. 144B)   28

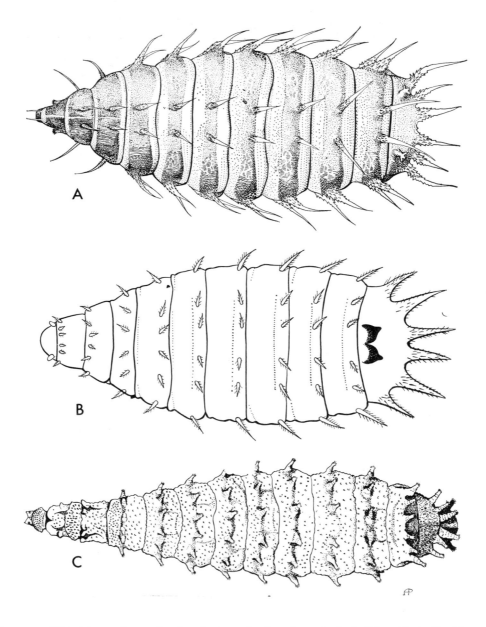

Fig. 143   Third instar larvae in dorsal view: A, *Fannia canicularis* (Muscidae); B, *Megaselia*
(Phoridae); C, *Chrysomya albiceps* (Calliphoridae).

26 Flattened larvae with filiform processes which are branched, at least basally, and may appear feathery, on the dorsal surface and sides of the segments. Posterior spiracles borne on stalks, each stalk with four lobes on which are found the three slits and the button (fig. 143A) . . . . . **MUSCIDAE,** *FANNIA*

 – More or less cylindrical larvae with short, or moderately short, unbranched lateral and dorsal tubercles on the segments 27

27 Small, dirty white, slightly flattened larvae, up to 4 mm. long, with short processes on the dorsal and lateral surfaces; posterior spiracles on brown, sclerotized tubercles, each with a narrow opening (fig. 143B) . . . . **PHORIDAE**

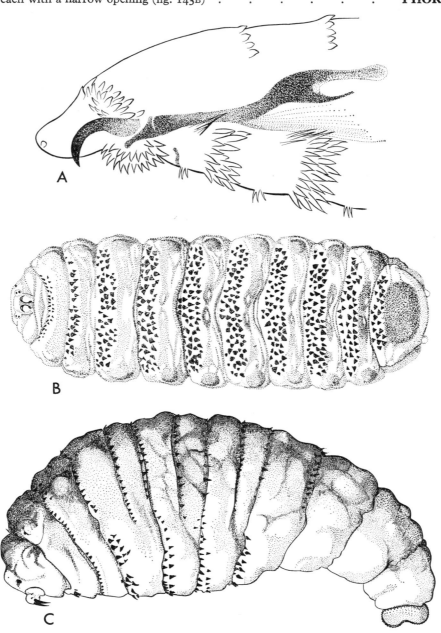

Fig. 144   Third instar larvae: A, *Oestrus ovis* (Oestridae), head and mouthparts lateral; B, *Oestrus ovis*, whole larva, ventral; C, *Dermatobia hominis* (Cuterebridae) lateral.

– Larger larvae, more nearly cylindrical, with larger, pointed, fleshy processes laterally and dorsally; posterior spiracles in a cleft on posterior face of anal segment, and consisting of flattened plates perforated by three slits (fig. 143C)

CALLIPHORIDAE, *CHRYSOMYIA albiceps*

28 Hind spiracles with a large number of small pores (fig. 149A, B) 29

–   Hind spiracles with slits, which may be straight, curved or sinuous, but never with numerous pores 31

29 Mouth-hooks rudimentary . . . . . . **HYPODERMATIDAE**

–   Mouth-hooks well-developed, strongly hooked (fig. 144A) 30

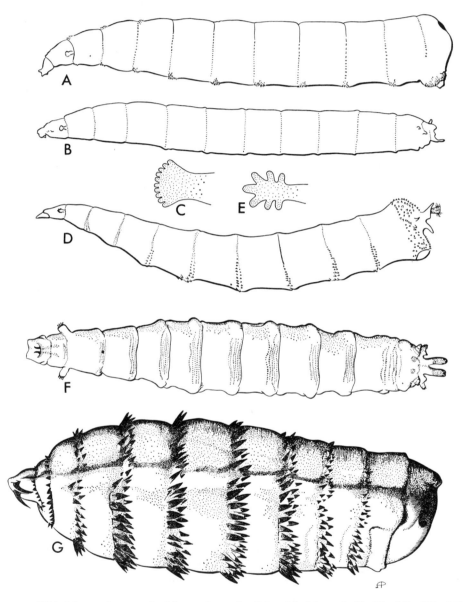

Fig. 145 Third instar larvae: A, *Musca domestica* (Muscidae) lateral; B, *Piophila* (Piophilidae) lateral; C, *Piophila*, anterior spiracle; D, *Sepsis* (Sepsidae) lateral; E, *Sepsis*, anterior spiracle; F, *Drosphila* (Drosophilidae) ventral; G, *Gasterophilus* (Gasterophilidae), lateral.

30 Body with weak spines, in limited areas only (fig 144B) . . . **OESTRIDAE**
 – Body with spines stronger and more evenly distributed (fig. 144C)
                                        CUTEREBRIDAE, *DERMATOBIA*
31 Posterior spiracles with one or two slits only (figs 137A, 146B). *First or second stage larvae of all families; generally impossible to identify, except by association with mature larvae, or with host.*
 – Posterior spiracles with three slits. *Third stage larva* 32
32 Slits of posterior spiracles strongly sinuous (fig. 149C–E) 33
 – Slits of posterior spiracles straight or arcuate 35
33 Slits of posterior spiracles each with three or more loops (fig. 149C). Posterior end of larva smoothly rounded (fig. 145A) . . . . . **MUSCIDAE**, *MUSCA*
 – Slits of posterior spiracles each with two S-shaped loops, and with setose tubercules ventrally 34

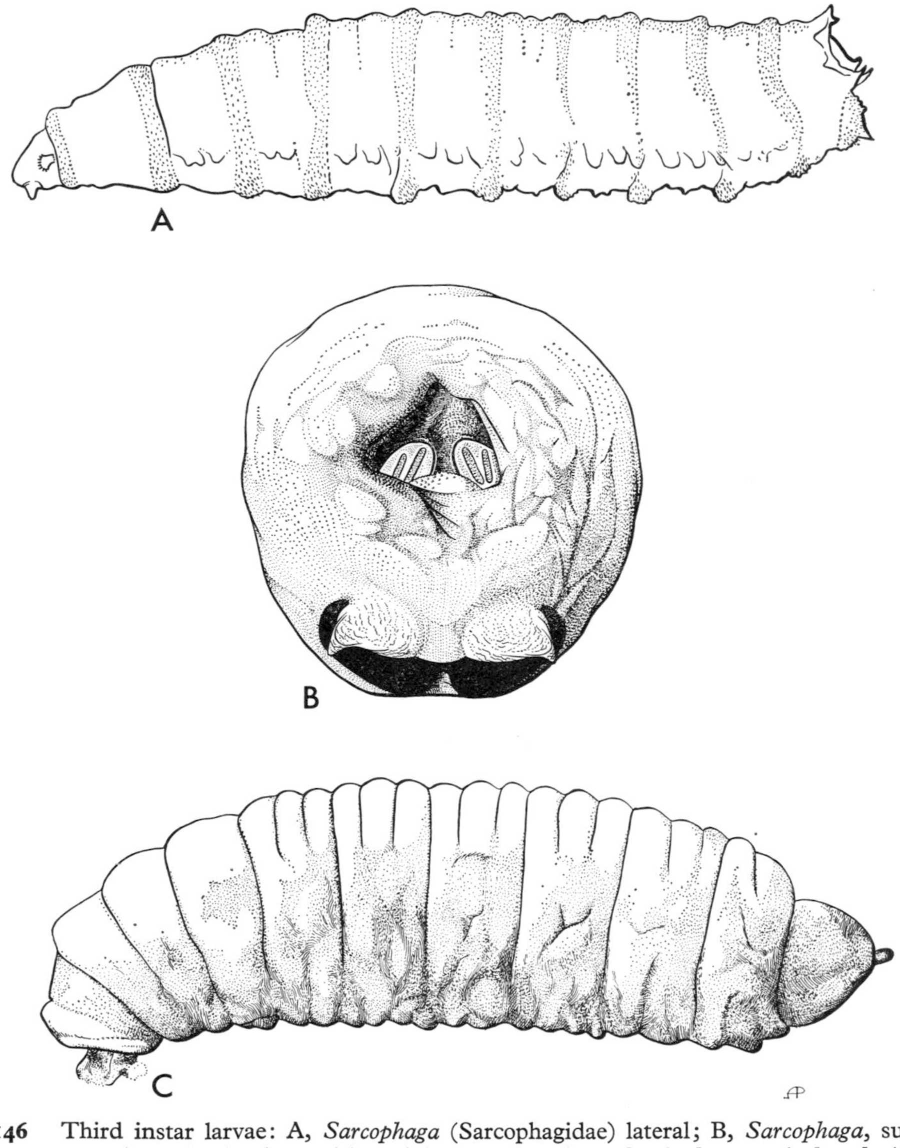

Fig. 146 Third instar larvae: A, *Sarcophaga* (Sarcophagidae) lateral; B, *Sarcophaga*, sunken posterior spiracles (second instar) in posterior view: C, *Auchmeromyia luteola* (Calliphoridae) lateral.

34   Posterior face of larva much deeper than broad, with three prominent setulose
     tubercules; slits of posterior spiracles not surrounding the button (fig. 149D)
                                                    MUSCIDAE, *SYNTHESIOMYIA*
 -   Posterior face of larva very little deeper than broad, and with moderately prominent
     setulose tubercles; slits of posterior spiracles surrounding button (fig. 149E)
                                                       MUSCIDAE, *STOMOXYS*
35   Anal segment with short, fleshy processes (fig. 145B–D)   36
 -   Without these processes projecting backward   38
36   Posterior spiracles situated upon two fleshy processes   37
 -   Posterior spiracles not situated upon two fleshy processes (fig. 145B)   **PIOPHILIDAE**
37   Anterior spiracle with lobes along the side of a central column. Cuticular hairs or
     spines present (fig. 145D,E)   .   .   .   .   **SEPSIDAE**
 -   Anterior spiracles with lobes confined to a fan at tip. Cuticular hairs or spines
     absent (fig. 145F)   .   .   .   .   **DROSOPHILIDAE**

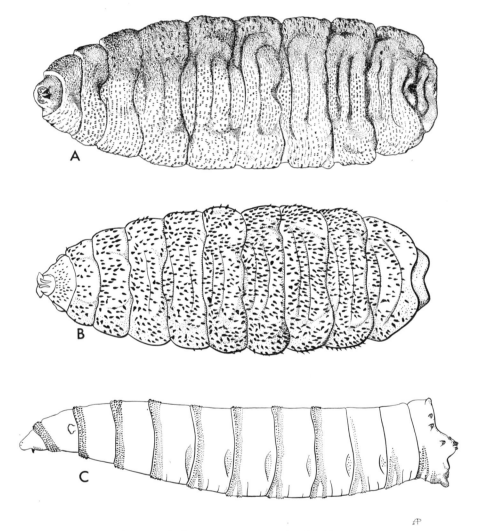

Fig. 147   Third instar larvae of Calliphoridae: A, *Cordylobia anthropophaga* (Tumbu-fly)
ventral (after Austen); B, *Cordylobia (Stasisia) rodhaini* ventral; C, *Cochliomyia* =
(*Callitroga*) *macellaria*, lateral.

MYIASIS

The term *myiasis* is used loosely for any kind of infestation of another living animal by larvae of Diptera.   Zumpt (1965) defines it more precisely as *the infestation of live human and vertebrate animals with dipterous larvae which, at least for a certain period, feed on the host's dead or living tissue, liquid body-substance, or ingested food.*

Myiasis has been classified into various categories, and from two opposing standpoints. For quick identification of the fly concerned it is useful to have a classification based upon the part of the body that is infested: whereas for an understanding of the biology of the fly as a guide to treatment, and to prevention of further attacks, as well as for the scientific study of evolution of flies, it is more informative to base a classification upon the degree of host parasite relationship.

Table 7   A Clinical Classification of Human Myiasis
(based upon Zumpt, 1965: xii and James, 1947: 6)

a. *Blood-sucking maggots.*   Larvae that attach themselves by their mouth-hooks to the human skin and suck blood.

    Auchmeromyia luteola (Calliphoridae)
    Tabanidae (Otsuru & Ogawa, 1959)

b. *Dermal and subdermal myiasis.*   Larvae that penetrate into the unbroken skin, or enter wounds, and form boils or other lesions, either at the site of entry or elsewhere.

    Sarcophagidae
    Calliphoridae
    Hypodermatidae
    Muscidae
    Tachinidae (*Mintho algira.*   James, 1947: 163)

c. *Nasopharyngeal myiasis*—infestion of the nasal fossae, frontal sinuses, pharyngeal cavities, eye-balls and eye-sockets. These larvae feed primarily on the mucous secretions, but may increase the flow of mucus by scraping with their mouth-hooks, thus leading by stages to wounds and tissue destruction.

Sarcophagidae
Calliphoridae
Cuterebridae
Gasterophilidae

Oestridae
Hypodermatidae
Phoridae
Muscidae

d. *Intestinal (enteric) myiasis.* This category includes such larvae as that of the Horse Bot-Fly, which is adapted to living in the intestine, and is therefore an *obligatory parasite;* and a variety of others that may be swallowed in food, or may invade the intestine *via* the rectum, and which are *facultative parasites.*

Sarcophagidae
Calliphoridae
Muscidae
Phoridae
Tipulidae
Psychodidae
Anisopodidae
Stratiomyidae (*Hermetia*)

Therevidae
Syrphidae
Drosophilidae
Ephydridae
Micropezidae
Piophilidae
Sepsidae

e. *Urinogenital myiasis.* There are no obligatory dipterous parasites in the urinogenital system: all are casual invaders, and are therefore *facultative parasites.*

Muscidae
Sarcophagidae
Calliphoridae
Anisopodidae
Scenopinidae

Larvae Concerned in Myiasis

In those families of which the adult flies are the medically important stage the eggs, larvae and pupae have already been briefly mentioned. There remain those families of which the larvae are important in medicine or hygiene, either accidentally, or as a normal part of their life-cycle.

The following section treats first those maggots and grubs that are actively concerned in myiasis of humans or of domestic animals, from which man may become infected.

Morphology of a maggot

Figure 136 shows a typical maggot, with the names of those few structures that are normally used for taxonomic purposes. There are three larval instars in this group of Diptera, but the third is much the longest. Third instar larvae are the most usual subjects for identification.

The body consists of eleven apparent segments, though both first and last are considered to be formed by the fusion of two segments. The integument is not sclerotized, and appears as a white skin, tough and wrinkled, with some areas of small spines that give a roughened appearance, like an unshaven chin; the spines are not always black and, if colourless, are difficult to see. Typically there is an anterior tract of spines and a posterior tract on each segment, but either of these may be incomplete dorsally, or entirely absent.

The pointed end of the maggot is anterior, and the broadly truncate end is posterior; the large posterior spiracles, deceptively like eyes, sometimes cause this end to be mistaken for the head.   Normally two pairs of spiracles can be seen: the anterior spiracles (asp) are placed laterally (except in Agromyzidae, where they are dorsal and close together), and may be quite simple, or branched in various ways.   Sometimes the form of the anterior spiracle is characteristic of the family, or even of the genus, but generally the anterior spiracles are less important taxonomically than the posterior ones.

The posterior spiracles (psp) are the most useful single character for identifying maggots.   In the mature larva (third instar) the hind spiracle normally has three slit-like openings, which are most commonly straight, or nearly so (fig. 136D).   The slits may be angled (fig. 150C) and in *Musca* and other Muscinae and Stomoxydinae the slits are highly convoluted (fig. 149C, E).   In some Tachinidae there may be more than three slits, and larvae of Oestridae have a large number of respiratory pores arranged in areas which occupy all the space of the spiracular disc (fig. 149A).

All these spiracles have a scar or button, which Keilin (1944) showed to be the remains of the spiracle of the second instar larva (fig. 137).   In practice the position of the button in relation to the spiracular slits, and to the surrounding peritreme; the peritreme itself, whether complete or broken; and the spacing apart of the two spiracles and their plates, with the orientation of the slits, are all important identification features.   In addition, the hind spiracles may be sunk in a deep cavity and normally concealed (at least in preserved specimens); or in a shallow cavity; on the flat or convex surface; or raised in various ways.

The *mouth-hooks* (figs 136A, Bm, 148) provide useful recognition features, though unfortunately they are not well systematized.   The presence or absence of the accessory oral sclerite (fig. 136A, os) distinguishes between *Calliphora* (present) and *Lucilia* (absent).   Gasterophilidae and Oestridae may have large prominently curved mouth-hooks for hanging on as well as for feeding, whereas in *Hypoderma* the mouth-hooks have almost vanished.

MUSCIDAE

A large and diverse family, both as adults (Chapter 5b) and as larvae.   Since only a very few species are of medical importance it is better, in the present context, to consider them separately.

Genus *Fannia*

The larvae are easily recognized by their flattened shape, with rows of fringed processes (fig. 143A).   They can be confused only with the larvae of some species of *Megaselia* (PHORIDAE, fig. 143B), but *Fannia* is much larger than *Megaselia* when fully grown. In both families the body processes appear to be a flotation device, enabling the larva to live in very dilute fluids without becoming immersed.

Figure 143A, B shows the larvae of *Fannia canicularis* and *Megaselia scalaris* side by side, for comparison.   The much longer and more elaborate processes of *Fannia* are distinctive of the genus, and also show good specific differences.   *Fannia* is a very large

X

genus, with a world-wide distribution, the larvae living generally in decaying organic matter of either vegetable or animal origin, habitats which are analysed by Chilcott (1960: 34–35). From this list it appears that the species recorded as causing myiasis—*canicularis, scalaris, manicata, incisurata*—are the most versatile, and occur in most or all of the types of habitat investigated. This versatility of breeding habitat is also a major factor in making the lesser house-fly *F. canicularis* an important nuisance to man.

Myiasis. *Fannia* larvae may be swallowed in food, may invade the urinary or other tracts of the body, and may infest wounds. The larvae will feed on liquid urine as, for example, in a soiled mattress in a baby's cot or pram.

Distribution. Originally probably Holarctic, but now cosmopolitan, spread by human agency.

Genus *Musca*

Musca domestica L. House-fly larvae are smooth maggots, with distinctive hind spiracles (fig. 149C). In practical cases the only likely confusion is between *Musca domestica* and *Stomoxys calcitrans*, and the two sets of hind spiracles are compared in fig. 149C, E.

Myiasis. Larvae have been reported from stools and vomits, and from the urinary passages. Even after allowing for all the possible mistakes of diagnosis some cases seem to be authentic. For example, in July 1952 a 2 yrs old infant in Lewisham Park Hospital, London, S.E.1, passed living House-fly larvae in its stool.

Wound infestations are also authenticated. In July 1952 larvae were taken from what was described as a 'dirty ulcer' on a man's leg at High Wycombe, Bucks., England. Other similar cases have occurred when a wound or an ulcer was covered in bandages and neglected.

Distribution. Cosmopolitan, spread by human agency. Place of origin doubtful, but possibly subtropical.

Genus *Stomoxys*

Stomoxys calcitrans L. Though the Stable-fly, or Biting House-fly, *Stomoxys calcitrans*, is less closely related taxonomically to *Musca domestica* than are some other genera, it is considered next because the two are readily confused both as adults and as larvae. Both larvae have convoluted hind spiracles, but in *Stomoxys* the three slits are less highly sinuous, and the button lies in the centre of the disc instead of on the inner periphery (fig. 149E).

The adult flies are extremely numerous around grazing stock and in farmyards, the larvae flourishing in organic materials with plenty of fibrous straw. Sometimes adults are abundant on beaches, both marine and freshwater—e.g. the Great Lakes of N. America —the larvae feeding in the debris along the strand line.

The larvae have been alleged to be involved in both intestinal and traumatic myiasis, but the importance of *S. calcitrans* as a myiasis-producer is disputed by Zumpt (1965 : 36). Probably it would be correct to say that this importance is slight and occasional.

Distribution. Cosmopolitan, spread by human agency. Its origin is entirely problematical, but it might have arisen on temperate steppes, in association with wild horses.

Genus *Muscina*

Muscina stabulans (Fallén).   The False Stable-fly is potentially dangerous because the adult is anthropophilic, and because the larva feeds in decaying organic matter, including any tainted or even strongly smelling food.

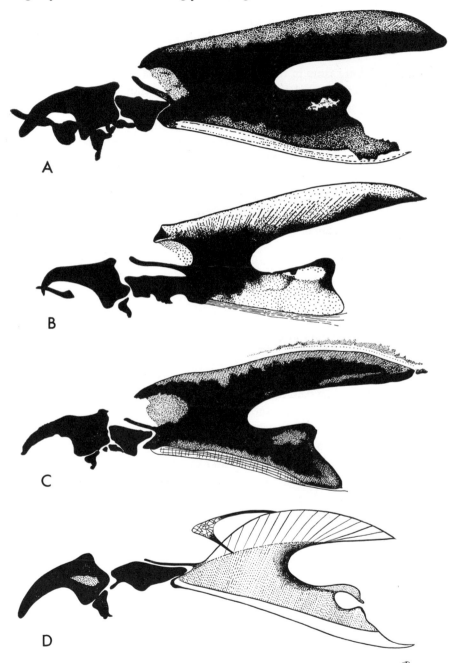

Fig. 148   Cephalopharyngeal skeletons of third instar larvae in lateral view: A, *Paralucilia wheeleri* (Calliphoridae); B, *Cynomyopsis cadaverina* (Calliphoridae); D, *Cochliomyia* (=*Callitroga*) *macellaria* (Calliphoridae); D, *Hylemyia* (Anthomyiidae) (after Thompson).

The distinctive character of the larva is the slightly boomerang-shaped slits of the hind spiracles (fig. 150C), set on a dense peritreme. According to Thomson (1937) the peritreme increases in density of sclerotization during the progress of the third instar, so that there is considerable variation between specimens of different ages.

Larvae of *Muscina* are among those discussed by Keilin (1917), which begin by being saprophagous, then in later larval life become carnivorous and attack other saprophagous larvae living in the same medium. This makes them particularly dangerous in myiasis because '. . . these larvae, thanks to their very dense integument, are resistant, and may live for a long time in the intestinal fluid of man, and, moreover, having their mouth-hooks better adapted to the act of piercing, they may provoke deep and dangerous wounds in the intestinal lining'. (Keilin, 1917: 420).

Taxonomically the species of *Muscina* are inadequately known. Besides *M. stabulans* two other species, *assimilis* (Fallén) and *pabulorum* (Fallén), are widespread in the Palaearctic and Nearctic Regions. The larvae of these two species are known to attack birds and sheep respectively (James, 1947: 137–138), but only *M. stabulans* is important in human myiasis. Many records exist of intestinal myiasis from larvae of *M. stabulans*; the adult fly readily enters houses, and readily oviposits on human food. James (1947) quotes description of damage by the larvae which bear out the comments made by Keilin (above).

Distribution. Cosmopolitan: probably carried round the world by man but by no means restricted to urban, or heavily populated areas.

CALLIPHORIDAE

Larvae of Calliphoridae are typical maggots, with mouth-hooks and spiracles of the general type shown in figs 148A–C, 149F–H etc. The hind spiracles have more or less straight lines, radiating from a button which lies in the line of the peritreme, though the peritreme may be weakened in the neighbourhood of the button, and thickened between the tips of the spiracular slits. The spiracular plates are surrounded by more or less heavily developed tubercles, and sometimes appear slightly sunken by comparison.

KEY TO THIRD STAGE LARVAE OF CALLIPHORIDAE

(adapted from James and others).

1 Larva grub-like, i.e. fleshy, and rounded at both ends   2
– Larva maggot-like, pointed at head, truncate posteriorly   4
2 Cuticle devoid of obvious spines, and with numerous fleshy grooves; anterior spiracles not projecting; posterior spiracles very widely separated, with three short, straight slits (fig. 150F)   .   .   .   .   ***AUCHMEROMYIA*** B. & B.
– Cuticle armed with obvious spines; anterior spiracles consisting of membranous stalks with finger-like processes; posterior spiracles not widely separated; slits highly convoluted   3

3   Spines of cuticle small, and often grouped into transverse rows of three or more; no bar of spiny processes behind the oral hooks; toothed plate bearing about six small yellow teeth; posterior spiracles small, with three short, sinuous slits (fig. 147A)

CORDYLOBIA Grünberg

–   Spines large, not grouped into transverse rows; a dark bar of spiny processes between the oral hooks; toothed plate bearing about ten yellow-brown teeth, arranged in two rows; posterior spiracles rather large, with three long, serpentine slits, of which at least one may show fragmentation into two parts (fig. 147B)

STASISIA Rondani

4   Peritreme of posterior spiracle incomplete and not enclosing button, which is sometimes poorly defined   5

–   Peritreme of posterior spiracle complete, though sometimes weakened in region of button   8

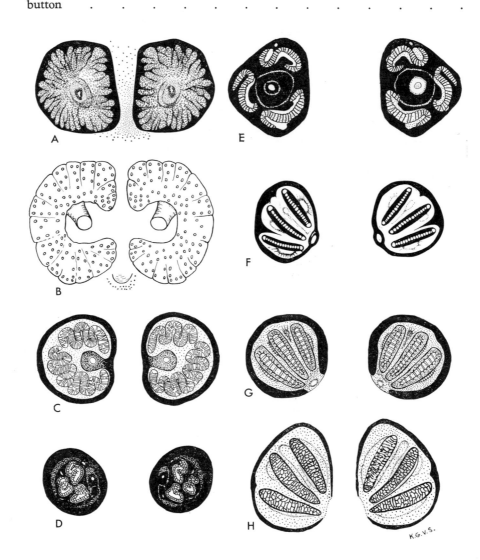

Fig. 149   Posterior spiracles of third instar larvae: A, *Oestrus ovis* (Oestridae); B, *Hypoderma lineatum* Hypodermatidae; C, *Musca domestica* (Muscidae) (after Zumpt); D, *Synthesiomyia nudiseta* (Muscidae) (after Siddons & Roy); E, *Stomoxys calcitrans* (Muscidae); F. *Lucilia sericata* (Calliphoridae); G, *Cochliomyia* (=*Callitroga*) *macellaria* (Calliphoridae); H, *Chrysomya bezziana* (Calliphoridae) (after Zumpt).

5   Posterior margin of segment 11 without dorsal spines; posterior spiracle without a definite button (fig. 149G) .   .   .   .   .   .   *COCHLIOMYIA* Brauer

–   Posterior margin of segment 11 with dorsal spines; button definite or not   .   .   6

6   Button indistinct (fig. 149H).   Old World tropics and subtropics   *CHRYSOMYA* R.-D.

–   Button distinct.   Palaearctic Region   7

7   Dorsal spines present on posterior margin of segment 10; larger tubercles on upper margin of posterior cavity distinctly longer than half width of one posterior spiracle (fig. 150A) .   .   .   .   .   *Protophormia terraenovae* R.-D.

–   Dorsal spines absent on posterior margin of segment 10; length of above-mentioned tubercles less than half width of one posterior spiracle (fig. 150B)
                                      Phormia regina (Meigen)

8   Accessory oral sclerite present (fig. 136A)   9

–   Accessory oral sclerite absent .   .   .   .   .   .   *LUCILIA* Meigen

9   Labial sclerite (mouth-hook) with toothlike apical portion longer than greatest depth of basal portion (fig. 136A) .   .   *Calliphora vicina* (=*erythrocephala*) R.-D.

–   Labial sclerite with toothlike apical portion as long as greatest depth of basal portion (fig. 148B) .   .   .   .   .   *Cynomyopsis cadaverina* R.-D.

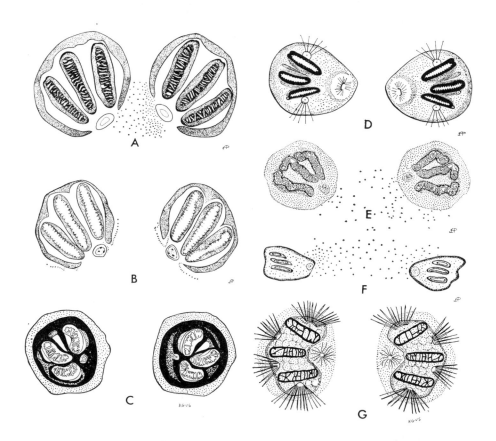

Fig. 150   Posterior spiracles of third instar larvae: A, *Protophormia terraenovae* (Calliphoridae); B, *Phormia regina* (Calliphoridae); C, *Muscina stabulans* (Muscidae) (after Zumpt); D, *Hydrotaea dentipes* (Muscidae); E, *Cordylobia anthropophaga* (Tumbu-fly, Calliphoridae); F, *Auchmeromyia* (Calliphoridae); G, *Ceratitis capitata* (Tephritidae).

Genus *Auchmeromyia*

Auchmeromyia luteola (Fabr.). Five species of *Auchmeromyia* are known, of which only one is associated with man. This is *Auchmeromyia luteola* (Fabr.), the blood-sucking larva of which is known as the Congo Floor Maggot, because it attacks people who sleep on the floor in tropical Africa.

The larva is recognized by the lack of cuticular spines, and the widely separated hind spiracles, each with 3 slits (figs 146C, 150F, Pl. 3A). At present no way is known of distinguishing the species in the larval stage; it is the association of the adult of *A. luteola* with human habitations that leads to the belief that this is the synanthopic species.

Zumpt (1965: 66) gives a key to adult *Auchmeromyia*. All five species, including *A. luteola*, may be found in and around the burrows of wart-hogs and ant-bears, along with the closely related *Pachychoeromyia praegrandis* (Austen). Whereas the adults feed on faeces and other fermenting materials, the larvae attach themselves to the skin, pierce, and feed on blood. Because adult flies are so numerous in comparison with the few specimens actually found in these burrows, Zumpt concludes that the true natural host must be some other, more obscure, burrowing animal.

It has often been said that *A. luteola* is a specific blood-sucker of man, but Zumpt questions this. Larvae have been found in association with domestic pigs, and adults at the entrance to the burrows of wart-hogs and ant-bears. Zumpt also mentions the sudden appearance of two adult flies in a tent at least 48 km (30 miles) from a human habitation.

For the biology of *A. luteola* see Garrett-Jones (1951).

Distribution. Throughout Africa, S. of the Sahara, and in the Cape Verde Islands, though not yet found in the drier parts of the Cape, nor in Madagascar.

Genus *Cordylobia* (including *Stasisia*)

The two genera *Cordylobia* and *Stasisia*, with one species each, have hitherto been separated as in the key on p. 311, but in 1953 Fain described *Cordylobia ruandae*, intermediate between the two in several respects. This led Zumpt (1965: 70) to reunite the two genera.

A useful comparison of the two species is made by Bertram (1938).

Cordylobia anthropophaga (Blanchard). The Tumbu-fly. Larvae (fig. 147A) are found in boil-like swellings beneath the skin of man and many other animals, domestic and wild. The eggs are laid in batches of 200–300 in sand or dust, especially if it is contaminated with urine or faeces, but only in shady places, not too wet. Eggs are not laid directly on the host.

The young larvae can live for as long as 9–15 days without food, but if opportunity occurs, they at once attach themselves to a host and burrow into the skin. The larvae remain *in situ*, feeding and moulting, for eight days then fall to the ground and pupate.

Infestations may occur on any exposed area of the body, and may involve one or more larvae, sometimes a score or so; but each is a separate invasion by an individual larva.

Distribution. Widely distributed in Africa, south of the Sahara. Like *C. rodhaini* (below) it occurs in many small rodents, but whereas *C. rodhaini* is principally a parasite of small antelopes (including duiker) and the Giant Rat, an important reservoir of *C. anthropophaga* is in monkeys and dogs.

Cordylobia (*Stasisia*) *rodhaini* Gedoelst. Often known under the name *Stasisia rodhaini*, and referred to in books as 'the larva of Lund', or as 'Lund's fly'; this name refers simply to the fact that Gedoelst (1909) based his original description on a larva extracted from the arm of a 'commandant Lund'.

C. rodhaini closely resembles *C. anthropophaga* as an adult, as a larva, and in its life-history, as far as the last is known. It is less frequently found in man (fig. 147B).

Distribution. Tropical Africa, but essentially the rain-forest areas. Besides small rodents, its main hosts are antelopes and the Giant Rat (cf. *C. anthropophaga* above).

Genus *Cochliomyia*

The New World 'screw-worms' have been known variously as *Callitroga* Brauer, 1833 or *Cochliomyia* Townsend, 1915, according to the passing whims of the International Code. They have become news in recent years as a result of the successful experiments in controlling them by releasing sterile males (for details, see *Rev. appl. Ent. B*).

There are four species, to which keys are given by James, 1947, and Cova Garcia, 1952 (see ch. 5C). Cases of myiasis involve the larvae of only two of these species, *C. macellaria* (Fabr.) (facultative) and *C. hominivorax* Coquerel (*americana* Cushing & Patton). The following table, comparing the larvae of these two species, is based upon Laake, Cushing & Parish, 1936.

	C. macellaria (Fabr.)	*C. hominivorax* Coquerel
Name	secondary screw-worm	primary screw-worm
Parasitology	facultative	obligatory
Eggs, number	250–1000	10–400
„ laid	in irregular groups, on hair or wool	tightly cemented to dry surface close to wound, dried pus or blood
„ appearance	cream-coloured, smooth surface; dorsal seam not cutting off a complete cap	white with reticulate surface, and with dorsal seam cutting off a complete cap
Larvae (1)	tracheal trunks not pigmented	tracheal trunks leading from hind spiracles pigmented, dark for length of 3–4 segments
(2)	posterior margin of segment II with spines only ventrally	posterior margin of segment II with complete ring of spines
(3)	hind spiracles smaller	hind spiracles bigger
(4)	anterior spiracles more often with 9–11 branches	anterior spiracles more often with 7–9 branches
(5)	ventral wall of pharynx ribbed	ventral wall of pharynx smooth

Cochliomyia hominivorax (Coquerel) (*americana* Cushing and Patton). The larva of this species is an obligatory parasite, which feeds in living tissue, head downward, with the posterior spiracles exposed. The larvae congregate in groups, and excavate a cavity or pocket. Although they can enter unbroken skin, they also infest any kind of wound, even a scratch. The destruction of tissues can be serious, and even fatal.

The average life cycle in summer in Texas is estimated at about 24 days, but considerable variation is attributable to the varying length of the pupal stage: 7–54 days, according to climatic conditions.

Distribution. From Illinois and S. Dakota right down to Argentina.

Cochliomyia macellaria (Fabr.). This is called the 'secondary screw-worm' (figs. 147C, 149H) because it is not an obligatory parasite, and is responsible for secondary infestation of wounds, as well as scavenging on dead tissues. Biologically, however, *macellaria* is evidently the more primitive species, not yet fully adapted to parasitism. Consistent with this status is the fact that *macellaria* is much the more numerous—Laake *et al.* give the relative abundance of *hominivorax* (as *americana*): *macellaria* as 1 : 590—and it completes its development in about half the time.

Distribution. Extends further north, to Quebec and Maine, and then southwards down to Argentina; it is also recorded from Ascension Island.

Paralucilia wheeleri (Hough). This is a little-known species, included by James (1947: 66) because the adult can be confused with *Cochliomyia*, and the larva has been reported from wounds. It is thought to be of little importance, though it sometimes happens that such apparently unimportant species are more abundant than was realized.

The larva is separated from that of *Cochliomyia* by the presence of an accessory oral sclerite (fig. 148A, D).

Distribution. Western U.S.: Washington down to New Mexico. Mexico.

Genus *Chrysomya*

This is a common and abundant genus of the Old World tropics, where it largely (but not entirely) replaces *Calliphora* and *Lucilia* of the temperate zone. It is also the Old World equivalent of the New World *Cochliomyia* (see above).

There are many species, and the taxonomy of the adults is not easy (see Zumpt, 1956 and Senior-White, Aubertin & Smart, 1940).

Zumpt (1965) offers the following key to larvae of the species that are known to be concerned in human myiasis:

KEY TO *CHRYSOMYA* LARVAE

1 Body provided with transverse rows of fleshy processes (fig. 143C) . . . 2
 – Body smooth; no fleshy processes except on last segment 3
2 Mature larvae reach a length of 18 mm. Segments with a great number of long processes; peritreme of posterior spiracle with a narrow opening and more or less distinctly forked at both ends . . . . . . a. *albiceps* (Wiedemann)
 b. *rufifacies* (Macquart)
 – Mature larva not more than 11 mm. long. Segments with fewer processes; peritreme of posterior spiracle with a broad opening, its ends not distinctly forked
 varipes (Macquart)

3 Segments with belts of strongly developed spines.   Anterior spiracles with 4–6
   branches   .   .   .   .   **bezziana** Villeneuve
– Segments with belts of moderately developed spines.   Anterior spiracles with 10–13
   branches   4
4 Posterior spiracles closely approximated, separated by about one-fifth of the diameter
   of a peritreme   .   .   .   .   a. **chloropyga** (Wiedemann)
                                                                                    ? b. **putoria** (Wiedemann)
– Posterior spiracles more remote from each other, separated by one-third to one-half
   of the diameter of a peritreme   .   .   .   .   .   a. **mallochi** Theowald
                                                                                    b. **megacephala** (Fabr.)

Chrysomya albiceps (Wiedemann), *Chrysomya rufifacies* (Macquart). These two species are bracketed together because their larvae are similar and distinct from all other Calliphorid larvae (fig. 143C); and also because it is debatable whether they may not be subspecies of one widely distributed species.

The larvae of both may cause secondary myiasis in sheep, but they are not known to attack man.

Distribution.   *C. albiceps* extends from the southern Palaearctic Region, throughout Africa, and sporadically in the Madagascan Region, to N.W. India.   In the Oriental and Australian Regions *C. albiceps* is replaced by *C. rufifacies*.

Chrysomya bezziana Villeneuve.   The Old World Screw-worm.   This is the most important species medically, since the larvae are obligatory parasites in wounds, and 'unlike all other *Chrysomya* species, never develop in carcasses or other decomposing matter' (Zumpt, 1965: 101).

Females are attracted to an exposed wound, however slight, on which they feed as well as oviposit.   Known hosts are man, cattle, water-buffalo, sheep, goats, horses, donkeys, dogs, camels and elephants.   Larval life lasts about 5–6 days, with much destruction of tissue; several hundreds of larvae have been reported from a single infestation.   Untreated wounds may rapidly become fatal.

Distribution.   Old World Tropics; tropical Africa, and most of Orient, but not Australia (see *C. mallochi*, below).

Chrysomya mallochi Theowald.   Before 1959 this species was known as *C. micropogon* Bigot, on the determination of Malloch.   According to Zumpt (1965: 102) records of *C. bezziana* from the Australian continent refer to this species.   The two species may meet, or even overlap, in New Guinea.

C. mallochi is mostly recorded as a secondary myiasis-producer, at any rate in the more southerly part of its range, but may be a primary parasite in the north.   A few cases of human myiasis, human and rectal, have been recorded.

Distribution.   New Guinea and Queensland.   James gives records of '*micropogon*' from all parts of Australia.

Chrysomya megacephala (Fabricius); *Chrysomya putoria* (Wiedemann); *Chrysomya chloropyga* (Wiedmann).

These are three species with rather similar habits.   The adult flies are numerous round any kind of exposed organic material, faeces or carrion.   They also take sugary foods, and thus are active in spreading any intestinal or other organism that can be spread by contact.

The larvae feed mainly in faeces and in carrion. They sometimes occur in wound myiasis in cattle, sheep and man, usually as a secondary infestation, with other larvae, sometimes alone.

Distribution

C. megacephala: Oriental and Australian Regions; China; Japan; Reunion; Mauritius. In Australia it occurs down the eastern seaboard. Not in Africa. *C. putoria* and *C. chloropyga*: Africa south of the Sahara, where they are possibly only sub-species; *putoria* is equatorial and West African; *chloropyga* occurs in the cooler highlands of central and eastern Africa, and in the south.

Genus *Protophormia*

Protophormia terraenovae (R.-D.). One of several steel-blue or 'black' blow-flies, that are easily confused with each other; in this case, particularly with *Phormia regina* (see below) and with *Protocalliphora azurea*, the larvae of which suck the blood of nestling birds.

The larva of *P. terraenovae* is a serious parasite of domestic animals—sheep, cattle, reindeer—and may sometimes cause death. It is not recorded from man.

Distribution. North temperate zone, sub-Arctic and Arctic, to within less than 960 km (600 miles) of the North Pole.

Genus *Phormia*

Phormia regina (Meigen). The Black Blow-fly. The larvae mostly feed in carrion, and may be very numerous. They may also become sheep maggots, especially in the warmer part of their range. Cases of human myiasis, both dermal and enteric, have been recorded in North America, though not, apparently, in the Old World (Zumpt, 1965: 82).

Distribution. Holarctic. Occurs also in the Hawaiian Islands and in Australia, but probably introduced there by man.

Genus *Cynomyopsis*

Cynomyopsis cadaverina (R.-D.). Another species of 'black' blow-fly, with larvae that feed in carrion, and which may be secondary invaders in wounds of domestic animals and man (fig. 148B).

Distribution. Nearctic. Reported by Enderlein to occur in Northern Europe, but Zumpt (1956, in Lindner, Flieg. pal. Reg. 190 (64i): 39) can only say that it 'ought to be found there'.

Genus *Lucilia* (inc. *Phaenicia*) Genus *Calliphora*

Greenbottles, bluebottles. These two genera are treated jointly, though of course they are abundantly distinct both as adults and as larvae. Their biology in general has been dealt with in Chapter 5C. The present chapter need only mention them in relation to myiasis (see also chapter 17).

Most, if not all, the larvae of the many species of *Lucilia* and *Calliphora* can feed in carrion, but they readily infect wounds.   Some species, especially of *Lucilia*, 'strike'-sheep—i.e. lay eggs in areas of the wool that are damp, soiled with urine or faeces, and subject to bacterial decay; and the larvae may lacerate the skin and start a festering sore. The principal species associated with sheep-strike are *L. sericata* Meigen and *L. cuprina* (Wiedemann), and the taxonomy of these is discussed in detail by Waterhouse & Paramonov (1950).

Wound myiasis in other animals, including man, is occasionally reported, but is not a regular habit.

The larvae of *Calliphora* are not as directly important in sheep-strike as those of *Lucilia*, but they are perhaps rather more frequently involved in wound myiasis in man. Considering the speed with which *Calliphora* locates even slightly tainted meat, it is surprising that such cases are not more frequent.   Intestinal myiasis is reported occasionally, as a result of swallowing meat containing maggots.   Again, the fact that such cases are not more frequent suggests that the larvae do not survive well in the human intestine.

Distribution.   Both genera probably originated in the North Temperate Region, and have been spread by human agency to the South Temperate Region.   The Ethiopian Region has one endemic species of *Calliphora*—*C. croceipalpis* Jaennicke—and two of *Lucilia*—*L. infernalis* (Villeneuve) and *L. cuprina* (Wiedemann): the last having later spread to the Orient and to Australia, where it is a major pest.

SARCOPHAGIDAE

Usually called Flesh-flies.   For many years this group of genera was regarded as a subfamily of Calliphoridae, but it is abundantly distinct.   The adult flies, in general, are grey and black, checkered or spotted, elongate, and conspicuously bristly.   The larvae are distinctive in having the hind spiracles hidden in a deep pit, and the spiracular slits nearly vertical, only slightly divergent (fig. 146A).

'The identification of *Sarcophaga* species must always be left to the specialist' (Zumpt, 1965: 103).   In this book, therefore, we shall merely mention the names of one or two species that are most relevant to human myiasis.   Within the family all degrees exist between breeding in faeces and obligatory parasitism.   It is also a family in which the deposition of larvae instead of eggs is the rule rather than the exception (see also chapters 5C and 17).

Genus *Sarcophaga*

Sarcophaga haemorrhoidalis (Fallén).   Breeds mainly in faeces, but also in decaying meat and other rotting food-materials.   It has been said to breed in the human intestine, but Zumpt (1965: 104) rejects this view, since *S. haemorrhoidalis* will larviposit on faeces as they are being deposited.   Many authentic cases exist, however, of larvae passing through the human intestine, causing disturbance, pain and nervous disorder. There is also direct evidence of injury to the intestinal wall.   Larvae of this species sometimes infest wounds, but rather rarely.

Distribution.   *S. haemorrhoidalis* is anthropophilic, and has been widely dispersed by human agency, though not uniformly so.   Its presence in Australia is doubtful.

Table 8 Table 8 lists other species of *Sarcophaga* recorded as producing human myiasis

Species	Authority James 1947	Zumpt 1965	Type of myiasis wounds	intestinal	Locality
herminieri R.-D.	+			+	N. America
sarraceniae Riley	+			+	N. America
placida Aldrich	+		+		Panama
plinthopyga Wied.	+		+		C. America
lambens Wied.	+		+		C. America
chrysostoma Wied.	+				Brit. Guiana
misera Walker (group)	+	+	+		widespread
crassipalpis Macquart	+	+	+		Balkans
bullata Parker	+		+	+	southern U.S.
striata (Fabr.)	+		+	+	southern Pal.
fertoni Villeneuve	+		+		Mediterranean
ruficornis (Fabr.)	+	+	+	+	India
peregrina (Fabr.)	+	+	+	+	Oriental; Australian
nodosa Engel	+	+	+		S. Africa
carnaria (L.)	+		+		Europe
albiceps Meigen	+	+	+		widespread
hirtipes Wied.	+	+		+	Africa
argyrostoma R.-D. (=barbata Thomp.)	+	+	+		Europe; S. America
tibialis Macquart		+	+		Africa; Mediterranean; Madagascar

Genus Wohlfahrtia

Biologically very similar to *Sarcophaga*, and with three species that are recorded as causing human myiasis.

Wohlfahrtia magnifica (Schiner). Larvae are obligatory parasites of warm-blooded animals, and never develop in carrion or other rotting materials. In man the larvae may be deposited in the ear, eye or nose, and cause extensive destruction of healthy tissue, sometimes causing death. Many human cases are recorded. Other hosts include horses, donkeys, cattle, water-buffalo, sheep, goats, pigs, camels, dogs, poultry and geese.

    Distribution. Southern Palaearctic and North Africa, but *not* Ethiopian Region.

Wohlfahrtia vigil (Walker). Apparently an obligatory parasite, dropping larvae on to the living skin, where they penetrate only if the skin is tender (e.g. young babies). Sometimes unsuccessful attempts to penetrate are diagnosed as impetigo. The larva remains in the dermal lesion and is easily removed.

Distribution.   Northern N. America, from Alaska down to Iowa, Ohio and Pennsylvania.

Wohlfahrtia nuba (Wiedemann).   Similar to *W. magnifica*, but smaller.   Causes wound myiasis in man and domestic animals.
   Distribution.   Saharan Region, from Senegal eastwards to Karachi.

GASTEROPHILIDAE, OESTRIDAE

Larvae of these two families are obligatory parasites of mammals, either in boils in the skin, in the mucus of head sinuses and throat, or in the intestine.   They are thus exclusively myiasis-producers, and Zumpt's book (1965) deals exhaustively with the species of the Old World.

   The classification of this group into families is debatable because of the difficulty of separating phylogenetic resemblances from those which result from convergence of habit.   It is fairly generally agreed that Gasterophilidae, the stomach bots of equines, and the bots of elephants, should be separated from the rest, which can then most conveniently be placed in one family Oestridae.

   No member of either family regularly produces myiasis in man, and the occasional infestation is accidental, usually caused by the first stage larva in its actively burrowing stage.   The horse bots (*Gasterophilus* spp.) lay their eggs either on hairs of the legs, where they are licked off (*G. intestinalis*) or directly on the lips (*G. haemorrhoidalis*), and the active larvae move through the mucous membrane of the mouth.   A few cases are reported of early larvae of *Gasterophilus* in the skin of man, especially on the face. The larvae of different species of *Gasterophilus* seem to differ in their ability to penetrate the human skin, and this is perhaps consistent with their different oviposition sites on their natural hosts.

   Invasions of the *human eye* by early larvae of *Gasterophilus* are recorded, and are not infrequently reported for the young larvae of *Oestrus ovis* and *Hypoderma* spp.   The larva of *Oestrus* usually dies without damaging the eye, but the larvae of *Hypoderma* are more dangerous, and may do considerable damage to the eye, or destroy it altogether. Semenov (1969) records penetration of the brain by *Hypoderma* with fatal results.

CUTEREBRIDAE

A family of New World flies, related to Calliphoridae and Oestridae, but apparently at a more primitive stage in the evolution of parasitism.   The larvae of Cuterebridae are dermal parasites of rodents, and *Cuterebra* is only rarely recorded as a human parasite (for summary see Rice & Douglas 1972), but *Dermatobia* has strong associations with man.

Dermatobia hominis (Linnaeus Jr.).   The association with man is frequent enough for this species to be called the Human Bot-fly; the larva, too, has many vernacular names in English, Spanish and Portuguese.

According to James (1947: 101): 'This is primarily a forest species; it occurs chiefly in the wooded tracts and forest margins of the lowlands and river valleys, but it may range up to an altitude of 3,000 ft.' *D. hominis* is famous for attaching its eggs to another biting arthropod—usually a fly or a tick—and most often to a day-flying mosquito. The larva remains within the chorion of the egg until it comes close to a warm-blooded animal, when the larva emerges and penetrates the skin, forming a boil. Larval life is long, from 6–12 weeks, and the larva is big and—in its second instar—flask-shaped (fig. 144C). The wound suppurates and the discharge may attract other myiasis-producing flies. An extensive and detailed account of the deliberate rearing of *D. hominis* in his own arm is given by August Busck (1912).

Though the attachment of the eggs to a carrier is spectacular, eggs may also be attached to leaves. The larva then emerges when a possible host brushes past (a device comparable with that found in adult fleas). Other hosts are cattle, dogs and birds.

Distribution: Neotropical, from Mexico to northern Argentina. Because of the long larval life, cases of infestation with *D. hominis* may appear in hospitals in any part of the world.

MAGGOTS IN FOOD

Intestinal myiasis caused by the accidental swallowing of larvae in food is covered in Chapter 18: Insects and Hygiene.

BIBLIOGRAPHY

BANKS, N. 1912. The structure of certain Dipterous larvae with particular reference to those in human foods. *Tech. Ser. Bur. Ent. U.S.* **22**: 44 pp.

BERTRAM, D. S. 1938. A note upon myiasis due to the larvae of *Cordylobia rodhaini* Gedeolst. *Ann. trop. Med. Parasit.* **32**: 431–433.

BRAUNS, A. 1954. *Terricole Dipterenlarven & Puppen terricoler Dipterenlarven*, 2 vols., pp. 179, 156. Berlin.

BRINDLE, A. 1961–. Taxonomic notes on the larvae of British Diptera. *Entomologist* **94**–. [Still appearing.]

—— 1963. Terrestrial Diptera larvae. *Entomologist's Rec. J. Var.* **75**: 47–62.

BRYCE, D. & HOBART, A. 1972. Biology and identification of the larvae of the chironomidae. (Diptera). *Entomologist's Gaz.* **23**: 175–217.

BURGESS, N. R. H. 1966. A case of human myiasis in London. *Trans. R. Soc. trop. Med. Hyg.* **60**: 432–433. [*Sarcophaga barbata*.]

BUSCK, A. 1912. On the rearing of *Dermatobia hominis* Linnaeus. *Proc. ent. Soc. Wash.* **14**: 9–12.

CHANDLER, A. E. F. 1968. A preliminary key to the eggs of some of the commoner aphidophagous Syrphidae (Diptera) occurring in Britain. *Trans. R. ent. Soc. Lond.* **120**: 199–218.

CHILCOTT, J. G. 1960. A revision of the Nearctic species of Fanniinae (Diptera: Muscidae). *Can. Ent., Suppl.* **14**: 1–295.

CHISWELL, J. R. 1954. A taxonomic account of the last instar larvae of some British Tipulinae (Diptera: Tipulidae). *Trans. R. ent. Soc. Lond.* **108**: 409–484.

COVA GARCIA, P. & SUAREZ, O. M.   1952.   Estudio sobre la morphología de la *Musca domestica*. *Revta. Sanid. Asist. soc., Caracas* **18**: 887–919.

C.S.I.R.O.   1970.   *The Insects of Australia*.   1829 pp.   Canberra.

DIXON, J. M., WINKLER, C. H., & NELSON, J. H.   1969.   Ophthalmomyiasis interna caused by *Cuterebra* larva.   *Trans. Amer. Ophth. Soc.* **67**: 110–113.

GARRETT-JONES, C.   1951.   The Congo floor maggot, *Auchmeromyia luteola* (F), in a laboratory culture.   *Bull. ent. Res.* **41**: 679–708.

GEDOELST, L.   1909.   *Cordylobia rodhaini* nova spec., Diptère congolaise à larve cuticole.   *Arch. Paras.* **13**: 538–547.

GREENBERG, B.   1971.   *Flies and disease*. 1. pp. 856.   Princeton.

GRUNIN, K. YA.   1953.   [Larvae of the gadflies of domestic animals of the U.S.S.R.]   *Opred. Faune SSSR* **51**: 1–123 [in Russian].

GUIMARÃES, J. H. & PAPAVERO, N.   1966.   A tentative annotated bibliography of *Dermatobia hominis*.   *Archos Zool. S. Paulo* **14**: 223–294.

HALL, D. G.   *The Blowflies of North America*.   pp. 477.   Baltimore.

HARTLEY, J. C.   1961.   A taxonomic of the larvae of some British Syrphidae.   *Proc. zool. Soc. Lond.* **163**: 505–573.

HENNIG, W.   1948–1952.   *Die larvenformen der Dipteren*, 3 vols, pp. 184, 458, 628.   Berlin.

HINTON, H. E.   1961.   How some insects, especially the egg stages, avoid drowning when it rains.   *Proc. S. Lond. ent. nat. Hist. Soc.* **196**: 138–154.

—— 1962.   Respiratory systems of insect egg-shells.   *Sci. Progr., Lond.* **50**: 96–113.

JAMES, M. T.   1947.   The flies that cause myiasis in Man.   *U.S.D.A. Misc. Pub.* **631**.   175 pp.   Washington.

JOHANSON, O. A.   1934–1937.   Aquatic Diptera.   *Bull. Cornell Univ. agric. Exp. Stn.*   Part I, Nematocera, excluding Chironomidae and Ceratopogonidae **164**: 71 pp.   Part II, Orthorrhapha, Brachycera and Cyclorrhapha **177**: 62 pp.   Part III, Chironomidae, subfamilies Tanypodinae, Diamesinae, Orthocladiinae **205**: 84 pp.   Part IV, Chironomidae, subfamily Chironominae, and Part V, Ceratopogonidae (Lillian C. Thomsen) **210**: 80 pp. [reprinted 1969].

KAMIMURA, K.   1967.   A case of human ocular myiasis due to [a larva of] the moth fly, *Psychoda alternata* [in Japanese]. [English summary.]   *Jap. J. sanit. Zool.* **18**: No. 4, 305–306.

KEILIN, D.   1915.   Recherches sur les larves de Diptères cyclorrhaphes.   *Bull. scient. Fr. Belg.* **49**: 14–198, 16 pls.

—— 1917.   Recherches sur les Anthomyides à larves carnivores.   *Parasitology* **9**: 325–450.

—— 1944.   Respiratory systems and respiratory adaptations in larvae and pupae of Diptera.   *Parasitology* **36**: 1–66.

KEILIN, D. & TATE, P.   1940.   The early stages of the families Trichoceridae and Anisopodidae (=Rhyphidae) (Diptera: Nematocera).   *Trans. R. ent. Soc. Lond.* **90**: 39–62.

KETTLE, D. A. & LAWSON, J. W. H.   1952.   The early stages of British biting midges *Culicoides* Latreille (Diptera: Ceratopogonidae) and allied genera.   *Bull. ent. Res.* **43**: 421–467.

KRIVOSHEINA, N. P. & MAMAEV, B. M.   1967.   [*Classification key to Diptera larvae of arboricole insects*.]   366 pp.   Moscow.   [In Russian.]

LAAKE, E. W., CUSHING, E. C. & PARISH, H. E.   1936.   Biology of the primary screw worm fly, *Cochliomyia americana*, and a comparison of its stages with those of *C. macellaria*. *Tech. Bull. U.S. Dep. Agric.* No. **500**, 24 pp.   Washington, D.C.

LEE, D. J.   1968.   Human myiasis in Australia.   *Med. J. Aust.* **1**: 170.

MAGY, H. I. & BLACK, R. J.   1962.   An evaluation of the migration of fly larvae from garbage cans in Pasadena, California.   *Calif. Vector Views* **9**: 55–59.

MALLOCH, J. R.   1971.   A preliminary classification of Diptera exclusive of Pupipara, based upon larvae and pupal characters, with keys to imagines in certain families. Part I.   *Bull. Ill. St. Lab. nat. Hist.* **12**: 161–407.

MURVOSH, C. M. & THAGGARD, C. W.   1966.   Ecological studies of the house-fly.   *Ann. ent. Soc. Am.* **59** (3): 533–547.

OKADA, T.   1968.   *Systematic study of the early stages of Drosophilidae*.   188 pp.   Tokyo.

OTSURU, M. & OGAWA, S. 1959. Observations on the bite of the Tabanid larvae in paddy-fields. *Acta. med. Biol., Niigata* **7**: 37–50.

PETERSON, A. 1957. *Larvae of Insects*. II. Coleoptera, Diptera etc. 416. pp. Columbus.

RAKUSIN, W. 1970. Ocular myiasis interna caused by the sheep nasal bot fly (*Oestrus ovis* L. Dipt., Oestridae). *S. Afr. Med. J.* **44**: 1155–1157.

RICE, P. L. & DOUGLAS, G. W. 1972. Myiasis in man by *Cuterebra* (Diptera: Cuterebridae). *Ann. ent. Soc. Am.* **65**: 514–516.

——, & GLEASON, N. 1972. Two cases of myiasis in the United States by the African Tumbu fly, *Cordylobia anthropophaga*. *Am. J. Trop. Med. Hyg.* **21**: 62–65.

ROBERTS, M. J. 1969. The feeding habits of higher Diptera larvae. *Entomologist* **102**: 99–106.

—— 1971. The structure of the mouthparts of some Calyptrate Diptera larvae in relation to their feeding habits. *Acta zool., Stockh.* **52**: 171–188 [*Calliphora*].

SEMENOV, P. V. 1969. [A case of the penetration of a larva of *Hypoderma lineatum* into the brain of a person] *Medskaya Parazit.* **38**(5): 612–3.

SENIOR-WHITE, R., AUBERTIN, D. & SMART, J. 1940. Diptera, VI. Calliphoridae, *Fauna Br. India*. 288 pp. London.

SKIDMORE, P. 1973. Notes on the biology of Palaearctic Muscids (1). *Entomologist*. **106**: 25–48.

SMART, J. 1936. Larvae of *Lucilia sericata* Mg. from a case of aural myiasis reported from Essex (Diptera). *Proc. R. ent. Soc. Lond.* (A) **11**: 1.

THOMSON, R. C. M. 1937. Observations on the biology and larvae of the Anthomyiidae. *Parasitology* **29**: 237–358.

WATERHOUSE, D. F. & PARAMONOV, S. J. 1950. The status of the two species of *Lucilia* (Diptera: Calliphoridae) attacking sheep in Australia. *Aust. J. Scient. Res. Ser. B.* **3**: 310–336.

ZUMPT, F. 1965. *Myiasis in animals and man*. 267 pp. London.

Y

7. SIPHONAPTERA
(Fleas)

by F. G. A. M. Smit

FLEAS, constituting the insect order Siphonaptera, are small wingless insects with a holometabolous metamorphosis (egg—larva—pupa—adult). Their bodies are streamlined, laterally flattened, hairy and shiny and vary in colour from yellowish brown to almost black. They are parasitic in the adult stage, sucking blood of warm-blooded animals (94% of the known species on mammals, 6% on birds). The elongate larvae lack eyes and legs and have rather sparse but strong setae; they are normally not parasitic but feed on organic matter which they find in the nest or dwelling-place of the host. The pupa is contained within a cocoon.

About 2000 different kinds of fleas, grouped into some 200 genera and 15 families, are known and probably another 500 await description.

LIFE HISTORY

The cycle of development, from egg through larval and pupal stages to adult, normally takes place in the nest or dwelling-place of the host and lasts at least three or four weeks. A female may produce several hundred eggs (fig. 151A) or more; a cat-flea, for instance, can under optimal conditions lay about 25 eggs a day for at least three or four weeks, in all some 800–1000 eggs during its lifetime. Most eggs are laid during the female's sojourn on the host or while in the latter's nest or lair and the majority will therefore land in the right places for the development of the larvae (fig. 151B) which usually hatch after about five days. With their chewing sucking mouth-parts the larvae feed on organic debris present in the nest or abode of the host. Blood is a nutritional requirement for larvae of a number of species. It is supplied by adult fleas which during feeding eject faeces consisting of the remnants of digested blood of a previous meal followed by droplets of virtually undigested blood (these mark the underwear or bed linen of flea-infested persons).

Some larvae even prod adult fleas to produce faecal blood which they then suck up. Moreover, larvae can be predators and scavengers and they may attack, kill and devour weak and small arthropods present in nest material and even feed on dead adults of their own kin. Larvae are sometimes found on host animals, e.g. in the coats of dirty dogs and cats, on human beings of unclean habits and on nestling birds. In south-east Australia and Tasmania the larvae of a certain species of flea actually live in burrows which they excavate in the skin of their marsupial hosts. After two or three weeks, during which the larva moults twice, the larva is fully grown and spins a cocoon of silk produced by the salivary glands. The viscosity of freshly spun silk causes dust and other fine particles to adhere to the cocoon which will thus be camouflaged. Two or three days later the cocooned larva, or prepupa, sheds its skin and is transformed into a pupa (fig. 151C).   The duration of the pupal stage depends on the ambient temperature but is usually one or two weeks.   The adult flea (fig. 151D, E), after emerging from the pupal case, requires a stimulus (usually vibration) to induce it to leave the cocoon and in the absence of such a stimulus can remain alive, but inactive, within the cocoon for long periods.   This peculiarity explains why the first person or animal to enter a dwelling or nest that had been uninhabited for a long time, may suddenly be attacked by innumerable fleas.

The adults of a number of species can copulate immediately after their emergence from the cocoon and egg-production may then begin after one or a few days; the females require nourishment before a batch of eggs is laid.   In such species there is often no apparent linkage between the breeding season of the host and that of the flea.   In other species the pattern of behaviour and development may be more or less completely interwoven with that of the—usually single—host species.

Fleas can fast for long periods and this enables them to spend a considerable time searching for a host after their original host has died or has vacated its nest, or to await patiently the return of a host to the nest.

Most fleas are not strictly host-specific, but are nest-specific, since they are to a large extent dependent on factors determined by the environment which governs their development, more so in fact than on the type of animal from which the adults obtain their food.   The adult's palate is not always very demanding and a good many fleas feed (or at least try to do so) on any available animal, though this will usually happen to be the one that built and occupies the nest.   While feeding on a wrong kind of host may be useful for the individual by keeping it alive and thereby prolonging its chances of coming into contact with a primary host, this may not be advantageous for the species as fertility can be seriously impaired by over-indulgence of foreign blood.

Whether a certain mammal or bird is a suitable host for fleas depends on the composition (food and shelter for larvae) and microclimate (temperature and humidity) of its nest or lair.   It follows that those mammals and birds which do not build or occupy nests, have nests or living-places unsuitable for the development of fleas or do not return regularly to some sort of home, cannot have fleas of their own unless these are strongly modified and adapted for a close association with the body of such hosts.

MEDICAL IMPORTANCE

The health of man can be adversely affected by fleas in the following ways:

(a) The perambulations, often rapid, by fleas on the skin can cause extreme annoyance and irritation, which can be aggravated by the difficulty of detection and capture of the culprits; during the hours of repose insomnia may result.   Many a traveller in regions where life is still unsophisticated has vivid memories of nights of flea torments.

(b) The after-effects of the repeated application of the flea's mouth-parts to the host's skin in the form of itching (pruritis), or possibly of an infection due to scratching, can be unpleasant.   The immediate traumatic effect caused by the flea's feeding apparatus is negligible.   When a flea has thrust its sucking-tube into the skin, blood is pumped up by dilatation of part of the oesophagus, but at the same time saliva is injected through a second extremely fine channel in the sucking-tube. The saliva prevents coagulation of blood and is the cause of the reaction of the skin (manifest as erythema and oedema) to the act of blood-sucking. A flea-puncture in a human skin may cause intense itching for one or more days and is characterized by a tiny dark spot (purpura pulicosa) —which may be visible for days—surrounded by a patch of swollen and reddish skin (roseola pulicosa) which as a rule disappears fairly quickly. The very first flea-puncture causes no observable skin reactions in a host but it induces hypersensitivity; when this host is then repeatedly feeding fleas over a long period the skin reactions are at first of a delayed type, then for some time there will be an immediate reaction followed by delayed reaction, the latter disappearing eventually, leaving an immediate reaction which finally will also be no longer apparent—the state of non-reactivity (immunity) has then been reached. The tissue-reaction to the act of feeding involving an antigen-antibody complex with accompanying chemical changes in a given stage of hypersensitivity appears to influence the development and survival of micro-organisms that may be injected by the flea into the skin.

(c) Following attacks by fleas (often cat-fleas, nowadays usually responsible for infestations of dwellings) certain individuals develop a mental condition, known as delusory parasitosis, which makes them suffer from imaginary ectoparasites. This condition is apparently not extremely uncommon and can be quite serious; doctors do not always seem to appreciate the nature of delusory parasitosis as those affected by it do not always receive a sympathetic response when consulting.

(d) The so-called jiggers, chigoes or sand-fleas constitute excellent examples of the extreme evolutionary modification a flea can undergo, structurally and habitually, in response to intensification of a need for a very close association with the body of the host. Ten species of such jiggers are known but only one of them, *Tunga penetrans* (Linnaeus), occurring in Central and South America and in tropical Africa, is of considerable medical importance. The larvae of these extremely small fleas develop in dry sandy soil at places frequented by the hosts; development from egg to adult takes about three weeks under favourable conditions. Freshly emerged adults are very agile and wander about till they find a host. The females of this jigger usually attach themselves to the feet of fairly large animals, man and pig being particularly suitable hosts. The soft skin between the toes or under toe-nails is especially favoured, but other parts of the foot may also be attacked, and hands as well, while heavy infestations may occur in the arms, particularly around the elbow, and in the genital region. The tiny female jigger buries herself with the aid of the strongly developed and serrated stilettos of the mouth-parts under the skin but the tip of her abdomen remains just outside the surface; these exposed terminal abdominal segments contain the large spiracles through which the flea breathes, as well as the anus and the opening of the oviduct. The abdomen gradually distends to reach the size and shape of a small pea (fig. 161c). This process of expansion, which takes 8–10 days, is not primarily a result of the maturation of eggs, which does not begin until the final stage of expansion. A total of several thousand eggs appear to be produced, and each of them is ejected so that it has a good chance of falling on the ground away from the host. When the embedded female dies she remains under the skin, where her presence often causes an inflammation which will eventually result in the dead jigger being expelled by ulceration. Neglected or improperly treated jigger-lesions often become secondarily infected with other organisms, sometimes causing loss of digits, septicaemia or tetanus.   The presence of the common jigger in some part of the human skin is referred to as tungosis.

(e) Several parasitic worms are associated with fleas.   One of the commonest tapeworms of dogs and cats is the double-pored tapeworm *Dipylidium caninum*, which is also a relatively uncommon human parasite.   Cat-fleas, dog-fleas and human-fleas are important intermediate hosts of this tapeworm.   Flea larvae (usually the large final instar ones) swallow eggs of this worm along with the organic matter which is found on the floor or ground in the haunts of dogs and cats.   The worm larvae hatching from the eggs in the midgut of the flea larva penetrate the gut wall and arrive in the body cavity where they remain during the pupal and adult stages of the fleas, increasing in size and becoming encapsulated infective larvae (cysticercoids).   An animal becomes infected by swallowing such infected fleas when, for example, licking its coat.   Similarly, such fleas do occasionally find their way into food or drink for human consumption and can thus cause helminthosis in man.   The rat tapeworm *Hymenolepis diminuta*, with rat fleas as intermediate hosts, is likewise a rather uncommon human parasite.

(f) As most fleas regularly commute between the bloodstream of one host and that of another, not necessarily related, animal, they are potential carriers (vectors) of micro-organisms that may be present in the blood of mammals and birds.   More often than not such organisms are harmful to the host animal and quite a number of species of fleas have now been incriminated as transmitters of aetiologic agents of diseases.   If man enters the mammal/bird—pathogen—flea cycle (the disease being a zoonosis) he may, if susceptible, contract the disease caused by the pathogen (the disease is then called an anthropozoonosis).   The role played by fleas as possible vectors of diseases transmitted from man to man (anthroponoses) is not very clear although e.g. interhuman transmission of plague, notably of the Black Death (bubonic plague) of the Middle Ages, by the human-flea and perhaps also by the cat-flea and dog-flea, may have been considerable.

Aetiologic agents of partially or wholly flea-borne zoonoses are protozoa, bacteria, rickettsiae and viruses.   No cases are known of fleas acting as vectors of protozoan anthropozoonoses although these insects may harbour quite a variety of protozoa (some of which are non-pathogens); the main diseases transmitted by fleas to man are bacterial or rickettsial.   For an annotated list of all micro-organisms that have been isolated from fleas, see Smit (1974).

Adult fleas become infected with micro-organisms by feeding on an infected host. Numerous natural infections of the alimentary canal of fleas are consequently detected but for a number of these the vectorial role of fleas is of no or only slight significance. No infection is known to be transstadial, i.e. if a flea larva feeds on infected material the pathogens succumb in the larva within a few days and are absent in the pupa.   Only adult fleas are therefore of direct medical importance.   Micro-organisms are either transmitted from host to host by the act of feeding (the usual modus for bacterial infections, e.g. by direct contamination from infected mouth-parts, or more indirectly by regurgitation of infected host blood but this occurs only when the bacteria multiply in the alimentary canal of the flea) or—especially in the case of rickettsiae—by their infected faeces (which may remain infective for a very long time—under certain conditions up to 3 years in the case of plague bacilli and 9 years for murine typhus rickettsiae) which may be licked up by grooming animals or reach mucous organs by some way or other. Salivary glands of fleas are hardly ever infected and it is therefore unlikely that microbes reach a host with the saliva injected into the skin during the flea's feeding process.

Fleas are known to be involved in the transmission of the following diseases (with the names of the aetiologic agents in brackets):

Bacterial infections:

PLAGUE (*Pasteurella pestis*). This is primarily an infection of rodents, especially of rats in which the disease usually develops into an acute and rapidly fatal septicaemia. When a rodent dies of the disease its infected fleas—the main vectors—will leave to attack other available animals and man may then be incidentally infected (normally most rodent fleas feed with reluctance on man). The bubonic form of plague is transmitted either by regurgitation of some of the mass of bacilli contained in the digestive tract of fleas (the proventriculus of which becomes usually temporarily blocked by them some time after infection) or by direct contamination from the mouth-parts. The incidence of human plague cases has markedly decreased since the last great Indian epidemic of 1898–1918 (when over 10 million people died). There is a vast reservoir of the plague bacillis in natural foci in many parts of the world, the bacillus now forming an integral part of ecosystems the equilibrium of which having doubtless initially been upset by the entrance of the pathogen. Plague flares up now and again and, although almost always curable if treated in time, it may become less easily preventable (apart from personal protection by vaccination) as rats and their fleas have shown an increased resistance to pesticides in certain regions.

TULARAEMIA (*Pasteurella tularensis*). A plague-like disease of mammals such as ground squirrels, water-rats, lemmings and other rodents, rabbits, hares, also affecting birds, and occurring only in the northern hemisphere. Ticks as well as certain flies appear to be the usual vectors but fleas may also play an important role. Naturally infected fleas have been found in Scandinavia, U.S.S.R. and U.S.A. In man tularaemia occurs in several clinical types, some resembling symptoms of plague.

PSEUDOTUBERCULOSIS (*Pasteurella pseudotuberculosis*). A disease of a large variety of mammals and birds; widespread, although apparently commoner in the Old World than in the Americas. In the Russian Far East a number of rat-fleas as well as a specific flea of the house-martin were found to be naturally infected. The disease symptoms in animals are either of a chronic type and resemble those of tuberculosis or of an acute form, which may occur in rodents, and which is more plague-like. In man the disease is relatively rare and may be typhoid-like with a high mortality rate or be relatively mild and simulating acute appendicitis.

ERYSIPELOID (*Erysipelothrix rhusiopathiae*). Man is usually infected through abrasions of the skin and the disease is a mild one with a low mortality rate. It is the cutaneous form of swine erysipelas which occurs in a variety of animals (especially pigs) and birds (mainly poultry) and it is not to be confused with human erysipelas which is caused by streptococci. The aetiologic agent of erysipeloid has been isolated from 12 species of fleas in Asiatic Russia and from four in European Russia.

LISTERIOSIS (*Listeria monocytogenes*). A world-wide disease of cattle, sheep, rabbits, guinea pigs, chickens and other animals. In man it may cause meningo-encephalitis. The mode of transmission appears to be unknown but it may be of some significance to note that several rodent fleas were found to be naturally infected in European and Asiatic Russia.

GLANDERS (*Malleomyces mallei*). A disease of horses and allied species, but also occurring in a variety of other mammals, including small rodents. Experimentally fleas have been shown to be suitable vectors among the latter.

MELIOIDOSIS (*Malleomyces pseudomallei*). Although world-wide in distribution, this disease is encountered especially in tropical and sub-tropical countries in rodents (rats), dogs, cats, sheep, goats, horses and pigs. In man it is a rare disease usually of acute character and a high mortality rate but chronic forms may last for years. Fleas have been experimentally shown to be good vectors.

BRUCELLOSIS (or Malta fever) (*Brucella melitensis*). Brucellosis *s.l.* is principally a world-wide disease of domestic animals, but the species of *Brucella* isolated from some fleas (*B. melitensis*) mainly infects goats and sheep in Mediterranean countries. In man this pathogen usually causes a febrile illness known as undulant fever and is normally acquired through contact with infected animals, or through infected milk. Several species of fleas tested in Malta and in various localities in the U.S.S.R. were found not to be involved in the transmission of this *Brucella* but from a certain flea in Turkey the organism was successfully isolated.

SALMONELLOSIS (*Salmonella enteritidis* and *S. typhimurium*). The two species of *Salmonella* isolated from fleas are the most important cause of bacterial food poisoning (gastro-enteritis). Among the various animals subject to *Salmonella* infections, rodents are extremely susceptible. Even though half-a-dozen species of rodent fleas have been found to be naturally infected in Asiatic Russia, the chances of man acquiring the organisms from such fleas would seem to be slight.

STAPHYLOCOCCUS AUREUS. In man, the principal host of this bacterium, acute pyogenic infections are caused, also acute osteomyelitis or septicaemia. A certain rodent flea in Asiatic Russia was found to be naturally infected.

Rickettsial infections:

MURINE TYPHUS (*Rickettsia typhi*=*R. mooseri*). A world-wide (but e.g. in North America only in the southern United States) disease of various rodents, especially of rats. In man this is clinically a milder disease than epidemic typhus (which is louse-borne). Fleas are the main vectors from rodent to rodent and rodent to man and naturally infected specimens are known from U.S.A., U.S.S.R., China and Thailand.

HAEMORRHAGIC NEPHROSO-NEPHRITIS (*Rickettsia pavlovskyi*). This rickettsial form of haemorrhagic fever occurs in eastern Asia, with a reservoir in wild rodents. A number of fleas have been found to be naturally infected in the U.S.S.R. Trombiculid mites perhaps also act as vectors.

BOUTONNEUSE FEVER (*Rickettsia conori*). The most important reservoir of this normally tick-borne disease is the dog. In south-eastern France the specific flea of the European rabbit was found to be infected and the presumption has been made that these fleas acquired the infection from voles (if so, a most accidental mode of infection).

Q-FEVER (*Coxiella burneti*). A cosmopolitan disease of wild and domestic animals. In man, who often acquires the infection through the respiratory tract, it causes usually a mild febrile illness resembling influenza. Although ticks are the normal vectors, at least one species of rodent flea carries a natural infection in Asiatic Russia.

Viral infections:

LYMPHOCYTIC CHORIOMENINGITIS. An acute but rarely fatal infectious disease of man, clinically rather variable, the virus of which occurring naturally in a large variety of mammals (especially mice), and in three species of rodent fleas in the U.S.S.R.

TICK-BORNE ENCEPHALITIS. This disease ranges throughout the Palaearctic and Oriental Regions; sheep and goats are among the principal hosts. In Asiatic Russia the 'spring-summer encephalitis' form produces a more severe illness in man than the milder forms known as West Russian and Central European encephalitis. Although ticks are the main vectors, the virus has been isolated from several species of fleas in Poland and the U.S.S.R. However, fleas are possibly not to be considered as important vectors in any practical epidemiological sense.

NEOPLASMS. Transmission of virus-induced neoplasms, including leukaemias, by fleas is currently being investigated.

It will be evident from these notes that there is an apparent preponderance of natural infections of fleas in eastern Europe and the U.S.S.R. This is presumably partly due to the fact that research on this subject has been more intense there than elsewhere.

IDENTIFICATION

It is mainly the fleas of terrestrial mammals, representing nearly 90% of the known species, that are of medical importance and such species figure largely in studies on e.g. plague transmission. Bat-fleas and bird-fleas are therefore not considered here (the European chicken-flea *Ceratophyllus gallinae* (Schrank) can occasionally be a nuisance).

Several species of fleas parasitic on synanthropic hosts have become virtually cosmopolitan and, being common and secondarily associated with man, they are of prime medical importance and as such are the most widely and frequently used species for experimental purposes. A separate and fully illustrated key for these common fleas is therefore provided here. It should be noted that like most other fleas, these common species have related forms which resemble them closely. It is not difficult to give a flea a wrong name! Apart from the classical example of misdeterminations with puzzling results by the Plague Commission in India in the beginning of this century

(failing to differentiate between some species of *Xenopsylla*, which was pardonable because flea taxonomy was still in its infancy), until fairly recently fleas of the genus *Pulex* in North America were determined as the human flea *P. irritans* Linnaeus as the differences between this taxon and the closely related *P. simulans* Baker had not been recognized.   Identification is not always facilitated by the fact that certain structures in many fleas are liable to show a fair amount of individual variation and the true nature of this kind of variation can only be recognized after study of long series of specimens. Moreover, structurally abnormal specimens may also be encountered.

It is not feasible to construct a single, simple, concise and foolproof key to the numerous other species of fleas that are known as potential or proven vectors (the list of which continues to grow).   The keys provided here only serve to identify the genus of such fleas, i.e. about a third of the genera of terrestrial mammal fleas.   In order to simplify determination, separate keys are given below for five large geographical areas: Eurasia (p. 344), Africa (p. 345), North America (p. 347), Central and South America (p. 356) and Australasia (p. 356).   Even so, keys to only some selected genera are most unsatisfactory and they should be used with care.   The world flea-fauna is still incompletely known and this is another trap for possible misdeterminations as even existing keys to species of most genera cannot be absolutely satisfactory.

PRESERVING AND MOUNTING

Collected fleas are kept, prior to mounting, in small glass tubes containing 70–80% alcohol; they can be stored indefinitely in this way.   Never use formalin as a preservative. The tubes can best be kept in well-sealed jars.   If storage is to be for many years, it is better to keep the specimens in a dry state: stick the specimens with paste or water-soluble gum to a narrow strip of card, after first having written the data on the reverse side of the strip; allow the paste or gum to dry and then place the strip in a tube and close firmly with a cork.

Most fleas can only be studied and identified satisfactorily if they are mounted properly on slides, but one may be able to identify less heavily sclerotized specimens without any previous treatment.

Pass the specimens (which should be kept in the same tube while changing fluids; pin the data-label on the cork) successively through the following liquids:

(*a*) 10% solution of potassium hydroxide (KOH) at room temperature (do not boil fleas); one or more (generally two) days, till the specimens are somewhat transparent.

(*b*) Water; a few minutes.

(*c*) 5–10% aqueous solution of glacial acetic acid; half an hour.

(*d*) Water (renew once); half an hour.

If permanent mounts are not desired, the specimens can, at this stage, be studied on a slide in water (under a coverslip) and afterwards returned to a tube containing alcohol or stuck on a strip of card.

(*e*) Put the specimens from one tube on a clean microscope-slide, a short distance from each other.   Remove with blotting paper any superfluous water round the specimens, but avoid letting them become dry.   Arrange with the aid of two fine mounted needles, under a dissecting microscope or a powerful lens, the legs in proper position, i.e. down-wards and free from each other. Put on a coverslip, or a quarter-piece of a slide, and let 96% or absolute alcohol run from a small pipette underneath the coverslip. Since the alcohol evaporates fairly quickly, more should be added from time to time. Very large and thick fleas should be covered with half a slide since the pressure of a coverslip would not be sufficient; leave for about half an hour. This flattening and leg-fixing process can be omitted (e.g. if too many specimens are involved) but it is an advisable procedure. Prior to straightening the legs, the abdomen of some specimens, e.g. gravid females, may have to be pierced with a fine needle laterally between sterna II and III, after which a gentle squeeze can remove most of the incompletely macerated contents.

(*f*) Scoop the specimens off the slide and place them in absolute alcohol; one hour or longer.

(*g*) Oil of cloves; at least for some hours, preferably a day or longer. If the specimens, which for the first hour or so will float on the surface of the oil, do not eventually sink, they should be pushed below the surface with a blunt needle, or the tube gently shaken.

(*h*) Xylol (renew once); about ten minutes (or longer).

Instead of (*g*) and (*h*) one can place the specimens overnight in a 1 : 1 mixture of terpineol and absolute alcohol, then in pure terpineol for a few hours at least. Some workers use only benzole for these two stages.

(*i*) Mount the specimens in Canada balsam, dissolved in xylol. It is customary, for the sake of uniformity, to mount every flea on its right hand side on the slide. In order to get a nice preparation, the specimen should be placed exactly on the centre of the slide; then put a clean coverslip on top of it, taking care that the specimen is beneath the centre of the slip. Now put a drop of fairly thin balsam on the slide, touching the edge of the slip; opposite this, place another drop of balsam so that the balsam will spread quickly underneath the coverslip; the space between coverslip and slide should be entirely filled with balsam but do not use more balsam than strictly necessary. While it is advisable not to place more than one flea under one coverslip, a large series of a single taxon permits the placing of e.g. 1♂ 1♀, or 2♂ 2♀, under one coverslip (this will also reduce required time, energy, materials and storage space).

(*j*) The slides can be dried on a hot-plate or in an oven at 80–90°C. for about half an hour (at room temperature it will take several weeks for the balsam to harden a little).

(*k*) When the slides are sufficiently hardened scratch off all balsam outside the coverslip, using a stiff razor blade. The slides may be ringed with Canada balsam if circular coverslips have been used; this ringing is not essential but it gives the slide a finished look and also protects the edge of the coverslip. Once ringed, the slides may be dried again at 80–90°C., but only for a few minutes.

(*l*) Label the slides and write (in clear script) the data from the tube label in Indian ink on the slide labels; after determination the name of the flea should be added. Store the slides in slide-boxes or cabinets.

One should always handle fleas with care and, for instance, lift them individually out of fluids with a lanceolate needle or a similar instrument, so as not to damage specimens through loss of setae.

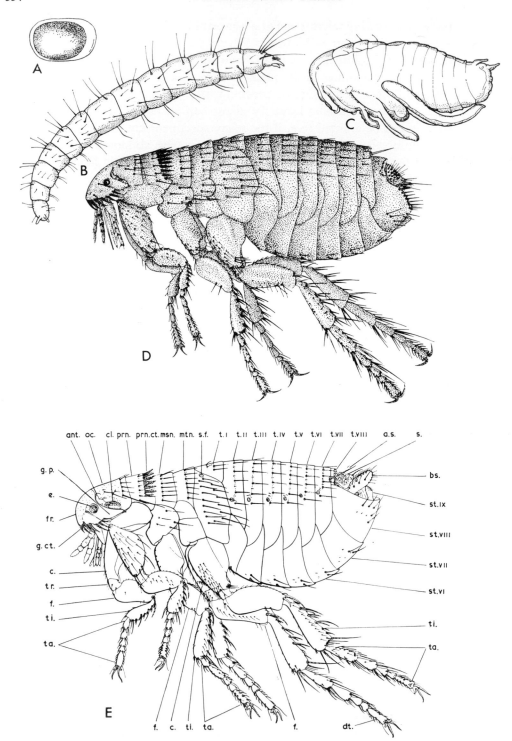

Fig. 151 *Xenopsylla cheopis* (after Patton & Evans): A, egg; B, larva; C, pupa   D, *Cteno-cephalides felis felis*, female (after Smit).   E, *Ctenocephalides canis*, male (after Smart);

Explanations of abbreviations used in figures: ac, acetabulum; ac.s, acetabular seta; an.sty, anal stylet; ant, antenna; ant.f, antennal fossa; a.s, antesensilial seta; b.c, bursa

GLOSSARY OF TECHNICAL TERMS USED IN THE KEYS

As in all insects, the body of fleas consists of three main divisions, head, thorax and abdomen, but unlike all other insects the body is laterally compressed and therefore fleas are studied, for the purpose of determination, from a side view; descriptions and figures of most of the paired structures generally relate to only one side (the left, as regards figures).

The taxonomically most important parts of the body are indicated in figs 151E, 152D, 156B, D, 157A, B, D, 158A, 159A. The following terms (only a fraction of the available terminology), arranged alphabetically, could not be avoided in the keys:

acetabular seta (fig. 157D, ac.s.)—one or two (sometimes more) setae, often long, placed on the basimere near or at its posterior margin and usually behind the acetabulum (fig. 157D, ac.).

aedeagus (fig. 155C)—the external part of the male intromittent organ.

anal sternum (fig. 169C, D)—the ventral plate of the small terminal abdominal segment.

anal stylet (figs 156B, 158A, an.sty.)—an elongate structure, bearing one long apical seta and often also one or a few preapical small setae, present in most female fleas on each side of the anal tergum, behind the sensilium.

antennal clava (fig. 151E, cl.)—the third, apical clublike segment of the antenna, subdivided into nine flagellomeres (except for Rhadinopsyllids and a number of Pulicids where one or more flagellomeres may have amalgamated).

antennal fossa (fig. 157B, ant.f.)—a deep groove on each side of the head in which the antenna rests.

antesensilial setae (figs 151E, 156B, D, 158A, a.s.)—one or more usually strong setae placed dorsally at or near the posterior margin of tergum VII; absent in *Chaetopsylla* and in ♂ *Rhadinopsylla*.

area spiculosa (fig. 168A)—a field of spicules on the inner side of the dorsal part of tergum VIII in males, present in many species of the family Ceratophyllidae.

basimere (figs 151E, 156D, 157D, bs.)—main part of paramere (*q.v.*).

bulga—see: spermatheca.

---

copulatrix; bs, basimere; bu, bulga of spermatheca; c, coxa; cl, clava of antenna; d.a, distal arm of sternum IX, ♂; d.b.c, ductus bursae; d.o, ductus obturatus; d.s, ductus spermathecae; dt, distitarsomere; e, eye; ep, epipharynx; f, femur; fr, frons; fr. tub, frontal tubercle; g.ct, genal ctenidium; g.p, genal process; hi, hilla of spermatheca; l, lacinia; la.p, labial palp; lam.m, lamina media of aedeagus; m, manubrium; max.p, maxillary palp; msem, mesepimeron; mses, mesepisternum; msn, mesonotum; mss, mesosternum; mtem, metepimeron; mtes, metepisternum; mtn, metanotum; mts, metasternum; oc, occiput; o.s, ocular seta; p.a, proximal arm of sternum IX, ♂; pl.r, pleural rod of mesothorax; prn, pronotum; prn.ct, pronotal ctenidium; s, sensilium; sp, spermatheca; spf, spiniform seta; s.f, spiracular fossa; st, sternum; sti, stipes; t. tergum; ta, tarsus; ti, tibia; tl, telomere; tnt, tentorial rod; tr, trochanter.

bursa copulatrix (fig. 158A, b.c.)—the copulatory pouch of the female, between vagina and ductus spermathecae.

coxa (fig. 151E, c.)—the basal segment of a leg.

ctenidium—a row of at least two spines (rarely reduced to only one spine).

distitarsomere (fig. 151E, dt.)—the apical claw-bearing section of a leg.

eye (figs 151E, 157B, e.)—eyes are variable in size and, particularly in subterranean nest fleas, can be absent.

frons (figs 151E, 157A, fr.)—the more dorsal and anterior area of the preantennal portion of the head.

frontal tubercle (fig. 157A, fr.t.)—a triangular projection at the frontal margin, very variable in shape and in structure; absent in a number of species.

gena—the ventral area of the preantennal part of the head.

genal ctenidium (figs 151E, 159A, g.ct.)—a comb of spines along the ventral marginal area of the gena (horizontal ctenidium, e.g. figs 160A–D, 164C) or along its posterior marginal area (oblique or vertical ctenidium, figs 159A, 164, 165).

genal process (figs 151E, 157B, g.p.)—the ventro-posterior part of the gena, its dorsal margin bordering the antennal fossa.

hilla—see: spermatheca.

labial palp (fig. 157A, la.p.)—usually consisting of five segments, in certain species fewer, rarely more.

lacinia (fig. 157A, l.)—the stiletto-shaped part of the maxilla; it has fine or coarse serrate margins and is as long as the epipharynx; the combination of the two laciniae plus the epipharynx forms the piercing-sucking tube.

lamina media (fig. 156D, lam.m.)—the median plate and in cleared specimens usually the only visible part of the aedeagal apodeme.

manubrium (fig. 156D, m.)—a forward projecting apodeme of the basimere.

mesonotum (figs 151E, 152D, msn.)—the dorsal sclerite of the second thoracic segment; it never bears a ctenidium.

mesopleuron—see: pleural rod.

metanotum (figs 151E, 152D, mtn.)—the dorsal sclerite of the third thoracic segment; this may have a ctenidium, but in most species it has not, or only a few small marginal spinelets.

metasternum (fig. 152D, mts.)—the ventral part of the third thoracic segment, mainly between the bases of the coxae, and anterior to the vertical pleural ridge.

metepimeron (fig. 152D, mtem.)—a large ventro-posterior sclerite of the third thoracic segment, containing dorsally the first abdominal spiracle; it has taken the place of the first abdominal sternum, which is lacking.

metepisternum (fig. 152D, mtes.)—the part of the third thoracic segment between the lateral metanotal area and the metasternum.

ocular seta (fig. 157B, o.s.)—the large seta placed nearest the eye, below or in front of it.

paramere (figs 155B, 157D, 160E, 160H, 166D, 167A–F)—a clasping organ joined to tergum IX in the male and consisting of a basimere with an anterior apodeme (manubrium) and one or two telomeres.

pleural rod (fig. 152D, pl.r.)—an internal rod which divides the mesopleuron into an anterior mesepisternum and a posterior mesepimeron.

pronotal ctenidium (figs 151E, 157A, 159A, prn.ct.)—a comb of spines along the posterior margin of the pronotum.

pronotum (figs 151E, 157A, prn.)—the dorsal sclerite of the first thoracic segment.

sensilium (figs 151E, 156B, D, s.)—a sharply defined area of tergum X, densely clothed with very short spicules; in this field are a number of pits (trichobothria) from each of which arises a long and very thin sensory seta.

spermatheca (figs 156B, 158A, sp.)—the semen receptacle of the female, consisting of a widened main part, the bulga (bu.), containing the orifice for the duct, and a terminal usually long and narrow sausage-shaped hilla (hi.); the latter bears in some species a sclerotized papilla at the apex.

spine—a well-developed, heavily sclerotized outgrowth of the cuticle, not arising from a distinct alveolus; a row of spines forms a ctenidium.

spiniform seta (fig. 159A, spf.)—a seta which is much thickened and often tapering abruptly to a fine point.

spiracular fossa (figs 151E, 156B, s.f.)—the cuticular pit of the body wall into which the abdominal spiracles are sunk.

sterna II–VII (figs 151E, 156B, D, st. II–VII)—the ventral halves of the abdominal segments; the first segment lacks a sternum (its place being taken by the large metepimeron; the basal sternum is therefore sternum II).

sternum VIII ♂ (figs 151E, 156D, 157D, 166A, 168D, E, st. VIII)—the size of this sternum is inversely correlated with that of tergum VIII.

sternum IX ♂ (figs 156D, 157D, st. IX)—a strongly modified sternum, usually L-shaped; the upright branch is the proximal arm (figs 156D, 157D, p.a.), the horizontal one is the distal arm (figs 156D, 157D, d.a.).

tarsus (fig. 151E, ta.)—the apical part of a leg, following the tibia, subdivided into five tarsomeres and with an apical pretarsus (consisting of two claws and an unguitractor which resembles a hive).

telomere (figs 156D, 157D, tl.)—a usually moveable process articulating with the inner side of the basimere; in Pulicoidea each basimere usually has two telomeres.

tentorial rod (fig. 159A, tnt.)—a thin rod of the tentorium, visible in members of some genera in front of the eye.

terga I–VII (figs 151E, 156B, D, t. I–VII)—the dorsal halves of the abdominal segments; these terga are not modified.

tergum VIII (figs 151E, 156B, D, t. VIII)—in the male this tergum can vary in size according to species; in the female it is always large and covers most or all of the small sternum VIII.

tibia (fig. 151E, ti.)—the segment of a leg between the femur and the tarsus.

trochanter (fig. 151E, tr.)—the short leg segment between coxa and femur.

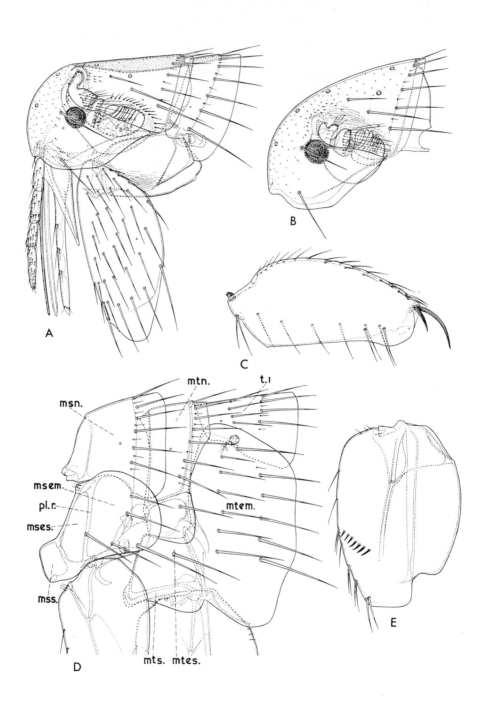

Fig. 152 *Xenopsylla cheopis*: A, head, prothorax and fore coxa, male (after Smit); B, head, female (after Smit); C, outer side of hind femur, female (original); D, mesothorax, metathorax and tergum I, male (after Smit); E, inner aspect of hind coxa, female (original).

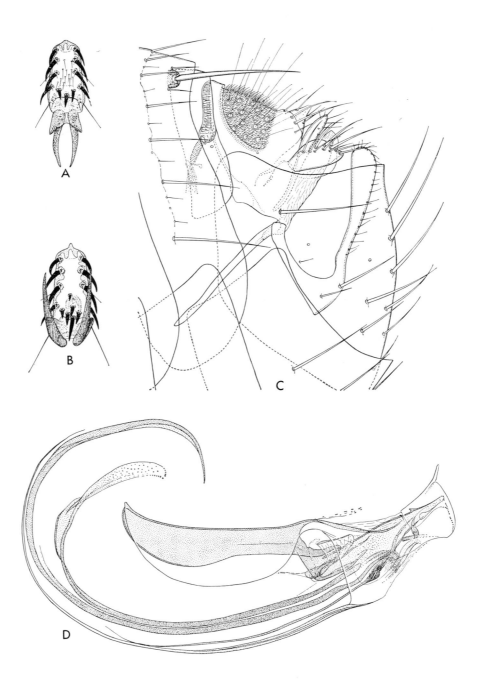

Fig. 153 Fore distitarsomere, male (originals): A, *Xenopsylla cheopis;* B, *Ctenocephalides felis damarensis.* C, D, *Xenopsylla cheopis*: C, terminalia, male (after Smit); D, phallosome (original).

Z

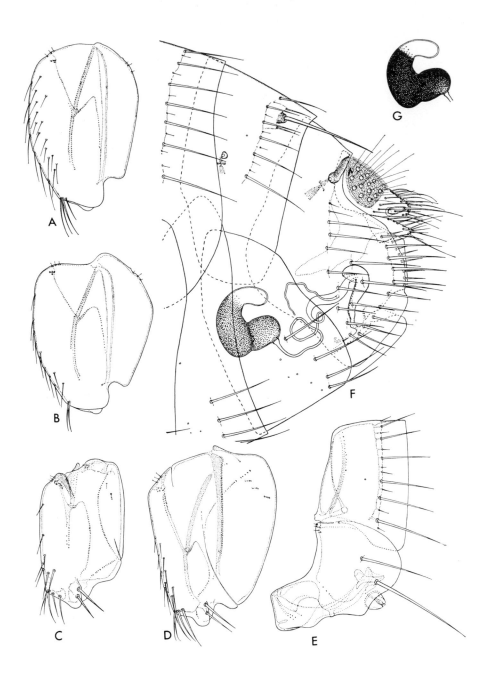

Fig. 154   Inner aspect of hind coxa, male (original): A, *Citellophilus simplex*; B, *Nosopsyllus fasciatus*.   C–D, outer side of mid coxa, female (originals): C, *Ctenocephalides felis felis*; D, *Nosopsyllus fasciatus*.   E, *Pulex irritans*, mesothorax, male (after Smit). F–G, *Xenopsylla cheopis* (after Smit).   F, terminalia, female; G, spermatheca (unbleached.)

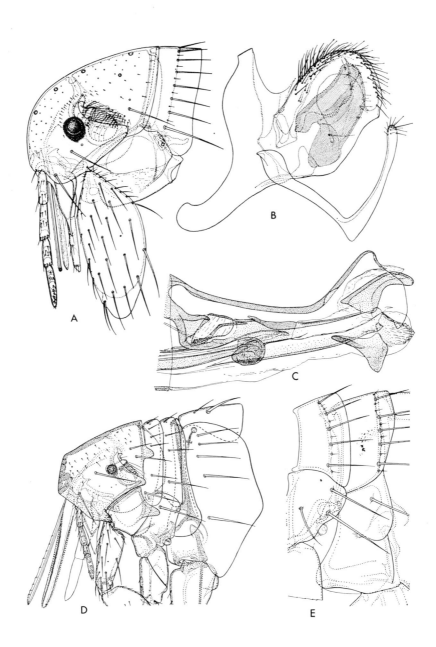

Fig. 155   *Pulex irritans* (after Smit): A, head, prothorax, and fore coxa, male; B, paramere and sternum IX; C, aedeagus.   D, *Echidnophaga gallinacea*, head, thorax and tergum I, female (after Smit).   E, *Synosternus pallidus*, mesothorax and metathorax, female (after Hopkins & Rothschild).

KEY TO COSMOPOLITAN SPECIES

1   Without ctenidia   2
-   At least pronotum with a ctenidium (e.g. figs 151D, E; 157A; 159A)   .   .   .   4

2   Mesopleuron with an internal rod (fig. 152D).   Antesensilial seta not marginal (figs 153C, 154F).   ♂ (fig. 153C) Basimere rather broad, with a straight or slightly concave apical margin and a number of slender setae; telomere with its tip a little curved downwards, never upwards (sometimes a little narrower than shown in the figure); sternum XI straight and widened towards the apex which is not much or hardly at all turned upwards; aedeagus as shown in fig. 153D, lamina media of aedeagus fairly broad, with a slightly concave dorsal margin and broadest preapically.   ♀ Sternum VII as in fig. 154F; bulga of spermatheca somewhat longer than broad, not broader than base of hilla, the ventral margins of bulga and hilla being about level (figs 154F, G)   .   .   .   .   *Xenopsylla cheopis*
-   Mesopleuron without an internal rod (fig. 154E)   3

3   Frons (fig. 155A) smoothly rounded; ocular seta placed below the conspicuous eye; one small pseudo-spinelet at the genal margin (rarely two, sometimes absent); post-antennal part of head with only one strong seta.   ♂ (fig. 155B) Basimere broad, covering the two telomeres.   Aedeagus as shown in fig. 155C.   ♀ (fig. 156A) Outline of sternum VII as in fig. 156A, but usually hardly visible in mounted specimens; spermatheca with a globular bulga and a curved hilla   .   *Pulex irritans*

-   Frons (fig. 155D) angulate; genal lobe directed backwards; laciniae very broad and coarsely serrated; post-antennal part of head with two setae and normally with a small lobe in the female only.   Thorax (fig. 155D) dorsally narrower than tergum I; distitarsomere (fig. 156C) with three pairs of stout, equally spaced, lateral setae and a smaller fourth pair, and with two apico-median plantar setae; pretarsal claws without a large basal projection.   ♂ Genitalia as in fig. 156D.   ♀ Terminal abdominal segments and genitalia as in fig. 156B   .   .   .   *Echidnophaga gallinacea*

4   With only a pronotal ctenidium (fig. 157A).   No setae in front of the row of three below the eye, but one or a few small ones above this row (fig. 157A).   None of the apical setae of the second hind tarsomere reaches the apical margin of the third tarsomere (fig. 158B).   ♂ Sternum VIII (fig. 157D) vestigial; tergum VIII (fig. 157C) strongly rounded from behind the last dorsal marginal seta; parameres and sternum IX as in fig. 157D.   ♀ (fig. 158A) Posterior margin of sternum VII slanting, not incurved; bursa copulatrix with a distinct forward curved duct and a strongly rounded apical part; hilla of spermatheca not markedly narrowed apicad   *Nosopsyllus fasciatus*
-   With a ctenidium on head as well as on pronotum (figs 159A, 160A–D)   .   .   .   5

5   Genal ctenidium vertical, consisting of four spines (fig. 159A); two of the setae near the frontal angle are spiniform.   Tibiae (fig. 159C) with a row of spiniform setae along the dorso-posterior margin, forming a so-called false comb.   ♂ Paramere and sternum IX as in fig. 159B.   ♀ Sternum VII and spermatheca as in fig. 159D   *Leptopsylla segnis*
-   Genal ctenidium horizontal, generally consisting of eight or nine spines (fig. 160A–D)   .   6

6   Head (fig. 160C, D) elongate, especially in the female, not strongly convex anteriorly.   Dorsal margin of hind tibia with six seta-bearing notches (fig. 160J).   ♂ (fig. 160H) Manubrium only a little dilated; paramere as in fig. 160H.   ♀ (fig. 160I) Apical part of hilla of spermatheca short   .   .   .   .   *Ctenocephalides felis felis*
-   Head (fig. 160A, B) strongly rounded anteriorly in both sexes.   Dorsal margin of hind tibia with eight seta-bearing notches (fig. 160G).   ♂ (fig. 160E) Manubrium with a dilated apex; paramere as in fig. 160E.   ♀ (fig. 160F) Apical part of hilla of spermatheca quite long   .   .   .   .   *Ctenocephalides canis*

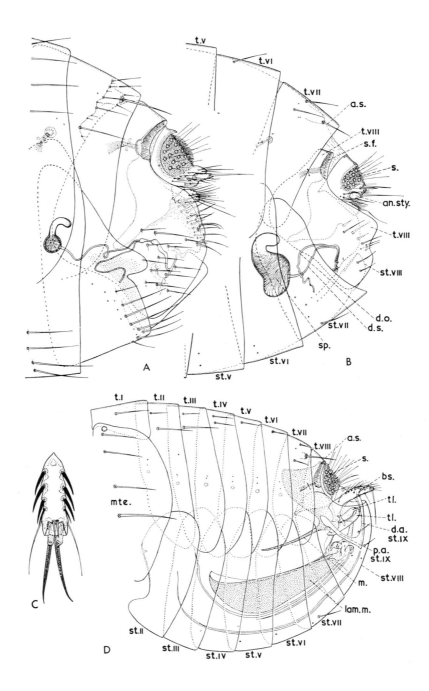

Fig. 156   A, *Pulex irritans*, terminalia, female (after Smit).   B–D, *Echidnophaga gallinacea* (after Smit): B, terminalia, female; C, hind distitarsomere, female; D, abdomen, male.

KEY TO SOME EURASIAN GENERA

1 Without ctenidia   2
– At least with a pronotal ctenidium   4
2 Lower part of frons with a transverse slit (fig. 162A)   .   .   .   ***COPTOPSYLLA***
– Frons entire   3
3 Frons smoothly rounded (fig. 152A, B); pleural rod of mesosternosome present (fig.
   152D)   .   .   .   .   ***XENOPSYLLA***
– Frons angulate (fig. 155D); pleural rod absent (fig. 155D)   .   ***ECHIDNOPHAGA***
4 Head without a ctenidium   5
– Head with a ctenidium   16
5 A rod-like structure between metepimeron and basal sternum (fig. 158E)   ***STIVALIUS***
– Without this rod   6
6 Tentorial rod visible in front of eye; upper seta of ocular row at or near margin of
   antennal fossa (fig. 162B)   7
– Tentorial rod absent; upper seta of ocular row in front of eye (fig. 157A, B)   .   .   10
7 First pair of plantar setae of distitarsomere placed between the members of the second
   pair (fig. 158J)   .   .   .   .   ***AMPHIPSYLLA***
– First pair of plantar setae at most slightly displaced mesad, but not between the
   members of the second pair (fig. 158K)   8
8 ♂ (fig. 167D) Telomere elongate, not very broad.   ♀ (fig. 170A) Bulga of spermatheca
   distinctly demarcated from hilla   .   .   .   .   ***PARADOXOPSYLLUS***
– ♂ Telomere usually more triangular.   ♀ Bulga either not sharply demarcated from
   hilla or not round   9
9 Ventro-posterior dark part of eye relatively large (fig. 162B).   ♂ (fig. 166A) Sternum
   VIII much reduced and modified; basimere with one acetabular seta.   ♀ (fig. 169A)
   Bulga of spermatheca fairly well demarcated from hilla by interruption of internal
   striae   .   .   .   .   ***OPHTHALMOPSYLLA***
– Ventro-posterior dark part of eye smaller (fig. 162C).   ♂ Sternum VIII (fig. 166E)
   unmodified; basimere (fig. 166D) with two acetabular setae.   ♀ (fig. 169F) Demarcation between bulga and hilla absent or obscure   .   .   .   ***FRONTOPSYLLA***
10 ♂ (fig. 167E) Telomere with a conspicuous ventro-posterior elongation.   ♀ (fig. 158C)
   Anal stylet with numerous setae; bursa copulatrix very large, wide and curved
   AMPHALIUS
– ♂ Telomere without such elongation.   ♀ Anal stylet with at most three or four lateral
   setae apart from a well-differentiated apical seta   11
11 Inner side of mid and hind coxae with thin setae laterally in the basal third (fig. 154A)   12
– Inner side of mid and hind coxae without such setae in at least the basal third (fig.
   154B)   13
12 Labial palp reaching to beyond apex of fore coxa.   ♂ (fig. 168B) Tergum VIII without
   an area spiculosa.   ♀ With three or four antesensilial setae   .   .   ***OROPSYLLA***
– Labial palp usually reaching at most to apex of fore coxa.   ♂ (fig. 168A) Tergum VIII
   with a dorsal area spiculosa.   ♀ With two antesensilial setae   .   ***CITELLOPHILUS***
13 ♂ (fig. 157D) Sternum VIII vestigial, usually without setae.   ♀ (fig. 158A) Bursa copulatrix with a distinct forward curved duct and a strongly rounded apical part
   NOSOPSYLLUS
– Sternum VIII much longer, usually with setae; bursa copulatrix not coiled   .   .   14
14 Spiracular fossa of tergum VIII much enlarged, especially in female (fig. 166B, C)
   MEGABOTHRIS
– This fossa not greatly enlarged   15

15 ♂ (fig. 168F) Sternum VIII long and narrow, with a long membranous process. ♀ (fig. 170F) Bulga of spermatheca longer than hilla which bears a distinct apical papilla   ***AMALARAEUS***

– ♂ (fig. 168E) Sternum VIII reduced and narrow, proximally triangular, its distal narrow part with or without setae. ♀ (fig. 170B) Bulga of spermatheca pyriform or cylindrical . . . . ***MONOPSYLLUS***

16 Terga II–VII with one row of setae (fig. 151D, E). Genal ctenidium horizontal, in the common species consisting of 8–9 spines (figs. 160A–D) . ***CTENOCEPHALIDES***

– Terga with more than one row of setae 17

17 Genal ctenidium of two spines 18

– Genal ctenidium of more than two spines 19

18 Bases of genal spines not or only slightly overlapping (fig. 163D). Tibiae without a false comb of spiniforms . . . . . . ***MESOPSYLLA***

– Base of anterior genal spine overlapping that of posterior spine (fig. 163F) ***NEOPSYLLA***

19 Tergum I with a well-developed ctenidium; genal ctenidium of 9–15 spines (fig. 165D) . . . . ***STENOPONIA***

– Tergum I without a ctenidium 20

20 Genal ctenidium of three spines, horizontal, the spines usually sharply pointed (fig. 164C) . . . . ***CTENOPHTHALMUS***

– Genal ctenidium of 4–8 more bluntly tipped spines, at least the hindermost spines vertical (fig. 165B) . . . . ***RHADINOPSYLLA***

KEY TO SOME AFRICAN GENERA

1 Pronotum without a ctenidium 2

– Pronotum with a ctenidium 5

2 Inner side of hind coxa without small spiniform setae, its ventro-anterior angle projecting downwards as a triangular tooth (fig. 161A). Sensilium (fig. 161B) with eight pits each side (jiggers or sand-fleas) . . . . . ***TUNGA***

– Hind coxa narrowing markedly below middle of posterior margin, with a row or patch of small spiniforms on inner side, its ventro-anterior angle not projecting downwards (fig. 152E). Pronotum dorsally shorter than mesonotum. No setae on genal process behind eye (fig. 155A). Sensilium with more than eight pits per side (figs 153C, 154F) 3

3 With a distinct separation between the small squarish lateral metanotal area and metepisternum (fig. 152D) . . . . ***XENOPSYLLA***

– Without this separation (fig. 155E) 4

4 African continent, Middle East . . . . . ***SYNOSTERNUS***

– Madagascar . . . . ***SYNOPSYLLUS***

5 Head without a ctenidium 6

– Head with a ctenidium 7

6 Frontal tubercle (fig. 162F) very large, triangular, situated in a deep cavity; a long slender spine immediately behind well-developed eye (fig. 162F). Anterior abdominal terga with marginal spinelets . . . . ***LISTROPSYLLA***

– Frontal tubercle small; no spine behind inconspicuous eye (fig. 163A). Terga without spinelets . . . . ***XIPHIOPSYLLA***

7 Terga with one row of setae. Genal ctenidium horizontal, in the common species consisting of 8–9 spines (fig. 160A–D) . . . ***CTENOCEPHALIDES***

– Terga with more than one row of setae 8

8   Head with oblique ctenidium of five dissimilar spines parallel with obliquely down-
    ward sloping frontal margin (fig. 165A)   .   .   .   .   .   .   ***DINOPSYLLUS***

–   Head with a different ctenidium   9

9   Genal ctenidium normally of two partially overlapping spines (fig. 163E) (sometimes
    reduced to one, or absent)   .   .   .   .   ***CHIASTOPSYLLA***

–   Genal ctenidium consisting of three spines   10

10 Genal ctenidium vertical, its spines bluntly tipped; two or three short spiniform
    setae near frontal angle (fig. 164A)   .   .   .   .   .   ***LEPTOPSYLLA***

–   Genal ctenidium horizontal, its spines usually sharply pointed; frons without
    spiniforms (fig. 164C)   .   .   .   .   ***CTENOPHTHALMUS***

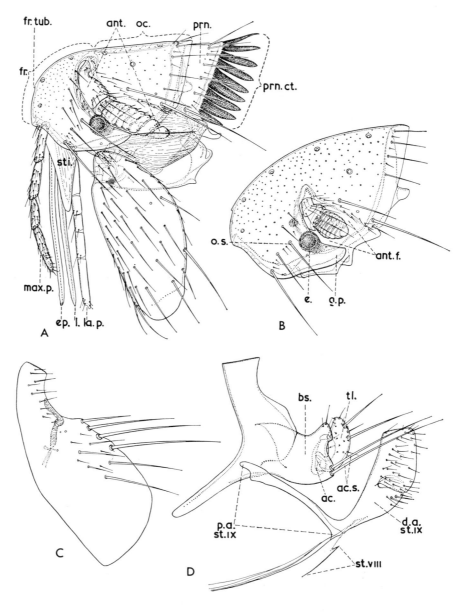

Fig. 157   *Nosopsyllus fasciatus* (after Smit): A, head, prothorax and fore coxa, male; B, head,
female; C, tergum VIII, male; D, paramere, sterna VIII and IX.

KEY TO SOME NORTH AMERICAN GENERA

1 Without ctenidia 2
– At least pronotum with a ctenidium 4

2 Frons without a tubercle (fig. 155A). Outer internal ridge of mid coxa absent (as in
    fig. 154C) . . . . ***PULEX***
– Frons with a tubercle. Outer internal ridge of mid coxa present (as in fig. 154D) . 3

3 Terga with one row of setae . . . . ***ANOMIOPSYLLUS***
– Terga with setae anterior to main row . . . . ***POLYGENIS***

4 Head without ctenidium 5
– Head with ctenidium 16

5 Terga II–VII with one row of setae . . . . . . ***HOPLOPSYLLUS***
– Terga with setae anterior to main row 6

6 Tentorial rod visible in front of eye (figs 162G, 163C). Metanotum without marginal
    spinelets 7
– No tentorial rod in front of eye. 8

7 Frons rounded; preantennal part of head with two rows of setae (fig. 163C). Hind
    coxa broad . . . . ***CATALLAGIA***
– Frons more angular; preantennal part of head with more than two rows of setae
    (fig. 162G). Hind coxa very narrow . . . . . ***STENISTOMERA***

8 Eye vestigial, pale (fig. 162D) . . . . . . ***DACTYLOPSYLLA***
– Eye large, dark 9

9 Basal sternum with a patch of lateral setae in upper half . . ***OPISOCROSTIS***
– This sternum with at most one or two lateral setae 10

10 Inner side of mid and hind coxa with thin setae laterally in the basal third (as in fig.
    154A) 11
– Inner side of mid and hind coxa without such setae in at least the basal third (as in
    fig. 154B) 13

11 ♂ (fig. 167A) Telomere long and slender, curved forwards; sternum VIII much reduced,
    without setae. ♀ (fig. 170C) Spermatheca with a constriction between bulga and
    hilla; anal stylet with only one long lateral seta . . . . ***DIAMANUS***
– ♂ Telomere shaped differently; sternum VIII much longer, with setae. ♀ Constriction
    between bulga and hilla absent or less distinct; anal stylet with two or more lateral
    setae 12

12 ♂ Sternum VIII rather short and fairly broad. ♀ With two or three antesensilial
    setae, bulga of spermatheca (fig. 170D) shorter than broad, hilla usually much
    longer than bulga, its papilla often small or absent . . . . ***THRASSIS***
– ♂ Sternum VIII narrow. ♀ With three or four antesensilial setae; bulga (fig. 170E)
    at least as long as wide, hilla at most a little longer than bulga and with a large
    papilla . . . . ***OROPSYLLA***

13 First pair of plantar setae of hind distitarsomere distinctly displaced mesad (fig.
    158K) 14
– First pair of plantar setae of hind distitarsomere not, or hardly at all, displaced . 15

14 ♂ Sternum VIII without setae; telomere (fig. 167C) with a row of four short black
    spiniforms fairly close together along posterior margin. ♀ (fig. 169C) Ventral margin
    of anal sternum distinctly angular near middle . . . . ***ORCHOPEAS***
– ♂ Sternum VIII with at least one seta; telomere with longer spiniforms. ♀ (fig. 169D)
    Ventral margin of anal sternum not so distinctly angular . . ***OPISODASYS***

15 ♂ (fig. 168D) Sternum VIII vestigial. ♀ (fig. 170G) Bulga of spermatheca at most as long as hilla . . . . **MALARAEUS**

 — ♂ Sternum VIII narrow and often reduced, proximally triangular, its distal part with or without a membranous process. ♀ Bulga of spermatheca pyriform or cylindrical **MONOPSYLLUS**

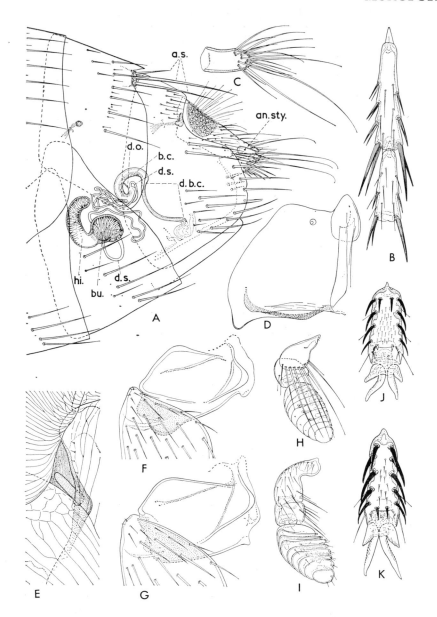

Fig. 158 *Nosopsyllus fasciatus* (after Smit): A, terminalia, female; B, hind tarsomeres II and III, male. C, *Amphalius runatus runatus*, anal stylet (original). D, *Delostichus talis*, metepisternum and metasternum, female (original). E, *Stivalius torvus*, rod-like structure between metepimeron and basal sternum, male (original). F–G, prosternosome and base of fore coxa, male (originals): F, *Rhopalopsyllus lugubris lugubris*; G, *Polygenis litargus*. H–I, antenna (originals): H, *Delostichus talis*, female; I, *Polygenis litargus*, male. J–K, hind distitarsomere (originals): J, *Amphipsylla rossica* female; K, *Orchopeas howardi howardi*, female.

16 Large fleas; genal ctenidium of 5–9 spines (fig. 164D). Pronotal ctenidium of
    numerous spines .   .   .   .   ***HYSTRICHOPSYLLA***

–   Smaller fleas; genal ctenidium of two spines   17

17 Genal spines parallel; frons angular, two or three spiniform setae near frontal angle
    (fig. 164B) .   .   .   .   ***PEROMYSCOPSYLLA***

–   Genal spines partially overlapping; frons without tubercle (fig. 163B)   ***MERINGIS***

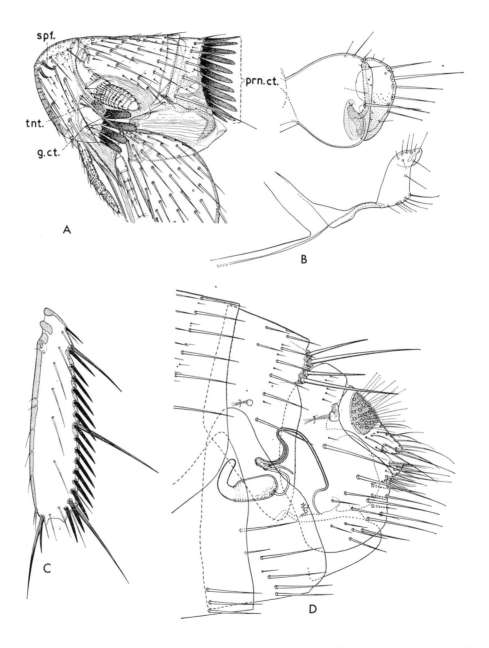

Fig. 159 *Leptopsylla segnis*: A, head, prothorax and base of fore coxa, male (after Smit);
B, paramere and sternum IX (original); C, hind tibia, female (original); D, terminalia,
female (after Smit).

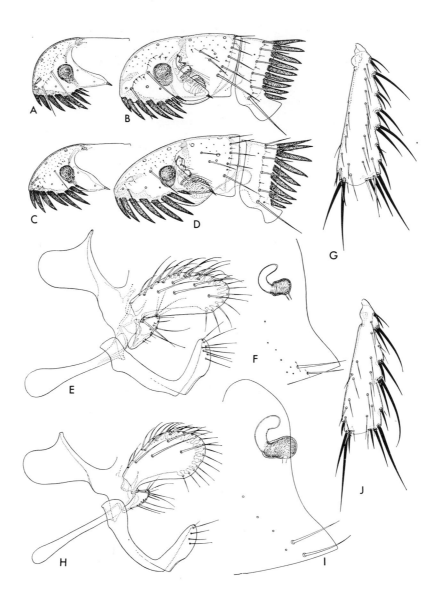

Fig. 160   A, *Ctenocephalides canis*: preantennal part of head, male; B, *idem*, head and pronotum, female.   C–D, *Ctenocephalides felis felis*: C, preantennal part of head, male;
D, *idem*, head and pronotum, female.   E–G, *Ctenocephalides canis*: E, paramere and
sternum IX; F, sternum VII and spermatheca; G, hind tibia, female.   H–J, *Ctenocephalides felis felis*: H, paramere and sternum IX; I, sternum VII and spermatheca;
J, hind tibia, female (all after Smit).

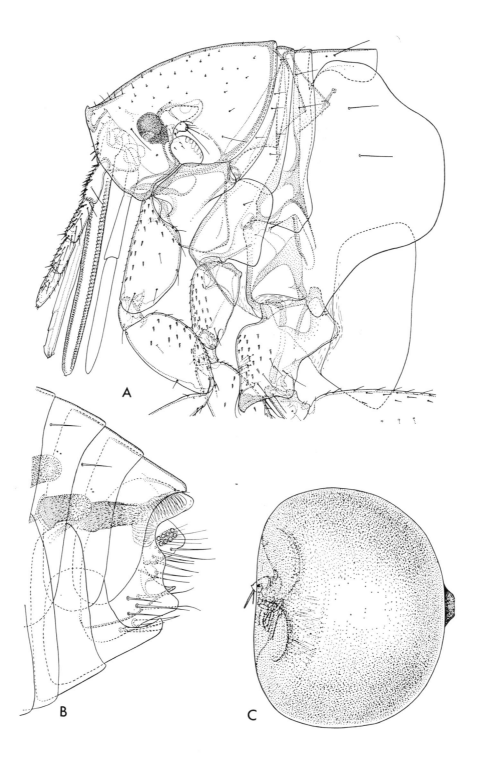

Fig. 161 *Tunga penetrans*, female (after Smit): A, head, thorax, tergum I and sternum II; B, terminalia, unexpanded female; C, fully expanded female (neosome) removed from skin.

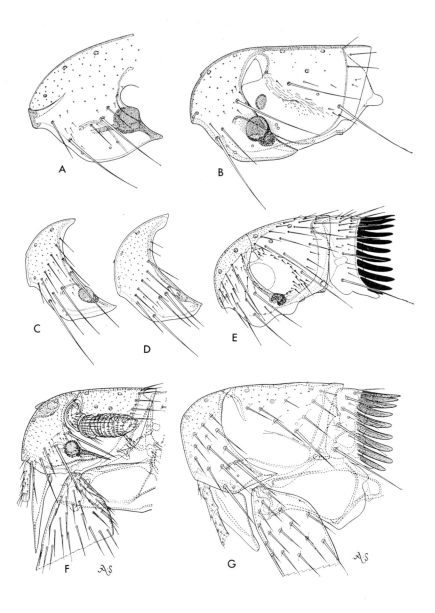

Fig. 162 A, *Coptopsylla lamellifer dubinini*, preantennal part of head, female (original). B, *Ophthalmopsylla volgensis palestinica*, head, female (after Smit). C–D, preantennal part of head, male (originals): C, *Frontopsylla wagneri*; D, *Dactylopsylla ignota ignota*. E, *Stivalius torvus*, head and pronotum, female (after Smit). F, *Listropsylla fouriei*, head and part of prothorax, male (after Hopkins & Rothschild). G, *Stenistomera macrodactyla*, head, prothorax and base of fore coxa, male (after Hopkins & Rothschild).

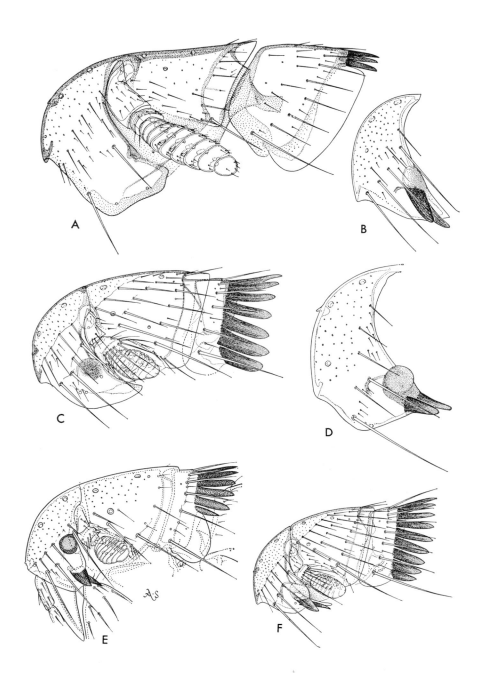

Fig. 163 A, *Xiphiopsylla daemonicola*, head and pronotum, male (after Smit). B, *Meringis shannoni*, preantennal part of head, male (original). C, *Catallagia dacenkoi*, head and pronotum, female (original). D, *Mesopsylla tuschkan andruschkoi*, preantennal part of head, male (original). E, *Chiastopsylla quadrisetis*, head and pronotum, female (after Hopkins & Rothschild). F, *Neopsylla setosa spinea*, head and pronotum, female (original).

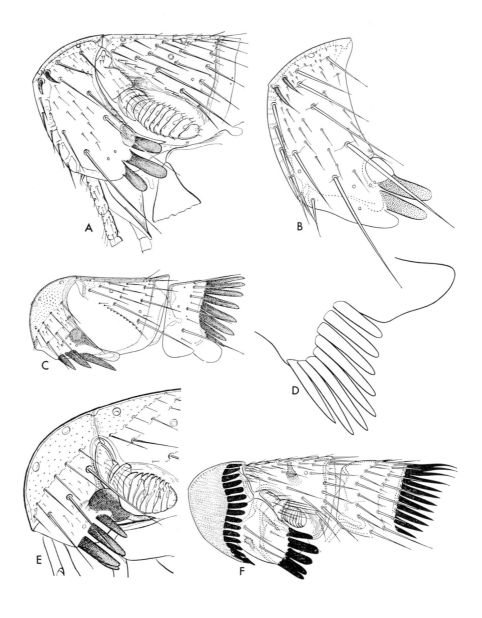

Fig. 164   A, *Leptopsylla algira algira*, head, male (after Jordan & Rothschild).   B, *Peromyscopsylla scotti*, preantennal part of head, male (after Hopkins & Rothschild).   C, *Ctenophthalmus calceatus cabirus*, head and pronotum, male (original).   D, *Hystrichopsylla occidentalis*, genal ctenidium (after Hopkins & Rothschild).   E, *Neotyphloceras crassispina crassispina*, head, female (after Rothschild).   F, *Sphinctopsylla inca*, head and pronotum, female (after Hopkins & Rothschild).

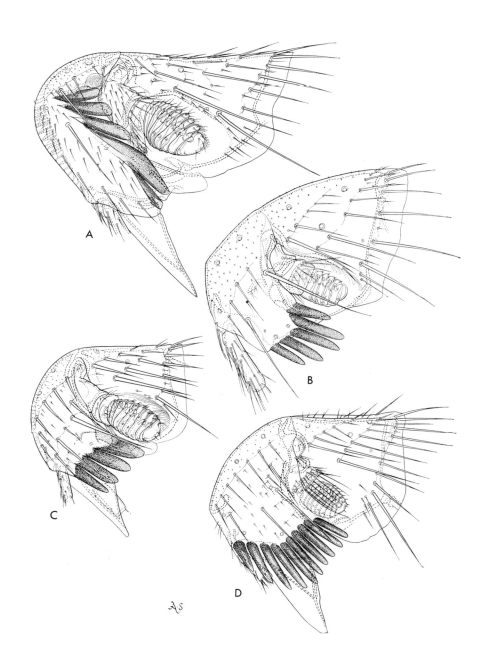

Fig. 165   A, *Dinopsyllus wansoni*, head, male.   B, *Rhadinopsylla li ventricosa*, head, female.
C, *Adoratopsylla intermedia intermedia*, head, male.   D, *Stenoponia sidimi*, head,
male (all after Hopkins & Rothschild).

AA

KEY TO SOME CENTRAL AND SOUTH AMERICAN GENERA

1 Pronotum without a ctenidium 2
– Pronotum with a ctenidium 7

2 Inner side of hind coxa without small spiniform setae (fig. 161A). Sensilium with eight sensory pits on each side (fig. 161B) 3
– Not so 4

3 Anterior apical angle of hind coxa projecting downward as a triangular tooth (fig. 161A) (Jiggers or sand-fleas) . . . . *TUNGA*
– Hind coxa without an apical tooth . . . . . . *HECTOPSYLLA*

4 Antennal clava symmetrical (fig. 158H); metasternum with a ventro-anterior projection (fig. 158D); labial palp of five segments . . . . . *DELOSTICHUS*
– Antennal clava asymmetrical (fig. 158 I); metasternum with ventro-anterior projection less well developed or absent 5

5 Prosternosome projecting downward between the coxae (fig. 158F) *RHOPALOPSYLLUS*
– Prosternosome without such downward projection (fig. 158G) 6

6 Labial palp not reaching beyond apex of fore coxa . . . *POLYGENIS*
– Labial palp extending to or beyond apex of fore trochanter . . . *TIAMASTUS*

7 Head without a ctenidium 8
– Head with a ctenidium 9

8 Terga II–VII with only one row of setae . . . . *HOPLOPSYLLUS*
– Terga II–VII with setae anterior to the main row . . . . *PLEOCHAETIS*

9 Terga II–VII with one row of setae . . . . . . *CEDIOPSYLLA*
– Terga with setae anterior to main row 10

10 Head with a long vertical anterior ctenidium and a shorter genal one (fig. 164F) . 11
– Head with only a genal ctenidium (figs 164E, 165C) 13

11 ♂ (fig. 168C) Basimere with an apical fringe of long setae. ♀ (fig. 169B) Base of hilla of spermatheca usually projecting well into lumen of the subcylindrical bulga *PLOCOPSYLLA*
– ♂ Basimere without this fringe of setae. ♀ Hilla not projecting into the bulga . . 12

12 ♂ (fig. 167B) Telomere with a blunt apex. ♀ (fig. 170H) Bulga of spermatheca elongate, with a tuberculoid invagination . . . . . *CRANEOPSYLLA*
– ♂ (fig. 167F) Telomere usually semilunar, with a tapering apex. ♀ (fig. 169E) Bulga globular, without invagination . . . . . *SPHINCTOPSYLLA*

13 Genal ctenidium of four uniformly spaced spines (fig. 165C) . . *ADORATOPSYLLA*
– Genal ctenidium also of four spines but the anterior (lowest) spine is largely covered by the second spine (fig. 164E) . . . . . . *NEOTYPHLOCERAS*

KEY TO SOME AUSTRALASIAN GENERA

1 Without ctenidia. Frons of head smoothly rounded (fig. 152A, B). *XENOPSYLLA*
– With a pronotal ctenidium. Eye situated above (not in front of) base of fore coxa (fig. 162E). A rod-like structure between metepimeron and basal sternum (fig. 158E) . . . . *STIVALIUS*

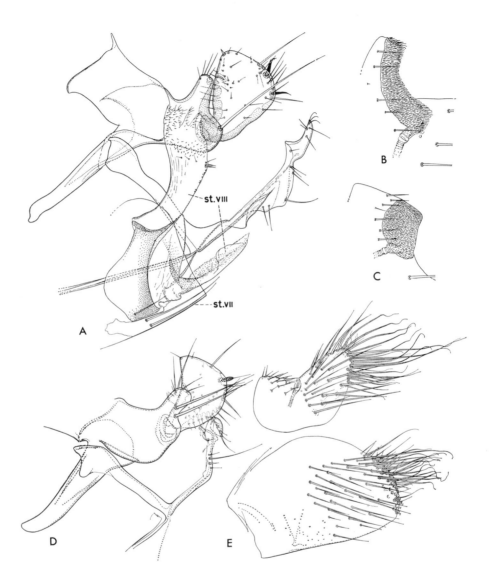

Fig. 166 A, *Ophthalmopsylla volgensis palestinica*, paramere and sterna VII–IX (after Smit). B–C, *Megabothris rectangulatus*, spiracular fossa of tergum VIII (originals): B, male; C, female. D, *Frontopsylla wagneri*, paramere and sternum IX (after Hopkins & Rothschild); E, *idem*, segment VIII, male (after Hopkins & Rothschild).

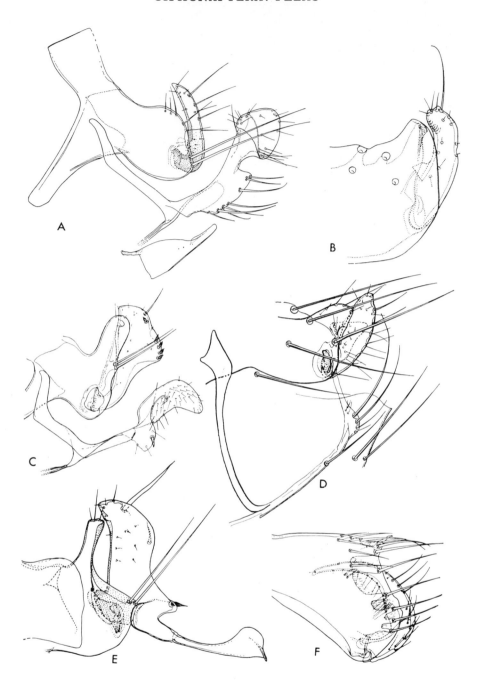

Fig. 167   A, *Diamanus montanus*, paramere, sterna VIII and IX.   B, *Craneopsylla minerva minerva*, basimere and telomere.   C, *Orchopeas sexdentatus pennsylvanicus*, paramere and sternum IX (after Jordan).   D, *Paradoxopsyllus teretifrons*, paramere, segment VIII and sternum IX (after Rothschild).   E, *Amphalius runatus runatus*, basimere and telomere (after Smit).   F, *Sphinctopsylla ares*, basimere and telomere (after Smit).

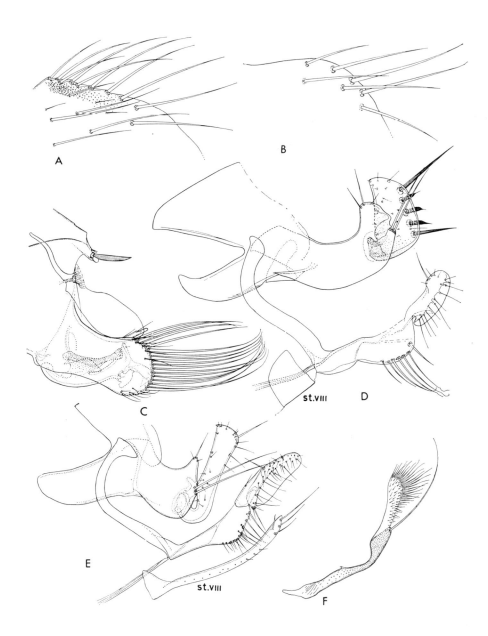

Fig. 168   Dorsal part of tergum VIII, male (originals): A, *Citellophilus tesquorum sungaris*;
B, *Oropsylla silantiewi*.   C, *Plocopsylla wolffsohni*, terga VII and VIII, basimere and
telomere (after Hopkins & Rothschild).   D, *Malaraeus telchinus*, paramere, sterna
VIII and IX (original).   E, *Monopsyllus anisus*, paramere, sterna VIII and IX (after
Smit).   F, *Amalaraeus penicilliger mustelae*, sternum VIII, male (after Smit).

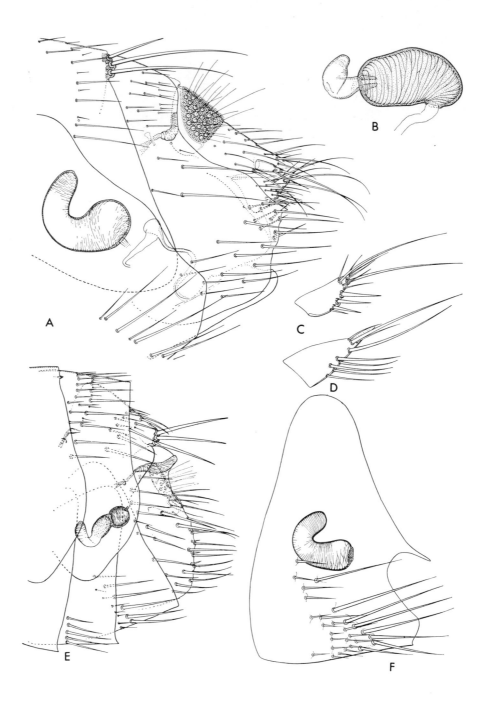

Fig. 169   A, *Ophthalmopsylla volgensis palestinica*, terminalia, female (after Smit).   B, *Ploco-psylla ulysses*, spermatheca (after Hopkins).   C–D, anal sternum, female (originals): C, *Orchopeas leucopus*; D, *Opisodasys keeni keeni*.   E, *Sphinctopsylla ares*, terminalia, female (after Hopkins & Rothschild).   F, *Frontopsylla wagneri*, sternum VII and spermatheca (after Hopkins & Rothschild).

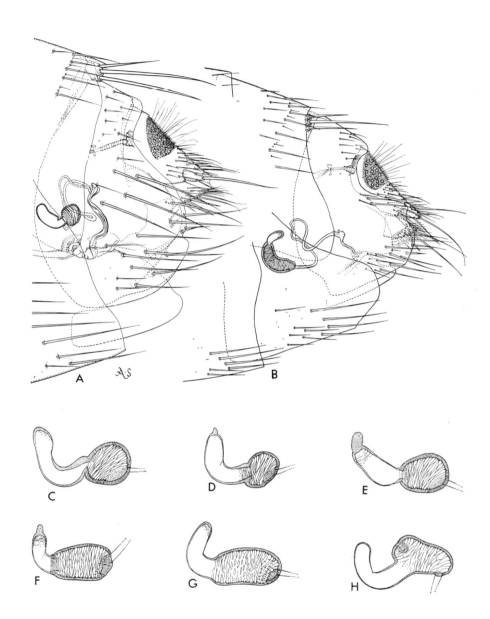

Fig. 170 A, *Paradoxopsyllus teretifrons*, terminalia, female (after Hopkins & Rothschild). B, *Monopsyllus anisus*, terminalia, female (after Smit).   C–H, spermatheca (originals): C, *Diamanus montanus*; D, *Thrassis acamantis acamantis*; E, *Oropsylla silantiewi*; F, *Amalaraeus penicilliger mustelae*; G, *Malaraeus telchinus*; H, *Craneopsylla minerva minerva*.

ALPHABETICAL LIST OF GENERA

The genera mentioned in the keys are listed alphabetically with the addition of the family name in brackets and also, where possible, a reference to a publication containing keys to species; 'Catalogue' stands for the Hopkins & Rothschild *Catalogue of the Rothschild collection of Fleas* (1953–1971). Distribution and host data are summarily given, followed by an enumeration of diseases and the fleas which, in the countries added, have been found to be naturally infected with the pathogens. The number of described species indicates the size of the genus (or subgenus); quite a number of flea species are divided into subspecies (geographical races) and the collective word 'forms' is used for the totals of species and subspecies.

Adoratopsylla (Hystrichopsyllidae) [*vide* Catalogue IV]
Distribution: South America; 2 subgenera: nominate subgenus with 3 species (5 forms), subgenus *Tritopsylla* with 2 species (4 forms).
Hosts: Didelphine marsupials.
Vector: Plague: *A.* (*T.*) *intermedia copha* (Jordan), Ecuador.

Amalaraeus (Ceratophyllidae) [*vide* Ioff & Scalon].
Distribution: Temperate and cold zones of Eurasia and North America; 5 species (14 forms).
Hosts: Microtine rodents.
Vectors: Haemorrhagic nephroso-nephritis: *A. penicilliger* ssp., Asiatic Russia [suspected vectorship]. Lymphocytic choriomeningitis: *A. penicilliger* ssp., Asiatic Russia.

Amphalius (Ceratophyllidae) [*vide* Ioff & Scalon].
Distribution: Eastern Asia and north-western North America; 3 species (4 forms).
Hosts: Picas (*Ochotona* spp.).
Vector: Erysipeloid: *A. runatus runatus* (Jordan & Rothschild), Asiatic Russia.

Amphipsylla (Leptopsyllidae) [*vide* Catalogue V].
Distribution: Mainly in colder regions of Eurasia but a few forms also in western North America; 28 species (39 forms).
Hosts: Murid rodents, especially microtines.
Vectors: Plague: *A. rossica* Wagner, European Russia; *A. primaris mitis* Jordan, Asiatic Russia and Mongolia. Erysipeloid and Salmonellosis: *A. primaris mitis* Jordan, Asiatic Russia.

Anomiopsyllus (Hystrichopsyllidae) [*vide* Catalogue III].
Distribution: North America; 13 species (15 forms).
Hosts: Wood rats (*Neotoma* spp.).
Vectors: Plague: *A. hiemalis* ssp., Texas; *A. nudatus* (Baker), western U.S.A.

Catallagia (Hystrichopsyllidae) [*vide* Catalogue III].
Distribution: Mainly in North America, a few species also in northern and eastern Eurasia; 13 species (15 forms).
Hosts: Murid rodents.
Vector: Plague: *C. decipiens* Rothschild, western U.S.A.

Cediopsylla (Pulicidae) [*vide* Catalogue I].
Distribution: Americas; 4 species (5 forms).
Hosts: Cottontail rabbits (*Sylvilagus* spp.).
Vector: Plague: *C. spillmanni* Jordan, Peru.

Chiastopsylla (Chimaeropsyllidae) [*vide* Catalogue II].
Distribution: South Africa; 14 species (16 forms).
Hosts: Murid rodents, especially Otomyinae.
Vector: Plague: *C. rossi* (Waterston).

Citellophilus (Ceratophyllidae) [*vide* Ioff, Mikulin & Scalon].
Distribution: Eurasia; 12 species (21 forms).
Hosts: Ground-dwelling Sciurid rodents (*Citellus* spp., some species on *Marmota* spp.).
Vectors: Plague: *C. lebedewi* ssp., Asiatic Russia; *C. tesquorum* sspp., Asiatic Russia.
Erysipeloid: *C. tesquorum* ssp., Asiatic Russia.

Coptopsylla (Coptopsyllidae) [*vide* Catalogue II].
Distribution: Mediterranean area to Central Asia; 15 species (19 forms).
Hosts: Gerbilline rodents.
Vectors: Plague: *C. bairamaliensis* Wagner and *C. lamellifer* sspp., Central Asia.

Craneopsylla (Stephanocircidae) [*vide* Catalogue II].
Distribution: South America; 1 species (2 forms).
Hosts: Rodents and marsupials.
Vector: Plague: *C. minerva* ssp., Argentina.

Ctenocephalides (Pulicidae) [*vide* Catalogue I].
Distribution: Mainly Africa and Eurasia, 2 species cosmopolitan; 9 species (12 forms).
Hosts: Especially on Carnivores, some African species on hares, hyraxes and ground-squirrels; on goats in Mediterranean and Oriental areas.
Vectors: Plague: *C. felis felis* (Bouché), cosmopolitan; *C. canis* (Curtis), cosmopolitan; *C. felis strongylus* (Jordan), Africa; *C. orientis* (Jordan), Oriental region.
Intermediate hosts for *Dipylidium caninum* [Cestoda]: *C. canis* (Curtis) and *C. felis felis* (Bouché), Europe and eastern U.S.A.; *C. felis strongylus* (Jordan), East Africa.

Ctenophthalmus (Hystrichopsyllidae) [*vide* Catalogue IV].
Distribution: Concentrated in Eurasia and Africa, a few species also in North and Central America and in Oriental Region; 118 species (204 forms).
Hosts: Mainly murid rodents, a few species on insectivores.
Vectors: Plague: *C. breviatus* Wagner & Ioff, European Russia; *C. calceatus cabirus* Jordan & Rothschild, East Africa; *C. dolichus* ssp., Central Asia; *C. phyris* Jordan, Central Africa; *C. pollex* Wagner & Ioff, European Russia; *C. congener secundus* Wagner, European Russia. Tularaemia: *C. teres* Ioff & Argyropulo, European Russia; *C. wladimiri* Isayeva-Gurvich, European Russia. Pseudotuberculosis: *C. congeneroides congeneroides* Wagner, Russian Far East. Erysipeloid: *C. teres* Ioff & Argyropulo, European Russia; *C. wladimiri* Isayeva-Gurvich, European Russia. Listeriosis: *C. orientalis* (Wagner), European Russia. Haemorrhagic nephroso-nephritis: *C. assimilis assimilis* (Taschenberg), Asiatic Russia; *C. agyrtes* ssp., European Russia; *C. orientalis*

(Wagner), European Russia; *C. solutus solutus* Jordan & Rothschild, European Russia. Lymphocytic choriomeningitis: *C. assimilis assimilis* (Taschenberg), Asiatic Russia. Tickborne encephalitis: *C. congeneroides congeneroides* Wagner, Russian Far East.

Dactylopsylla (Ceratophyllidae).
Distribution: North America; 2 subgenera: nominate subgenus with 15 species (20 forms), subgenus *Foxella* with 4 species (14 forms).
Hosts: Murid rodents.
Vector: Plague: *D. (F.) ignota* ssp., southern U.S.A.

Delostichus (Rhopalopsyllidae) [*vide* Johnson].
Distribution: South America; 5 species.
Hosts: Rodents.
Vector: Plague: *D. talis* (Jordan), Argentina.

Diamanus (Ceratophyllidae).
Distribution: Eastern Asia and western North America; 2 species.
Hosts: Ground-dwelling Sciurid rodents (*Citellus* spp.).
Vector: Plague: *D. montanus* (Baker), western U.S.A.

Dinopsyllus (Hystrichopsyllidae) [*vide* Catalogue IV].
Distribution: Africa; 2 subgenera: nominate subgenus with 22 species, subgenus *Crypto-ctenopsyllus* with 1 species.
Hosts: Rodents.
Vectors: Plague: *D. (D.) ellobius* (Rothschild), South Africa; *D. (D.) lypusus* Jordan & Rothschild, Central and East Africa.

Echidnophaga (Pulicidae) [*vide* Catalogue I].
Distribution: Eurasia (mainly in warmer areas), Africa, Australia, 1 species cosmo-politan; 18 species (19 forms).
Hosts: Rodents, marsupials, carnivores, warthogs, 1 species on bats, the cosmopolitan species on birds (especially chickens) as well as on rats.
Vectors: Plague: *E. gallinacea* (Westwood), U.S.A.; *E. oschanini* Wagner, Asiatic Russia and Mongolia.   Murine typhus: *E. gallinacea* (Westwood), southern U.S.A.
Intermediate host for *Dipylidium caninum*: *E. larina* Jordan & Rothschild, East Africa.

Euhoplopsyllus (Pulicidae) [*vide* Catalogue I].
Distribution: Eastern Asia and Americas; 3 species (8 forms).
Hosts: Leporidae (hares and rabbits).
Vectors: Plague: *E. andensis* (Jordan), Peru; *E. glacialis affinis* (Baker), western and southern U.S.A.; *E. manconis* (Jordan), Peru.

Frontopsylla (Leptopsyllidae) [*vide* Catalogue V].
Distribution: Eurasia; 4 subgenera, nominate subgenus only thus far of medical impor-tance, with 13 species (27 forms).
Hosts: Nominate subgenus: rodents.
Vectors: Plague: *F. (F.) semura* Wagner & Ioff, European Russia.   Erysipeloid: *F. (F.) luculenta* ssp., Asiatic Russia; *F. (F.) wagneri* Ioff, Asiatic Russia.   Listeriosis: *F. (F.) luculenta* ssp., Asiatic Russia.   Salmonellosis: *F. (F.) luculenta* ssp., Asiatic Russia.

Hectopsylla (Pulicidae) [*vide* Catalogue I].
Distribution: South America; 2 subgenera: nominate subgenus with 9 species, subgenus *Rhynchopsyllus* with 2 species.
Hosts: Nominate subgenus: mostly rodents, 2 species on birds; subgenus *Rhynchopsyllus* on bats.
Vectors: Plague: *H. (H.) eskeyi* Jordan, Peru; *H. (H.) suarezi* Fox, Ecuador.

Hoplopsyllus (Pulicidae) [*vide* Catalogue I].
Distribution: North and Central America; 2 species.
Hosts: Ground-dwelling Sciurid rodents (*Citellus* spp.) and Leporidae.
Vector: Plague: *H. anomalus* (Baker), western U.S.A.

Hystrichopsylla (Hystrichopsyllidae) [*vide* Catalogue III].
Distribution: Eurasia, North and Central America; 2 subgenera: nominate subgenus with 1 species (2 forms), subgenus *Hystroceras* with 13 species (15 forms).
Hosts: Mainly rodents.
Vectors: Plague: *H. (Hystroceras) dippiei* ssp. and *H. (Hystroceras) linsdalei* Holland, western U.S.A.

Leptopsylla (Leptopsyllidae) [*vide* Catalogue V].
Distribution: Eurasia, Africa, North and Central America; 2 subgenera: nominate subgenus with 9 species (18 forms), subgenus *Pectinoctenus* with 6 species.
Hosts: Murine rodents.
Vectors: Plague: *L. (L.) aethiopica* ssp., Africa. Erysipeloid, murine typhus and tick-borne encephalitis: *L. (L.) segnis* (Schönherr), Asiatic Russia, South-East U.S.A. and China, and Poland respectively. Salmonellosis: *L. (P.) pavlovskii* Ioff, Asiatic Russia.

Listropsylla (Hystrichopsyllidae) [*vide* Catalogue IV].
Distribution: Africa; 9 species (11 forms).
Hosts: Rodents.
Vector: Plague: *L. dorippae* (Rothschild), southern Africa.

Malaraeus (Ceratophyllidae).
Distribution: North America; 3 species.
Hosts: Murid rodents.
Vectors: Plague: *M. sinomus* (Jordan) and *M. telchinus* (Rothschild), western U.S.A.

Megabothris (Ceratophyllidae) [*vide* Ioff & Scalon].
Distribution: Eurasia and North America; 17 species (21 forms).
Hosts: Especially microtine rodents, 1 species on Sciurid rodents.
Vectors: Plague: *M. abantis* (Rothschild) and *M. clantoni clantoni* Hubbard, western U.S.A. Tularaemia: *M. rectangulatus* (Wahlgren), northern Europe. Pseudotuber-culosis: *M. calcarifer* (Wagner), Russian Far East. Lymphocytic choriomeningitis: *M. rectangulatus* (Wahlgren), Asiatic Russia. Haemorrhagic nephroso-nephritis: *M. turbidus* (Rothschild), European Russia.

Meringis (Hystrichopsyllidae) [*vide* Catalogue III].
Distribution: North America; 16 species.
Hosts: Dipodomyine rodents.
Vector: Plague: *M. shannoni* (Jordan), western U.S.A.

Mesopsylla (Leptopsyllidae) [*vide* Catalogue V].
Distribution: Mediterranean Region and Asia; 5 species (14 forms).
Hosts: Dipodid rodents (jerboas).
Vectors: Plague: *M. apscheronica* Wagner & Argyropulo and *M. tuschkan* ssp., Central Asia.

Monopsyllus (Ceratophyllidae) [*vide* Johnson, 1961; Ioff & Scalon].
Distribution: Eurasia and North America; 22 species (28 forms).
Hosts: Sciurid and murine rodents.
Vectors: Plague: *M. eumolpi* ssp., western U.S.A., *M. exilis* (Jordan), southern U.S.A., *M. wagneri* (Baker), western U.S.A. Pseudotuberculosis, Erysipeloid and Listeriosis: *M. anisus* (Rothschild), Asiatic Russia. Tick-borne encephalitis: *M. indages indages* (Rothschild), Asiatic Russia.

Neopsylla (Hystrichopsyllidae) [*vide* Catalogue III].
Distribution: Eurasia and North America; 33 species (44 forms).
Hosts: Murid rodents.
Vectors: Plague: *N. bidentatiformis* (Wagner), Asiatic Russia and Mongolia; *N. inopina* Rothschild, western U.S.A.; *N. mana* Wagner, Mongolia; *N. setosa* ssp., European Russia. Pseudotuberculosis: *N. bidentatiformis* (Wagner), Asiatic Russia. Erysipeloid: *N. bidentatiformis* (Wagner) and *N. pleskei* ssp., Asiatic Russia. Salmonellosis: *N. bidentatiformis* and *N. pleskei orientalis* Ioff & Argyropulo, Asiatic Russia. Staphylococcus aureus: *N. pleskei orientalis* Ioff & Argyropulo, Asiatic Russia. Q-fever: *N. pleskei* ssp., Asiatic Russia.

Neotyphloceras (Hystrichopsyllidae) [*vide* Catalogue IV].
Distribution: South America; 2 species (4 forms).
Hosts: Rodents and marsupials.
Vector: Plague: *N. rosenbergi* (Rothschild), Ecuador.

Nosopsyllus (Ceratophyllidae) [*vide* Ioff & Scalon; Ioff, Mikulin & Scalon].
Distribution: Mainly Eurasia, a few species in Africa, 1 species cosmopolitan (and another near-cosmopolitan); 4 subgenera: nominate subgenus with 27 species (31 forms), subgenus *Gerbillophilus* with 17 species (26 forms), subgenera *Nosinius* and *Penicus* each with 1 species.
Hosts: Rodents.
Vectors: Plague: *N. (G.) aralis* ssp., Asiatic Russia; *N. (N.) consimilis* (Wagner), European Russia; *N. (N.) fasciatus* (Bosc), U.S.A.; *N. (G.) laeviceps* ssp., European Russia; *N. (N.) mokrzeckyi* (Wagner), European Russia; *N. (N.) nilgeriensis* (Jordan & Rothschild), India; *N. (G.) tersus* (Jordan & Rothschild), Asiatic Russia; *N. (G.) turkmenicus* ssp., Asiatic Russia. Pseudotuberculosis: *N. (N.) consimilis* (Wagner), European Russia; *N. (N.) fasciatus* (Bosc), Russian Far East. Erysipeloid: *N. (N.) fasciatus* (Bosc), Russia.
Intermediate host for *Hymenolepis diminuta*: *N. (N.) fasciatus* (Bosc), England, Argentina, Australia.

Ophthalmopsylla (Leptopsyllidae) [*vide* Catalogue V].
Distribution: South-eastern Europe, Asia; 3 subgenera, only nominate subgenus of medical importance, with 3 species (15 forms).

Hosts: Rodents, especially Dipodids.
Vectors: Plague: *O. (O.) volgensis* sspp., European and Asiatic Russia. Listeriosis: *O. (O.) volgensis* ssp., European Russia. Salmonellosis: *O. (O.) kukuschkini* Ioff, Asiatic Russia.

Opisocrostis (Ceratophyllidae).
Distribution: North America; 6 species (7 forms).
Hosts: Ground-dwelling Sciurid rodents.
Vectors: Plague: *O. bruneri* (Baker), *O. hirsutus* (Baker), *O. labis* (Jordan & Rothschild) and *O. tuberculatus* ssp., western U.S.A.

Opisodasys (Ceratophyllidae).
Distribution: North America; 7 species (9 forms).
Hosts: Mainly arboreal Sciurid rodents.
Vector: Plague: *O. keeni nesiotus* Augustson, western U.S.A.

Orchopeas (Ceratophyllidae).
Distribution: North and Central America; 10 species (20 forms).
Hosts: Arboreal Sciurid rodents and Murid rodents.
Vectors: Plague: *O. leucopus* (Baker) and *O. neotomae* Augustson, southern U.S.A.; *O. sexdentatus* ssp., western U.S.A.

Oropsylla (Ceratophyllidae) [*vide* Ioff & Scalon].
Distribution: Asia and North America; 6 species (7 forms).
Hosts: Ground-dwelling Sciurid rodents.
Vectors: Plague: *O. idahoensis* (Baker), western U.S.A.; *O. ilovaiskii* Wagner & Ioff, European Russia; *O. rupestris* (Jordan), western U.S.A.; *O. silantiewi* (Wagner), Asiatic Russia and Mongolia. Erysipeloid: *O. silantiewi* (Wagner), Asiatic Russia and Mongolia.

Paradoxopsyllus (Leptopsyllidae) [*vide* Catalogue V].
Distribution: Asia; 18 species.
Hosts: Rodents.
Vectors: Plague: *P. curvispinus* Miyajima & Koidzumi, Asiatic Russia and Japan; *P. dashidorzhii* Scalon, Mongolia; *P. teretifrons* (Rothschild), Asiatic Russia.

Peromyscopsylla (Leptopsyllidae) [*vide* Catalogue V].
Distribution: Eurasia, North and Central America; 19 species (24 forms).
Hosts: Murid rodents.
Vector: Plague: *P. hesperomys adelpha* (Rothschild), southern U.S.A.

Pleochaetis (Ceratophyllidae) [*vide* Johnson].
Distribution: Americas; 16 species (18 forms).
Hosts: Murid rodents.
Vectors: Plague: *P. dolens quitanus* (Jordan) and *P. equatoris* (Jordan), Peru.

Plocopsylla (Stephanocircidae) [*vide* Catalogue II].
Distribution: South America; 14 species.
Hosts: Murid rodents.
Vector: Plague: *P. hector* Jordan, Ecuador.

Polygenis (Rhopalopsyllidae) [*vide* Johnson].
Distribution: South-eastern North America, Central and South America; 41 species (50 forms).
Hosts: Mostly rodents, also marsupials.
Vectors: Plague: *P. brachinus* Jordan, Peru; *P. gwyni* (Fox), south-eastern U.S.A.; *P. litargus* (Jordan & Rothschild), Ecuador, Peru; *P. platensis cisandinus* (Jordan), Argentina.

Pulex (Pulicidae) [*vide* Catalogue I].
Distribution: Americas, 1 species cosmopolitan; 2 subgenera: nominate subgenus with 3 species, subgenus *Juxtapulex* with 3 species.
Hosts: Various large coarse-coated mammals such as pig, canids, mustelids, deer, tapir and peccary; 1 species also on man.
Vector: Plague: *P. (P.) irritans* Linnaeus, U.S.A.   [*P. (P.) simulans* Baker may actually be of more importance].   Erysipeloid: *P. (P.) irritans* Linnaeus, Asiatic Russia and Mongolia.
Intermediate host for *Dipylidium caninum*: *P. (P.) irritans* Linnaeus, Italy, Switzerland.

Rhadinopsylla (Hystrichopsyllidae) [*vide* Catalogue III].
Distribution: Eurasia, North Africa, North and Central America; 5 subgenera: nominate subgenus with 6 species; subgenus *Ralipsylla* with 2 species (5 forms), subgenus *Micropsylla* with 1 species (2 forms), subgenus *Micropsylloides* with 1 species and subgenus *Actenophthalmus* with 42 species (48 forms).
Hosts: Mostly rodents.
Vectors: Plague: *R. (Rh.) cedestis* Rothschild, European Russia; *R. (Ral.) li ventricosa* Ioff & Tiflov, Asiatic Russia; *R. (Rh.) ucrainica* Wagner & Argyropulo, European Russia.

Rhopalopsyllus (Rhopalopsyllidae) [*vide* Johnson].
Distribution: Central and South America; 6 species (13 forms).
Hosts: Mostly rodents.
Vector: Plague: *R. sp. indet.*, Argentina.

Sphinctopsylla (Stephanocircidae) [*vide* Catalogue II].
Distribution: South America; 6 species.
Hosts: Rodents.
Vector: Plague: *S. mars* (Rothschild), Peru.

Stenistomera (Hystrichopsyllidae) [*vide* Catalogue III].
Distribution: North America; 3 species.
Hosts: Murid rodents.
Vectors: Plague: *S. alpina* (Baker), western U.S.A.; *S. macrodactyla* Good, southern U.S.A.

Stenoponia (Hystrichopsyllidae) [*vide* Catalogue III].
Distribution: Eurasia, North Africa, North America; 16 species (28 forms).
Hosts: Murid rodents.
Vectors: Plague: *S. conspecta* Wagner, Asiatic Russia; *S. tripectinata* ssp., Asia Minor and European Russia; *S. vlasovi* Ioff & Tiflov, Asiatic Russia.

Stivalius (Pygiopsyllidae).
Distribution: Africa, South-East Asia, Australasia; 60 species.
Hosts: Mostly rodents, some on marsupials, single species on insectivores and birds.
Vectors: Plague: *S. ahalae* (Rothschild), South-East Asia; *S. cognatus* Jordan & Rothschild, Java.

Synopsyllus (Pulicidae) [*vide* Catalogue I].
Distribution: Madagascar; 5 species.
Hosts: Rodents and insectivores.
Vector: Plague: *S. fonquerniei* Wagner & Roubaud.
Intermediate host of *Hymenolepis ?diminuta: S. fonquerniei* Wagner & Roubaud.

Synosternus (Pulicidae) [*vide* Catalogue I].
Distribution: Africa and south-western Asia; 6 species (7 forms).
Hosts: Mostly rodents.
Vector: Plague: *S. pallidus* (Taschenberg), West Africa.

Thrassis (Ceratophyllidae) [*vide* Stark].
Distribution: Western North America; 4 subgenera: nominate subgenus with 4 species (12 forms), subgenus *Nomadopsylla* with 2 species (8 forms), subgenus *Pandoropsylla* with 2 species (3 forms) and subgenus *Thrassoides* with 3 species (6 forms).
Hosts: Mostly ground-dwelling Sciurid rodents.
Vectors: Plague: *T. (T.) acamantis* sspp., *T. (Thrassoides) arizonensis* (Baker), *T. (N.) bacchi* ssp., *T. (Thrassoides) fotus* (Jordan), *T. (T.) francisi* ssp., *T. (P.) pandorae* ssp., *T.(P.) petiolatus* (Baker), *T. (T.) stanfordi* Wagner.

Tiamastus (Rhopalopsyllidae) [*vide* Johnson].
Distribution: South America; 5 species.
Hosts: Mainly rodents.
Vector: Plague: *T. cavicola* (Weyenbergh), Ecuador, Peru.

Tunga (Pulicidae) [*vide* Catalogue I].
Distribution: Americas, Africa, East Asia; 9 species.
Hosts: Edentates, man, domestic animals, murid rodents.
Causing tungosis: *T. penetrans* (Linnaeus), Central and South America, Africa.

Xenopsylla (Pulicidae) [*vide* Catalogue I].
Distribution: Throughout warmer parts of the Old World; 1 species world-wide; 68 species (76 forms).
Hosts: Mainly Murid rodents.
Vectors: Plague: *X. astia* Rothschild, Oriental region; *X. brasiliensis* (Baker), Central Africa; *X. buxtoni* Jordan, Asia Minor; *X. cheopis* (Rothschild), Eurasia, Africa, Americas [mainly between latitudes 45°N. and S.]; *X. conformis* ssp., Asiatic Russia; *X. eridos* (Rothschild), South Africa; *X. gerbilli* ssp., Asiatic Russia; *X. hirsuta* ssp., South Africa; *X. hirtipes* Rothschild, Asiatic Russia; *X. nubica* (Rothschild), northern Africa; *X. nuttalli* Ioff, Asiatic Russia; *X. philoxera* Hopkins, South Africa; *X. phyllomae* de Meillon, South Africa; *X. piriei* Ingram, South Africa; *X. skrjabini* Ioff, Asiatic Russia; *X. versuta* Jordan, South-West Africa; *X. vexabilis* Jordan, Australasia, Hawaii.

Brucellosis: *X. conformis* ssp., Asia Minor. Melioidosis: *X. cheopis* (Rothschild), North Africa. Erysipeloid: *X. cheopis* (Rothschild), Russia. Murine typhus: *X. cheopis* (Rothschild), southern U.S.A., European and Asiatic Russia.
Intermediate host of *Hymenolepis diminuta: X. cheopis* (Rothschild), Australia, Argentina, Mexico.

Xiphiopsylla (Xiphiopsyllidae) [*vide* Catalogue II].
Distribution: East Africa; 7 species (8 forms).
Hosts: Murid rodents.
Vector: Plague:*X. lippa* ssp., central Africa.

BIBLIOGRAPHY

HAESELBARTH, E. 1966. Order: Siphonaptera. *In:* F. Zumpt, The Arthropod parasites of vertebrates in Africa south of the Sahara (Ethiopian Region). *Publ. S. Afr. Inst. med. Res.* **13** (52): 117–212.

HIRST, L. F. 1953. *The conquest of Plague.* 478 pp. Oxford.

HOLLAND, G. P. 1949. The Siphonaptera of Canada. *Tech. Bull. Dept. Agric. Can.* **70**: 1–306.

—— 1964. Evolution, classification and host relationships of Siphonaptera. *A. Rev. Ent.* **9**: 123–146.

HOPKINS, G. H. E. & ROTHSCHILD, M. 1953. *An illustrated catalogue of the Rothschild collection of fleas (Siphonaptera) in the British Museum (Natural History).* London. Vol. I. Tungidae, Pulicidae. 361 pp.

——, —— 1956. *Op. cit.* Vol. II. Coptopsyllidae, Vermipsyllidae, Stephanocircidae, Macropsyllidae, Ischnopsyllidae, Chimaeropsyllidae, Xiphiopsyllidae. 445 pp.

——, —— 1962. *Op. cit.* Vol. III. Hystrichopsyllidae (part). 560 pp.

——, —— 1966. *Op. cit.* Vol. IV. Hystrichopsyllidae (concluded). 549 pp.

——, —— 1971. *Op. cit.* Vol. V. Leptopsyllidae, Ancistropsyllidae. 529 pp.

HUBBARD, C. A. 1947. *Fleas of Western North America.* 533 pp. Iowa.

IOFF, I. G. & SCALON, O. I. 1954. *Handbook for the identification of the fleas of eastern Siberia, the Far East and adjacent regions.* Moscow. 275 pp. [in Russian].

IOFF, I. G., MIKULIN, M. A. & SCALON, O. I. 1965. *Handbook for the identification of fleas of central Asia and Kazakhstan.* Moscow. 370 pp. [in Russian].

IOFF, I. G. & TIFLOV, V. E. 1954. *Keys to the Aphaniptera of south-east U.S.S.R.* 201 pp. Stavropol'. [in Russian].

JOHNSON, P. T. 1957. A classification of the Siphonaptera of South America, with descriptions of new species. *Mem. ent. Soc. Wash.* **5**: 1–299.

—— 1961. A revision of the species of *Monopsyllus* Kolenati in North America (Siphonaptera: Ceratophyllidae). *Techn. Bull. U.S. Dep. Agric.* **1227**: 1–69.

LINK, V. B. 1955. A history of plague in the United States. *Publ. Hlth Monogr.* **26**: 1–120.

MEILLON, B. DE, DAVIS, D. H. S. & HARDY, F. 1961. *Plague in southern Africa.* I. *The Siphonaptera (excluding Ischnopsyllidae).* 280 pp. Pretoria.

POLLITZER, R. 1954. Plague. *Monograph Ser. W.H.O.* **22**: 1–698. [Identification of fleas, Smit, pp. 648–682.]

—— 1966. *Plague and plague control in the Soviet Union.* 478 pp. New York.

ROSICKÝ, B. 1957. Blechy—Aphaniptera. *Fauna Č.S.R.* **10**: 1–439.

SAKAGUTI, K. 1962. *A monograph of the Siphonaptera of Japan.* 255 pp.. Osaka.

SCALON, O. I. 1970. Order Siphonaptera (Aphaniptera, Suctoria)—Fleas. *In:* G. Ya. Bei-Bienko, *Keys to the insects of the European part of the U.S.S.R.* **5.** Leningrad. pp. 799–844 [in Russian].

SMIT, F. G. A. M. 1957. Siphonaptera. *Handbk. Ident. Br. Insects.* **1** (16): 1–94.

—— 1974. An annotated list of pathogens, parasites and enemies of fleas (in press).

Smit, F. G. A. M. & Wright, A. M.   1974.   Siphonaptera.   Catalogue of type data, 1758–1972, with code-number list, host list, gazetteer and bibliography (in the press).

Stark, H. E.   1970.   A revision of the flea genus *Thrassis* Jordan 1933 (Siphonaptera: Ceratophyllidae).   *Univ. Calif. Publ. Ent.* **53**: 1–184.

Traub, R.   1950.   Siphonaptera from Central America and Mexico.   *Fieldiana, Zool.* **1**: 1–127.

Wu, L. T., Chun, J. W. H., Pollitzer, R. & Wu, C. Y.   1936.   *Plague, a manual for medical and public health workers.*   547 pp.   Shanghai.

8. HEMIPTERA
(Bugs)

by M. S. K. Ghauri

THE Hemiptera have sucking mouth-parts, possess two pairs of wings and exhibit incomplete metamorphosis. The anterior pair of wings are most often of harder consistency than the posterior pair, either uniformly so (Homoptera) or with the apical portion more membranous than the remainder (Heteroptera). The mouth-parts are piercing and suctorial and the palpi are atrophied; the labium is a dorsally grooved sheath receiving two pairs of bristle-like stylets (modified mandibles and maxillae).

Almost all Hemiptera in all their stages of development possess sucking mouth-parts (fig. 171) (the exception being adult males of Coccoidea, which have atrophied mouth-parts). The order includes plant-bugs, assassin-bugs, bed-bugs, cone nose-bugs or kissing-bugs, leafhoppers, cicadas, green-fly (Aphides) and mealy-bugs and scale insects. It is divided into two sub-orders; the Heteroptera, which includes all the species usually known as true bugs, and the Homoptera, which includes the cicadas, the leafhoppers, froghoppers, green-flies, scale insects etc. In most of the Heteroptera, as the name suggests, the consistency of the fore-wing is not uniform, the anterior part being leathery and the posterior membranous. In the Homoptera the texture of the fore-wing is uniform; the difference in consistency of fore- and hind-wings in Heteroptera is much more evident than in the case of Homoptera. The Heteroptera fold their wings flat on the back with the apices superimposed while the Homoptera carry them raised like a roof over their backs.

The great majority of the Hemiptera suck up the juices of plants of various kinds which they obtain by inserting their mouth-parts into the tissues of the plants. It should be noted that the formidable 'beak' or rostrum on the underside of the head is really only the sheath (formed of the labium) in which lie the other mouth-parts in the form of fine piercing stylets. This sheath folds up when the piercing parts are thrust into the tissues of the plant or, in the case of the blood-sucking species, their hosts' skin (fig. 171).

The Hemiptera have three stages in their life-cycle; egg, larva and adult (fig. 175). The larval form is also known as a nymph or neanide. The pupal stage is usually absent and the adult stage is preceded by larval stages which resemble it.

The eggs of Heteroptera are laid singly and there are five nymphal instars.

The Sub-order Heteroptera is subdivided into Gymnocerata and Cryptocerata.

The antennae of Cryptocerata are small and inconspicuous and are concealed in a socket between the gena and the compound eye or closely pressed underside of the head. These are truly aquatic bugs and their antennae are modified to suit a mode of life in water.

In Gymnocerata the antennae are long and conspicuous, capable of being moved about freely in front of the head. The Gymnocerata include land-bugs and water surface dwellers.

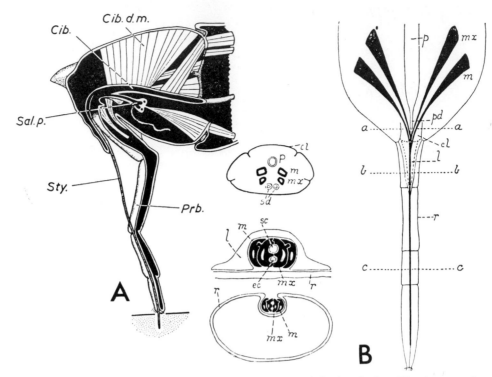

Fig. 171 Diagram of mouthparts and adjacent regions of the head of an Hemipterous insect: A, section (lateral view) of head of *Graphosoma italicum* in feeding position; cib, cibarium; cib.d.m, cibarial dilator muscles; prb, proboscis (labium); sal.p, salivary pump; sty, stylets; B, section in frontal view and on transverse sections at points a–a, b–b and c–c (magnifications not uniform); cl, clypeus; ec, ejection canal with salivary duct; l, labrum; m, mandible; mx, maxilla; p, pharynx; pd, cibarial pump; r, rostrum; sd, salivary ducts; sc, suction canal with pharyngeal duct (after Imms).

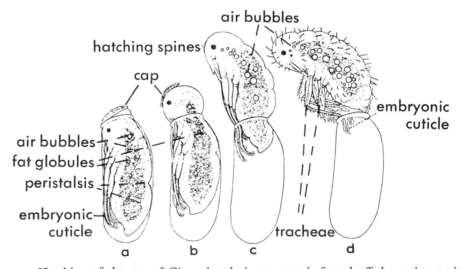

Fig. 172 Hatching of the egg of *Cimex lectularius*: a, cap is forced off, b; active peristalsis continues as head is distended and bulges from the egg; c, air is swallowed, hatching spines are visible on front of head; d, embryonic cuticle has split and slipped backward, allowing bristles to stand erect, trachaea have filled with air (after Sikes and Wigglesworth, 1931).

KEY TO FAMILIES OF HETEROPTERA

The following key to families of Heteroptera is based mainly on Reuter (1912) and is adapted from Imms (1957). Although at present only two families of Heteroptera, Cimicidae (bed-bugs) and Reduviidae (cone nose-bugs, including Triatominae) are known to contain species of medical importance, there is some possibility that as a result of future research, members of certain other families treated below might be incriminated as indirectly causing human ailments. These suspected bugs belong to the genera *Oxycarenus* of the family Lygaeidae and *Lethocerus* of the family Belostomatidae.

Some terms used in the keys, peculiar to Heteroptera and not explained in the general Introduction (Ch. 1) are defined here.

clavus: The sharply pointed anal area of the hemelytra, next to the scutellum when folded.
connexivia: The prominent abdominal margin.
corium: The elongate middle section of the hemelytra.
hemelytra: The anterior wing, the basal half of which is thickened and the apical half membranous. Sometimes called the elytron or tegmen.
rostrum: A jointed beak-like sheath formed by the labium and enclosing the stylets or trophi.

1 Antennae shorter than head, usually concealed; eyes present . . . **2**
 - Antennae usually much longer than head; if shorter, then eyes absent . . . **10**

2 Front tarsi spatulate, 1-segmented; rostrum 1- or 2-segmented . . **CORIXIDAE**
 - Front tarsi not spatulate, rarely 1-segmented; rostrum 3- or 4-segmented . . **3**

3 Head completely fused with prothorax, the boundary indicated by a shallow impression . . . **HELOTREPHIDAE**
 - Head free or at most partially fused with prothorax . . . **4**
 - Ocelli present . . . **5**
 - Ocelli absent . . . **6**

5 Fore legs raptorial; antennae concealed . . . **GELASTOCORIDAE**
 - Fore legs normal; antennae free . . . **OCHTERIDAE**

6 Fore coxae inserted at front of prosternum; hind tarsi with distinct claws . . **7**
 - Fore coxae inserted at back of prosternum; hind tarsi without claws . . **9**

7 Membrane without veins . . . **NAUCORIDAE**
 - Membrane with reticulate venation . . . **8**

8 Hind tibiae flattened and fringed for swimming; respiratory funnel short **BELOSTOMATIDAE**
 - Hind tibiae simple; respiratory funnel usually long . . . **NEPIDAE**

9 Rostrum 4-segmented; abdomen with ventral furrow in which lies median carina **NOTONECTIDAE**
 - Rostrum 3-segmented; abdomen without furrow or carina . . . **PLEIDAE**

10 Eyes present . . . **11**
 - Eyes absent . . . **49**

11 Pulvilli usually absent; if present then either meso- and meta-pleura divided or clavus membranous or hemelytra reticulate or base of membrane densely reticulo-punctate; tibiae simple apically . . . **12**
 - Pulvilli usually present; if absent then tibiae with apical membranous lobe; meso- and metapleura simple; hemelytra with coriaceous clavus, never reticulate or with reticulo-punctate membrane . . . **37**

12 Hemelytra of macropterous forms with clavus which meets that of opposite side behind scutellum to form distinct claval commissure 13

– Clavus of macropterous forms not exceeding scutellum to form commissure; body flattened, with head produced and bearing pointed antennal tubercles . . 36

13 First two antennal segments very short, last two long and pilose, third segment swollen basally 14

– Second antennal segment usually equal to or longer than third which is never swollen basally 15

14 Head directed backwards between fore coxae . . . **SCHIZOPTERIDAE**

– Head more or less porrect . . . . **DIPSOCORIDAE**

15 Meso- and metapleura divided or if, rarely, simple then clypeus triangular, broader apically; cuneus present in macropterous forms 16

– Meso- and meta-pleura simple 20

16 Clypeus with parallel or sub-parallel sides 17

– Clypeus broadened apically; ocelli absent; micropterous . . . **CIMICIDAE**

17 Ocelli absent in both sexes; tarsi usually 3-segmented . . . . **MIRIDAE**

– Ocelli present in males, sometimes absent in females, in which case these are micropterous with swollen abdomen and 2-segmented tarsi 18

18 Tarsi 3-segmented . . . . **ANTHOCORIDAE**

– Tarsi 2-segmented 19

19 Membrane with divided semicircular basal cell; male genitalia asymmetrical; female with eighth sternite entire . . . . . . **ISOMETOPIDAE**

– Membrane with one rectangular basal cell; male genitalia symmetrical; females with swollen abdomen and divided eighth sternite . . . . **MICROPHYSIDAE**

20 Hind coxae hinged; rostrum 3-segmented 21

– Hind coxae rotatory 25

21 Ocelli absent; eyes small; hemelytra reduced . . . . **AEPOPHILIDAE**

– Ocelli present; eyes large and projecting 22

22 Cuneus present . . . . **VELOCIPEDIDAE**

– Cuneus absent 23

23 Fore tarsi 1-segmented, middle and hind tarsi 2-segmented . . **LEOTICHIIDAE**

– All tarsi 3-segmented 24

24 Ocelli on pedunculate tubercle . . . . **LEPTOPODIDAE**

– Ocelli not on pedunculate tubercle . . . . **SALDIDAE**

25 Claws ante-apical: hemelytra homogeneous 26

– Claws apical 27

26 Middle femora markedly exceeding end of abdomen; vertex without longitudinal groove . . . . **GERRIDAE**

– Middle femora scarcely, if at all, exceeding end of abdomen; vertex with longitudinal groove . . . . **VELIIDAE**

27 Pulvilli absent, rarely with membranous appendage between claws, in which case clavus and membrane confluent 28

– Pulvilli present; tarsi 2-segmented 35

28 Prosternum without stridulatory groove 29

– Prosternum with stridulatory groove 34

29 Ocelli absent; body linear with long, apically dilated head . **HYDROMETRIDAE**

– Ocelli usually present, if absent then head not dilated apically . . . 30

30 Head constricted basally and behind eyes; hemelytra entirely membranous
ENICOCEPHALIDAE

–	Head not thus constricted; hemelytra not entirely membranous	31
31	Clavus membranous or sub-membranous and confluent with veinless membrane .	32
–	Clavus coriaceous and distinct	33
32	Rostrum 4-segmented; antennae usually 5-segmented . . . . **HEBRIDAE**
–	Rostrum 3-segmented; antennae 4-segmented . . . . . **MESOVELIIDAE**
33	Tarsi 2-segmented . . . . **JOPPEICIDAE**
–	Tarsi 3-segmented . . . . **NABIDAE**
34	Antennae filiform, often apically slender, geniculate . . . **REDUVIIDAE**
–	Last antennal segment clavate or fusiform; fore legs stoutly raptorial **PHYMATIDAE**
35	Head bifid anteriorly; ocelli present in macropterous forms . . . **PIESMIDAE**
–	Head not bifid anteriorly; ocelli absent; body and hemelytra densely reticulate **TINGIDAE**
36	Head widened posteriorly and enclosing eyes; trochanters distinct **DYSODIIDAE**
–	Head not widened behind prominent eyes; trochanters fused with femora **ARADIDAE**
37	Membrane usually with five or fewer veins; if veins numerous and branching then ocelli absent; scutellum normal	38
–	Membrane with many veins and ocelli present, if otherwise scutellum unusually large	43
38	Pulvilli absent; tibiae with apical membranous lobe . . **THAUMASTOCORIDAE**
–	Pulvilli present; tibiae without lobe	39
39	Ocelli absent . . . . **PYRRHOCORIDAE**
–	Ocelli present	40
40	Membrane without distinct veins; elongate forms with long antennae **COLOBATHRISTIDAE**
–	Membrane with distinct veins	41
41	Antennae geniculate; head constricted in front of eyes . . **BERYTIDAE**
–	Antennae not geniculate; head not thus constricted	42
42	Membrane with four veins which form three large preapical cells then branch **HYOCEPHALIDAE**
–	Membrane usually with four or five veins which never form preapical cells **LYGAEIDAE**
43	Scutellum smaller, never reaching membrane; antennae 4-segmented **COREIDAE**
–	Scutellum larger, rarely not reaching membrane, in which case antennae 5-segmented	44
44	Hemelytra about twice as long as abdomen, folded beneath enlarged scutellum at rest; tarsi 2-segmented . . . . **PLATASPIDAE**
–	Hemelytra usually normal; if elongate then tarsi 3-segmented	45
45	Connexivia of six abdominal tergites visible; tibiae spinose . . . **CYDNIDAE**
–	Connexivia of seven abdominal tergites visible; tibiae rarely spinose . . .	46
46	Head and body with lamellate lateral expansions; antennae 3-segmented **PHLOEIDAE**
–	Lamellate lateral expansions absent; antennae 4- or 5-segmented . . .	47
47	Base of antenna concealed by side-margins of head	48
–	Base of antenna not concealed . . . . **UROLABIDAE**
48	Body concave ventrally; base of corium not reaching sides of body to form epipleura **APHYLIDAE**
–	Body flat or convex ventrally; base of corium reaching sides of body and forming epipleura . . . . **PENTATOMIDAE**
49	Broadly oval, flat, apterous forms resembling woodlice and living in termites' nests **TERMITAPHIDIDAE**
–	Oblong forms with vestigial hemelytra, parasitic on bats . . **POLYCTENIDAE**

REDUVIIDAE (INCLUDING TRIATOMINAE) (CONE NOSE-BUGS, KISSING-BUGS AND ASSASSIN-BUGS)

China *et al.* (1950) define this family thus: 'Head with a transverse sulcus dividing the head into two distinct lobes, rarely obsolete (Triatominae and *Xenocaucus*) in which case prosternum with a distinct stridulatory groove. Rostrum (fig. 173B) usually robust and three segmented, strongly curved below the head (straight in Triatominae) so that the apex comes in contact with the stridulatory furrow, rarely slender and straight and apparently two segmented (Tegeinae) or with stridulatory furrow absent (*Amulius*, *Ectinoderus*) in which case the membrane with a large cell with a distinct apical "spur" bounded by veins M and Cu and a smaller inner cell bounded by veins Cu and Ist A. Aedegus with a distinct rod-like basal plate-extension ending in two struts supporting the phallosoma, rarely with the two struts fused to form a flattened plate-like sclerite (Vesciinae). Metathoracic scent glands opening into hind coxal cavities and without the visible orifices on metapleura or metasternum which are characteristic of many other families.'

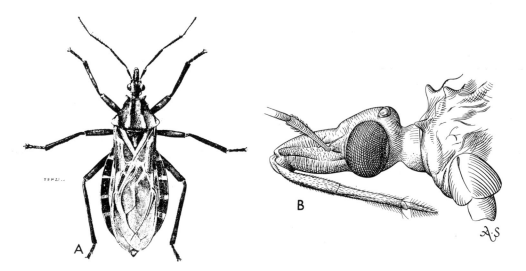

Fig. 173　*Panstrongylus megistus*, a cone nose-bug: A, adult (from Castellani & Chalmers); B, head from the side (from Smart).

There are more than 20 subfamilies of Reduviidae. The subfamily Triatominae is most important medically and includes those species of which both sexes feed exclusively by sucking the blood of vertebrate animals. They are popularly known as cone nose-bugs and occur in the nests of birds and mammals and in the dwellings of man and his domesticated animals. The majority of the 80 or so described species occur in the Western hemisphere with an Indian species and 6 species of *Triatoma* in South-East Asia. Their size ranges up to 40 mm. in length. They may be recognized by reference to fig. 173 and pl. 7.

Since Stål's treatment of the Acanthaspidinae (=Reduviinae, *s.lat.*) there has been a continual culling out of peculiar types that Stål lumped together in his rather unnatural group.   Jaennel (1919) first separated out the Cetherinae and Triatominae as tribes and Pinto (1926 and 1927) erected separate families of *Sphaeridops* and for the Triatomas. Unfortunately, the family name Triatomidae has found extensive use in parasitological literature but there is absolutely no foundation for such a status, no characters were given by Pinto to separate it from such Reduviinae as *Physoderes*, and hemipterists have not generally accepted the name.   *To accept the family Triatomidae would require the elevation of twenty-three other subfamilies to family rank.*

The Triatominae are characterized by elongate head, rarely constricted behind the eyes, the ocelli present except in rare brachypterous froms, located behind compound eyes, usually on oblique elevations at postero-lateral angles of the cylindrical head; under surface of head without a buccal groove; a rostrum three-segmented and straight whose tip fits into a longitudinal, cross-striated prosternal stridulatory groove; antenna four segmented, apical antennal segments not thicker than the remaining segments, the former inserted at apices of the latter, second antennal segment not subdivided; pronotum constricted at or near the middle, with well-differentiated clavus, corium, and membrane and without greatly modified front legs; cubitus simple, not forming an extra four- to six-angled cell between corium and membrane, the latter with more than one close cell; anterior coxae shorter, usually less than twice as long as broad and not extending beyond apex of head, front femora not tremendously dilated, the stink gland passing laterally into a groove which runs along the internal face of the emarginated edge of the metacoxal cavity and communicates with the exterior by a small pore; base of abdomen without a trichome, dorsal abdominal scent gland opening absent in nymphs as well as in adults; usually broad insects; the tibial fossae are usually present on the front and middle tibiae of adult males and absent in females, their occurrence is variable in some genera, since they are so generally present in other Reduvioidea their absence in Triatominae should be considered as a specialization or degeneration because Triatominae no longer have to catch and hold small prey.

The Triatominae are closest to the typical genus *Reduvius* of the Reduviinae in type of egg and in most other characters.   These two subfamilies are closer to each other than to any of the other subfamilies of Reduviidae.   Reduviinae have a shorter, more curved head, a shorter, stouter and curved rostrum, and three dorsal abdominal scent gland openings.   Although the Old World genus *Physoderes* (Reduviinae) has a longer, more cylindrical head with a long, straight rostrum, similar to Triatominae, it has a distinct transverse impression behind the eyes.

There are fifteen genera in Triatominae according to the majority of taxonomists, but Usinger (1944) insists rightly on the validity of separate status for *Mestor* Kirkaldy as distinct from *Panstrongylus* Berg.

Most of the genera are distributed in South and Central America, but there is one, *Paratriatoma* Barber found only in California (North America) and the Oriental genus *Linshcosteus* Distant is so far known only from South India.   The largest genus is *Triatoma* Laporte, which is also the most widely distributed and incidently contains the most important vector species.

Apart from *Triatoma*, the other genera significant as vector of Chagas' disease are, *Rhodnius* Stål, *Panstrongylus* Berg, *Eratyrus* Stål, *Cavernicola* Barber and *Dipetalogaster* Usinger.   These are distributed in South and Central Americas.   With the help of the following key, the above-mentioned six genera can be readily separated.

KEY TO GENERA OF VECTORS OF CHAGAS' DISEASE

1 Ocelli not elevated, located in or just behind arcuate impressed suture on posterior lobe of head.   Clavus (see p. 375) and corium (see p. 375) smooth, the veins almost invisible .   .   .   . *CAVERNICOLA* Barber

— Ocelli located on postero-lateral, rounded elevations, directed laterally or antero-laterally.   Clavus and corium with one or more distinctly elevated veins   .   .   2

2 Clypeus more or less widened apically.   Genae strongly divergent anteriorly.   Head with a lateral callosity behind each eye   .   .   .   .   . *RHODNIUS* Stål

— Clypeus distinctly narrowed apically.   Genae sub-parallel or slightly convergent anteriorly.   Head without postocular callosities   3

3 Apex of scutellum long, acute, upturned   .   .   .   .   . *ERATYRUS* Stål

— Apex of scutellum long or short, tapering or bulbous, but never acute and upturned, base without distinct basal tubercles   4

4 Ventral abdominal margins extending laterad to connexivial margins, the abdominal margins thus appearing double-plated.   Very large, 38 mm. long
DIPETALOGASTER Usinger

— Ventral abdominal plates never extending to abdominal margins.   Size less than 38 mm.   5

5 Head short, broad, the antennae inserted well behind middle of anteocular region close to the eyes (fig. 173)   .   .   .   .   . *PANSTRONGYLUS* Berg

— Head longer, subcylindrical or convex above, the antennae inserted at or near the middle of anteocular portion (pl. 7)   .   .   .   . *TRIATOMA* Laporte

In South and Central America and the south-western United States, several species of *Triatoma, Rhodnius, Panstrongylus* and *Eratyrus* have been reported infected with *Trypanosoma cruzi* Chagas, the cause of Chagas' disease in man which is often a zoonosis (see table below).   In nature most of the species of these bugs are found associated with wood-rats, armadillos, ground-squirrels, skunks and large iguanas, but many also inhabit human dwellings.   Infection from Triatominae is transferred to human beings through faecal droplets of the insect.   Other trypanosomes occur naturally in South American Triatominae and *Trypanosoma rangeli* Tejera has been recovered from man, but is non-pathogenic (Pifano, 1954).

Unlike most species of Triatominae, *Paratriatoma hirsuta* Barber has never been found infected with *Trypanosoma cruzi* Chagas in nature.   The report of infected *P. hirsuta* by Hall (1953) appears to be a misquote.   Swezey (1963) reports a case of severe reaction to the bite of *Paratriatoma*, which shows that although members of this genus inflict bites on human beings, they are non-vectors as far as known.

The Triatominae live in forest as well as under desert conditions.   In deserts the bugs live in places which are protected against excessive heat, *e.g.* nests of wood-rats. Originally the Triatomine bugs were non-domestic, but some species have now become totally domesticated while others have populations both in forests and in human dwellings, and even some populations are found in the process of moving from forests which have been cleared for domestic use, to houses.

One species of Triatominae, *Triatoma rubrofasciata* (de Geer) has become world-wide in distribution aided, presumably, by rats associated with man.   In the Far East *T. rubrofasciata* (de Geer) bites man but is not known to transmit human disease (Kalshoven, 1970).

The various species differ in flight habits and in their attraction to lights. In the U.S.A. many *Triatoma rubida uhleri* Neiva have been collected in flight during the daytime 15 ft above ground in summer. Some species are less active than others, for example Wood & Wood (1938) noted that *T. protracta* (Uhler) is a relatively sluggish species which does not fly readily and which bites hesitantly in contrast to the very active *T. r. uhleri* with its long legs, ready flight, and prompt biting reaction.

Petana (1971) reported capturing of *T. dimidiata* (Latr.) during night between 7–11 p.m. on the structure and roof of a 45 ft high wooden tower in the forest of British Honduras. The adults being winged are able to fly considerable distances but the nymphs are forced to stay in or near the place of their birth until they become adults. The rainy season forces the bugs to accumulate in caves and other protected places whereas in the dry season their population tends to spread out far more.

The Triatominae are dependent for their existence upon animal blood. As has been observed in the laboratory, the contact feeding time may range from 2 minutes 50 seconds (*Paratriatoma hirsuta*) to 29 minutes 10 seconds (*T. protracta*). Higher temperature (93°F.) probably speeds up the feeding process compared to low temperature (83°F.). The feeding may be continuous for short or long periods or with one or more than two intervals. The success of transmission of causative agents of Chagas' disease depends on the creation of a wound as a result of biting by the bugs and its contamination with infective faeces of the bugs. When the weather is not warm and the sleeping persons' bodies are mostly covered, the bugs feed on exposed parts of body such as mouth, nose and mucus membrane of eyes, but in hot weather, when not much cover is necessary during sleep, other parts of the body are equally attacked.

The defaecation rate is also important in relation to transmission of Chagas' disease. Some species defaecate more often; *T. rubida uhleri* tends to deposit immediately and abundantly at short intervals (1, 9, 17, 26, 32, 43, 50 minutes) during the first hour after the blood-meal; *T. protracta* appears to be a less efficient environmental contaminator, withholding faecal deposits for hours after meals (Wood, 1951).

The bugs can live without food for long periods. During this long fast the trypanosomes are accumulated in the gut.

During mating the male takes a dorsal position to one side of the female, revolving the genital capsule to effect coition.

Eggs may be laid within 9 to 28 days after emergence of the adult. The total number of eggs laid may vary from 96 up to 701 according to the species involved and the prevailing temperature. 701 eggs were laid in 240 days with 26 blood-meals at 30°C. by *Rhodnius prolixus*, but the average is much less and the number of eggs laid under natural conditions may not reach these figures very often because of changing temperature and irregular blood-meals.

The Triatominae have very simple eggs without an elaborate circum-micropylar process and without an enlarged cap region. Special features are confined to details in sculpturing of chorion and cap. The eggs are slightly asymmetrical, elongate-oval to oval with more or less constricted 'neck' region flaring slightly to receive the cap. The eggs may be white, dull or shining. The eggs may be laid in a mass, fastened together, and to a support by a secretion, or laid singly either adhering to the substrate or perfectly free. These variations may be due to the different species involved, but some slight variations may occur within the same species or even in the same female.

The incubation period depends, of course, upon prevailing temperature, and the species involved.   The duration of the egg stage may vary from 7 to 60 days.   Eggs show reddish eye spots.   In some species white eggs become pink as development progresses. Some species have eggs with a thinner chorion in which case the embryo is visible.

Nymphs moult after one feeding but they must engorge quite fully with the abdominal integument stretched and the abdomen oval to start the moulting process.   After a full meal, no additional blood is taken until the moult is completed.   Insufficiently engorged nymphs may live for many months without moulting.   Nymphs take six to twelve times their body weight whereas adults only about three times.

Nymphs of Triatominae have smaller eyes than adults and lack ocelli and differ from other reduviine nymphs by lacking dorsal abdominal scent gland openings.

There are five nymphal instars.   The nymphal period may last from 120 days to more than 2 years, in different species.   Most species take 1 year to complete their life-cycle.

Petana (1971) noted a very interesting camouflage habit of nymphs, which is common to many living in old world deserts.   Nymphs living on the ground were found carrying a hard shield of soil on top of their bodies and were only noticeable on moving about; in contrast, nymphs on cave walls and in wall crevices carried little or no soil on their bodies.

Our knowledge of the parasites and predators of Triatominae is inadequate, hence very little is known about the biological control methods of these vectors.   There is one Hymenopterous parasite, *Teleonomus fariai* Costa Lima (Scelionidae) which is a promising egg parasite of *Triatoma* spp. and *Panstrongylus megistus* (Burmeister).   In a single host egg this can develop 10–16 adult offspring.   There is a Hymenopterous predator also, *Solenopsis* sp. (Formicidae), which feeds on *Triatoma infestans* (Klug).   Members of two non-triatomine genera of Reduviidae, *Reduvius* and *Opisthacidius* also feed on Triatominae.   Cannibalism amongst the vector species themselves is known.

There are 10 species of mites of the genus *Pimeliaphilus* parasitizing triatomines in rodent nests, human habitations and poultry houses.   Some of these are found in the U.S.A., Mexico, Chile and Argentina, but their potentialities for the control of bugs are not known.

The control of triatomine bugs has become quite possible by the use of organic insecticides, DDT, BHC, Dieldrin etc.   Houses in five townships in Venezuela were sprayed with a 2.5% suspension of Dieldrin at 1 g insecticide per square metre.   Three applications at 50 day intervals, beginning in February, gave highly effective control for 3 years after the last spraying.   In this case replacing straw thatching by aluminium or zinc roofing material was found unnecessary.   Similarly in Brazil, limited housing improvements and house-spraying with BHC at 500 mg gamma isomer per square metre completely killed off the bugs.

The construction of rural dwellings with walls of concrete blocks also contributes to the elimination of the bugs; also paint should be applied to the interior and exterior of the dwellings to form a substrate for the insecticide deposits.

Dogs, not any other domestic animal like cats, donkeys, goats, cows, horses etc., are considered to be the most important reservoir of Chagas' disease.   Their removal will certainly eliminate an important factor in the spread of the disease.

Light traps are also suggested as a means of reducing the population by long term removal of dispersing adults.

However, the control of Triatominae with chemical insecticides led to the conclusion that whereas their use considerably reduces the population of the bugs, no species has been eradicated even by repeated treatment. In fact in view of the seriousness of the

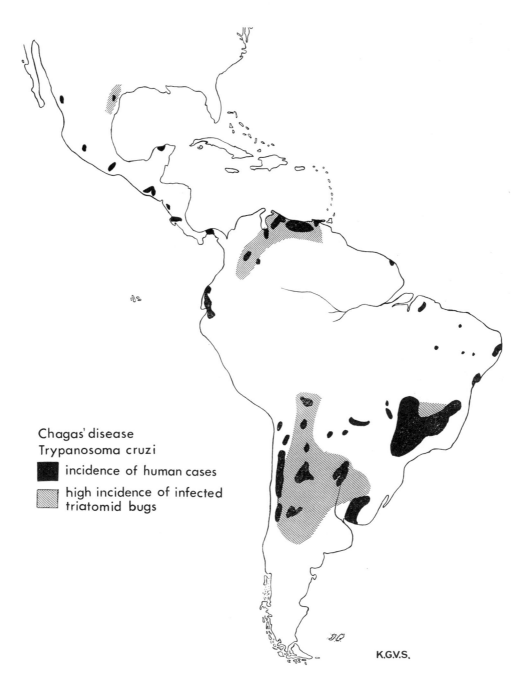

Fig. 174 The distribution of *Trypanosoma cruzi* (Chagas' disease) (modified from map by Dias, 1951, with additions).

problem, the aim should be eradication of the vectors.   The outdoor breeding of some of the species which infest houses (*Rhodnius prolixus* Stål and *Triatoma infestans* (Klug)) should be examined to assess its probable bearing on the likelihood of effecting eradication.   In this respect the sex attractant shown to be present in *Triatoma phyllosoma pallidipennis* Stål gives a clue for another line of approach for the control of these bugs. The possibilities of an integrated programme including the use of parasites and predators and chemicals have yet to be fully investigated.

Table 9   Hemiptera found naturally infected with *Trypanosma cruzi*
(adapted from Usinger, 1944)

Species	Distribution
Cavernicola pilosa	Brazil
Dipetalogaster maximus	Mexico
Eratyrus cuspidatus	Panama
Nesotriatoma flavida	Cuba
Panstrongylus geniculatus	Brazil, Panama
„ *herreri*	Peru
„ *megistus**	Brazil
„ *rufotuberculatus*	Venezuela
Rhodnius pallescens	Panama
„ *prolixus**	Venezuela
*Triatoma barberi**	Mexico
„ *dimidiata**	Panama through Central America into Mexico
„ *gerstaeckeri*	Texas
„ *hegneri*	Mexico
„ *infestans**	S. Brazil, Uruguay, Paraguay, Argentina, Chile and S. Bolivia
„ *lecticularis*	Texas
„ *longipes*	Arizona
„ *maculata**	Venezuela
„ *maculipennis*	San Salvador
„ *phyllosoma* & subspp.	Mexico
„ *protracta*	Arizona, California, New Mexico
„ *protracta woodi*	Texas
„ *recurva*	Arizona
„ *rubida*	Mexico
„ *rubida uhleri*	Arizona, Texas
„ *rubrofasciata*	Brazil
„ *sanguisuga ambigua*	Texas
„ *sanguisuga sanguisuga*	Texas

*Proven vectors for man.

In Panama *Rhodnius pallescens* Barber, which inhabits human dwellings, prefers certain hiding places and occurs particularly in cracks in walls near beds, crevices and folds of bed clothes, beneath sleeping mats and in crevices in the bedstead, behind clothes hanging on nails or pegs in the wall and in roof thatch immediately over the sleeping platforms in the loft.

Outside houses, pig pens and chicken houses harbour the bugs.

Estimates made by the World Health Organisation in 1967 showed that probably some 7 000 000 people harbour *T. cruzi* infections. While less than 10% of these cases may ultimately die from Chagas' disease, the long term economic and social effects of the disease in its chronic form must be comparable to the effects of sleeping sickness in Africa.

Bugs become infected from reservoir animals, from man, or from others of their own species by feeding on engorged nymphs.

The incidence of infection in nature is high (fig. 174). Usinger (1944) found that 43% of 4181 *Triatoma* bugs examined were infected and Wood (1942) found that 25% of 816 *Triatoma protracta* were infected in California.

The relationship between the incidence of infection in *Triatoma* and man appears to depend on the habits of the former. Not all *Triatoma* species occur near human habitation and others fly off soon after feeding so that contamination from the faeces is less likely. Species found to be naturally infected are listed in table 9, the proven vectors for man being indicated with an asterisk. A bibliography of Chagas' Disease is given by Olivier, *et al* (1972).

Various viruses have been recovered from Triatominae. *Triatoma infestans* has been found to transmit murine typhus in Brazil (Violle & Sauter, 1938) and other species have been found to carry equine encephalitis viruses (Kitselman & Grundmann, 1940; Lépine *et al.*, 1941) (see also Chapter 19).

CIMICIDAE (BED-BUGS)

Members of the Cimicidae are small, oval, dorsoventrally flattened insects with very short hemelytra, rostrum lying in a ventral groove, ocelli absent, tarsi 3-jointed and are parasites of mammals and birds.

Bed-bugs are mahogany brown and, when seen on a wall at a distance, look rather like mobile brown lentils. Close inspection will show that they are rather rounded in outline, with the head somewhat sunken back into the front of the thorax, measuring 4–5 mm. in length and 3 mm. across. They are without functional wings though the fore-wings remain represented in the adults by two small brown pads on the dorsal surface of the thorax. As in other Hemiptera, metamorphosis is incomplete and the young resemble their parents in general appearance though smaller in size (fig. 175, pl. 6). They suck blood in all stages of their development. The proboscis is not very apparent since it is held flat along the underside of the head and thorax.

Bed-bugs infest the habitations of man. Closely related species attack birds (such as pigeons and house martins), bats and some rodents and are found in the nests and sleeping-places of these animals. Bed-bugs live in cracks and crevices in the structure, behind wallpaper and in beds and other furniture. They approach man, when he is sleeping, only for the purpose of obtaining a meal of blood. The eggs are laid in the crevices and cracks where the bugs live. It is important to remember these points since consideration of them will show that men themselves do not normally carry the infestation from one place to another but that furnishings from infested premises removed to clean premises are likely to result in an infestation unless steps are taken to disinfect the articles removed. Heavily infested premises have a characteristic odour.

There is experimental evidence to show that bed-bugs are capable of transmitting a variety of diseases, but there is no satisfactory evidence that they are the normal vectors of any under natural conditions. Their bites are a severe nuisance and cases of iron deficiency caused by excessive feeding of bed-bugs on infants have been reported in India (Venkatachalam and Belavadi, 1962; in Usinger, 1966).

There are 3 human bed-bugs: *Leptocimex boueti* (Brumpt) on bats and man locally in West Africa; *Cimex hemipterus* (F.) (fig. 176) attacking man, chickens, and rarely bats throughout the Old and New World tropics; and *Cimex lectularius* L. (figs 175, 177, pl. 6) associated with man, bats, chickens, and occasionally other domesticated animals over most of the world.

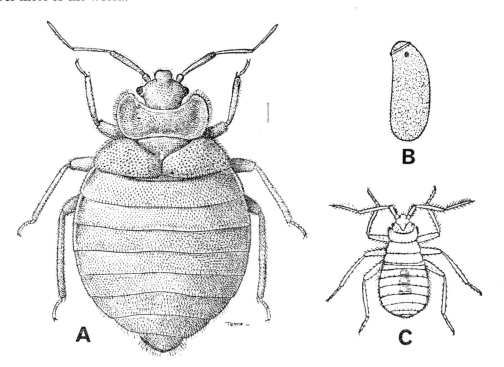

Fig. 175   *Cimex lectularius*, the common bed-bug: A, dorsal view of male adult; B, egg; C, newly hatched nymph (not all to same scale).

The two bed-bugs of importance are the *Cimex* species. Both are very similar in general appearance but they can be easily separated by the examination of certain details (figs. 176, 177). In *Cimex hemipterus* (fig. 176) or the Tropical bed-bug the prothorax (1st segment of the thorax, the one immediately behind the head) is rather less expanded laterally and the extreme margins are not so flattened as in the common bed-bug or *Cimex lectularius* (fig. 177), the colour is darker, the head is shorter and not so broad and the abdomen not so orbicular. Of these characters the first is probably the most easily appreciated and best to use.

The common bed-bug is now called *Cimex lectularius* Linneaus, but has been included in the genera *Clinocoris* and *Acanthia* in the literature. Similarly the tropical bed-bug, *Cimex hemipterus* Fabricius appears in the literature as *Cimex rotundatus* Signoret, *Clinocoris rotundatus*, and *Acanthia rotundatus*.

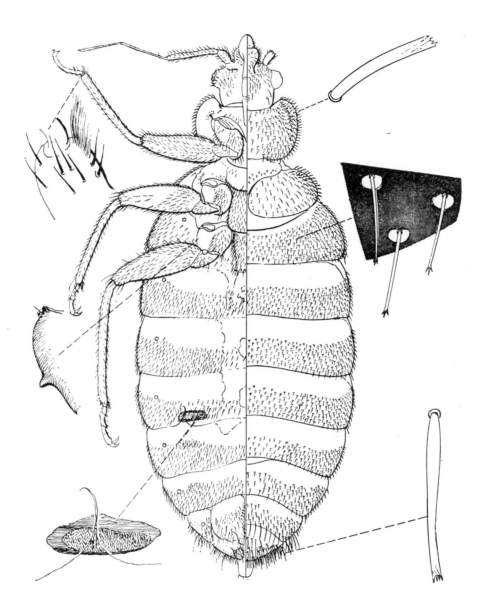

Fig. 176 *Cimex hemipterus*, female, the tropical bed-bug (from Usinger & Ferris, 1960).
CC

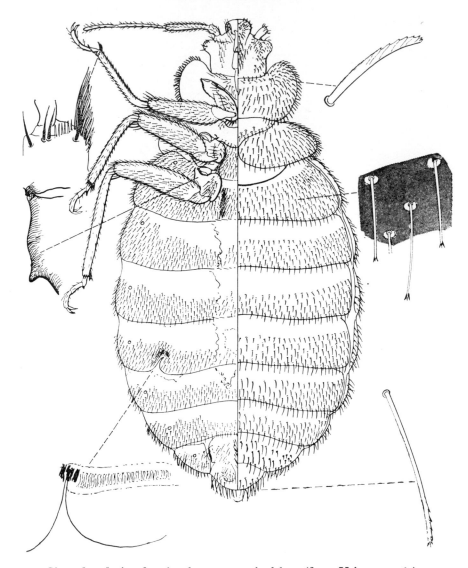

Fig. 177   *Cimex lectularius*, female, the common bed-bug (from Usinger, 1960).

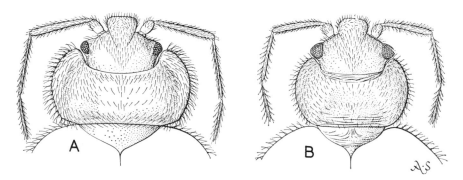

Fig. 178   Bed bugs: A, head and prothorax of *Cimex lectularius*, the common bed-bug; B, head and prothorax of *Cimex hemipterus*, the tropical bed-bug (from Smart).

POLYCTENIDAE

The Polyctenidae are a small family of Hemiptera closely related to Cimicidae. They are permanent ectoparasites of bats. Polyctenidae are provided with ctenidia or one or more combs, rostrum is 3-segmented, antennae and tarsi 4-segmented, eyes wanting, hemclytra short and of uniform consistency and devoid of a membrane. They are viviparous, the embryo remaining in the ovarioles where they gradually mature. The young are born at an advanced stage but differ very considerably from the adults and two postnatal nymphal instars occur.

Nothing of medical importance is known at present about Polyctenidae, but they are included here because of their close relationship with Cimicidae.

ANTHOCORIDAE

Small, elongate-oval, flattened insects in which ocelli are present, rostrum 3-segmented, metathorax with scent-gland opening, tarsi 3-segmented, and hemelytra with cuneus and embolium. The Anthocoridae are related to Cimicidae. They are mostly predacious on other insects but *Lyctocoris campestris* (Fabricius) occurs in association with human habitations (granaries, thatched roofs etc.) and it and some other species (e.g. *Anthocoris kingi* Brumpt and *Anthocoris nemorum* (Linnaeus)) are known to bite man occasionally.

BELOSTOMATIDAE (GIANT WATER-BUG OR TOE-BITERS)

Antennae 4-segmented; posterior legs adapted for swimming, the tibiae flattened and fringed with hairs. Membrane reticulate; abdomen with 2 retractile apical appendages.

Belostomatidae includes the largest members of Hemiptera. In habits they are very rapacious, feeding upon small fish, tadpoles, young frogs and insects. On occasion they may 'bite' unsuspecting bathers in fresh water with their fore-legs, which are modified as raptorial organs for catching their prey and holding it while they suck its juices. This unfortunate habit has earned for them the common name of Toe-Biters. Apart from the immediate pain from the 'bite' not much is known at present about their medical importance. The Toe-Biters are included in the genus *Lethocerus*.

LYGAEIDAE (CHINCH-BUGS, COTTON STAINERS)

Small, dark or brightly coloured forms, ocelli almost always present. Antennae inserted well down on sides of head. Rostrum 4-segmented. Thoracic gland openings present. Membrane with 4–5 veins. Coxae rotary, tarsi 3-segmented; pulvilli present.

Apart from a few Lygaeids which are predacious on mites etc., most of them are plant feeders.   Although most of them are harmless to health, the cotton stainers (*Oxycarenus*) are suspected, at present, to cause industrial disease to the workers in cotton ginning industry.   The dead insects are mixed with the unginned cotton in large numbers. During the process of ginning their bodies are broken into small pieces and also the oil contents are volatilized.   It is presumed that these impurities in the air being breathed by the factory workers might prove to be causing respiratory troubles similar to pollen and dust allergy.

MISCELLANEOUS

Some other plant and water bugs sometimes pierce the skin of man with their mouth-parts (see Usinger, 1934), causing pain and irritation.

The Pentatomid bugs, *Carbula pedalis* Bergroth, *Agonoscelis versicolor* (Fabricius), *Aethemenes chloris* Westwood, *Piezodorus teretipes* Stål and others have been found to enter houses in Nigeria during the dry season causing much annoyance and is some cases were so troublesome that the occupants had to abandon them and in one instance the only remedy was to burn the house to the ground.

APHIDIDAE (HOMOPTERA) (GREENFLY, PLANT LICE)

Aphids sometimes occur in vast numbers and may when fragmented, become inhalant allergens.

BIBLIOGRAPHY

ABALOS, J. W. & WYGODZINSKY, P.   1951.   *Las Triatominae argentinas (Reduviidae, Hemiptera)* [Argentinian Triatominae].   178 pp.   Tucamán.

ALVES, U. P. & NODA, J.   1964.   Os transmissores da doença de Chagas da Região de Sorocaba, Estado de São Paulo, Brasil (Hemiptera, Reduviidae).   *Archos Hig. Saúde públ.* **29**: 14–57.

ARAGÃO, J. M. B., AGUIRRE, G. H., LEAL, J. M. & SERAFIM, E.   1955.   Contribuição ao conhecimento da distribuição geográfica dos tritomíneos domiciliários e seus índices de infecção natural por *Schizotrypanum cruzi*, no Estado da Bahia.   [The geographical distribution and indices of infection with *Trypanosoma cruzi* of triatomids in houses in the State of Bahia.] *Revta bras. Malar. Doenç. Trop.* **7**: 409–421.

ARAGÃO, M. B.   1961.   Aspectos climáticos da doença de Chagas.   II. Area de ocorrência *do Panstrongylus megistus* (Burmeister, 1835).   [Climatic aspects of Chagas's disease. II. The distribution of *Panstrongylus megistus*.]   *Revta bras. Malar. Doenç. Trop.* **13**: 171–193.

ARAGÃO, M. B. & DIAS, E.   1956.   Aspectos climáticos da doença de Chagas.   I. Considerações sôbre a distribuição geográfica do *Triatoma infestans*.   [Climatic aspects of Chagas's disease. I. Observations on the geographical distribution of *T. infestans*.]   *Revta bras. Malar. Doenç. Trop.* **8**: 633–641.

ARNOLD, H. L. Jr. & BELL, D. B.   1944.   Kissing bug bites.   *Hawaii Med. J.* **3**: 121–122.

BARRETTO, M. P.   1967.   Estudos sôbre reservatórios e vectores silvestres do *Trypanosoma cruzi*.   XIX. Inquérito preliminar sôbre triatomíneos silvestres no sudeste do Estado de Goiás, Brasil (Hemiptera, Reduviidae).   [Studies on wild reservoirs and vectors of *T. cruzi*. XIX. Preliminary survey on wild triatomids in the south-east on Goiás State, Brazil. *Revta Inst. Med. Trop. S. Paulo* **9**: 313–320.

BICE, D. E.   1966.   The incidence of *Trypanosoma cruzi* in *Triatoma* of Tucson, Arizona. *Revta Biol. Trop.* **14**: 3–12.

BRASIL, A.   1970.   Consideracões sôbre a evolucão clinica da doença de Chagas.   *Revta Ass. Méd. Bras* (V) **16**: 57–64.

BUSTAMANTE, F. M. 1957. Distribuição geográfica dos transmissores da doença de Chagas no Brasil e sua relação com certos fatôres climáticos. Epidemiologia e profilaxia da enfermidade. [Geographical distribution of the vectors of Chagas's disease in Brazil and its relationship to certain climatic factors. Observations on epidemiology and control.] *Revta bras. Malar. Doenç. Trop.* **9**: 191–211.

CARVALHO, A. G. & BARBOSA, J. A. 1957. Zoogeografia de Triatominae neotropicais (Hemiptera, Reduviidae) [Zoogeography of neotropical Tritominae]. *Revta Goiana Med.* **3**: 181–196.

CHAGAS, C. 1909. Nova tripanozomiaze humana. Estudos sobre a morfolojia e o circlo evolutivo do *Schizotrypanum cruzi* n. gen., n. sp., ajenta etiologico de nova entidade morbida do homen. [The new human trypanosomiasis. Studies on the morphology and developmental cycle of *S. cruzi* n. gen., n. sp., the cause of a new disease in man.] *Mems Inst. Oswaldo Cruz* **1**: 159–218. [In Portuguese and German.]

—— 1921. American trypanosomiasis. Study of the parasite and of the transmitting insect. *Proc. Inst. Med. Chicago* **3**: 220–242.

—— 1935. Summula dos conjecimentos actuaes sôbre a trypanosomiasis americana. [Review of present knowledge of American trypanosomiasis.] *Mems Inst. Oswaldo Cruz* **30**: 387–416.

CHINA, W. E., USINGER, R. L. & VILLIERS, A. 1950. On the identity of *Heterocleptes* Villiers, 1948 and *Hydrobatodes* China & Usinger, 1949. *Rev. Zool. Bot. Afr.* **43**: 336–344.

COURA, J. R. & PETANA, W. B. 1967. American trypanosomiasis in British Honduras. II. The prevalance of Chagas's disease in Cayo District. *Ann. trop. Med. Parasit.* **61**: 244–249.

COVA GARCIA, P. & SUAREZ, M. A. 1959. Estudio de los triatominos en Venezuela. [Study of tritomid bugs in Venezuela.] *Publnes Div. Malar. Caracas*, No. **11**, 209 pp.

DAVIS, N. T. 1966. Contribution to the morphology and phylogeny of the Reduvioidea (Hemiptera: Heteroptera). Part III. The male and female genitalia. *Ann. ent. Soc. Am.* **59**: 911–924.

DEL PONTE, E. 1961. Los propagadores de la enfernedad de Chagas-Mazza. [The vectors of Chagas's disease.] *Revta Fac. Med. Tucumán* **3**: 282–290.

DUNN, L. H. 1934. Attempts to transmit *Trypanosoma cruzi* Chagas with ticks of the genus *Ornithodoros*. *Am. J. Trop. Med.* **14**: 283–289.

—— 1934. Notes on the reduviid bug *Eratyrus cuspidatus* Stål, naturally infected with *Trypanosoma cruzi* Chagas, found in Panama. *Am. J. trop. Med.* **14**: 291–292.

FRAZIER, C. A. 1970. *Insect Allergy.* 493 pp. St. Louis.

GALVÃO, A. B. 1955. Chave para as espécies brasileiras do gênero *Triatoma* Laporte, 1833 (Hemiptera, Reduviidae) [Key to the Brazilian species of *Triatoma*]. *Revta bras. Malar. Doenç. Trop.* **7**: 343–345.

—— 1956. Chave ilustrada para adultos das espécies brasileiras do gênero *Panstrongylus* Berg, 1879. [Illustrated key to the adults of Brazilian species of *Panstrongylus*.] *Revta bras. Malar. Doenç. Trop.* **8**: 421–432.

GALVÃO, A. B., MELLO, L. R., FERREIRA NETO, J. A. & LEAL, H. 1961. Sôbre a distribuição geográfica e infecção natural do *Rhodnius domesticus* Neiva & Pinto, 1923. [The geographical distribution and natural infection of *R. domesticus*.] *Revta bras. Malar. Doenç. Trop.* **13**: 57–60.

HALL, R. P. 1953. *Protozoology.* 682 pp. New York.

HERNANDES DE PAREDES, C. & PARADES-MANRIQUE, R. 1949. Un caso de infección humana por *T. rangeli.* [A case of human infection by *Trypanosoma rangeli.*] *Revta Fac. Med. Univ. Colomb.* **18**: 343–375.

IMMS, A. D. 1957 (reprinted 1964). *A general textbook of entomology*, 9th edn., revised by O. W. Richards, and R. G. Davies. 886 pp. London.

JAENNEL, R. 1919. Insectes Hémiptères. iii. Henicocephalidae et Reduviidae. Voy. Alluaud et Jaennel. pp. 133–313. Paris.

KAGAN, I. G., NORMAN, L. & ALLAIN, D. 1966. Studies on *Trypanosoma cruzi* isolated in the United States: a review. *Revta Biol. trop.* **14**: 55–73. [RAE '68.]

KALSHOVEN, L. G. E. 1970. Observations on the blood-sucking Reduviid *Triatoma rubrofasciata* (De Geer) in Java. *Ent. Ber., Amst.* **30**: 41–47.

KITSELMAN, C. M. & GRUNDMANN, A. W.   1940.   Equine encephalomyelitis virus isolated from naturally infected *Triatoma sanguisuga* (Le Conte).   *Tech. Bull. Kansas Agric. exp. Sta.* **50**: 1–15.

LEON, J. R.   1949.   El *Trypanosoma rangeli* observado en seres humanos en Guatemala [*T. rangeli* in human beings in Guatemala].   *Publnes Inst. Invest. Cient.* No. **4**, 19 pp.

LÉPINE, P., MATHIS, M. & SAUTER, V.   1941.   Infestation expérimentale de *Triatoma infestans* par le virus de l'encephalomyélite equine, type Venezuala.   *Bull. Soc. Path. exot.* **34**: 115.

LEWIS D. J.   1958.   Hemiptera of Medical Interest in the Sudan Republic.   *Proc. R. ent. Soc. Lond.* (A) **33**: 43–47.

LUCENA, D. T. DE.   1965.   Intraducao do *Triatoma infestans* (Klug, 1834) em Pernambuco. *Revta bras. Malar. Doenç. trop.* **17**: 407–414.

—— 1970.   Estudos sõbre a doenca de Chagas no Nordeste do Brasil.   *Revta bras. Malar. Doenç. trop.* **22**: 3–173.

MACHADO, H. & PINTO, O. S.   1952.   Contribuição ao conjecimento da distribuição geográfica dos triatomídeos domiciliários e de seus índices de infecção natural no Estado do Ceará, Brasil.   [Geographical distribution of domestic triatomids and their index of natural infection with *Trypanosoma cruzi* in Ceará State, Brazil.]   *Revta bras. Malar. Doenç. trop.* **4**: 157–170.

MACHADO-ALLISON, C. E. & PEREZ, J. R.   1967.   *Chipos* (Hemiptera, Reduviidae, Triatominae). Cuardernos Cientificos Ser. 1, No. 2, pp. 33 +ills.   Caracas.

MARSDEN, P. D.   1968.   South American trypanosomiasis (Chagas's disease).   *PANS Manual* **14**: 177–188.

McKENNY-HUGHES, A. W. & JOHNSON, C. G.   1942.   *The Bed-bug, its habits and life history and how to deal with it*, 5th edn.   British Museum (Natural History), Economic series No. **5**. 20 pp.   London.

MAZZA, S. & JÖRG, M. E.   1938.   Tercera nota sõbre Triatomidae (Hemipt. Het. Reduvioidea) argentinos [Argentinian Triatomidae].   *Publnes Misión Estud. Patol. reg. argent. Jujuy* No. **36**: 26–58.

MILES, M. A. & ROUSE, J. E.   1970.   Chagas's Disease (South American Trypanosomiasis). A bibliography.   *Trop. Dis. Bull. Suppl.* **1970**.   209 pp.   London.

MILLER, N. C. E.   1956 [reprinted 1971].   *The Biology of the Heteroptera.*   162 pp.   London.

OLIVIER, M. C., OLIVIER, L. J. & SEGAL, D.   1972.   A bibliography of Chagas's Disease (1909–1969).   *Index-cat. med. Vet. Zool. sp. Publs.* **2**: 1–633.

PESSOA, S. B.   1959.   Biologia dos triatomíneos.   [Biology of triatomids.]   *Revta Goiana Med.* **5**: 3–11.

PETANA, W. B.   1971.   American trypanosomiasis in British Honduras.   X: natural habitats and ecology of *Triatoma dimidiata* (Hemiptera, Reduviidae) in the El Cayo and Toledo Districts and the prevalence of infection with *Trypanosoma* (*Schizotrypanum*) *cruzi* in the wild-caught bugs.   *Ann. trop. Med. Parasit.* **65**: 169–178.

PHILIP, C. B.   1942.   Mechanical transmission of rabbit fibroma (shope) by certain Haematophagous bugs.   *J. Parasit.* **28**: 395–398.

PIFANO, C. F.   1954.   Nueva trypanosomiasis humana de la región neotrópica producida por el *Trypanosoma rangeli*, con especial referencia a Venezuela.   [New human trypanosomiasis of the neotropical region produced by *T. rangeli*, with special reference to Venezuela.] *Arch. venez. Patol. trop. Parasit. méd.* **2**: 89–120.

PINTO, C.   1926.   Classificao dos Triatomideos.   *Cienc. med.* **4**(9): 485–490.

—— 1927.   Classification de genres d'Hémiptères de la famille Triatomidae (Reduvidioidea). *Bolm biol. Lab. Parasit. Fac. Med. S. Paulo* fasc. **8**: 103–114 [also in Portuguese: 1927, *Revta méd. S. Paulo* **12**: 271–281].

PIPKIN, A. C.   1968.   Domiciliary reduviid bugs and the epidemiology of Chagas's disease in Panama (Hemiptera: Reduviidae: Triatominae). *J. Med. Ent.* **5**: 107–124.

RODENWALDT, E.   1959.   Geographie der Chagas-Krankheit.   *Ztschr. Tropenmed. Parasit.* **10**: 1–5.

ROMANA, C.   1961.   Epidemiologia y distribucion geografica de la Enfermedad de Chachgas. *Bol. oficina. San. Panamer.* **51**: 390–403.

RYCKMAN, R. E. 1971. The genus *Paratriatoma* in western North America (Hemiptera: Reduviidae). *J. Med. Ent.* **8**(1): 87.

RYCKMAN, R. E., FOLKES, D. L., OLSEN, L. E., ROBB, P. L. & RYCKMAN, A. E. 1965. Epizootology of *Trypanosoma cruzi* in south-western North America. Parts I–VII. *J. Med. Ent.* **2**: 87–108.

RYCKMAN, R. E. & RYCKMAN, A. E. 1966. Reduviid bugs. In Smith, C. N. Ed. *Insect colonization and mass production.* pp. 183–200. New York.

SERVICE, M. W. 1961. The problem of stink bugs in Northern Nigeria. *Nigerian Field* **26**: 183–188.

SWEZEY, R. L. 1963. 'Kissing bug' bite in Los Angeles. *Archs intern. Med.* **112**: 977–980.

TOBIE, E. J. 1965. Biological factors influencing transmission of *Trypanosoma rangeli* by *Rhodnius prolixus*. *J. Parasit.* **51**: 837–841.

USINGER, R. L. 1934. Bloodsucking among phytophagous Hemiptera. *Can. Ent.* **66**: 97–100.

—— 1944. The Triatominae of North and Central America and the West Indies and their public health significance. Public Health Bulletin No. **288**, 81 pp. Washington, D.C.

—— 1966. *Monograph of Cimicidae* (Hemiptera-Heteroptera). 585 pp. Baltimore.

USINGER, R. L., WYGODZINSKY, P. & RYCKMAN, R. E. 1966. The biosystematics of Triatominae. *A.-Rev. Ent.* **11**: 309–330.

VENKATACHALAM, P. S. & BELAVADI, B. 1962. Loss of haemoglobin iron due to excessive biting by bed-bugs. *Trans. R. Soc. trop. Med. Hyg.* **56**: 218–221.

VIOLLE, H. & SAUTER, J. 1938. Transmission du typhus murin par un hémiptère brésilien (*Triatoma infestans*). *C. R. Soc. Biol. Paris* **127**(13): 1276–1278.

WOOD, F. D. & WOOD, S. F. 1938. On the distribution of *Trypanosoma cruzi* Chagas in the Southwestern United States. *Am. J. trop. Med.* **18**: 207–212.

WOOD, S. F. 1942. Observations on vectors of Chagas's disease in the United States. 1. California. *Bull. Calif. Acad. Sci.* **41** (2 pt): 61–69.

—— 1951. Importance of feeding and defecation times of insect vectors in transmission of Chagas's disease. *J. econ. Ent.* **44**: 52–54.

WOOD, S. F. & WOOD, F. D. 1961. Observations on vectors of Chagas's disease in the United States. III. New Mexico. *Am. J. trop. Med. Hyg.* **10**: 155–165.

WOODY, N. C. & WOODY, H. B. 1955. American trypanosomiasis (Chagas's disease), first indigenous case in the United States. *J.A.M.A.* **159**: 676–677.

WORLD HEALTH ORGANIZATION. 1960. Chagas's disease. Report of a study group, Washington, D.C., 7–11 March, 1960. *Wld Hlth Org. Techn. Rep. Ser.*, No. **202**, 21 pp.

—— 1969. Trypanosomiasis in Africa and America, 1959–1968. *Wld Hlth Stat. Rep.* **22**: 635–703.

9. PHTHIRAPTERA
(Lice)
by Theresa Clay

THE human-louse, *Pediculus humanus* Linn. (fig. 179A), is found in two forms, the head-louse and the body-louse which can be considered as unstable environmental subspecies of the one species. The morphological differences between them are slight and variable and many specimens cannot be assigned to one or other of the subspecies. The pubic-louse (fig. 180) is a different species, *Phthirus pubis* (Linn.).*

These species are sucking-lice (Anoplura) which have mouth-parts adapted for piercing the skin and sucking blood; they spend their whole life history from egg to

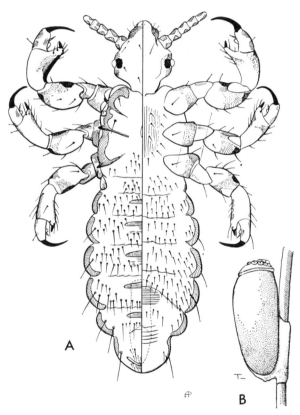

A   B

Fig. 179 *Pediculus humanus* f. *capitis*, the head-louse: A, (adult ×35) (after Ferris); B, egg attached to human hair (×60).

*Correct name of head-louse: *Pediculus humanus capitis* De Geer (synonyms: *americanus, angustus, cervicalis, maculatus, pubescens*). Body-louse: *Pediculus humanus humanus* Linn. (synonyms: *albidior, chinensis, corporis, marginatus, nitritarum, nigrescens, tabescentium, vestimenti*). Pubic-louse: *Phthirus pubis* Linn. (synonyms: *Pediculus pubis, Phthirus inguinalis*).

395

adult on the host, away from which they cannot survive for more than a short time. Other species of sucking-lice are found on mammals, and the pig-louse (*Haematopinus suis* Linn.), has been reported as living on humans, but such infestations are rare and unlikely to persist for any length of time.   In addition most birds and some mammals are parasitized by biting- or chewing-lice (Mallophaga) which feed on feathers and tissue fluids and may temporarily be found on a person handling infested animals.

The head-louse is not confined to the head but may also be found on hairs in other parts of the body, while the body-louse is found only on the body or attached to clothing in contact with the body.   These lice vary in colour from greyish to brownish and in length from 2.00–3.25 mm in the males and 2.40–4.20 mm in the females.   The eggs (or 'nits') of both lice are similar (fig. 179B), those of the head-louse being attached to hairs, and those of the body-louse usually to clothing.   The nymphs which hatch from the eggs resemble the adults in general appearance and also suck blood.   It takes two to four weeks, depending on temperature, for the eggs to hatch, the nymphs to pass through three stages and the adults to reach sexual maturity.

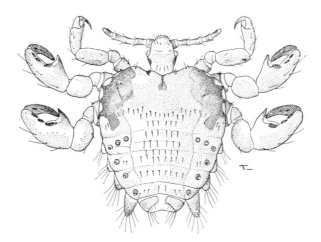

Fig. 180   *Phthirus pubis*, the pubic-louse or 'crab' (×30) (from Smart).

The human-louse, which is cosmopolitan in its distribution, is of great importance as the vector of Exanthematous Typhus Fever (caused by *Rickettsia prowazeki*) and louse-borne Relapsing Fever (caused by the Spirochaete *Borellia* (= *Spirochaeta*) *recurrentis*).

Lice are spread by direct contact or by some article of clothing or bedding recently used by a lousy person; they tend to leave the body when it is suffering from high fever, probably an important factor in the spread of louse-borne diseases.   Although infestations of two or more thousand lice are known, these are rare, most infestations being less than a hundred and many chronically lousy people have only a dozen or so lice. The presence of lice on the body is called *pediculosis* and persons habitually carrying heavy infestations develop hard pigmented skin, a condition known as 'vagabond's disease' (*morphus errorum*).   A number of common names have been applied to the human-louse: grey-backs, cooties, vermin and among troops during the Second World War the popular name was 'mechanized dandruff'.

The pubic or crab-louse (fig. 180) can easily be distinguished from the body-louse by its size (1.25–2.00 mm) and characteristic appearance. These lice cause infestations known as *pediculosus pubis* or *phthiriasis* in which they are found attached to hairs in the pubic and perianal regions; other parts of the body may also be infested including armpits, beard, eyebrows and eyelashes, but the scalp only rarely. The egg, which is attached to a hair, is similar to that of the body-louse but smaller. The pubic-louse, cosmopolitan in distribution, is not known to transmit any disease.

BIBLIOGRAPHY

ANON. 1969. *Lice.* Econ. Ser. British Museum (Nat. Hist.), No. **2A.**, 23 pp. London.

ANON. 1969. *Recommended Methods for Vector Control.* Appendix to 17th Rept. Expert Committee on Insecticides. W.H.O., Geneva. (Up-to-date information on control.)

BUSVINE, J. R. 1966. *Insects and hygiene.* 467 pp. London.

BUXTON, P. A. 1947. *The louse.* 164 pp. London.

CUSHING, E. C. 1957. *History of entomology in World War II.* 117 pp. Washington, D.C. [Good account of wartime typhus epidemic in Naples.]

FERRIS, G. F. 1951. The sucking lice. *Pacif. cst Ent. Soc. Mem.* 320 pp. San Francisco.

GIROUD, P. 1966. Le typhus épidémique peut-il se conserver en dehors du pou de l'homme son véritable reservoir? *Bull. Wld Hlth Org.* **35**: 119–122.

HOPKINS, G. H. E. 1949. The host-associations of the lice of mammals. *Proc. zool. Soc. Lond.* **119**: 387–604.

PHILIP, C. B. 1965. Epidemic typhus: preliminary evidence of *Rickettsia* zoonoses in Latin America. Res. Activities PAHO in selected fields, 1964–1965, Res. a/2A: 63–68.

PHILIP, C. B. & IMAM, I. Z. E. 1967. New concepts in the epidemiology of typhus fever. *In:* First International Conference on Vaccines against Viral and Rickettsial Diseases of Man. PAHO Scientific Publication No. **147**. pp. 517–522.

SPARROW, H. 1958. Etude du foyer éthiopien de fièvre recurrente. *Bull. Wld Hlth Org.* **19**: 673–710.

WEYER, F. 1960. Biological relationships between lice and microbiological agents. *A. Rev. Ent.* **5**: 405–420.

ZINSSER, H. 1935. *Rats, lice and history.* 301 pp. London.

10. DICTYOPTERA
(Cockroaches and Praying Mantises)
by D. R. Ragge

THE only members of this order that are of any medical importance are the cockroaches (sub-order Blattodea) associated with man.   These rather flattened, sometimes beetle-like, rapidly moving insects are familiar to most people; they may be easily distinguished from beetles by their very flexible, thread-like antennae.   The species associated with man attack stored food and infest premises used for storing, preparing and cooking food, such as bakehouses and kitchens, as well as sewers and rubbish dumps.   They are known to carry pathogenic viruses, bacteria and helminths, and to act as intermediate hosts for such pathogens as the nematode *Gongylonema pulchrum* Molin (gullet worm) and the acanthocephalid *Moniliformis moniliformis* Bremser; they are also capable of causing allergic dermatitis.   More than a dozen species have some degree of medical importance, but the following six species, all with world-wide distributions, are the principal vectors (figs 181–185): the common cockroach ('Blackbeetle') (*Blatta orientalis* L.), the American cockroach (*Periplaneta americana* (L.)), the Australian cockroach (*P. australasiae* (Fab.)), the German cockroach ('Steamfly', 'Shiner', 'Croton Bug') (*Blattella germanica* (L.)), the brown-banded cockroach (*Supella supellectilium** (Serville)) and the Madeira cockroach (*Leucophaea maderae* (Fab.)).

In view of the recent appearance of a number of important works on the biology of cockroaches (all listed in the bibliography), it has been thought inappropriate to give more than a very brief account here.   For comprehensive information on the medical and economic importance of cockroaches, readers are referred in particular to the excellent reviews of Roth and Willis (1957, 1960).

KEY TO THE ADULTS OF SIX MEDICALLY IMPORTANT SPECIES OF COCKROACH

1 Fore wings well developed, reaching at least the tip of the abdomen . . . 2
– Fore wings absent or poorly developed, not reaching the tip of the abdomen . . 6
2 Total length (to the tips of the fore wings) more than 18 mm. 3
– Total length (to the tips of the fore wings) less than 17 mm. 5
3 General colour greyish-brown, the pronotum and fore wings patterned as in fig. 185
Leucophaea maderae (**Fab.**)
– General colour reddish-brown to dark brown, the fore wings not patterned as in fig. 185 4
4 Fore wings with a pale yellow stripe along the basal part of the anterior margin (fig. 182A) . . . . *Periplaneta australasiae* (**Fab.**)

---

* This name has been shown by Princis (1960, *Atti Soc. ital. Sci. nat.* **99**: 193) to be a junior synonym of *Supella longipalpa* (Fab.); it has been thought better, however, to retain the well-known name in this book.

– Fore wings without a pale yellow stripe along the basal part of the anterior margin (fig. 181D) . . . . ***Periplaneta americana*** (L.)

5 Pronotum with two conspicuous longitudinal dark bands. Fore wings uniform in colour . . . . ***Blattella germanica*** (L.)

– Pronotum without dark bands (brown with translucent lateral margins). Fore wings dark basally and pale distally in the male, and dark with pale bands in the female

Supella supellectilium (Serville)

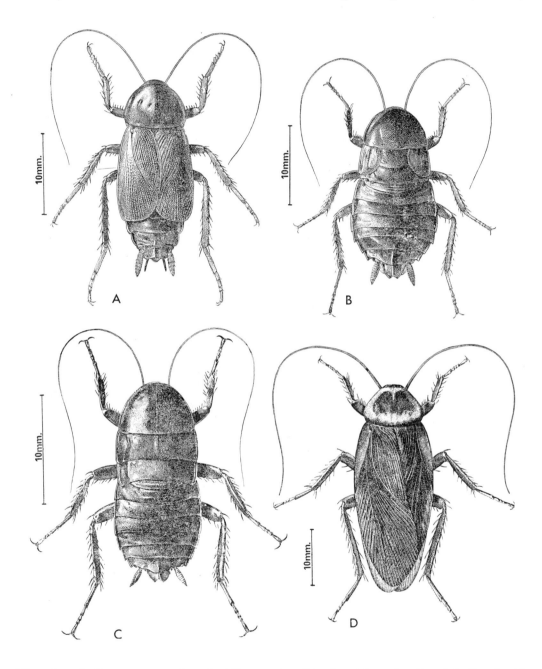

Fig. 181 A–C, common cockroach (*Blatta orientalis*): A, adult male; B, adult female; C, nymph; D, American cockroach (*Periplaneta americana*), adult male.

6 Legs reddish-brown to dark brown. Pronotum uniformly opaque. Total length more than 15 mm. . . . . *Blatta orientalis* **L.**

– Legs straw-coloured. Pronotum with translucent lateral margins. Total length less than 14 mm. . . . . . *Sufella supellectilium* **(Serville)**

Young cockroaches are similar to the adults but lack wings (figs 181C, 182C).

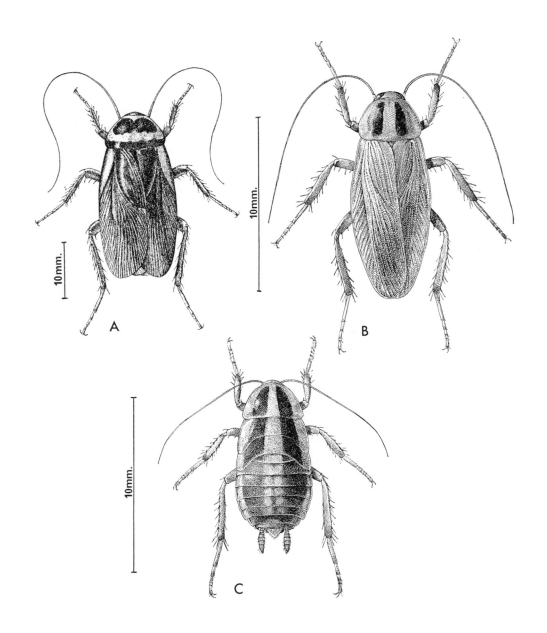

Fig. 182 A, Australian cockroach (*Periplaneta australasiae*), adult female; B–C, German cockroach (*Blattella germanica*); B, adult male; C, nymph.

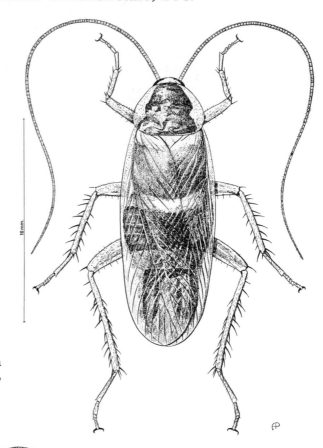

Fig. 183 Brown-banded cockroach
(*Supella   supellectilium*),
adult male.

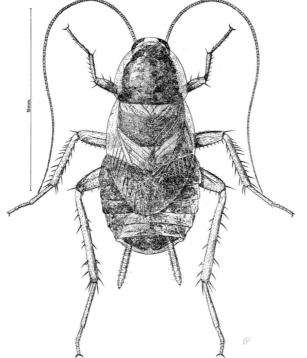

Fig. 184 Brown-banded cockroach
(*Supella   supellectilium*),
adult female.

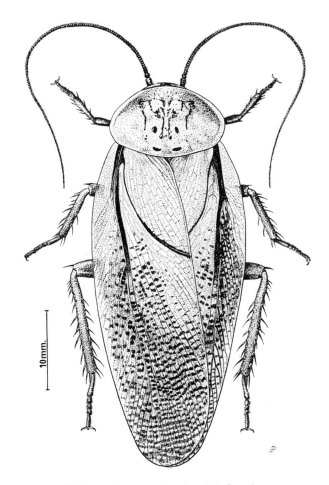

Fig. 185 Madeira cockroach (*Leucophaea maderae*), adult female.

BIBLIOGRAPHY

BURGESS, N. R. H., MCDERMOTT, S. N. & WHITING, J. *In press.* The vector status of cockroaches, initial results using *Blatta orientalis*. *Trans. R. Soc. trop. Med. Hyg.*

CAMERON, E. 1961. *The cockroach (Periplaneta americana L.).* 111 pp. London.

CORNWELL, B. P. 1968. *The cockroach.* Vol. I. 391 pp. London.

GANGWERE, S. K. 1967. Biting in Orthoptera and their allies. *Eos Madr.* **42**: 363–382.

GUTHRIE, D. M. & TINDALL, A. R. 1968. *The biology of the cockroach.* 408 pp. London.

MINISTRY OF AGRICULTURE, FISHERIES & FOOD. Advisory Leaflet No. **383**, Cockroaches. 7 pp.

RAGGE, D. R. 1965. *Grasshoppers, crickets and cockroaches of the British Isles.* 299 pp. London.

ROTH, L. M. & WILLIS, E. R. 1957. The medical and veterinary importance of cockroaches. *Smithson. misc. Collns* **134**: 147 pp.

ROTH, L. M. & WILLIS, E. R. 1960. The biotic associations of cockroaches. *Smithson. misc. Collns* **141**: 470 pp.

DD

11. LEPIDOPTERA
(Butterflies and Moths)
by Kenneth G. V. Smith

SOME moth and butterfly caterpillars, especially Arctiidae, Lymantriidae, Saturnidae and Nymphalidae possess 'nettling' or urticating hairs which can cause dermatitis of varying intensity (pl. 9B).   These hairs are usually connected to a gland which releases poison into the skin *via* punctures made by the hairs.   The poison has the property of liberating histamine (Vallette & Huidobro, 1954; Valle *et al.*, 1954).   The intensity of the irritation varies with the species of caterpillar and the sensitivity of the victim, but the symptoms are usually transitory.   The irritation is more severe when the hairs reach a mucous membrane and in the eye of man may give rise to nodular conjunctivitis (see Watson & Sevel, 1966).   An unidentified Saturniid larva from Venezuela (pl. 9B) injects a powerful anticoagulant if handled or even brushed against and serious bleeding results (Arocha-Piñango & Layrisse, 1969; Marsh & Arocha-Piñango, 1971).

If inhaled caterpillar hairs may cause dyspnoea and if ingested may give rise to erucic stomatitis.   A recent pharmacological analysis of extracts from these urticating hairs has been carried out by Frazer (1965).

Urticating hairs may also become attached to the cocoon when the larva pupates and may sometimes become attached to the adult moth.   The hairs of some species may retain their urticating properties long after being shed by the caterpillar and persons working with entomological collections may thus be affected.

Some caterpillars eject secretions when handled and these may cause transitory symptoms such as irritation or inflammation of the skin.   A useful list of urticating Lepidoptera is included in James & Harwood (1969) with a survey of the literature. A list of other papers on these topics is given in the references.

Sometimes Lepidopterous larvae may accidentally occur in the gut of man and such infestation is known as scholechiasis.   The larvae may be ingested with vegetable and salads or stored food products.   These cases are extremely rare and there are few reliable data on the subject.

Moths of the families Pyralidae (pl. 8), Geometridae and Noctuidae include species that frequent the eyes of various mammals where they feed upon eye discharges. Bänziger (1966) recorded the first association of these moths with man in South-East Asia.   Bänziger & Buticker (1969) found that of 20 species recorded from mammals in Thailand, 6 troubled human beings.   Slight pain is caused by this feeding habit and the eye may become inflamed.   Although these eye-frequenting moths have recently been shown to be incapable of penetrating live tissue and thus sucking blood, they may well be vectors of mammalian epidemic keratoconjunctivitis and other eye diseases.

The Noctuid moth *Calyptra eustrigata* (Hmps.) has been shown (Bänziger, 1968) to pierce the skin of mammals, including man, and is therefore a potential vector of disease (pl. 9).

The rat tapeworm *Hymenolepis diminuta* Rudolphi which occasionally affects man is known to have the meal-moth *Pyralis farinalis* (L.) as an intermediate host.

BIBLIOGRAPHY

AITKEN, AUDREY D. 1963. A key to the larvae of some species of Phycitinae (Lepidoptera, Pyralidae) associated with Stored products, and some related species. *Bull. ent. Res.* **54**: 175–188, figs 1–24.

ALLARD, H. F. & ALLARD, H. A. 1958. Venomous moths and butterflies. *J. Wash. Acad. Sci.* **48**: 18–21.

AROCHA-PIÑANGO, C. L. & LAYRISSE, M. 1969. Fibrinolysis produced by contact with a caterpillar. *Lancet*, 19 April **1969**: 810–812.

BÄNZIGER, H. 1966. First records of eye-frequenting Lepidoptera from Man (preliminary communication). [Unpublished World Health Organization document *WHO/EBL/66.81*, 13 pp. Geneva.]

—— 1968. Preliminary observations on a skin-piercing blood-sucking moth (*Calyptra eustrigata* (Hmps.) (Lep., Noctuidae)) in Malaya. *Bull. Ent. Res.* **58**: 159–163.

—— 1970. The piercing mechanism of the fruit-piercing moth *Calpe* [*Calyptra*] *thalictri* Bkh. (Noctuidae) with reference to the skin-piercing blood-sucking moth *C. eustrigata* Hmps. *Acta tropica* **27**: 54–88.

—— 1971. Bloodsucking moths of Malaya. *Fauna.* **1**: 5–16.

BÄNZIGER, H. & BUTICKER, W. 1969. Records of eye-frequenting Lepidoptera from Man. *J. Med. Ent.* **6**: 53–58.

BEARD, R. L. 1963. Insect toxins and venoms. *A. Rev. Ent.* **8**: 1–18.

BERCOWITZ, S. 1945. Caterpillar dermatitis. *U.S. Army Med. Dept.* **4**: 464–467.

BISHOPP, F. C. 1923. The puss caterpillar and the effects of its sting on man. *Dep. Bull. U.S. Dep. Agric.* No. **288**.

BÜCHERL, W. & BUCKLEY, E. E. 1972. *Venomous animals and their venoms.* 3. *Venomous invertebrates.* 560 pp. London.

CAFFREY, D. J. 1918. Notes on the poisonous urticating spines of *Hemileuca oliviae* larvae. *J. econ. Ent.* **11**: 363–367.

CHEVERTON, R. L. 1936. Irritation caused by the processionary caterpillar. *Trans. R. Soc. trop. Med. Hyg.* **29**: 555–557.

CLEMENTS, A. N. 1951. On the urticating properties of adult Lymantridae. *Proc. R. ent. Soc. Lond.* (A) **26**: 104–108.

DAVIS, R. A. 1947. Notes on urticating lepidopterous larvae becoming of some local medical importance. *Proc. R. ent. Soc. Lond.* (A) **22**: 3–4.

FAUST, B. E. C. & RUSSELL, P. F. 1964. Craig & Faust's Clinical Parasitology. 1099 pp. Philadelphia. [See Table 13, p. 827.]

FOOT, N. C. 1922. Pathology of the dermatitis caused by *Megalopyge opercularis*, a Texan Caterpillar. *J. Exp. Med.* **35**: 737–753.

FRAZER, J. F. D. 1965. The cause of irritation produced by larval hairs of *Arctia caja* (L.) (Lepidoptera: Arctiidae). *Proc. R. ent. Soc. Lond.* (A) **40**: 96–100.

GILMER, P. M. 1925. A comparative study of the poison apparatus of certain lepidopterous larvae. *Ann. ent. Soc. Am.* **18**: 203–239.

GOETH, H., BRETT, R. & WEIDNER, H. 1968. 'Butterfly itch', eine Schmetterling-sclermatose an bord eines Tankers. *Ztschr. Tropenmed. Parasit.* **18**: 5–16.

GOLDMAN, L., SAWYER, F., LEVINE, A., GOLDMAN, J., GOLDMAN, S. & SPINNANGER, J. 1960. Investigative studies of skin irritations from caterpillars. *J. Invest. Derm.* **34**: 67–79.

GUSMAO, H. H., FORATTINI, O. P. & ROTBERG, A. 1961. Dermatite provacado por lepidópteros. do gênero *Hylesia*. *Revta. Inst. Med. trop. S. Paulo* **3** (3): 114–120 [with English summary]

HILL, W. R., RUBENSTEIN, A. D. & KOVACS, J. 1948. *J. Am. Med. Ass.* **138** (10): 737–740.

JAMES, M. T. & HARWOOD, R. F. 1969. Herms's '*Medical Entomology*' 6th edn. London.

KATZENNELLENBOGEN, I. 1955. *Dermatologica* **3**: 99

KEEGAN, H. L. 1963. Caterpillars and moths as public health problems *In* '*Venomous and poisonous animals of the Pacific Region*', ed. Keegan, H. L.

KEMPER, H. 1958. Experimentelle untersuchungen über die Wirkung von Raupenhaaren auf die Menschliche Haut. *Proc. 10th Internat. Cong. Entomol.* **3**: 719–723.

KEPHART, C. F. 1914. The poison glands of the larva of the brown-tail moth. *J. Parasit.* **1**: 95–103.

KNIGHT, H. H. 1922. Observations on the poisonous nature of the white-marked tussock moth. *J. Parasit.* **8**: 133–135.

LUCAS, T. L. 1942. Poisoning by *Megalopyge opercularis*. *J. Amer. Med. Ass.* **119**: 877–880.

MARSH, N. A. & AROCHA-PIÑANGO, C. L. 1971. Observations on a Saturniid moth caterpillar causing severe bleeding in man. *Proc. R. ent. Soc. Lond.* **36** (2): 9–10.

McGOVERN, J. P., et al. 1961. *Megalopyge opercularis;* observations on its life history, natural history of its sting in man a report of an epidemic. *J. Amer. Med. Ass.* **175**: 1155.

McMILLAN, C. W. & PURCELL, W. R. 1964. Hazards to health: the puss caterpillar. *New Engl. J. Med.* **271**: 147–149.

MICKS, D. W. 1952. Clinical effects of the sting of the puss caterpillar on man. *Texas Reports on Biol. & Med.* **10**: 399–405.

MILLS, R. G. 1923. Observations on a species of cases of dermatitis caused by a Liparid moth, *Euproctis flava* Bremer. *China Med. Jour.* **37**: 351–371.

PERRY, H. M. & POOLE, L. T. 1935. A common caterpillar injurious to man. *Jl. R. Army med. Cps* **65**: 217.

RANDEL, H. W. & DOAN, G. B. 1956. Caterpillar urticaria in the Panama Canal Zone in Buckley, E. E. & Porges, N. (eds.) 'Venoms'. *Amer. Assoc. Adv. Sci.* **44**: 111–116.

TYZZER, E. E. 1907. The pathology of the brown-tail moth dermatitis. *J. med. Res.* **16**: 43–64.

VALETTE, G. & HUIDOBRO, H. 1954–1956. Pouvoir histaminotiberateur du venim de la chenille processionaire du pin (*Thaumatopoea pityocampa* Schiff). *C. r. Séanc. Soc. biol.* Paris **148**: 1605–1607, **150**: 658–661.

VALLE, J. R., PICARELLI, Z. P. & PRADO, J. L. 1954. Histamine content and pharmacological properties of crude extracts from setae of urticating caterpillars. *Arch. int. Pharmacodyn.* **98**: 324–334.

WATSON, P. G. & SEVEL, D. 1966. Ophthalmia nodosa. *Brit. J. ophthal.* **50**: 209–217.

WEIDNER, H. 1936. Beitrage zu einer Monographie der Raupen mit Gifthaaren. *Z. angew. Ent.* **23**: 432–484.

12. HYMENOPTERA
(Ants, Bees and Wasps)
by I. H. H. Yarrow

A GREAT many of the members of this order can inflict a severe sting and, in the case of some ants, painful bites. The stinging apparatus is the modified ovipositor, which is used to inject venom into the victim's flesh. Wasps and ants extract their stings after use and though honey bees try to do so they rarely succeed as the barbs on the lancets hold the sting so firmly in place that the bee must literally tear itself away to escape. It must, however, be remembered that there are many Hymenoptera which, though furnished with long 'tails' and wearing a frightening wasp-like aspect, are yet quite harmless. The 'tail' is, in these cases, a greatly produced ovipositor by means of which the egg is placed in some otherwise inaccessible situation; it is not a sting, though some Ichneumonidae (e.g. genera *Ophion*, *Netelia* (fig. 187C) and *Pimpla instigator*) can use it as such.

The formic acid-squirting ants are of little importance but as they usually bite before squirting their attacks can be quite painful; the stinging ants (Dorylinae, Ponerinae, Myrmicinae) can produce some very unpleasant effects, even when only quite tiny species are involved and multiple stingings may induce anaphylactic response when larger and more powerful species are concerned. Investigations into the chemistry of ant venoms has shown that there is considerable variation throughout this large group of insects and that some of them are certainly proteinaceous. Some ants have been suspected or proved to be unwitting carriers of disease and their attraction to filthy substances as well as to human food makes them undesirable visitors to kitchens and food stores. Ants have bitten coal-miners working 1900 feet below ground.

For most people, to be stung by bee or wasp is of little significance, being accompanied by short-lived local swelling and itching easily soothed by cooling with ice-pack, cold water, blue bag, diluted vinegar, Eau de Cologne or antihistamine ointment. In cases of multiple stinging from an enraged bee hive or wasps' nest or even a single sting inside the mouth, medical assistance should be sought immediately. Some people, usually bee keepers, become desensitized to bee stings but others may react more and more violently until they become so hypersensitive that their lives are in serious danger. For the violently allergic, a desensitization process has been developed consisting of a two month-long series of daily injections of gradually increasing dosages of bee or wasp venom in an extract of whole insect, followed by monthly 'booster' injections; the results of this process, though temporarily satisfactory, are not permanent and may have to be repeated each year for the rest of the sufferer's life. For this reason deliberate desensitization is not recommended except for those who lose consciousness within five minutes of being stung. Persons who may be approaching such an advanced stage of hypersensitivity are advised to carry a supply of adrenaline for immediate self injection. Bee and wasp venoms are very complicated substances composed of proteins including enzymes with haemolytic and spreader effects, and histamine which is responsible for the smarting and swelling near the site of the sting and in sensitized subjects for the release of tissue histamine throughout the body causing burning and other sensations far removed from the site of the sting.   In addition there are many other components of bee and wasp venom none of which is formic acid as used to be supposed.   Wasp venom differs from bee venom by containing a much higher proportion of histamine and also 5-hydroxytryptamine (serotonin) as well as other substances not found in bee venom.   Basically, however, bee and wasp venom are alike in that the chief toxic element is a protein supported by the enzymes phospholipase A and hyaluronidase and the more superficial aspects of being stung due to histamine.   It follows that identical first-aid measures may be used for stings from either kind of insect.   It has been found, however, that the constituents of wasp venom vary to some extent according to the type of wasp and desensitization extracts may have to be prepared accordingly.

The technique of desensitization is still in an early stage of development but there are already indications of an advance towards replacement of the tedious multi-injection method mentioned above by a single injection of whole insect extract held in an emulsion from which it is only slowly released.

The bees, wasps and ants referred to above live socially in colonies which they will defend with great vigour and self-sacrifice (*Apis, Bombus, Vespa, Vespula, Polistes*, etc.). The majority of stinging Hymenoptera, however, live as single units, though sometimes gregariously, and many of them, for instance the solitary bees, probably never use their stings at all, while the various solitary wasps use them only in self defence and to provide food for themselves or their larvae.   Humans are rarely stung by such insects and little is known about their venom.   The specific identity of any of the Hymenoptera which inflict stings or bites is seldom of importance except in cases of such extreme allergy that desensitization is necessary.

The stingless social bees of the tropics (*Melipona, Trigona*, etc.) can be very aggressive though the effects of their mass biting (and in some Neotropic species squirts of caustic fluid) are rarely more than exceedingly unpleasant.

The tiny stingless 'sweat-bees' of the genus *Trigona* are much attracted by human perspiration and can be intensely annoying in the African savannas because of the large numbers in which they will hover around the body and alight on the skin, especially of the face and hands.   At times they are an aggravating pest to anyone walking or working in the bush, and are then sometimes confused with Simuliidae because of their slight superficial likeness to *Simulium damnosum* (q.v., Chapter 3b).

The honey of hive-bees, the wild honey-bees of Africa and Asia, bumblebees and the tropical 'stingless' bees may be toxic for a variety of reasons though usually associated with the forage plants and if hive-bees are not provided with drinking water they will join the wild-bees to refresh themselves from any source of liquid, however putrid and may then contaminate their honey with infected matter from sewage effluents etc.

The introduction of the tropical African honeybee into South America ostensibly to boost, through hybridization the honey production of the long-established European bees has proved devastating not only for agriculture, but for man and his domestic animals as well.   Normally aggressive, but with due care and respect manageable in its native Africa, this bee has run riot in Brazil where it was originally introduced in 1956 and from where it has now spread into neighbouring parts of the Continent.   Hybridization has completely swamped the desirable feature of the comparatively docile European bees and in their place has produced a bee whose excessive viciousness has reduced a once valuable and important honey harvest almost to nil. Research aimed at taming this viciousness, which takes the form of mass attack on the slightest provocation, is in progress, but in the meantime reports of the death through multiple stingings of humans and domestic animals who have inadvertently approached a nest or a swarm, if not exaggerated, are extremely alarming.

BIBLIOGRAPHY

BEARD, R. L. 1963. Insect toxins and venoms. *A. Rev. Ent.* **8**: 1–18 [120 references].

BEATSON, S. H. Pharaoh's Ants as pathogen vectors in Hospitals. *Lancet.* Feb. 19th. 1972: 425–427.

BÜCHERL, W. & BUCKLEY, E. E. 1972. *Venomous animals and their venoms.* 3. Venomous invertebrates. 560 pp. London.

FLUNO, J. A. 1961. Wasps as enemies of man. *Bull. ent. Soc. Am.* **7**: 117–119.

FRANKLAND, A. W. 1963. Treatment of bee sting reactions. *Bee World* **44**: 9–12.

FRAZIER, C. A. 1969. *Insect Allergy.* 493 pp. St. Louis [good bibliography].

MARSHALL, T. K. 1957. Wasp and bee stings. *Practitioner* **178**: 712–722.

NOGUEIRO-NETO, P. 1964. The spread of a fierce African bee in Brazil. *Bee World,* **45**: 119–121.

PARRISH, H. M. 1959. Deaths from bites and stings of venomous animals and insects in the United States. *Arch. Int. Med.* **104** (2): 198–207.

PERLMAN, F. 1962. Treatment for severe reactions to the bites and stings of Arthropods. *Med. Times* **9**: 813.

SCRAGG, F. F. R. & SZENT-IVANY, J. J. H. 1965. Fatalities caused by multiple hornet stings in the territory of Papua and New Guinea. *J. Med. Ent.* **2**: 309–313.

VECHT, J. VAN DER. 1957. The Vespinae of the Indo-Malayan and Papuan areas (Hymenoptera, Vespidae). *Zool. Vehr. Leiden.* **34**: 1–83. [pp. 66–69 aggressiveness and effect of stings.]

WEBER, N. 1939. The sting of the ant *Paraponera clavata.* *Science.* **89**: 127–128.

—— 1959. The stings of the Harvesting Ant, *Pogonomyrmex occidentalis* Cresson, with a note on populations (Hymenoptera). *Ent. News.* **70**: 85–90.

WIGGLESWORTH, V. B. 1965. *Principles of Insect physiology,* 6th ed. pp. 546–548. London.

13. COLEOPTERA AND OTHER INSECTS
(Beetles, Mayflies, Caddisflies)
by Kenneth G. V. Smith

INVASION of the living body by beetles or their larvae is called canthariasis and there are numerous cases in the literature, a summary of which is provided by Théodorides (1948–1951). The majority of these cases appear to be accidental, but Senior-White (1920) and Fletcher (1924) record adults of the dung-beetles (Scarabaeidae) *Onthophagus bifasciatus* F., *O. unifasciatus* F. and *Caccobius vulcanus* F. (fig. 186) from the bowels of young children in Ceylon and lower Bengal. Fletcher suggested that these beetles enter the anus when the child is asleep causing diarrhoea and griping. Other cases of intestinal infestation involve species normally found in foodstuffs such as the larvae of larder-beetles (Dermestidae) or meal-worms (Tenebrionidae). It is probable that in these cases eggs or small larvae were ingested with raw or imperfectly cooked food. Very few cases of urino-genital infestation by Coleoptera are on record (Archibald & King, 1919) and only one or two cases of nasal canthariasis have been reported (Ligett 1931; Hurd, 1954).

The larvae of some Dermestidae (fig. 186E) have urticating hairs which penetrate the skin. This stimulates the liberation of histamine in the affected area and various forms of dermatitis result. If inhaled the hairs can cause dyspnoea and dockers have been affected when unloading a cargo heavily infested with Dermestid larvae. If ingested the hairs may give rise to erucic stomatitis.

Some species of the family Meloidae, or blister-beetles, are an important source of cantharidin used medicinally. The best known species is the Spanish-fly, *Lytta vesicatoria* (L.) (fig. 186A). Other species include *Mylabris cichorii* (L.) and *Epicauta hirticornis* (Haag-Rutenberg) of India. The body-fluids of these beetles produce blisters if rubbed on the skin. In the family Oedemeridae, *Sessinia collaris* (Sharp) and *S. decolor* Fairm., known as coconut-beetles in the Gilbert Islands, also cause severe blistering.

Some species of the genus *Paederus* (Staphylinidae) cause urticaria and blistering. Some very small Staphylinidae of the genera *Atheta* and *Oxytelus* fly in numbers during summer and may enter the eye causing a burning sensation and thus be a hazard to motorists and cyclists.

Body-fluids from the larvae of the Chrysomelid beetle *Diamphidia nigroarrata* Stål (=*simplex* Peringuey) are used by South African bushmen as a lethal arrow poison, death resulting from general paralysis.

So many diverse families have been involved in canthariasis that there is little point in providing keys. Adults are easily recognizable, as beetles and the larvae may be distinguished from the larvae or maggots of Diptera by the possession of three pairs of small distinct 'thoracic' legs, each terminating in a pointed claw, and the presence of a distinct head capsule. The larvae of Lepidoptera, while possessing a distinct head capsule and thoracic legs, also have pseudopods on segments 3–6 and 10 with the exception of the Geometridae or 'loopers' which have pseudopods only on segments 6 and 10.   The majority of Diptera do not have a complete head capsule and lack true legs, only a very few possessing fleshy pseudopods.

Some beetles serve as intermediate hosts to various parasites affecting man.   The tapeworm *Hymenolepis diminuta* Rudolphi is common in rats and is occasionally found in man.   Its intermediate hosts are stored products beetles, including *Tenebrio molitor* L.   *Gongylonema pulchrum* Molin, a nematode of rats, develops in various Scarabaeid

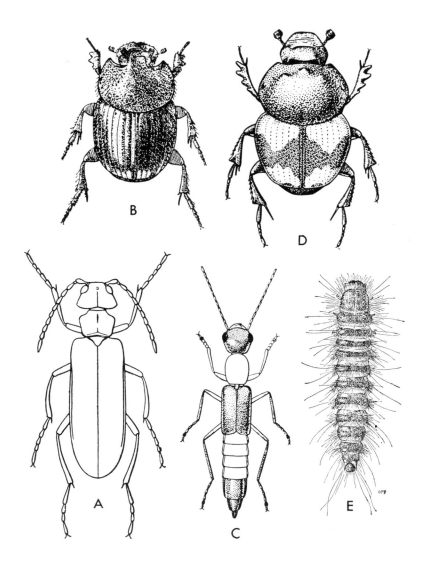

Fig. 186   A, *Lytta vesicatoria*, the Spanish-fly; B, *Caccobius vulvanus* (Scarabaeid); C, *Paederus sabaeus* (Staphylinidae); D, *Onthophagus bifasciatus* (Scarabaeidae); E, larva of *Dermestes lardarius* (Dermestidae).

beetles of the genus *Aphodium* and may on occasion occur in man, where it burrows in the mucous layers of the mouth and upper digestive tract.   The Acanthocephalid worm *Macroacanthorhynchus hirudinaceus* Pallas (=*Echinorhynchus gigas* Bloch) develops in the larvae of cockchafers, occasionally in pigs and rarely in man.   *Moniliformis moniliformis* Bremser, another Acanthocephalid, found in man, has the beetle *Blaps mucronata* Latreille (Tenebrionidae) as in intermediate host.

Some coprophagous and necrophagous beetles such as *Dermestes*, *Attagenus* and *Anthrenus* spread the virulent spores of anthrax (*Bacillus anthracis*).

BIBLIOGRAPHY

ANONYMOUS. 1967. Scarabiasis. *J. trop. Med. Hyg.* **70**: 49–50.

ARCHIBALD, R. G. & KING, H. H.   1919.   A note on the occurrence of a coleopterous larva in the urinary tract of man in the Anglo-Egyptian Sudan.   *Bull. Ent. Res.* **9** (3): 255–256.

BROWN, S. G.   1960.   Report of cantharidin poisoning due to blister beetle.   *Br. Med. J.* **11**: 1290.

FLETCHER, T. B.   1924.   Intestinal Coleoptera.   *Indian Med. Gaz.* **59**: 296–297.

GILGLIOLI, M. E. C.   1965.   Some observations on blister beetles, family Meloidae, in Gambia, West Africa.   *Trans. R. Soc. Trop. Med. Hyg.* **59**: 657–663.

HALL, M. C.   1929.   Arthropods as intermediate hosts of Helminths.   *Smithson. misc. Collns* **81**, no. 15: 1–77.

HINTON, H. E.   1945.   A monograph of the beetles associated with stored products.   Vol. 1 443 pp.   London.

HINTON, H. E. & CORBETT, A. S.   1972.   Common insect pests of stored food products.   *Econ. Ser. Br. Mus. nat. Hist.* No. 15 [under revision].

HURD, P. D.   1954.   'Myasis' resulting from the use of the aspirator method in the collection of insects.   *Science, N.Y.* **119**: 814–815.

LEHMANN, C. F., PIPKIN, J. L. & RESSMAN, A. C.   1955.   Blister beetle dermatosis.   *A.M.A. Archs. Derm.* **71**: 36.

LEWIS, D. J.   1958.   Coleoptera of medical interest in the Sudan Republic.   *Proc. R. ent. Soc. Lond.* (A) **33**: 37–42.

LIGETT, H.   1931.   Parasitic infestations of the nose.   *J. Amer. Med. Assoc.* **96** (19): 1571–1572.

ROBERTS, J. I. & TONKING, H. D.   1935.   Note on East African vesicant beetle, *Paedrus crebipunctatus* Epp.   *Ann. trop. Med. Parasit.* **29**: 415–420.

SENIOR-WHITE, R. A.   1920.   On the occurrence of Coleoptera in the human intestine.   *Ind. J. Med. Res.* **7**: 568–569.

SWARTS, W. B. & WANAMAKER, J. F.   1946.   Skin blisters caused by vesicant beetles.   *J. Am. med. Ass.* **131**: 594.

THÉODORIDES, J.   1948.   Les Coléoptères parasites accidentales de l'homme.   *Annls Parasit. hum. comp.* **23**: 348–363.

—— 1950a.   Les Coléoptères parasites accidentales de l'homme et des animaux domestiques. *Annls. Parasit. hum. comp.* **25**: 69–76.

—— 1950b.   The parasitological, medical and veterinary importance of Coleoptera.   *Acta Trop.* **7**: 48–60.

—— 1951.   Notes sur les Coléopteres d'importance médicale.   *Acta Trop.* **16**: 101–105.

OTHER INSECTS

Mass emergences of some aquatic insects such as Ephemeroptera (mayflies, fig. 187A) and Trichoptera (caddisflies, fig. 187B) result in large numbers of exuviae.   These cast skins are fragmented and wind-borne and become inhalent allergens.   The emerging insects themselves may become a nuisance or hazard if in sufficient numbers (Corbet,

1966, Seshadri, 1955). Allergy to these and other insects is reviewed by Frazier (1970) and other useful references are given below. Some earwigs (Dermaptera) can draw blood with their 'pincers' (Bishopp, 1961); *Anisolabis* carries the Cestode *Hymenolepis diminuta*.

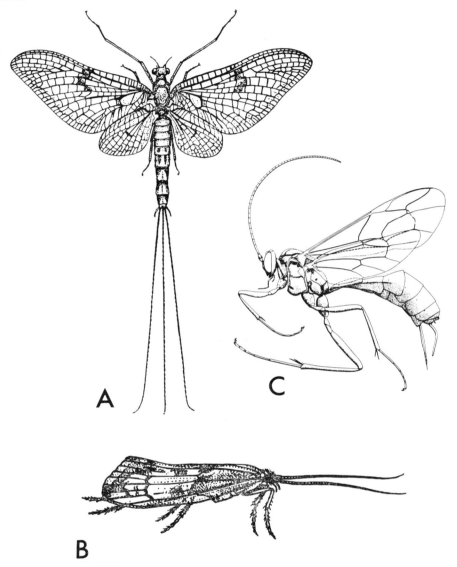

Fig. 187 A, a mayfly (Ephemeroptera); B, a caddisfly (Trichoptera); C, *Netelia*, an ichneumon-fly (Hymenoptera—after Townes).

BIBLIOGRAPHY

BISHOPP, F. C. 1961. Injury to man by earwigs. *Proc. ent. Soc. Wash.* **63**: 114.

CORBET, P. S. 1966. A quantitative method for assessing the nuisance caused by non-biting aquatic insects. *Can. Ent.* **98**: 683–687.

CORBET, P. S., SCHMID, F. & AUGUSTIN, C. L. 1966. The Trichoptera of St. Helen's Island, Montreal. 1. *Can. Ent.* **98**: 1284–1298.

FRAZIER, C. A. 1970. *Insect allergy.* 493 pp. St. Louis.

SESHADRI, A. R. 1955. An extraordinary outbreak of caddis-flies (Trichoptera) in the Meltur-dam township area in Salem district, South India. *South Ind. J. Entom.* **3**: 337–340.

14. ARACHNIDA
(Scorpions, Spiders, Ticks, etc.)
by J. G. Sheals

LIVING members of the phylum Arthropoda include representatives of five well defined subphyla: the Onychophora, Crustacea, Pycnogonida, Insecta-Myriapoda and Chelicerata. The Chelicerata are in turn made up of two classes: Merostomata and Arachnida, and while the merostomes (king-crabs) are aquatic and equipped with external gills, the arachnids, with the exception of certain mites which have become secondarily adapted to life in water, are essentially terrestrial.

Arachnids are sometimes referred to loosely as insects but they differ fundamentally in structure and organization from the true Insecta. For example, antennae, wings and compound eyes are never found, and the body is never divided into the familiar head, thorax and abdomen. In the arachnids the body (see for example figs 188, 190) can be conveniently divided into two regions, the prosoma (or cephalothorax), which is almost invariably unsegmented, and the opisthosoma (or abdomen), which is sometimes segmented. The adult prosoma normally bears six pairs of appendages. The most anterior of these, the chelicerae, are essentially feeding organs. In most of the arachnids the chelicerae are chelate-dentate, that is to say pincer-like, although as a result of adaptation for various methods of feeding they can exhibit a considerable variety in structure, and they may, for example, take the form of curved claws or long piercing stylets. The basal segments of the second pair of appendages, the pedipalps, may perform a masticatory function, while the distal segments are usually tactile although they are often modified to form raptorial organs. The pedipalps may also be used for cleaning the chelicerae after feeding, and in some mites for example, certain distal segments carry inwardly directed comb-like setae for this purpose. The remaining four pairs of appendages generally serve as walking legs, although the first pair, and less frequently the first and second pairs, of legs may be tactile in function. Moreover, in some subclasses the basal segments of the first two pairs of legs are modified to form masticatory organs (gnathobases).

SCORPIONES (SCORPIONS)

Scorpions are relatively large heavily sclerotized arachnids in which the prosoma is covered dorsally by a compact shield or carapace (fig. 188). This shield bears a pair of median eyes (ocelli) and a group of two to five smaller ocelli on each lateral margin. The massive pedipalps are chelate. Four pairs of walking legs are present, and the basal segments of the first two pairs form masticatory organs. The opisthosoma is divided into a broad seven-segmented mesosoma or pre-abdomen, and a slender five-segmented tail-like metasoma or post-abdomen. The latter terminates in a sting. On the ventral side of the mesosoma, immediately behind the genital opercula, lies a pair of comb-like organs known as pectines (fig. 189B, C). The function of these organs is uncertain although Stahnke (1956) states that they 'are used by the scorpion much as a blind man would use a cane'. Scorpions respire by means of four pairs of lung-books, and paired respiratory openings are located ventrally on mesosomal segments III–VI. Scorpions, which are an ancient group known from the Silurian period, are carnivorous and occur most commonly in the warmer parts of the world, although they are not found in New Zealand.

Amongst many accounts of the systematics of scorpions those of Kraepelin (1899) and Werner (1935) are perhaps the most comprehensive. Vachon (1952) has monographed the North West African species. The six families can be separated with the following key:

KEY TO SCORPION FAMILIES

1 Sternum inconspicuous, made up of two small transverse plates (fig. 189A) **BOTHRIURIDAE**

    (Australia and South America.)

– Sternum prominent, pentagonal or triangular in outline, at least half as long as broad, frequently longer than broad (fig. 189B, C) 2

2 Tarsi of walking legs with an exterior and interior pedal spur (fig. 189D) . . . 3

– Tarsi of walking legs with only an exterior pedal spur (fig. 189E), or with no spurs, sternum pentagonal in outline 5

3 With two pairs of lateral eyes . . . . **CHACTIDAE**

    (Southern Europe, North Africa, Asia, southern U.S.A., Central and South America.)

– With three to five pairs of lateral eyes 4

4 Sternum generally triangular in outline, median lamellae of pectines never rounded (fig. 189C), without trichobothria (sensory setae) on the ventral surface of the pedipalpal femur . . . . **BUTHIDAE**

    (Wide distribution, mainly circum-tropical and circum-subtropical, but extending into temperate regions, notably southern Europe.)

– Lateral margins of the sternum almost parallel, median lamellae of the pectines often rounded (fig. 189B), at least one sensory seta on the ventral surface of the pedipalpal femur . . . . **VEJOVIDAE**

    (Greece, Asia, southern U.S.A., Central and South America.)

5 With a tooth-like tubercle below the sting . . . **DIPLOCENTRIDAE**

    (Middle East, Sokotra, Mexico.)

– Without a tooth-like tubercle below the sting . . . . **SCORPIONIDAE**

    (Wide circum-tropical and circum-subtropical distribution.)

Scorpions are nocturnal and they feed principally on insects and other arthropods. They do not attack man spontaneously and even the most dangerous species can be allowed to walk over the back of the hand with little risk. Accidents occur commonly when scorpions secrete themselves in clothing and when dark secluded corners are being cleared of rubbish. Their medical significance varies considerably and is dependent on their habits and venom potency rather than on their size.

Although published accounts of the nature of scorpion venom and its effects on man are by no means consistent, it would appear that in general two types can be distinguished.

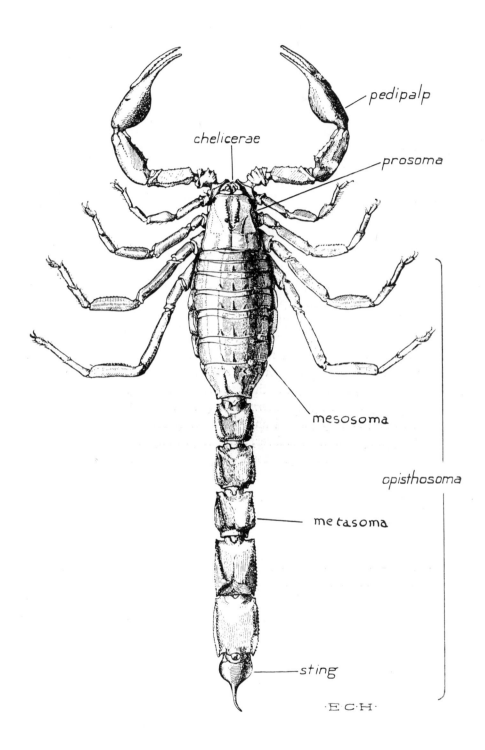

Fig. 188 A scorpion, *Androctonus australis*, labelled to show the main anatomical features (×1.4).

EE

The first type produces a local reaction of varying severity with only mild or with no systemic effects. Barring an unusual hypersensitivity there is very little danger of death from stings of scorpions with venom of this type. The second type of venom is a neurotoxin and its effects can be lethal, particularly in children. Although all the scorpions with powerful neurotoxic venoms belong to the family Buthidae, within this family the nature of the venom varies independently of other characters, so that in a given genus some species can be dangerous and others comparatively harmless.

Many scorpions, for example the yellow-tailed scorpion of Provence, *Euscorpius flavicaudis* (De Geer) (Chactidae), are quite innocuous, and their sting, a mere pinprick, is of little consequence. Similarly the venom of the large Javanese scorpion, *Heterometrus cyaneus* (Koch) (Scorpionidae) lacks potency, and it has been observed that the entire venom content of a telson of an allied species, *H. gravimanus* (Pocock), did not constitute a lethal dose for a mouse (Keegan *et al.*, 1964). Moreover, in some species, notably *Opisthocanthus lepturus* (Peters) (Scorpionidae), the venom apparatus is reduced and the sting is apparently never used.

In southern U.S.A. the venom of certain species of *Vejovis* and *Hadrurus* (Vejovidae) and that of *Diplocentrus whitei* (Gervais) (Diplocentridae) can produce a local swelling with a varying amount of pain, but although the pain may travel some distance from the site of the sting, the symptoms are generally of short duration. Occasionally a severe sting of the stripe-tailed scorpion, *Vejovis spinigerus* Wood, a typical desert species of the southwestern U.S.A., may evoke a systemic reaction in the form of a prickly sensation around the mouth and a partial paralysis of the tongue (Stahnke, 1966). Other species, for example buthids such as *Buthus occitanus* (Amoreux)* in parts of southwest Europe and *Lychas marmoreus* (C. L. Koch) of Australia, are more harmful. Their sting is very painful and is generally followed by a local swelling, temporary partial paralysis and some fever, but, except in hypersensitive persons, the consequences are not serious and the symptoms disappear within two days.

Dangerous Buthids

Amongst the Buthidae with neurotoxic venom, *Androctonus australis* (L.), the fat-tailed scorpion of the Atlas mountains and northern Sahara (fig. 188), the Middle-Eastern *Leiurus quinquestriatus* (Hemprich and Ehrenberg) (pl. 10B) and certain species of *Centruroides* (pl. 10C) from southern U.S.A. and Mexico are often cited as the most dangerous. In particular, the venom of *A. australis* is, drop for drop, almost as toxic as that of a cobra and is capable of killing a dog in seven minutes (Millot and Vachon, 1949).

Some of the symptoms evoked by the stings of these buthids resemble those of strychnine poisoning. Baerg (1929) has described the symptoms produced by the sting of the Durango scorpion of Mexico (*Centruroides suffusus* Pocock) (pl. 10C). Immediately

* There appears to be a considerable biological difference between the south-western European populations of *B. occitanus* and those from other regions. The venom of the North-African and Middle-Eastern populations is much more toxic to man, and in these regions a number of deaths (mainly in young children) have been attributed to the sting of this species (see for example Bouisset & Larrouy, 1962; and Wahbeh, 1965).

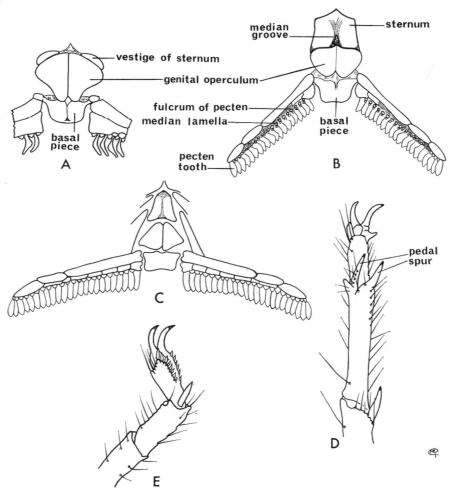

Fig. 189   The sternal region (A–C) and the terminal segments of leg IV (D–E) of certain scorpions: A, *Bothriurus* sp.; B, *Vejovis spinigerus*; C, *Centruroides sculpturatus*; D, *Androctonus australis*; E, *Scorpio maurus* (A–C based on Stahnke, 1966).

following the sharp pain of the sting there is a feeling of numbness or drowsiness and an itching sensation in the mouth, nose and throat.   Initially there is an excessive production of saliva and this, together with a feeling that is described as a sensation of a ball of hair in the throat, induces rapid swallowing.   The tongue is sluggish and the muscles of the jaw are contracted so that it is difficult or impossible to give medicine through the mouth.   There is also a disorder of movement in the arms and legs.   A rapid rise in temperature is accompanied by a reduction of the salivary secretion and by a scarcity of urine.   The senses of touch and sight are affected; the hair feels rigid, the face bulky and vision is veiled.   Strong light is unpleasant to patients, and luminous objects, such as a candle, appear to be surrounded by a red circle.   There may be a haemorrhage of the stomach, intestines and lungs.   The convulsions which follow come in waves, and increase in severity for an hour and a half to two hours, or, in severe cases, until a fatal result.   When the case ends in death, respiration stops a full minute before the pulse ceases to beat.   Patients surviving for three hours are generally considered to be out of danger, but death may occur six to eight hours after the sting.

Well documented accounts of serious scorpionism refer almost exclusively to North Africa, the Middle East, South Africa, Brazil, Trinidad, Mexico or to the state of Arizona in the U.S.A.*   To a great extent this undoubtedly reflects the distribution of dangerous buthids producing neurotoxic venom, although it would seem that in many regions the problem has yet to be properly appraised.   In India, for example, reports on the effects of scorpion stings are to some extent conflicting, and reliable data on the identity of species involved in accidents are lacking.   Caius and Mhaskar (1932), while reporting a number of fatalities in children (attributed mainly to species of *Buthus*), concluded that scorpions in the subcontinent were no more dangerous to human beings than bees or wasps.   Basu (1939) reported that during the period 1928–1937, 19 scorpion-sting cases were treated at the Calcutta Medical College Hospital.   Five cases (all children) were fatal.   It appears that a scorpion involved in one accident may have been *Buthus tamulus* (Fabricius) but whether this species was responsible for one of the fatalities is not clear.   More recently Mundle (1961), without noting the species involved, reported that out of 78 cases of scorpion sting seen over a period of 14 years in Bombay State, 23 were fatal and 9 of these fatalities were in adults.   On the other hand Roantree (1961), whilst noting that the annual numbers of scorpion-sting cases treated in a hospital in southern India were between 800 and 1000, reported no fatalities, and suggested that many of Mundle's fatal cases may have been due to snake bites.

In a review of scorpionism in North Africa, Balozet (1964) reported that over a period of 17 years, 20,164 cases of scorpion sting in the southern Algerian territories were sufficiently serious to warrant medical intervention.   Of these, 386 were fatal, a mortality rate of 1.9 per cent.   In North-West Africa as a whole *Androctonus australis* is considered to be responsible for 80 per cent of the accidents and 95 per cent of the fatalities, although in this region four other buthids, namely *Androctonus aeneas* C. L. Koch, *A. amoreuxi* (Audouin and Savigny), *Buthus occitanus* and *Buthacus arenicola* (Simon) are also considered to be dangerous.

The most widespread of the medically important scorpions of the Mediterranean region is *Buthus occitanus* which is represented by dangerous subspecies and varieties throughout most of North Africa and in Israel and Jordan (see for example Wahbeh, 1965; Vachon, 1966).   Other dangerous buthids occurring in this region are *Androctonus crassicauda* (Olivier) (pl. 10A) and *Leiurus quinquestriatus* (Hemprich and Ehrenberg) (pl. 10B).   *A. crassicauda* is common in the south eastern provinces of Turkey (Goren, 1950) and its distribution extends eastwards to India and southwards through Syria, Israel and Jordan to Arabia.   It is not found in Egypt.   *L. quinquestriatus*, on the other hand, which has a venom four to five times more toxic to rats than *A. crassicauda* (Tulga, 1964), is widespread in the eastern part of North Africa and its distribution extends through Israel and Syria to eastern Turkey, and through the Sinai peninsula to the shores of the Red Sea.   In an early account of *L. quinquestriatus* in Egypt, Wilson (1904) noted that it was common throughout the country but especially so in Upper

---

* Unsupported statements citing the Manchurian scorpion, *Buthus martensi* Karsch, as a dangerous species have appeared in a number of text-books and articles, although Kubota (1918), who studied the effects of the venom of this species on frogs and guinea-pigs, noted that: 'The bite of the Manchurian scorpion is not fatal even in the case of children, which is very different from the case of the Mexican and Egyptian varieties.'

Egypt. It was frequently found in houses and he considered that it was probably responsible for the numerous scorpion-sting fatalities amongst children in this region. Todd (1909) observed that fatalities were particularly numerous in the town of Assouan, where, over the seven-year period 1901–1907, the annual death rate due to scorpion sting was 0.64 per 1000 of the population. This represented 1.6 per cent of the total death rate. More recently Shulov (1939) and Efrati (1949) have reported fatalities due to this species amongst children in Israel.

Grasset, Schaafsma and Hodgson (1946) observed that in South Africa scorpion stings, although painful and responsible in some cases for severe symptoms of intoxication, are seldom fatal, even among children whose symptoms are more severe. They reported that the few scorpion-sting fatalities that occurred amongst children appeared to be due mainly to species of *Parabuthus* and noted that the toxic action of the venom of these scorpions was similar to that of *B. occitanus* and *L. quinquestriatus* in North Africa.

In Brazil, two species of the genus *Tityus*, *T. serrulatus* Lutz-Mello and *T. bahiensis* (Perty), are extremely dangerous and both have been responsible for fatalities in adults as well as in children. *T. serrulatus* is commonly found in centres of population and frequently invades houses and outbuildings, while *T. bahiensis*, which geographically is rather more widely distributed, is essentially a scorpion of fields and plantations (see for example Bücherl, 1953).

Waterman (1938) reported that in Trinidad 698 scorpion sting cases were treated in one hospital in a cane-growing district during the five-year period 1929–1933. Thirty-three of these cases were fatal—a case mortality rate of 4.7 per cent—and in a later paper (Waterman, 1939) it was implied that the species involved in these accidents was *Tityus trinitatis* Pocock (pl. 10D).

An annual scorpion-sting fatality rate of well over one thousand has been reported from Mexico (Mazzoti and Bravo-Becherelle, 1963). Fatalities are most frequent in the one- to four-year-old age group, and the principal offending species are *Centruroides noxius* Hoffman and *C. suffusus* Pocock (pl. 10c), although in this region other species of *Centruroides*, including *C. limpidus* Karsch, are also considered to be dangerous. From the public health viewpoint scorpions are regarded as the most important venomous animals and in some places, during certain months, scorpion sting has been listed as the most important cause of death.

Two very dangerous *Centruroides* species, *C. sculpturatus* Ewing and *C. gertschi* Stahnke, are found in the United States although their distribution is restricted to the southern half of Arizona and to small areas in neighbouring states. Stahnke (1963) reported that during the period 1929 to 1948 these scorpions were responsible for twice as many deaths in Arizona than all other venomous animals combined. Although most of the deaths attributed to scorpion sting in this state have been in children under 17 years of age, recorded fatalities include a 44-year-old hypersensitive female and a 72-year-old male.

Treatment of Scorpion Sting

Symptomatic medication has been reported to be of little value for the treatment of scorpion stings of the neurotoxic type, and it can even be harmful, for certain drugs, notably morphine, can synergize the venom (Stahnke, 1963). However, local pain can be relieved safely with xylocaine (Stahnke, 1966).

Antivenins, which are produced in most areas where scorpion stings are important, are very effective when administered early and in sufficient quantity.   Several species of scorpions have certain venom components in common but while some antivenins will neutralize a heterologous venom—that is to say the venom of a species other than that against which the antivenin was prepared—others are quite specific in their action. For example, an antivenin prepared against the venom of *Androctonus australis* was found to give protection in white mice against the venoms of *Centruroides noxius* and *C. limpidus* (Sergent, 1949).   Again, Whittemore *et al.* (1961) found that Turkish *A. crassicauda* antivenin neutralized the venoms of *A. australis*, *B. occitanus*, *Tityus serrulatus* and *T. bahiensis*, and Tulga (1964) reported that the same antivenin neutralized the venom of *L. quinquestriatus*.   On the other hand Grassett *et al.* (1946) found that *Tityus* antivenin gave no appreciable protection to white mice against *Parabuthus* venom, and it is interesting to find that while *A. crassicauda* antivenin neutralizes the venom of a number of species (and indeed in the case of *A. australis* venom, more effectively than *A. australis* antivenin), the venom of *A. crassicauda* is effectively neutralized only by the homologous antivenin (Whittemore *et al., op. cit.*).

Ligature-cryotherapy treatment, at least as a first-aid measure, is advocated by Stahnke (1956, 1966) not only for scorpion stings but also for the stings or bites of other venomous animals.   With this technique a ligature (that is to say a tight tourniquet) is applied at once between the site of the sting or bite and the body, but as near the point of entrance of the venom as possible.   The envenomized member is then immersed in iced water to well above the point of ligation.   After a period of ten minutes the ligature is removed and the member is kept immersed in the iced-water for at least two hours.   The hypothermia decreases the absorption rate of the venom and localizes its action.   The hypothermia must be started at once, otherwise the ligature must be released every 10–15 minutes.   Care must be taken not to traumatize the tissue with the ligature as such injury not only increases the substrate for venom action but also releases autolytic enzymes.   If ice is not immediately available, refrigerating agents such as Freon or ethyl chloride may be used, although these should not be sprayed directly on to the skin. The hypothermia must be discontinued gradually, and this can be accomplished by removing the ice and allowing the water to come up gradually to room temperature. It has been found empirically that two or three hours of hypothermia prevents serious untoward effects from scorpion and most spider venoms, and anaphylaxis can also be prevented within this time period.

ARANEAE (SPIDERS)

In spiders the prosoma is joined to the opisthosoma (or abdomen) by a narrow pedicel (fig. 190).   The chelicerae have hook-like movable digits which carry the ducts of poison glands, and in the male the terminal segments of the pedipalps are modified to form secondary sexual organs which contain an apparatus for taking up sperm and transferring it to the female.   The genital orifice, which in the female is usually covered by a rudimentary scelorotized plate (the epigyne), is located antero-ventrally on the sac-like opisthosoma.   One to four pairs of spinnerets, which carry the openings of the abdominal silk glands, are situated ventrally or postero-ventrally on the opsithosoma.

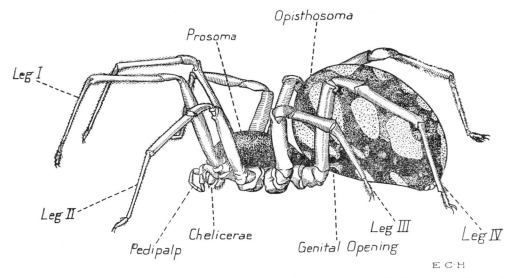

Fig. 190 A female spider, *Latrodectus mactans tredecimguttatus*, labelled to show the main anatomical features (×5.5).

General systematic accounts of spiders are given by Berland (1932) and Petrunkevitch (1928) and a useful survey is given by Millot (1949). Two major groups can be distinguished: the Orthognatha in which the chelicerae work in a plane parallel to the long axis of the body and in which respiration is effected by means of two pairs of lung-books, and the Labidognatha in which the chelicerae work in a plane at right angles to the long axis of the body and in which, with very few exceptions, respiration is effected by means of a single pair of lung-books and tracheae. The Orthognatha are made up of two suborders, the Liphistiomorphae and Mygalomorphae. In the former, which have a very restricted distribution in South-East Asia, the opisthosoma is segmented. The Mygalomorphae have an unsegmented opisthosoma and occur mainly in the warmer parts of the world. Most of the so-called bird-eating spiders belong to this group. The Labidognatha are made up of a single suborder, the Araneomorphae, which are cosmopolitan in distribution.

The vast majority of spiders are completely harmless, either because their chelicerae are too feeble to pierce the skin of larger animals, or because their venom is completely innocuous to man and domestic animals. On the whole they do not deserve the exaggerated fear and loathing which they inspire in many people and very few species are actually dangerous to man. During the Middle Ages a legend grew up in southern Europe around the 'tarantula' whose bite was reputed to give rise to a disorder known as tarantism which caused, and could only be cured by, frenzied dancing to the point of exhaustion (hence tarantella). This manifestation is now thought to have been a form of epidemic hysteria. Moreover, the terms tarentula and tarantula must be used with some caution. *Tarantula* is used scientifically as a genus-group name for certain whip-scorpions (Amblypygi), and both tarentula and tarantula are used somewhat loosely throughout the world as a common name for any large spider. The tarentula or tarantula *spider* of Taranto (and other parts of southern Europe) is *Lycosa tarentula* (Rossi). This species is apparently quite harmless, although as will be noted below, in southern Europe serious general symptoms not very different from those described for tarentism can be caused by the bite of the inconspicuous 'malmignatte', a species of the genus *Latrodectus*.

Some species of spiders can be regarded as being moderately harmful as their bites can lead to quite severe local symptoms. Some large araneomorph species such as *Chiracanthium punctorium* (Villers) and most of the large mygales belong to this category, although the latter, with few exceptions, are not aggressive, and in vivaria have to be provoked considerably before they can be induced to bite.

Latrodectism

Latrodectism, a condition caused by the bite of spiders belonging to the genus *Latrodectus*, is common in agricultural regions throughout the warmer parts of the world and is characterized by epidemic outbreaks (Bettini, 1964). There is some disagreement amongst arachnologists on the status of certain *Latrodectus* species and in the most recent revision of the genus Levi (1959) recognizes only six species: *L. mactans* (Fabricius) (pl. 11A), *L. geometricus* C. L. Koch, *L. curacaviensis* (Müller), *L. pallidus* O. P.-Cambridge, *L. hystrix* Simon and *L. dahli* Levi. To maintain reference with older usage Levi recognized five subspecies of *L. mactans* although this subspecific classification was regarded as being artificial.

L. mactans, the most toxic species, occurs in warm areas throughout the world, and *L. geometricus*, although probably native to Africa, now has a world-wide tropical and subtropical distribution. In the United States it occurs only in Florida. *L. curacaviensis* has a wide distribution in the Americas from southern Canada to Patagonia, although it is apparently absent from Mexico and Central America and is more common in temperate regions. *L. pallidus*, which ranges from Libya into Azerbaidjan, is predominantly a Middle-Eastern species. *L. hystrix* has been recorded from Aden and Yemen, and *L. dahli* from Iran and from the island of Sokotra.

The common name black widow is used fairly generally for *Latrodectus* spiders in the United States where the most widespread representative of the genus is the nominate subspecies of *L. mactans*, although in Florida *L. geometricus* is sometimes referred to as the brown or grey widow. In southern Europe a subspecies of *L. mactans*, *L. mactans tredecimguttatus* (Rossi) is known as the 'malmignatte', and another subspecies, *L. mactans hasselti* Thorell, is known in Australia as the red-backed spider and in New Zealand as the katipo. In South Africa the subspecies *L. mactans cinctus* Blackwall is known as the knoppie-spider or button-spider.

Latrodectus venom is a neurotoxin and was used by certain American Indian tribes as an arrow poison (Thorp and Woodson, 1945). The chain of symptoms following a *Latrodectus* bite is very characteristic.* The bite itself may not be felt and there is usually little evidence of a lesion, although a slight local swelling around two tiny red spots may occur. Sometimes this develops into a low triangular-shaped erythematous

---

* If Levi's (1959) classification is followed only the bites of the nominate and other subspecies of *L. mactans* and those of *L. curacaviensis* are likely to cause serious symptoms. The symptoms evoked by the bites of *L. geometricus* and *L. pallidus* are said to be relatively mild, although illness due to the bites of *L. geometricus* has been reported (see for example Finlayson, 1956). There appear to be no data on the toxicity of *L. hystrix* and *L. dahli*.

swelling covering a large area. Pain, usually at the site of the bite, occurs almost immediately and reaches its maximum intensity in one to three hours. Excruciating pain in the inguinial region and generalized aching of the body, especially in the legs, are common reactions, and rigidity and tonic spasm of the large muscle groups, particularly those of the abdomen, are a constant feature. Other reactions are shock, low fever, leucocytosis, nausea, severe headache, raised blood-pressure, difficulty in breathing and profuse perspiration. The condition is self-limiting and in most cases symptoms disappear in two or three days.

Latrodectism has received most attention in the southern U.S.A. although even here some authorities suggest that, while over a period of time the number of poisoning cases is large, bites are relatively uncommon as the spiders are shy and retiring and are not aggressive except perhaps when guarding their eggs. On the other hand it has been argued that the number of reported cases may represent only a fraction of the total, as missed diagnoses have been frequent. In the U.S.A., *L. mactans mactans* populations occur commonly in and around buildings, and people are apt to be bitten when spiders clinging to rubbish or clothing are accidently squeezed. The spiders are also reputed to be a hazard in outdoor privies where they rush out in response to vibrations set up in the web. Thorp and Woodson (1945) have tabulated cases of black-widow bite in the U.S.A. for the period 1826–1943. They recorded a total of 1,291 cases and 55 deaths— a case fatality rate of 4.26 per cent. The problem is clearly most serious in California, where for this period, 578 cases and 32 fatalities were recorded. More recently Parrish (1963) reported 65 spider-bite fatalities in the U.S.A. for the period 1950–1959. Sixty-three of these were attributed to *Latrodectus* species and two to *Loxosceles reclusa* (see below).

In certain circumstances and for various reasons the trend of endemic latrodectism may slowly increase (Bettini, 1964). For example, in the case of *L. mactans mactans* in the U.S.A. and Argentina the spider has gradually spread over wider areas and has become more closely associated with man, but in the case of *L. mactans hasselti* in Australia, the increase may be more apparent than real, for, according to Wiener (1961), physicians have recntley been more willing to report cases. Bettini points out that the epidemiological picture is complicated by several factors. Some spider populations, for example *L. mactans tredecimguttatus* in Europe and those of *L. mactans mactans* in certain parts of South America, live in fields, while others, such as *L. mactans mactans* in North America, inhabit outbuildings and have closer contact with man. This difference has considerable significance, for in the case of the 'field' populations farmers constitute the class most frequently bitten (in Italy 70 per cent) and the yearly peak in the number of cases occurs at harvest time. In the case of 'domestic' spider populations the victims may belong to any class and the yearly peak depends more on spider densities than on increased man-spider contact. Another complicating feature is that some species such as *L. geometricus* and *L. pallidus*, which induce relatively minor symptoms in man, may be sympatric with the more poisonous ones. For example, *L. geometricus* is sympatric with *L. mactans mactans* in Florida, and *L. pallidus* is sympatric with *L. mactans tredecimguttatus* in Israel.

Several epidemics of latrodectism have been reported and most of these have occurred in Europe. Bettini (*op. cit.*) studied cases in central Italy during the epidemic which occurred in Italy and Yugoslavia over the period 1938–58. Of the 946 cases studied only two were fatal and in the non-fatal cases the commonest duration of hospital stay was two days.

The pain from *Latrodectus* poisoning may be relieved by relaxing the muscle spasm with intravenous injections of calcium salts (Gilbert & Stewart, 1935), and Bogen (1956) comments that the value of this treatment has been abundantly confirmed. Calcium gluconate is actually used in preference to the chloride, because with the gluconate there is less danger of tissue necrosis from accidental extravasation. Relief is rapid and the treatment may be repeated as required. Magnesium sulphate has also been used with good effect in many cases (Bogen, *op. cit.*).

Latrodectus antivenins, which are available in most areas inhabited by these spiders, are effective, but the results are not so dramatic as those obtained from the calcium-salt therapy. The antivenins are apparently paraspecific in action. McCrone & Netzloff (1965), for example, found that the venoms of four species known to occur in North America (*L. mactans*, *L. variolus* Walckenaer, *L. bishopi* Kaston and *L. geometricus*)* contained several common antigens, and, in experiments with mice, showed that a commercial antivenin ('Lyovac'), prepared against the venom of *L. mactans* or *L. variolus*, neutralized the lethal effects of the venom of all four species.

Stahnke (1966) has reported successful results with ligature-cryotherapy treatment (see p. 424).

Necrotic Arachnidism

In the U.S.A. a condition known as necrotic arachnidism or loxoscelism is caused by the bite of the brown recluse spider, or violin spider, *Loxosceles reclusa* Gertsch & Mulaik (fig. 191). This species occurs predominantly in Missouri, Arkansas and Oklahoma, but it is also found from Texas east to Georgia and from Illinois south to Louisiana. It is commonly found in and around houses and outbuildings (Baerg, 1959), but situations such as crevices in rocks on and below cliff-faces may form its natural habitat (Lancaster, cited by Dillaha *et al.*, 1964). The venom of *L. reclusa* can produce a gangrenous slough near the site of envenomation (Atkins *et al.*, 1958). There may be a mild transitory stinging at the time of the bite, but patients may be unaware that they have been bitten. Mild to severe pain begins after two to eight hours and a thick weal with necrosis of tissue usually forms at the site of the bite. This necrotic area soon becomes dark and dry, and after 7–14 days separates from the surrounding tissue to leave an open ulcer which may persist for several weeks. The healing time is directly proportional to the size of the ulcer, and when there are large ulcers skin grafts may be required.

In addition to the cutaneous lesions the bite of *L. reclusa* can evoke severe systemic reactions, particularly in young children, and grave haemolytic anaemia and thrombocytopaenia attributable to its bite have been responsible for two deaths in children (records cited by Dillaha *et al.*, 1964).

In South America a condition similar in many respects to North American loxoscelism is caused by the bite of *Loxosceles laeta* (Nicolet). In a report of forty cases of

---

* Levi (1959) considered *L. variolus* Walckenaer and *L. bishopi* Kaston to be synonymous with, respectively, *L. mactans* (Fabricius) and *L. curacaviensis* (Müller).

arachnidism caused by this species, Schenone & Prats (1961) distinguished two forms of the condition: a localized cutaneous form manifested by a severe necrotic reaction at the site of the bite, and a relatively uncommon viscerocutaneous form which is often fatal, and which is characterized by haemolytic anaemia and haematuria.

While a number of drugs such as phentolamine hydrochloride and antihistamines have been advocated in the treatment of North American loxoscelism, Dillaha *et al.* (1964) favoured the prompt administration of corticosteroids. In Chile, Schenone & Prats (1961) treated localized cases of necrotic arachnidism due to *L. laeta*, with satisfactory results, by administering antihistamines intramuscularly or subcutaneously. The pain and oedema decreased within 24 hours and the course of the condition was shortened. For severe cases with haemolysis these authors advocated the early use of ACTH and corticosteroids. In Peru, Vellard (1954) obtained satisfactory results in some patients with a specific antivenin.

A cutaneous necrosis can also be caused by the venom of certain wolf spiders (family Lycosidae) although in the case of these spiders the necrotic action of the venom appears to be limited and systemic symptoms do not occur. Necrosis due to the bites of *Lycosa* species, especially *L. raptoria* Walckenaer, has been reported in South America (Vellard, 1936; 1966).

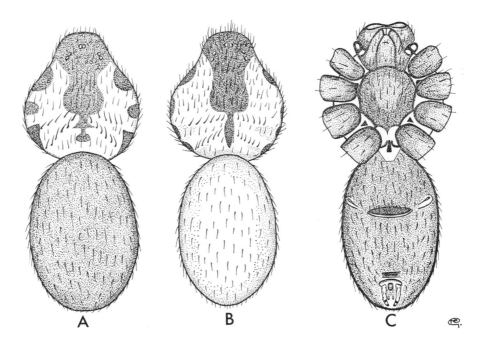

Fig. 191   The brown recluse spider, *Loxosceles reclusa*: A, dorsum of female; B, dorsum of male; C, venter of female (after Gertsch, 1958).

Other Dangerous Spiders

Bites of the Brazilian *Phoneutria fera* (Perty) can result in very painful symptoms and the venom acts on both the central and peripheral nervous systems. This species is very aggressive and is sometimes brought into Europe in cargoes of fruit. In Brazil several hundred accidents involving this species occur annually (Bücherl, 1956) although fatal cases are rare and are most likely to occur in children under four years of age. An effective antivenin is available.

In Australia several fatalities, mainly in children, have been caused by the funnel-web spider (see for example McKeown, 1952; Wiener, 1961) and the species generally implied by this common name is the mygalomorph *Atrax robustus* O.P.-Cambridge (pl. 11B). When another species of *Atrax* was discovered in Australia, *A. robustus* became known as the Sydney funnel-web spider because the crannies and crevices of the sandstone rocks in the vicinity of Sydney were its only known habitat. However, this species is now known to extend northwards until it reaches the territory of the North Coast funnel-web spider (*A. formidablis* Rainbow) and in recent years it has also been found in the far east of Victoria. There is a possibility that a mygalomorph recently collected on the Great Dividing Range east of Melbourne may prove to be yet another species of *Atrax* (Garnet, 1968).

All *Atrax* species should be regarded as dangerous. According to Garnet (*op. cit.*) the symptoms of envenomation include pain at the site of the bite—soon extending to other parts of the body—profuse sweating, delirium and respiratory failure. An antivenin for *Atrax* venom has not been produced since none of its components is antigenic.

ACARI (Mites and Ticks)

The Acari are the most heterogeneous subclass of the Arachnida and cannot be defined precisely in terms of a small set of characters. The body (fig. 192) is always divisible into two regions, the gnathosoma (comprising the mouth parts) and the idiosoma, and in many groups the latter is further divided by a transverse furrow, located between legs II and III, into an anterior propodosoma and a posterior hysterosoma. The opisthosoma (that is to say the region behind the legs) is never clearly separated from the anterior part of the body. The chelicerae are often pincer-like, but they may be variously modified. Similarly, the palps may be simple, or modified to form raptorial organs. Two, three or four pairs of legs may be present in the adult stage. The genital and anal orifices vary in position although commonly the former is situated between coxae IV, and the latter is subterminal. Respiration is cutaneous or effected by tracheae, which in some orders open to the exterior by paired stigmata.

The Acari are unique amongst the Arachnida in having a six-legged larva (the plant-feeding Eriophyoidea, which have only four legs in all postembryonic stages, are exceptional). Following the larval instar there may be as many as three active, morphologically different, octopod nymphal stages before the mite attains sexual maturity, although in many groups the number of nymphal stages is reduced. In some Prostigmata, for example, two of the active nymphal stages are replaced by inactive resting stages, the nympho-chrysalis and imago-chrysalis. On the other hand in argasid ticks (Metas-

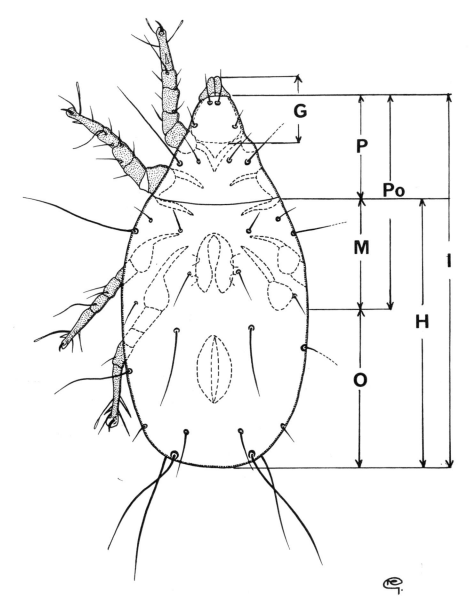

Fig. 192   Divisions of the body of an acarid mite: G, gnathosoma; H, hysterosoma; I, idiosoma; M, metapodosoma; O, opisthosoma; P, propodosoma; PO, podosoma (much enlarged).

tigmata) the nymph undergoes several moults, and in some cases there may be as many as eight nymphal instars.   However, the existence of morphological differences between all the nymphal instars in this family has not been established.

General systematic accounts of mites are given by Vitzthum (1943) and Baker & Wharton (1952), and an introductory account of their morphology, biology and higher classification is given by Evans, Sheals & Macfarlane (1961).   In the higher classification the nature of the respiratory system and the position of the respiratory openings provide useful diagnostic features.   Seven orders can be recognized: Notostigmata (=Opilioacarida), Tetrastigmata (=Holothyroidea), Metastigmata, Mesostigmata, Cryptostigmata, Astigmata and Prostigmata. All except the Notostigmata and Cryptostigmata include species of medical importance.*

Acari-Tetrastigmata

The Tetrastigmata are large heavily sclerotized reddish mites. The tracheal system opens by means of a pair of stigmata situated near coxae III, but additionally a pair of pores leading to a system of air sacs is located posterior to the stigmata. The Tetrastigmata are thought to be carnivorous and have been recorded from New Guinea, Australia, New Zealand and from islands in the Indian Ocean. In Mauritius *Holothyrus coccinella* Gervais is evidently common. It secretes an irritant poison and is sometimes swallowed by geese and ducks with fatal results. For this reason it is called 'touille canard' by the inhabitants, and children have been reported to suffer ill-effects by putting the mite into their mouths (Hirst, 1922).

Acari-Metastigmata (Ticks)

Ticks are the largest acarines. Cosmopolitan in distribution, all are external parasites of terrestrial vertebrates, and, with very few exceptions, all postembryonic stages feed on the blood and tissue fluids of the host. The dorso-ventrally flattened body (fig. 193B, C) is sac-like in outline and the respiratory openings are located behind coxae IV or laterally above coxae III–IV. Each stigma lies on a sclerotized plate and a peritreme may be present. A conspicuous component of the mouth parts is the long hypostome which is armed with recurved teeth. The digits of the chelicerae are directed externally and work in a horizontal plane.

General systematic accounts of many metastigmatid genera are given in the volumes of an incompleted series entitled *Ticks: A Monograph of the Ixodoidea* (Nuttall *et al.*, 1908; Nuttall & Warburton, 1911, 1915; Robinson, 1926 and Arthur, 1960), while systematic accounts of regional tick faunas are given by Pomerantzev (1950, translation 1959) (U.S.S.R.); Anastos (1950) (Indonesia); Hoogstraal (1956) (Sudan); Theiler (1962) (Africa south of the Sahara); Yeoman & Walker (1967) (Tanzania) and Elbl & Anastos (1966) (Central Africa).

There are two major families of ticks and these can be separated as follows:

Dorsal scutum present, covering the whole of the dorsal surface in the male, but relatively small and covering only the anterior part of the dorsum in the female. Mouth parts terminal, visible from above (fig. 193A). Stigmata situated behind coxae IV (fig. 193B, C) .   .   .   .   .   .   *IXODIDAE* (hard ticks)

Without dorsal scutum. Mouth parts ventral, not visible from above (fig. 193D). Stigmata usually situated between coxae III and IV (fig. 199B) *ARGASIDAE* (soft ticks)

A third family, Nuttallielidae, represented by a single rare species in South West Africa, has some characteristics in common with the Ixodidae and others in common with the Argasidae. This 'intermediate' family is of no medical significance.

* Some cryptostigmatid species are important in veterinary medicine since they function as intermediate hosts for anoplocephaline cestodes.

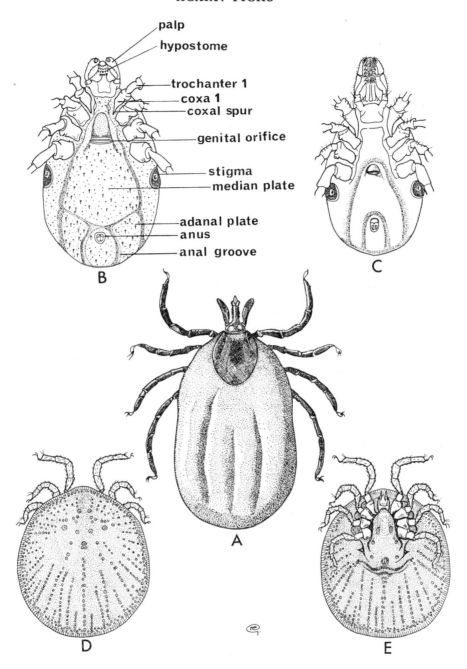

palp
hypostome
trochanter 1
coxa 1
coxal spur
genital orifice
stigma
median plate
adanal plate
anus
anal groove

Fig. 193 A–C, the sheep tick *Ixodes ricinus*, A, dorsum of an engorged female; B–C, B, venter of male; C, venter of female (all ×23); D–E, *Argas vespertilionis*, female: D, dorsum; E, venter (after Hoogstraal, 1958).

The development cycle of ticks follows the general acarine pattern. All have a hexapod larva, but whereas the Ixodidae have only one nymphal instar certain Argasidae may have up to eight. The Ixodidae may be grouped into three categories according to the number of host individuals required to complete the life-cycle. In one-host ticks, such as *Boophilus*, the larval, nymphal and adult stages engorge on the same animal and both moults take place on the host. Two-host ticks, for example certain species of *Rhipicephalus*, require two host individuals to complete the life-cycle. The larva engorges and moults on the first host, but after engorgement the nymph drops to the ground and a new host is sought by the adult. Most species of *Ixodes* are three-host ticks. Each instar requires a different host individual and each ecdysis takes place on the ground. The argasids are multi-host ticks, and nymphs as well as adults may take several short blood meals. Moreover, the argasids may be regarded as 'habitat ticks' for while the larvae may remain attached to the host for several days, unlike the Ixodidae, the nymphs and adults are rarely carried away from the hosts' nests (Nuttall, 1911).

Ticks and Disease

While the discomfort and irritation caused by their bites is far from negligible, in human medicine ticks are important largely because of their activity as vectors of pathogenic organisms. Transovarian and trans-stadial transmission has been observed to a greater or lesser extent in the case of all the major categories of organisms carried by ticks. For example, in the case of *Dermacentor andersoni* Stiles (fig. 194), the principal vector of Rocky Mountain spotted fever rickettsiae (*Rickettsia rickettsi*), organisms ingested by the larvae may be passed to the nymph, thence to the adult which in turn may transmit the pathogen transovarially to the larvae of the next generation.

The majority of ticks are essentially parasites of wild animals and man must be regarded as an incidental host, both of the ticks and the organisms which they transmit. Tick-transmitted organisms and wild hosts generally exist in a state of equilibrium, and infection in these hosts rarely leads to overt disease. Thus wild animals and ticks form a reservoir of infection, and man plays no part in the maintenance of this enzootic cycle. Indeed as far as the transmitted organisms are concerned man is a dead end. For example, in the United States and other parts of North and South America, leporids, rodents, other small mammals and birds are hosts of the rickettsiae of Rocky Mountain spotted fever. The infection in the wild hosts is evidently very transitory, but the cycle is maintained by several tick species, including *Dermacentor andersoni* Stiles. The larvae and nymphs of this species feed on small animals but the adults favour larger mammals, and man, if accidently bitten by an infected tick, may suffer a rickettsial infection of great severity. *D. andersoni* has a wide host range, but ixodid ticks with a restricted host range such as *Haemaphysalis leporispalustris* Packard and possibly *Dermacentor parumapertus* Neumann may also be important in maintaining the natural cycle.

However, at least as far as rickettsiae are concerned, this general picture of the natural history of tick-borne infections, although substantially correct, may be a gross over-simplification (Bertram, 1962). Again taking Rocky Mountain spotted fever as an example, in some areas domestic animals and argasid as well as ixodid ticks may be implicated. Moreover, it is possible that the disease in man has its origin in a complex of inapparent rickettsial infections, not necessarily with the characteristics typical of *R. rickettsi*. During the course of successive passages between different vertebrate and tick hosts the rickettsial agents may undergo changes in the direction of *R. rickettsi*, possibly as a result of selection acting on mutant forms, with, eventually, a degree of pathogenicity for man.

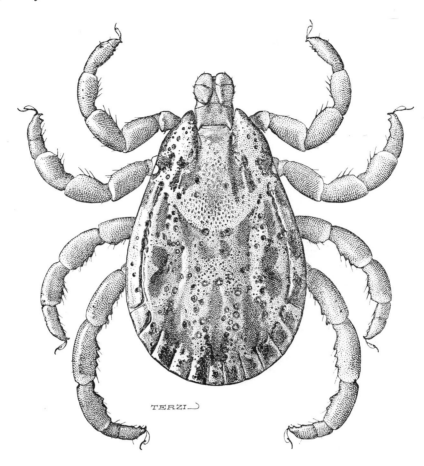

Fig. 194   The Rocky Mountain wood tick, *Dermacentor andersoni*, male (×18).

Viruses

The tick-borne viruses causing human disease fall into two main categories (Hoogstraal, 1966).   Firstly the large and varied Russian spring-summer complex in antigenic group B (RSS subgroup) which are vectored only or almost entirely by ticks.   Secondly, arboviruses in antigenic group A, other subgroups of B or ungrouped viruses for which ticks play either a definite or more or less incidental role as hosts.   The Russian spring-summer (RSS) complex includes Powassan encephalitis, Negishi encephalitis, Langat encephalitis, Kyasanur forest disease (KFD), Omsk haemorrhagic fever (OHF), Russian spring-summer encephalitis in the strict sense (designated RSSE to distinguish it as an entity in the RSS complex), Central European tick-borne encephalitis (TE), louping ill and a 'related' neurotropic virus isolated in East Germany. The arbovirus infections in the second category are Sindbis (a virus of antigenic group A), Japanese B and West Nile (which together form a separate subgroup of antigenic group B) and the following

FF

antigenically independent viruses: Colorado tick fever, Quaranfil fever, Crimean-Congo haemorrhagic fever (CHF-Congo),* Kemerovo tick fever and Tribec. Additionally ticks may play a role in the epidemiology of other human viral infections, for example St Louis encephalitis, eastern equine encephalitis, western equine encephalitis and lymphocytic choriomeningitis. Host and distributional data for the principal tick-borne viruses causing human disease are given in Table 10.

Table 10 Principal tick-borne viruses causing human disease

Virus	Group	Main tick hosts†	Main vertebrate hosts	Distribution
Powassan encephalitis	B	*Dermacentor andersoni, Ixodes marxi, I. cookei, I. spinipalpis*	Leporids, rodents, other small mammals	Canada, U.S.A.
Langat encephalitis	B	*Ixodes granulatus*	Rodents	Malaysia
Kyasanur Forest disease (KFD)	B	*Haemaphysalis spinigera, H. turturis, H. papuana kinneari, H. cuspidata, H. kyasanurensis, H. minuta, H. wellingtoni*	Rodents, other small mammals, forest monkeys	India (Mysore State)
Omsk haemorrhagic fever (OHF)	B	*Dermacentor reticulatus, D. marginatus, Ixodes persulcatus, I. apronophorus*	Rodents, other small mammals, domestic mammals, birds	South-western Siberia; distribution in other parts of Asia not clear, possibly occurs in eastern Europe in association with *Ixodes ricinus*
Russian spring-summer encephalitis (RSSE)	B	*Ixodes persulcatus, I. lividus, Haemaphysalis japonica douglasi, H. concinna, H. neumanni, Dermacentor silvarum, D. marginatus*	Small mammals, birds	Northern and temperate regions of Asia, North and Central Europe

* The aetiological agents of Crimean haemorrhagic fever (CHF) and a haemorrhagic disease known from the Central Asian republics [Central Asian haemorrhagic fever (CAHF) or Uzbekistan haemorrhagic fever] have been shown to be similar if not identical. Additionally, it has been shown that CHF and a virus isolated from febrile patients in the Congo and Uganda (Simpson *et al.*, 1967) are closely related if not identical (Casals *et al.*, 1970).

† It should be noted that not all the species listed transmit viruses to man; many are important merely because of their role in the maintenance of natural cycles.

Central European tick-borne encephalitis (TE)	B	*Ixodes ricinus, Dermacentor marginatus*	Leporids, rodents, other mammals, birds	European U.S.S.R., Scandinavia to Balkans
Louping ill	B	*Ixodes ricinus*	Sheep, cattle, deer, various small mammals, red grouse	Northern England, Scotland, Ireland
Colorado tick fever	—	*Dermacentor andersoni, D. occidentalis, D. parumapertus, Haemaphysalis leporispalustris, Otobius lagophilus*	Leporids, rodents	Canada, U.S.A.
Quaranfil fever	—	*Argas arboreus, A. reflexus hermani*	Birds	Egypt
Crimean-Congo haemorrhagic fever (CHF-Congo)	—	*Hyalomma marginatum marginatum, H. anatolicum anatolicum, H. marginatum rufipes, H. truncatum, Rhipicephalus rossicus, Amblyomma variegatum, Boophilus decoloratus, B. microplus*	Small mammals, domestic mammals and probably birds	Bulgaria, U.S.S.R. (Kazakhstan, Uzbekistan, Turkmenistan, Tadzhikstan, Crimea, lower Don and Volga valleys and flood plains), W. Pakistan, Uganda, Kenya, Nigeria, Congo
Kemerovo fever	—	*Ixodes persulcatus*	? Rodents	Siberia (Kemerovo)

Rickettsiaceae

The role of ticks in the transmission of human rickettsial infections is reviewed by Bertram (1962) and Hoogstraal (1967). The main human rickettsial diseases with which ticks are associated are the related spotted fevers, tick-bite fevers and tick-typhus fevers which can be grouped together as tick-borne rickettsioses. These include Rocky Mountain spotted fever (*Rickettsia rickettsi*), Siberian tick typhus (*Rickettsia sibericus*), Queensland tick typhus (*Rickettsia australis*) and boutonneuse fever (*Rickettsia conori*). A more appropriate name for Rocky Mountain spotted fever might be American spotted fever since the disease has been recorded from almost every state in the U.S.A. It has also been recorded from Canada and Central America and from several South American countries where it is known by various names including São Paulo fever. Other names for the typhus-like disease caused by *R. conori* are tick-borne typhus, Kenya typhus, South African tick-borne fever, Marseilles fever and Indian tick typhus. Host and distributional data for the spotted fever group of rickettsiae are given in Table 11.

Several species of ticks have also been found infected with the rickettsiae of Q-fever or nine-mile fever (*Coxiella burneti*), but, as Bertram points out, 'tick-borne' would be a misnomer for this disease since transmission of the infection commonly occurs without any association with ticks.   Actual transmission to man by tick bite has yet to be demonstrated, but there is evidence to suggest that ticks are important in maintaining enzootic cycles.

Table 11   Tick-borne human rickettsial diseases of the spotted fever group

Rickettsia	Main tick hosts*	Vertebrate hosts	Distribution
Rickettsia rickettsi (Rocky Mountain spotted fever)	*Dermacentor andersoni, D. occidentalis, D. parumapertus, D. variablis, D. albipictus, Haemaphysalis leporispalustris, Rhipicephalus sanguineus, Amblyomma americanum, A. brasiliense, A. cajennense, A. striatum, Ornithodoros nicollei, O. rudis, O. rostratus, Otobius lagophilus*	Leporids, rodents, other mammals, birds	North, Central and South America
Rickettsia siberica (Siberian tick typhus)	*Dermacentor marginatus, D. nuttalli, D. reticulatus, D. silvarum, Haemaphysalis concinna, H. japonica douglasi, H. punctata, Hyalomma asiaticum, Rhipicephalus sanguineus*	Leporids, rodents, other small mammals, birds	Asiatic Russia (Pacific Far East to Armenia)
Rickettsia conori (Boutonneuse fever)	*Rhipicephalus sanguineus, R. appendiculatus, Amblyomma hebraeum,* numerous other ixodids including other *Rhipicephalus* and *Amblyomma* spp., and *Haemaphysalis, Hyalomma, Boophilus, Ixodes* and *Dermacentor* spp.	Leporids, rodents, dogs, other mammals, birds	Africa, European Mediterranean, Middle East and Southeast Asia
Rickettsia australis (Queensland tick typhus)	? *Ixodes holocyclus*	Small marsupials	Queensland

Bacteria

Ticks appear to be the primary transmitters of *Pasteurella tularensis*, the aetiological agent of tularaemia or rabbit fever (reviewed Arthur, 1962).

* It should be noted that not all the species listed transmit rickettsiae to man.   Many are important because of their role as reservoirs and because of their role in the maintenance of natural cycles.

Spirochaetaceae

The spirochaetes of human relapsing fever are transmitted by species of *Ornithodoros* (family Argasidae). Aspects of transmission are reviewed by Varma (1962). Tick-borne relapsing fevers occur in the Old World and in the Americas, and three large endemic areas can be recognized. Firstly the Central African region which extends from Kenya southwards through Uganda and Tanzania to the Transvaal and Cape Province. The causal organism in this region is *Spirochaeta duttoni*. The second endemic region, the Mediterranean, extends from Portugal through Spain, North and North-West Africa southwards to Dakar. Eastwards this region extends along the Mediterranean and includes most of Turkey, Iran, Arabia and Transcaucasia. The eastern limit of this region was once thought to be Russian Central Asia but foci are now known to exist in Kashmir and Sinkiang. In the western part of the Mediterranean region the causal organism is *Spirochaeta hispanica* while in northern Libya and eastwards the causal organism is *Spirochaeta persica*. The third endemic area covers the warmer parts of America and extends from central and western U.S.A. through Mexico to northern Argentina. In the U.S.A. there appear to be three distinct species of human relapsing fever spirochaetes: *S. turicatae*, *S. parkerii*, and *S. hermsii*, while in Central and South America the causal organism is *S. venezuelense*.

Other Diseases

In 1962 a haemorrhagic disease, Himalayan haemorrhagic disease (HHD), occurred in primary jungle in the eastern Himalayan foothills among Indian troops who had been extensively bitten by ticks reported to be nymphs of *Dermacentor auratus* Supino and of *Hyalomma marginatum* subsp. After four to six months' service in the region, 78 soldiers contracted the disease and nine died. Certain *Toxoplasma*-like and other organisms were suggested as agents but the findings were inconclusive (Singh *et al.*, 1965; Hoogstraal, 1967*a*).

Tick paralysis, an acute intoxication, may be caused by the bites of rapidly engorging female ticks of certain species. It is characterized by fever and an acute ascending paralysis. If the ticks are removed promptly, recovery follows in a few days, but if they are not, death may occur from respiratory paralysis. Tick paralysis is thought to be caused by a toxin secreted by the salivary glands during the period of rapid egg development. It occurs chiefly in Oregon and British Columbia although it has been recorded from other areas. In the Pacific North West it is usually associated with *Dermacentor andersoni*, and the condition is most frequently observed in children under two years of age.

Family Ixodidae

The main genera of Ixodidae (hard ticks) may be separated with the key given below. Genera of outstanding medical importance are marked with an asterisk. Others are important in veterinary medicine and possibly also as reservoirs of human pathogens.

1	Without eyes	2
–	With eyes	4
2	Anal groove distinct and extending anteriorly around anus (fig. 193B, C)	*IXODES**
–	Anal groove faint or if distinct not extending anteriorly around anus (fig. 195B) .	3
3	Palpi short and conical. Palpal segment II at least as wide as long and projecting laterally beyond basis capituli. Scutum inornate (fig. 195)	*HAEMAPHYSALIS**
–	Scutum ornate or inornate, second palpal segment about twice as long as broad (on reptiles)	*APONOMMA*
4	Palps much longer than basis capituli, palpal segment II much longer than wide (figs. 196, 197)	5
–	Palps about as long as basis capituli, palpal segment II about as long as wide . .	6
5	Scutum inornate. Males with adanal and subanal plates. Marginal festoons present or absent, if present irregular and partly coalesced (fig. 196) .	*HYALOMMA**
–	Scutum with coloured ornamentation. Males lacking adanal and subanal plates. Marginal festoons present, regular and not coalesced (fig. 197)	*AMBLYOMMA**
6	Marginal festoons present	7
–	Marginal festoons absent	10
7	Scutum usually ornate, basis capituli rectangular dorsally. Ventral plates absent in both sexes (fig. 198B)	*DERMACENTOR**
–	Scutum usually inornate, basis capituli variously shaped. Ventral plates present or absent	8
8	Ventral plates present in the male (fig. 198D)	*RHIPICEPHALUS**
–	Ventral plates absent in both sexes	9
9	Basis capituli hexagonal dorsally	*RHIPICENTOR*
–	Basis capituli rectangular dorsally	*ANOCENTOR*
10	Male with a median pre-anal plate continued backward as two long spines on each side of the anus. Segments of leg IV in the male greatly enlarged	*MARGAROPUS*
–	Male without a pre-anal plate. Adanal and accessory plates absent	*BOOPHILUS**

* Of outstanding medical importance

Ixodes

The genus *Ixodes* has a world-wide distribution and many of its species are proven vectors and reservoirs of human pathogens. This is the largest of the metastigmatid genera and useful systematic treatments of regional faunas are given by Cooley and Kohls (1945) (North American species) and by Arthur (1963, 1965) (British and African species, respectively).

I. ricinus (L.) (fig. 193A–C) is the most common tick in the British Isles. It is principally a parasite of sheep and cattle, but its host range is extremely wide and it will probably parasitize almost any lizard, bird, or mammal. In the British Isles it transmits, among other livestock diseases, louping-ill, an encephalomyelitis of sheep caused by a neurotropic virus belonging to the Russian spring-summer group. Human cases of louping-ill have been recorded (Smith, 1962; Varma, 1964) and the disease also occurs in cattle.

On the Eurasian land mass the distribution of *I. ricinus* extends eastwards to about 50–55°E. longitude. The southern boundary of its distribution, which in Iran lies at about 35°N. latitude, extends through the mountains of Turkey, Albania, Italy, Sardinia and the Pyrenees. The northern limit of its distribution lies at about 65°N. latitude (Arthur, 1966). In continental Europe, *I. ricinus* is the principal vector of Central

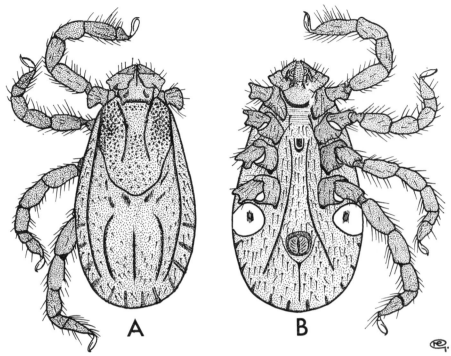

Fig. 195 The yellow dog tick, *Haemaphysalis leachi*, female: A, dorsum; B, venter (after Hoogstraal, 1958).

European tick-borne encephalitis (TE). The Czechoslovakian Tribeč virus (which has some antigenic properties in common with the Siberian Kemerovo virus) has been isolated from this tick, and in southwestern U.S.S.R. it is associated with 'Bukovinian haemorrhagic fever', a disease which may be a variant of Omsk haemorrhagic fever (reviewed Hoogstraal, 1966).

Giroud *et al.* (1965) present evidence to suggest that, in France, *I. ricinus* and *I. hexagonus* Leach (a species known in the British Isles as the hedgehog tick) harbour *Rickettsia conori*, the aetiological agent of boutonneuse fever. *R. conori* has been isolated from *I. ricinus* in Belgium (Jadin *et al.*, 1969) and in France from *Dermacentor marginatus* Sulzer and *D. reticulatus* (Fabricius) (p. 445). Thus, in addition to the principal vector (*Rhipicephalus sanguineus* (Latreille)), these *Ixodes* and *Dermacentor* species may play an important role in the maintenance of the natural cycle of boutonneuse fever in Europe.

The taiga tick, *I. persulcatus* Schulze, is associated with taiga and other forest types in Eurasia. Its distribution extends south of 65°N. eastwards from northern Germany to the Pacific coast and near-by islands. Under continental climatic conditions it reaches as far south as Lake Balkash, but under maritime conditions it extends to 32°N in the island of Kyushu, Japan (Arthur, 1966). This tick is the principal vector of Russian spring-summer encephalitis (RSSE). It is also a reservoir and vector of Omsk haemorrhagic fever (OHF) and a vector of Siberian Kemerovo fever. The larvae and nymphs of this species parasitize a variety of small forest mammals and birds, whilst the adults tend to feed on larger wild and domestic mammals (reviewed Hoogstraal, 1966).

I. apronophorus Schulze appears to play an important role in the circulation of Omsk haemorrhagic fever (OHF) virus in nature (reviewed Casals *et al.*, 1970). This species does not bite man, but spends its entire life in the burrows of the narrow-skulled vole, *Microtus (Stenocranius) gregalis*, a rodent whose population numbers exploded before the 1945–48 outbreak of OHF and declined in 1947 prior to the gradual disappearance of human OHF infections transmitted by ticks.

Natural infections of Powassan virus have been found in *I. spinipalpis* Hadwen & Nuttall taken from *Peromyscus* mice in Connecticut and South Dakota, and in Canada isolates of virus and seasonal serological evidence showed that young groundhogs (*Marmota monax*) acquire active summer infections of Powassan from the tick parasites, *I. cookei* Packard. Similar evidence showed that red squirrels (*Tamaisciurus hudsonicus*) acquired infections from *I. marxi* Banks (reviewed Hoogstraal, 1966). The evidence of serological and other studies indicates that leporids and rodents and their tick parasites are the main vectors of this virus, but since the ticks involved rarely bite him, man is unlikely to be implicated.

Other medically important *Ixodes* species include *I. lividus* Koch (a parasite of sand martins (*Riparia riparia*)) from which RSSE virus has been isolated, and *I. holocyclus* Neumann which is a suspected vector of Queensland tick typhus.

Haemaphysalis

The genus *Haemaphysalis* is made up of about 90 species, several of which are important vectors. Three species, namely *H. concinna* Koch, *H. punctata* Canestrini & Fanzago and *H. japonica douglasi* Nuttall & Warburton, have been implicated as vectors and reservoirs of Siberian tick typhus (reviewed Hoogstraal, 1967). *H. concinna* is widely distributed in the forests of Eurasia from the Atlantic Ocean to the islands of the Sea of Japan and it also occurs in the southern republics of the U.S.S.R. *H. punctata* is also fairly widely distributed, ranging through most of Europe and the north-western corner of Africa and south-western Asia. *H. japonica douglasi* occurs in the southern districts of the far eastern section of the U.S.S.R. and in Korea and China. *H. concinna* may also be the chief vector of Russian spring-summer encephalitis in areas such as the Krasnoyarsk region of western Siberia, where it is the dominant tick species, and the RSSE virus has also been recovered from *H. japonica douglasi* and from *H. neumanni* Doenitz—a species with a distribution rather similar to that of *H. japonica douglasi*.

Although rarely biting man, *H. leporispalustris* (Packard), the rabbit tick, which is widely distributed in North and South America, is a vector of the rickettsiae of Rocky Mountain spotted fever, and it may be important in spreading this infection, as well as *Pasteurella tularensis* (the causal agent of tularaemia or rabbit fever), amongst animal hosts. The virus of Colorado tick fever has also been isolated from this species.

H. leachi (Audouin) (fig. 195), the yellow dog tick of Africa, has been found infected with the rickettsiae of Q-fever (Hoogstraal, 1956) and is an important urban vector of boutonneuse fever in southern Africa (Bertram, 1962; Hoogstraal, 1967).

The genus *Haemaphysalis* has recently received considerable attention in relation to the epidemiology of Kyasanur Forest disease. As well as species of other genera, the tick fauna of the Kyasanur Forest area in Mysore includes fourteen species of the genus

Haemaphysalis and all stages of this haemaphysalid fauna are keyed by Trapido *et al.* (1964). Isolations of KFD virus from pooled samples collected in nature have been made from the following seven species: *H. spinigera* Neumann, *H. turturis* Nuttall & Warburton, *H. papuana kinneari* Warburton, *H. kyasanurensis* Trapido, Hoogstraal & Rajgopalan, *H. minuta* Kohls and *H. wellingtoni* Nuttall & Warburton (reviewed Hoogstraal, 1966). The most important vector of KFD is considered to be *H. spinigera*.

Hyalomma

The majority of *Hyalomma* species are large ticks, and the genus, which is made up of about twenty species, is widely distributed in Asia, southern Europe and Africa.

H. aegyptium (L.) (fig. 196), the large tortoise tick, which is frequently imported into Britain with pet-shop tortoises from southern Europe, is completely innocuous and has not been incriminated either as a vector or reservoir of organisms pathogenic in warm-blooded animals.

H. marginatum marginatum Koch and *H. anatolicum anatolicum* appear to be the principal vectors of CHF-Congo virus in Asia. It is interesting to note that birds migrating southward from the U.S.S.R. to central Africa frequently carry nymphs of *H. marginatum marginatum* while birds journeying northwards from Africa commonly carry *H. marginatum rufipes* Koch. Either the migrating birds or their tick passengers

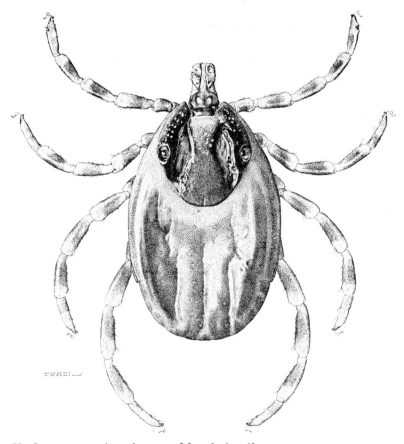

Fig. 196 *Hyalomma aegyptium,* dorsum of female ($\times 7\frac{1}{2}$).

may be responsible for disseminating the CHF-Congo virus and possibly other pathogenic organisms over very wide geographical areas (Hoogstraal *et al.*, 1961; Casals *et al.*, 1970).

Several *Hyalomma* species, including *H. marginatum*, *H. dromedarii* Koch and *H. detritum* Schulze, are natural reservoirs of Q-fever, and *H. detritum* is a possible host of CHF-Congo virus. The latter species is also known to harbour *Pasteurella pestis*, the causative organism of plague, for several days after feeding on an infected animal.

In south-western Kirgizia, foci of Siberian tick typhus and Q-fever are associated with red-tailed jirds (*Meriones erythrourus*), and here *H. asiaticum* Schulze & Schlottke parasitizing the jirds in their burrows are infested with *Rickettsia siberica* (Hoogstraal, 1967).

The bites of *H. truncatum* Koch, a species confined to the Ethiopian faunal region, may cause tick paralysis in man.

Amblyomma

Amongst species of this genus, *A. americanum* (L.), the lone star tick of southern U.S.A., will bite man in all its stages. It is an important vector of Rocky Mountain spotted fever and possibly of tularaemia (Cooley & Kohls, 1944). The same authors note that *A. cajennense* (Fabricius) is a vector of Rocky Mountain spotted fever in Central America and Brazil.

A. hebraeum Koch (fig. 197) is an important vector of boutonneuse fever in southern Africa, and *A. variegatum* (Fabricius), which has a very wide African distribution, has been found naturally infected with the rickettsiae of Q-fever (Hoogstraal, 1956).

Dermacentor

This genus, which is made up of about thirty species, occurs in Asia, Europe, Africa and in North and Central America. A general systematic account of the genus is given by Arthur (1960) and the species occurring in the United States are dealt with by Cooley (1938).

The best known and most important species is *D. andersoni* Stiles, the Rocky Mountain wood tick (fig. 194). This three-host tick is widely distributed throughout the western part of North America and plays an important role as a vector of Rocky Mountain spotted fever and Colorado tick fever. Powassan or a closely related virus has been isolated from *D. andersoni* in Colorado and South Dakota (Thomas *et al.*, 1960), and this tick can cause tick paralysis in both man and cattle.

D. variabilis Say, the American dog tick, which is widely distributed in North America east of the Rocky Mountains, is the principal if not the only vector of Rocky Mountain spotted fever in the central and eastern parts of the United States. It also transmits tularaemia. The immature stages of this tick feed almost exclusively on leporids and rodents, but the adults prefer larger mammals. Dogs are the most favoured hosts, but other domestic mammals and man are also freely attacked.

D. parumapertus Neumann is a parasite of leporids and rodents in semi-arid regions of the south-western United States. Although this tick does not bite man, it serves as a reservoir of Colorado tick fever and tularaemia, and it is known to be capable of transmitting the rickettsiae of Rocky Mountain spotted fever.

D. occidentalis Marx, the western dog or Pacific tick, is common in California and in the southern part of Oregon. It has been found naturally infected with the virus of Colorado tick fever and is a possible natural carrier of Rocky Mountain spotted fever rickettsiae (Parker, 1937).

D. albipictus (Packard), the moose tick, is distributed widely in North America. It is a one-host tick which feeds only during the winter, and is principally a parasite of larger wild and domestic mammals. This species has been implicated as a vector of *Rickettsia rickettsiae* in the state of Washington, where a man bitten by a tick recently detached from a dead elk subsequently contracted spotted fever (Phillip & Kohls, 1952).

Dermacentor species are also an important component of the Eurasian tick fauna (Hoogstraal, 1966, 1967). *D. reticulatus* (Fabricius) (= *D. pictus* of Russian authors) (fig. 198) is widely distributed, extending westward from northern Kazakhstan to the Atlantic, although in the British Isles it is known only from S.W. England and from Cardiganshire. The immature stages of this species feed on a variety of small mammals and they are occasionally found on birds. The adults are commonly found on larger wild and domestic mammals. This tick is a vector of Siberian tick typhus, and it almost certainly transmits Omsk haemorrhagic fever (OHF) virus to man, although it should be noted that this disease can be transmitted to man by contact with infected

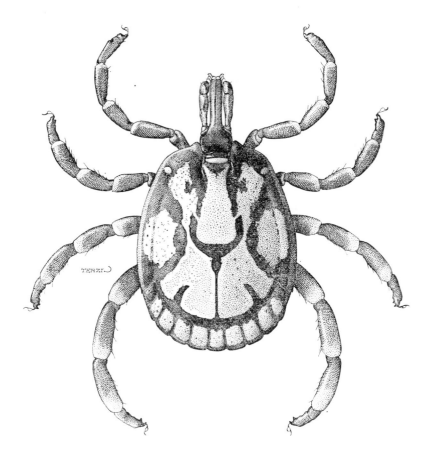

Fig. 197 The bont tick, *Amblyomma hebraeum*, dorsum of male (× 10).

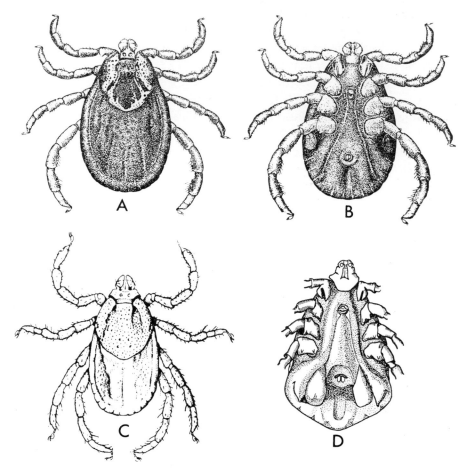

Fig. 198 A–B, *Dermacentor reticulatus* female: A, dorsum; B, venter; C–D, *Rhipicephalus sanguineus*: C, dorsum of female; D, venter of male (×*c*.12) (A–C after Pomerantzev, 1950).

muskrats—indeed OHF first attracted attention as a clinical disease shortly before World War II at about the time when American muskrats were introduced into Siberian waters for the fur industry.

The distribution of *D. marginatus* Schulze is rather similar to that of *D. reticulatus*, although this species has not been recorded from the British Isles. The host range of the two ticks is also rather similar. Omsk haemorrhagic fever virus has been isolated from *D. marginatus* (Casals *et al.*, 1966), and it is regarded as a secondary host of the virus of Central European tick-borne encephalitis (TE) in nature (Hoogstraal, 1966). *D. marginatus* appears to be a vector of Russian spring-summer encephalitis (RSSE) virus in certain parts of Kazakhstan, and it is also a reservoir and vector of Siberian tick typhus. In France, *Rickettsia conori*, has been isolated from this tick and from *D. reticulatus* (Giroud *et al.*, 1965*a*).

Amongst other medically important species of *Dermacentor*, *D. nuttalli* Olenev and *D. silvarum* Olenev have been reported as vectors and reservoirs of Siberian tick typhus, and in the easternmost part of the U.S.S.R. *D. silvarum* has been found naturally infected with RSSE virus.

Rhipicephalus

Rhipicephalus species are widely distributed in the Old World, but only one species of this genus, the cosmopolitan kennel tick or brown dog tick, *R. sanguineus* (Latreille) (figs 24–25), is found in America. The genus, which is dealt with in detail by Zumpt (1950) and Hoogstraal (1956), is made up of about 60 species, and several are important as reservoirs and vectors of a variety of animal and some human pathogens.

A number of species with a wide distribution in Africa, notably *R. appendiculatus* Neumann, *R. evertsi* Neumann and *R. simus* Koch, have been incriminated as vectors of boutonneuse fever, although there can be little doubt that the principal vector of this disease is *R. sanguineus*. This species, which is also a vector of Rocky Mountain spotted fever in the warmer parts of the Americas, has an extremely wide host-range, and is probably the most widely distributed tick. It is found in practically all countries between 50°N. and 35°S., and it is known to spread very rapidly once established in a new area.

Boophilus

Members of this genus, which are all one-host ticks, are chiefly important as pests of livestock. However, isolations of CHF-Congo virus have been made from *B. decoloratus* (Koch) in Nigeria and from *B. microplus* (Canestrini) in West Pakistan (reviewed Casals *et al.*, 1970). In Ethiopia serological and immunological tests have provided evidence of infection of adult *B. decoloratus* with *Rickettsia conori*, and in Brazil, *B. microplus* is considered to be a vector of Rocky Mountain spotted fever (reviewed Hoogstraal, 1967).

Family Argasidae

The Argasidae (soft ticks) attain their greatest abundance in dry regions and when they extend into humid areas they tend to select drier niches. The family is made up of four genera, three of which are of medical importance. They can be separated as follows:

1 Peripheral region of the body usually structurally different from the dorsum. Lateral suture usually present (fig. 193D) . . . . **ARGAS**
- Peripheral region of the body similar in structure to the dorsum. Lateral suture absent (fig. 199) 2
2 Adults with a granular integument, hypostome vestigial. Hypostome of nymphs normal. Integument of nymphs beset with spines . . . . **OTOBIUS**
- Integument of nymphs and adults mamillated or tuberculated, devoid of spines. Hypostome never vestigial 3
3 Hypostome broad at base and scoop-like. On bats . . . . **ANTRICOLA**
- Hypostome variable, never scoop-like . . . . . **ORNITHODOROS**

Argas

Members of this genus are most commonly associated with bats and birds although some species are found on reptiles and to a lesser extent on mammals other than bats. An account of the more important species is given by Hoogstraal (1956).

A. reflexus (Fabricius), the pigeon argas, appears to be a Middle or Near Eastern tick which has spread northwards into Europe and south-western Russia and eastward to

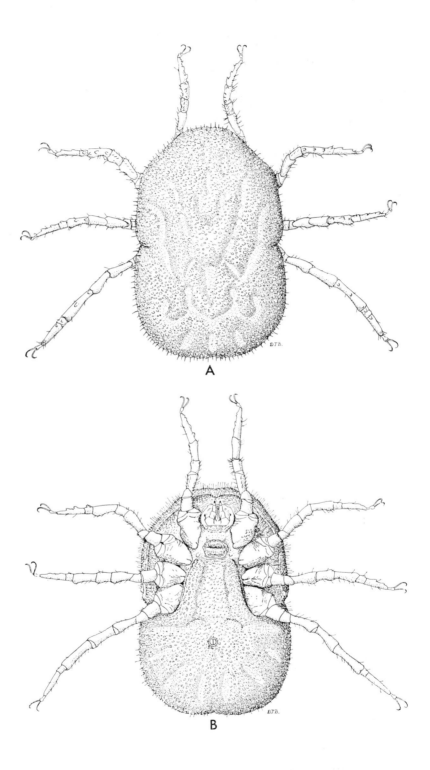

Fig. 199   The eyeless tampan, *Ornithodoros moubata*, female: A, dorsum; B, venter ($\times c.7\frac{1}{2}$).

India and other parts of Asia. There are also some records of this species from South America. Although pigeons are its main hosts (heavy infestations frequently cause death through exsanguination) a number of other hosts, both avian and mammalian, may be attacked. Man is frequently attacked, with painful sequelae, especially in the vicinity of unoccupied pigeon cotes. In Egypt, West Nile and Quaranfil viruses have been isolated from the subspecies *A. reflexus hermanni* Audouin (Hoogstraal, 1966 (review); Taylor *et al.*, 1966).

A. persicus (Oken), the fowl argas, is a serious pest of domestic poultry especially in tropical and subtropical regions, although this species is now established in most parts of the world between 40°N. and 30°S. It is a vector of a number of avian diseases including avian spirochaetosis. It may attack man, but accounts of the frequency and ferocity of attacks on humans have probably been exaggerated. A closely related species, *A. arboreus* Kaiser, Hoogstraal and Kohls, which is very numerous in rookeries of the common cattle egret (*Bubulcus ibis ibis*) near Cairo, has been found naturally infected with Quaranfil virus (Kaiser, 1966). High transmission rates of Quaranfil virus have been obtained in experimental studies with this tick (Kaiser, 1966*a*).

Argas vespertilionis (Latreille) (fig. 193D, E) has a wide Old World distribution and a wide host-range amongst bats. Both nymphs and adults readily bite man and cause a mild itching which may persist for several weeks.

Otobius

This genus is made up of two species, *O. megnini* (Dugès) and *O. lagophilus* Cooley and Kohls.

O. megnini, the spinose ear tick, is a serious pest of cattle in the warmer parts of North and South America and in Africa and India. The life-cycle is not typical of the Argasidae (Herms, 1917). Larvae enter the ears of cattle, sheep, horses and dogs, and, attaching themselves to the tender skin inside, feed for 7–14 days. They moult inside the ear, and the emerging nymphs, which have a spinose skin, feed (again in the ear) for 1–7 months. They then drop to the ground and climb up structures such as fence posts and buildings to moult. The adults, which do not feed, mate and lay eggs over a period of six months. The hosts pick up the larvae by rubbing up against structures in which they are hidden. There are several records of the occurrence of the nymphs in the ears of man.

O. lagophilus, which is a parasite of leporids in the western parts of North America and in Mexico, has been found infected with the rickettsiae of Rocky Mountain spotted fever and with the virus of Colorado tick fever.

Ornithodoros

Ornithodoros species are important as vectors of the spirochaetes which cause relapsing fevers, and in the case of the New World spirochaetes there is evidence of a tick-spirochaete specificity. For example, Brumpt (1933) showed that the spirochaetes of Texan relapsing fever (*S. turicatae*) were transmitted only by their natural vector, *O. turicata* (Dugès), and Davis (1956) found that when different species of *Ornithodoros* were fed on patients, only one, the natural vector, picked up the infection and was subsequently able to transmit it during feeding.

The only known vector of *Spirochaeta duttoni*, the causal organism of Central African relapsing fever, is the eyeless tampan, *O. moubata* (Murray) (Hoogstraal, 1956). However, Walton (1962) regarded Murray's *moubata* as a super-species complex made up of four species, and he retained the name *moubata* for one of them. *O. moubata* (Murray) (fig. 199), in the restricted sense, is a species of arid conditions and its wild and domestic populations (the latter being those inhabiting human dwellings) may be subspecies, races or demes. It has a wide distribution in Central, East and southern Africa. Heavy populations of the 'wild' form, which tend to be larger than domestic specimens, have been found in burrows of large mammals such as porcupines and warthogs, and it is probable that this tick did not invade human dwellings until very recent times. A second species of the complex is *O. porcinus* Walton, and a domestic subspecies of this is considered to be primarily a human parasite of (relatively) moist conditions. The main host of the 'wild' subspecies, *O. porcinus porcinus* Walton, is the warthog.

O. tholozani Lalboulbène and Mégnin, the Persian relapsing-fever tampan, which occurs in India, Iran, Russian Turkestan, the Middle East and North Africa, is a vector of *S. persica*, the causal organism of relapsing fever in the eastern part of the Mediterranean endemic region. The causal organism in the western part of this region, *S. hispanica*, is vectored by *O. erraticus* (Lucas). This species, which occurs in nature as two so-called forms, large and small, is found in Iran, Turkey and throughout much of the African and European Mediterranean area, and in West and East Africa.

In southern, central and western U.S.A., the three causal organisms of relapsing fever, *S. turicatae*, *S. parkerii* and *S. hermsii*, are named after the species of *Ornithodoros* which transit them. *O. turicata* (Dugès) is widely distributed in the southern United States and also occurs in Mexico. It readily attacks man in all its stages and its bite can be painful. *O. parkeri* Cooley, and *O. hermsi* Wheeler, Herms and Meyer, which are primarily parasites of rodents, are widely distributed in the Rocky Mountain and Pacific Coast states. *O. rudis* Karsch is considered to be the most important vector of the spirochaete of relapsing fever (*S. venezuelense*) in Panama, Colombia, Venezuela and Ecuador, although *O. talaje* (Guérin-Méneville) may also be a spirochaete vector in Central and South America.

The eyed tampan, *O. savignyi* (Audouin), is distributed locally throughout the arid parts of North, East and South Africa. It is not a vector of relapsing fever, but it is a serious pest of camels and cattle, which may be killed through exsanguination. It frequently bites man with very painful sequelae. After parenteral infection this species has been found to transmit West Nile and Sindbis viruses when biting suckling mice (reviewed Hoogstraal, 1966).

Ornithodoros coriaceus Koch occurs in Mexico and in California where it is common in the more mountainous coastal counties. This tick is known in Mexico as the 'talaja' and in California as the 'pajaroello'. Although this species has not been implicated as a vector of disease organisms its bites can be painful and it is widely feared (Nuttall *et al.*, 1908; Herms, 1969; Waldron, 1962). The bites were described by Nuttall *et al.* as being intolerably sharp and painful and they noted that an intermittent irritation persisted after four months. Eight months after the bite was inflicted there remained a nodule which occasionally itched. A rather similar description of the bites is given by Herms who also noted that the lesions exuded a clear lymph from beneath a scab which remained in evidence for two or three months. Waldron, however, on the basis of data collected from Forest Service personnel, noted that the bites were not invariably painful—indeed in most cases no pain was felt at the time of attachment—and that the reaction to the bites appeared to be a direct allergic response on the part of the host to the saliva of the tick.

Acari-Mesostigmata

In this order of mites the stigmata are located dorso-laterally in the region of coxae II–IV. Associated with the stigmata are long anteriorly directed grooves called peritremes whose function is not fully understood. The order contains over 50 families, and from the medical point of view the most important is the family Dermanyssidae to which all the species listed below belong. Taxonomic accounts of this family are given by Strandtmann and Wharton (1958) and by Evans and Till (1965, 1966).

Dermanyssus gallinae (DeGeer) (fig. 200A, B)
Commonly known as the red mite of poultry, this species is an obligatory blood-sucking parasite with a wide host-range amongst wild and domestic birds. In temperate regions it is an important pest of poultry; turkeys and fowls being the most susceptible to attack. *D. gallinae* is nocturnal; the mites engorge themselves quickly and leave the birds to hide in dark crevices and cracks in perches. It may also attack man, causing a severe irritation, and it may cause a skin ailment resembling strophulus or animal scabies (Cremer and Morrien, 1962). In urban areas minor infestations of dwelling houses, due to a migration of mites from abandoned nests and roosting sites, are not uncommon. Smith *et al.* (1948) have reported that *D. gallinae* is capable of transmitting the virus of St. Louis encephalitis amongst chickens, but its exact role in the epidemiology of this disease is not known.

Liponyssoides sanguineus (Hirst) (fig. 200C, D)
From the medical point of view *L. sanguineus* may be the most important of all the dermanyssid mites. First described in 1914 as a parasite of rats in Egypt, it has since been found parasitizing rats and mice in the U.S.A. Medical interest was aroused with the recovery of *Rickettsia akari*, the aetiological agent of rickettsial pox, from specimens taken from buildings in New York where human cases of the disease had occurred (Huebner *et al.*, 1946). More recently Zdrodovskij (1964) has cited this mite as an apparent vector of rickettsial pox in the Ukraine, and the rickettsiae of Q-fever (*Coxiella burneti*) have been isolated from specimens feeding on laboratory animals (Zemskaya and Pchelkina, 1968).

Ornithonyssus bacoti (Hirst) (fig. 201A, B)
Commonly known as the tropical rat mite, this species is cosmopolitan in distribution but in temperate areas it occurs mainly in sea ports. Although primarily a parasite of rats, there are numerous records of it biting man. Bites are painful and result in the formation of an irritant papule up to 18 mm ($\frac{3}{4}''$) in diameter. Hopla (1951) has shown that under experimental conditions *O. bacoti* can become infected with *Pasteurella*

GG

tularensis. In southern U.S.A. and in Chile, murine typhus rickettsiae have been recovered from specimens, but the mite is believed to be of little importance in the epidemiology of this disease (Worth and Richard, 1951).

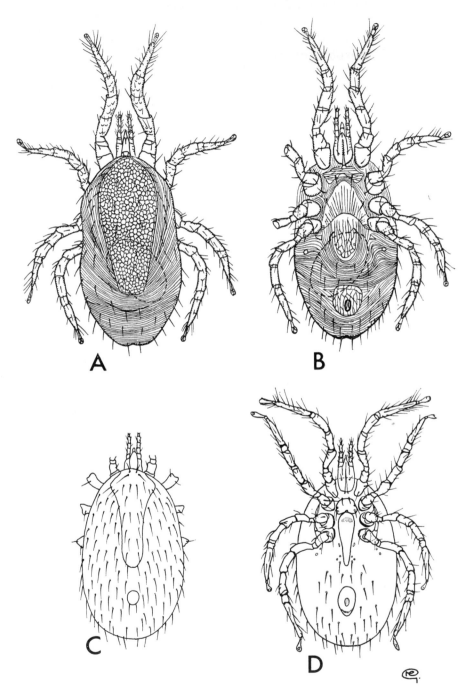

Fig. 200 A–B, the red mite of poultry, *Dermanyssus gallinae*, female: A, dorsum; B, venter (×90); C–D, *Liponyssoides sanguineus*, female: C, dorsum; D, venter.

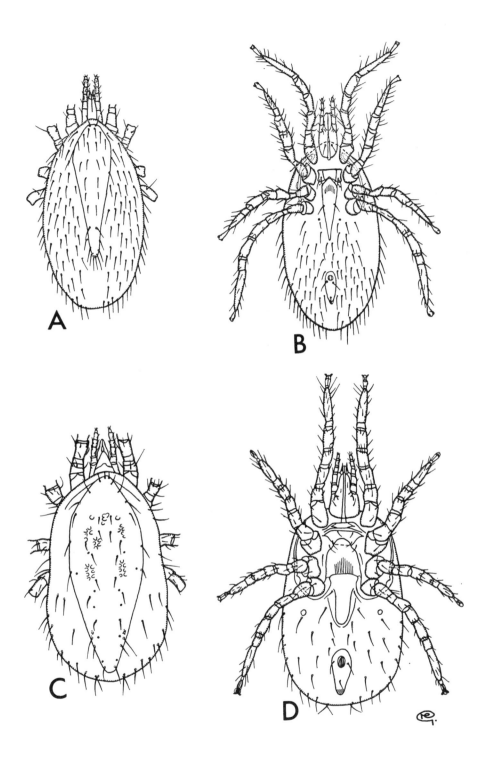

Fig. 201 A–B, the tropical rat mite, *Ornithonyssus bacoti*, female: A, dorsum; B, venter (both much enlarged); C–D, the northern fowl mite, *Ornithonyssus sylviarum*, female: C, dorsum; D, venter (×100).

Ornithonyssus sylviarum (Canestrini and Fanzago) (fig. 201C, D)

This species, which is known commonly as the northern fowl mite, is a blood-sucking ectoparasite with a wide host-range amongst wild birds and domestic poultry. Unlike *D. gallinae*, this mite spends the whole of its life-cycle on the host. It is a serious pest of domestic poultry in temperate regions, and it will bite man, causing some irritation. In the U.S.A. the viruses of western equine encephalitis and St. Louis encephalitis have been isolated from specimens taken from wild birds' nests (Reeves *et al.*, 1955).

Ornithonyssus bursa (Berlese) (fig. 202)

O. bursa, the tropical fowl mite, is an important blood-sucking pest of domestic poultry in tropical and subtropical regions. This mite will also bite man, causing some discomfort.

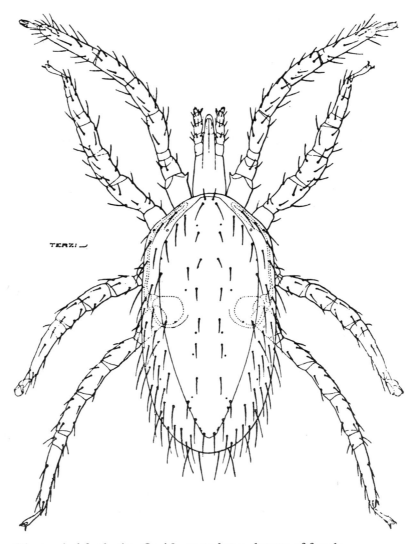

Fig. 202　　The tropical fowl mite, *Ornithonyssus bursa*, dorsum of female.

Androlaelaps casalis (Berlese)

This cosmopolitan species is a common component of the mite fauna of the nests of birds and small mammals. In Britain it is frequently found in enormous numbers in broiler-chicken houses. While it may take blood from abraded skin areas, it appears to be a general feeder and can subsist, for example, on chicken faeces and on farinaceous material. It has also been observed feeding on the eggs of *D. gallinae*. It is reputed to cause irritation in man, possibly by biting (see for example, Hill and Gordon, 1945, who refer to this mite as *Hypoaspis fenilis* (Mégnin)). *A. casalis* is amongst nine dermanyssid species associated with rodents and birds which Zemskaya and Pchelkina (1968) report as having been found infected with the rickettsiae of Q-fever (*Coxiella burneti*).

Additionally a number of other dermanyssid species normally associated with birds or small mammals have been found infected, either naturally or under experimental conditions, with human pathogens. For example, Alifanov *et al.* (1961) isolated the virus of Omsk haemorrhagic fever from the following dermanyssids combed from experimentally infected water voles (*Arvicola terrestris*): *Laelaps muris* (Ljungh), *Hyperlaelaps amphibia* Zachvatkin, *Hirstionyssus isabellinus* (Oudemans) and *Androlaelaps fahrenholzi* (Berlese) (cited as *Haemolaelaps glasgowi*). The virus was also recovered from *Ondatralaelaps multispinosa* (Banks) combed from experimentally infected musk-rats (*Ondatra zibethica*). Again a number of dermanyssids have been found infected with the virus of Russian spring-summer encephalitis (reviewed Naumov *et al.*, 1963), but, as these authors point out, the role of these mites in the epidemiology of the disease has yet to be properly elucidated.

Acari-Astigmata

Members of this order are weakly sclerotized mites, which, as adults, range in size from about 200–1200 μm. The palps are two-segmented and lie closely adpressed to the ventral components of the mouth-parts (infracapitulum). With few exceptions the chelicerae are pincer-like, and the idiosoma, which carries relatively few setae, is often divided by a groove, the sejugal furrow, into a propodosoma and hysterosoma. Stigmata and tracheae are never present. The order includes mycetophagous forms and detritus feeders as well as parasites. Three families are of medical importance.

Acaridae (*s. lat.*)

These mites are best known as pests of stored food products, but several species have been implicated as a cause of a contact dermatitis in persons handling infested material. While the mites may be a primary cause of the dermatitis, it has been suggested (Baker *et al.*, 1956), that a hypersensitivity is necessary and that the dust in the infested material might also be a contributory factor. The more familiar 'stored product itches' and the mites alleged to cause them are: copra itch (*Tyrophagus putrescentiae* (Schrank)), (figs 37–38); wheat-pollard itch (*Suidasia nesbitti* Hughes); bakers' itch (*Acarus siro* L.); dried-fruit dermatitis (*Carpoglyphus lactis* (L.)) and grocers' itch (*Glycyphagus domesticus* (DeGeer)). Grocers' itch is often used as a generic term for all these conditions.

Species of the genus *Glycyphagus*, particularly *G. destructor* (Schrank) (fig. 203C), are

often very abundant in hay, and Voorhorst, Spieksma and Varekamp (1969) present
evidence to suggest that the allergenic factor in hay causing rhinitis, itching eyes and
wheezing in sensitized persons is due mainly to the allergen produced by *Glycyphagus*
mites.   Evidence from skin-reaction tests also suggested that an identical allergen may
be produced by *Goheria fusca* (Oudemans), a species frequently found in large numbers
in flour.

Additionally, infestations of the urinary and intestinal tracts by certain Acaridae and
Pyroglyphidae (see below) have been reported, and these infestations have been associated
in some cases with a clinical picture suggesting an irritation of the urinary tract and with

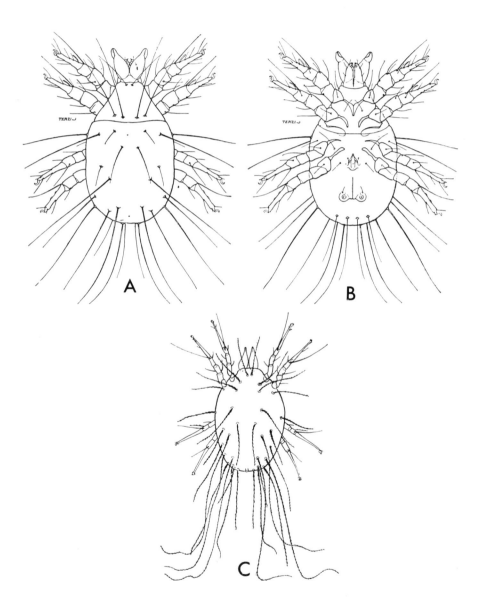

Fig. 203   A–B, *Tyrophagus putrescentiae*, male: A, dorsum; B, venter (×*c.*90); C, *Glycyphagus
destructor*, female (×55).

intestinal upsets (see for example Mekie, 1926; Hinman and Kampeier, 1934, and Sasa, 1950). Species of Acaridae and Pyroglyphidae as well as certain prostigmatid mites have also been recovered from the sputum of patients with lung disorders (Carter *et al.*, 1944; Taboada, 1954). However, Sasa (1964), reviewing the relationship between human acariasis and mites in stored food and drugs in Japan, concluded that there was 'no reliable proof whether the mites were actually parasitic in human tissues and caused the disease, or were the accidental contaminants of the human excrement or of the laboratory equipment'.

Pyroglyphidae

This family is made up of five genera, of which one, *Dermatophagoides* Bogdanov (=*Mealia* Trouessart), is of outstanding importance in human medicine. *Dermatophagoides* species are found on the skin surface of mammals and birds and they are common components of the mite fauna of nests. It has been hypothesized that their principal food is skin detritus, but some species at least appear to be fairly general feeders since they are frequently found in large numbers in stored food products such as fish meal and poultry and pig-rearing meal (see for example Hughes, 1961). They are also commonly found in dust from granaries and dwelling houses.

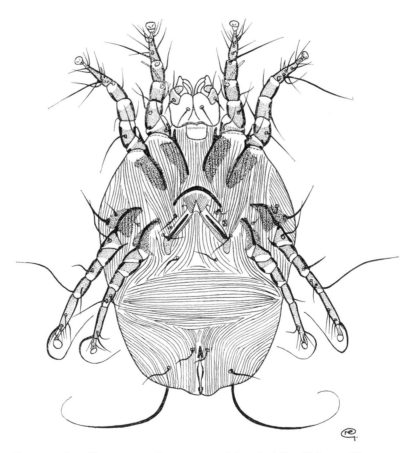

Fig. 204   *Dermatophagoides pteronyssinus*, venter of female (after Fain, 1966).

Dermatophagoides mites have been reported to cause a severe dermatitis in man (see for example Traver, 1951; Baker *et al.*, 1956; Dubinin *et al.*, 1956).* Traver (*op cit.*) reported infestations of the scalp which persisted for at least seven years. The mites evidently burrow into the dermis and only occasionally appear on the surface of the skin. Symptoms of attack include an intense irritation, the appearance of small red papules and a thickening of the epidermis over the affected areas.

Voorhorst, Spieksma-Boezma and Spieksma (1964) found that among the many constituents of house-dust causing allergic reactions of the respiratory tract (asthma and rhinitis) in persons sensitized to house dust, *D. pteronyssinus* (fig. 204) produced the most potent allergen. In Britain this species has been found to occur frequently in mattress dust and it can become airborne during bedmaking (Cunnington and Gregory, 1968). Maunsell, Wraith and Cunnington (1968) examined 186 samples of dust from houses in London, the Home Counties and South Wales—176 from the homes of bronchial asthmatics and 10 from the houses of non-allergic persons. *D. pteronyssinus* was found in 82 per cent of the samples and 94 per cent of the house-dust-sensitive patients gave skin reactions to *D. pteronyssinus* extracts. More recently, Voorhorst, Spieksma and Varekamp (1969), after considering the evidence from a long series of surveys and experiments, concluded that *D. pteronyssinus* 'produces a very potent allergen which, on the basis of the skin-test method, cannot be distinguished from the house-dust allergen, and that this species of mite must be considered as the cause of the specific allergenic potency of house-dust'.

In Japan, Miyamoto *et al.* (1968) found that a close antigenic identity existed between house-dust and *Dermatophagoides farinae*. This species has been found to occur in houses in Europe although it is evidently less common in dust samples than *D. pteronyssinus*. In the U.S.A., on the other hand, *D. farinae* seems to be the more frequently encountered species. For example, in Ohio, Larson *et al.* (1969) found that all the adult *Dermatophagoides* mites in the house-dust samples which they collected were *D. farinae*. In a comparison of the skin reactions of 16 atopic patients, Voorhorst *et al.* (1969) could detect no differences between the extracts of cultured *D. farinae* and *D. pteronyssinus*.

Another pyroglyphid mite, *Euroglyphus maynei* (Cooreman), has also been found to occur very frequently in house-dust in Europe. Voorhorst *et al.* (1969) found that this mite produced an allergen which gave identical skin reactions when compared in equivalent test fluids with extracts of house-dust and cultured *D. pteronyssinus*.

Sarcoptidae

The sarcoptids are skin parasites of warm-blooded animals. Some species cause mange in mammals and others are responsible for 'scaly leg' and loss of plumage in birds. The scabies or itch mite, *Sarcoptes scabiei* (L.) (fig. 205), infests mammals, and although no constant morphological differences have been detected between the forms occurring on different mammalian hosts, considerable physiological differences must exist since mites taken from one host species will not usually establish themselves permanently on another. Thus, at present, it is considered that each known mammalian host has its own biological race of *S. scabiei*.

The human itch mite, *S. scabiei f. scabiei* has been monographed by Mellanby (1943). The main infective stage is probably the newly fertilized female, and having found a new host, the mite selects a site for burrowing. The most favoured areas are the hands and wrists, although the elbows, feet and scrotal area are also frequently selected. The mite takes about an hour to bury itself in the cornified epithelium, and its burrows never penetrate below this layer. During the process of burrowing the mite fixes itself firmly to the substratum by means of ambulacral suckers, and uses its chelicerae as well as cutting edges located on the tibiae of the first pair of legs to cut a channel into the skin. Egg-laying begins within a few hours of the commencement of burrowing, the eggs being laid at the rate of two or three per day for a period of up to two months. The eggs hatch in three or four days and the larvae move out of the burrows to wander freely over the skin before finding shelter, and probably also food, in the hair follicles. Here they moult to become nymphs. There are two nymphal instars and both are found in the follicles. Males are comparatively rare and are found in short burrows less than 1 mm in length. The fertilized females also excavate short burrows in which they remain for two or three days. Fertilization probably takes place on the surface of the skin and the whole life-cycle may be completed in 10–14 days.

In most cases of scabies the mite population is very low, often in the region of six adult mites per patient. Infestation by the mite causes a severe itching, which is often worst at night, and a peculiar rash characterized by erythematous patches and follicular papules. Although partly due to the parasite, the rash is caused to a greater extent by the patient's scratching. There is evidence to suggest that, before the symptoms arise, the patient must first be sensitized to the mites, for cases have been observed in which

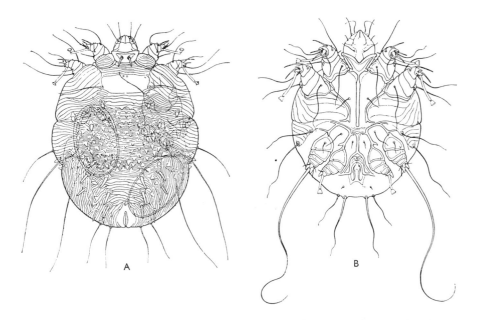

Fig. 205   The itch mite, *Sarcoptes scabiei*: A, dorsum of female; B, venter of male (×115).

the mites were present for several weeks without causing the least sensation. In persons with no previous history of scabies, mite activity can proceed for about a month before producing a general rash and symptoms of itching. After this period drastic changes occur; the erythematous patches appear around the burrows and severe itching commences.

Man can also become infested by races of *Sarcoptes scabiei* from domestic animals (see for example Chakravorty *et al.* 1953), but such infestations are usually short lived. Infestations of this sort, which have been called 'animal scabies', produce a complex of symptoms resembling those of typical scabies, but they can be distinguished from the latter by the absence of mite burrows in the epidermis and by the sites of infestation which are generally the points of contact with the infested animal.

In a comparatively rare, but well-defined, type of scabies in man, known as Norwegian or crusted scabies, the skin over the mite-infested areas is covered with thick keratotic crusts, and, in contrast to typical scabies, enormous numbers of mites are present. In this form of scabies itching is much less pronounced and is not a distressing symptom. Norwegian scabies was at one time thought to be due to a distinct variety of *S. scabiei*, *S. scabiei f. crustosae* Fürstenberg, which, although closely resembling the typical species, was said to differ from it in relation to the lengths of certain dorsal spines (Fürstenberg cited by Buxton, 1921; Canestrini and Kramer, 1899). However, Buxton (*op. cit.*), in a discussion of the aetiology of this condition, largely discounted this view and argued that as Norwegian scabies was such a rare condition, a distinct host-specific causative mite would be unlikely to survive. He also noted that a secondary infection of typical scabies by moulds could not be dismissed as a possible cause. Norwegian scabies has also been attributed to infestations by races of *S. scabiei* from other animals, but again recent observations (for example those of Burks *et al.*, 1956 and Pirila *et al.*, 1967) fail to support this view. Mellanby (1943) suggested that the condition might arise in patients who had developed an immunity to the effects of the mites so that large numbers could be present without symptoms of itching. On the other hand, Burks *et al.* (*op. cit.*) present evidence to suggest that Norwegian scabies develops in patients who *fail* to develop the *Sarcoptes* antibody, and the absence of this antibody is the reason for the unrestricted growth of the mite population as well as for the absence of itching, which in typical scabies is thought to occur as a result of the sensitization of the skin to the mite antigen.

The acaricide most frequently used in the treatment of scabies is benzyl benzoate. A 25 per cent emulsion is painted over the body from the neck down with a soft flat paint brush about two inches wide and the patient allowed to dry in a warm room for 10–15 minutes. When properly used this substance is extremely effective and one treatment is usually sufficient to kill all the mites. However, it may not be well tolerated by young children. Sulphur ointment BPC and certain organic sulphur preparations, in particular dimethyl diphenyl disulphide ('Mitigal') and tetraethylthiuram mono-sulphide ('Tetmosol'), are also effective. Being greasy, sulphur ointment is rather unpleasant to use and it is liable to produce a sulphur dermatitis. Two treatments, either on succeeding days within the period of one week, have been recommended, and approximately three ounces of the ointment are required per treatment. 'Mitigal' is more pleasant to use, but it also is liable to produce a dermatitis although to a lesser extent than inorganic sulphur. 'Tetmosol' rarely produces a dermatitis and is particularly valuable for the treatment of scabies in young children. It is slower in action than benzyl benzoate and three or more treatments, 24 hours apart, have been recommended.

Acari-Prostigmata

The Prostigmata is the most heterogeneous of all the orders of the Acari, and its components include aquatic species (mainly freshwater, but some marine), phytophagous, mycetophagous, predatory and parasitic terrestrial species, as well as free-living terrestrial species which are parasitic in the larval stage. The chelicerae may be pincer-like but they are often modified in various ways to become, for example, hook-like structures or piercing stylets. Similarly, the structure of the pedipalps is extremely variable, and in many families the tibia and tarsus of this appendage are opposable and form a 'thumb-claw' complex. When present, the tracheae open through paired stigmata situated between the chelicerae or further back on the dorsal surface of the propodosoma. Peritremes may be present. This order contains over 50 families, one of which, the Trombiculidae, is of outstanding importance in human medicine. Two other families, the Pyemotidae and Demodicidae, are of relatively minor importance. Additionally, members of other families have on occasion been associated with minor skin disorders. For example, *Cheyletiella parasitivorax* (Mégnin) (family Cheyletidae), which commonly occurs on the fur of cats and rabbits and which probably feeds mainly on epizooic species such as the fur mites, has been found in association with various skin lesions in man and other animals. Cooper (1946), after reviewing the literature, considered that there was insufficient evidence to attribute the injuries directly to this mite, although Hughes (cited by Evans *et al.*, 1961) has observed *C. parasitivorax* with its chelicerae actually embedded in the host.

Pyemotidae

The pyemotids are lightly sclerotized mites in which the palps are reduced and the chelicerae stylet-like. *Pyemotes ventricosus* (Newport) (fig. 206), the grain itch mite, which is primarily a parasite of insects infesting grain, straw and other stored products, will bite man (Swan, 1934; Booth and Jones, 1952). Agricultural workers are chiefly affected but industrial workers and others have been attacked through occupational or incidental exposure to infested materials. Infestation rates in humans reaching epidemic proportions have occurred in the mid-western part of the U.S.A. (Booth and Jones, *op. cit.*). Attacks appear to be most frequent in hot weather. All parts of the body may be affected, and, although the mites do not burrow into the skin, their bites cause a severe skin eruption. The weals vary in size and form but are usually characterized by a raised whitish area with a small central vesicle marking the site of the bite. Rubbing and scratching may break the vesicles and lead to secondary infections. On its normal hosts *P. ventricosus* has a curious life-cycle. As the fertilized female feeds the opisthosoma becomes enormously distended (fig. 206B) and the eggs develop and hatch within this sac. The young mites complete their development within the parent and do not emerge until they are sexually mature. At first males are produced and these remain attached to the parent, from which they obtain sustenance, becoming grouped around the genital orifice (fig. 206B). Later females are produced and these are fertilized by the males as soon as they are born.

Booth and Jones (*op. cit.*) note that the local application of acaricides such as sulphur and benzyl benzoate is a valuable prophylactic measure. They also point out that, as the eruption in grain itch is self limiting once the contactant material is removed, the purpose of active treatment is to relieve pruritus and to prevent or eliminate secondary bacterial infection. A lotion of warm water and vinegar was the treatment credited as giving the greatest relief by chronically infested workers in a factory manufacturing strawboard. Swan (*op. cit.*) found that a saturated solution of picric acid in 90 per cent alcohol was effective in relieving the irritation.

Demodicidae

Commonly known as follicle mites, members of the genus *Demodex* burrow into the hair follicles of mammals and feed on subcutaneous secretions, particularly sebum (Spickett, 1961). A taxonomic and biological survey of the genus is given by Hirst (1919). *D. folliculorum* (Simon) (pl. 12) inhabits the hair follicles and sebaceous glands of man, particularly around the nose and eyelids. Infestations of the scalp have also been reported (Miskjian, 1951). It is probable that most persons harbour this mite and infestations are not generally considered to be important. However, there are reports of the mites being associated with conditions such as acne rosacea, blepharitis and impetigo contagiosa (Hirst, *op. cit.*) and more recently Russell (1962) has suggested that the mites may play a contributory role in the complex aetiology of rosacea. Although treatment is rarely necessary it can be noted that a case of scalp demodicidosis was easily cured with an ointment containing 10 per cent sulphur and 5 per cent Peruvian balsam (Miskjian, *op. cit.*).

Trombiculidae

Larval trombiculids, commonly known as chiggers or red bugs, are ectoparasites of vertebrate and to a lesser extent of arthropod hosts, but the nymphs and adults are free-living predators feeding mainly on soil arthropods. Since Trombiculidae are most easily collected as larvae, most of the currently recognized species have been described in the larval stage only, the nymphs and adults being either unknown or uncorrelated with the larvae. Over 700 species have been described and about 20 of these are important either as a cause of dermatitis (trombidiosis or scrub-itch) in man or as vectors of human pathogens. The family has been monographed by Wharton and Fuller (1952) and the genera and subgenera of the Far East are keyed by Vercammen-Grandjean (1968).

The eggs of chiggers are usually laid in the soil, and, on hatching, the larvae crawl on the surface of the soil or lurk in the ground vegetation until they find a suitable host. The larvae tend to select attachment sites on parts of the body where the skin is thin. In the case of small mammals favoured sites are the ears or around the genitalia, while in man they are most frequently found around the ankles, in the groin and in the armpits, but attachment also occurs in regions constricted by tight clothing. The mode of

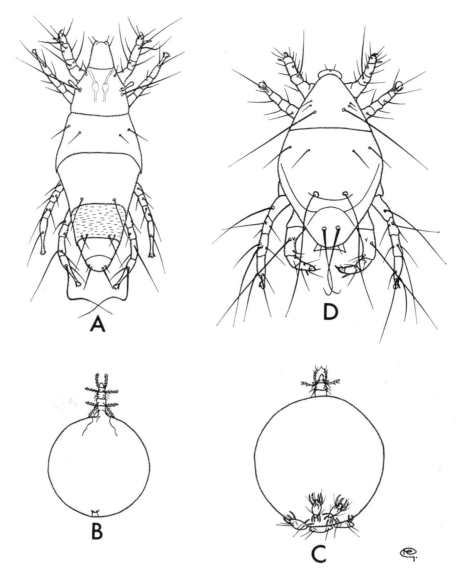

Fig. 206 The hay-itch mite, *Pyemotes ventricosus*: A, dorsum of female (×*c*.315); B, dorsum of gravid female (×35); C, males grouped around genital opening of gravid female; D, dorsum of male (×*c*.315).

feeding on vertebrate hosts has been studied in some detail (Jones, 1950; Wharton and Fuller, 1952). The mites insert their scimitar-like chelicerae into the surface layers of the skin and the histiolytic action of their salivary secretion leads to the formation in the hosts' tissue of a feeding tube (cytostome or stylosome) which penetrates the lower strata of the epidermis as far as the malpighian layer and sometimes into the underlying dermis. The mites feed by sucking back lymph and the contents of disintegrated cells through the stylosome. No blood is taken. After engorgement from a single feed the larvae move into the soil to enter a quiescent phase before moulting to become nymphs. The latter live freely as predators and eventually enter another quiescent phase before undergoing their final moult.

Scrub-itch in man, which is the result of an allergic reaction to the saliva of the chigger, can be caused by about 15 trombiculid species. The majority of these belong to the genera *Eutrombicula* and *Schoengastia*, although in Britain the condition is caused by *Neotrombicula autumnalis* (Shaw) (fig. 207A), the familiar harvest mite. This species also has a wide distribution in continental Europe. The larvae, which first appear in July, are most abundant in early September. In Britain the most favoured natural host of this species is the rabbit. The most common species attacking man in the western hemisphere is *Eutrombicula alfreddugesi* (Oudemans) which has a distribution ranging from Canada through the U.S.A. and the West Indies to South America. In the U.S.A. trombidiosis is also caused by *E. splendens* Ewing, a species which is widely distributed along the Gulf and Atlantic coasts and which also occurs northwards and westwards to Ontario, Michigan and Minnesota. *E. batatas* (L.), which is an important dermatitis-inducing species in tropical America, has also been recorded from southern U.S.A. In South East Asia, Australia and the Pacific islands scrub-itch is commonly caused by *E. wichmanni* (Oudemans), *E. hirsti* Sambon and by various species of *Schoengastia*.

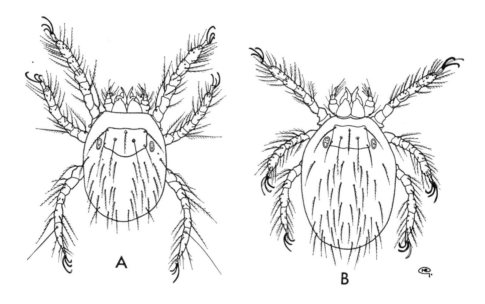

Fig. 207   A, the harvest mite, *Neotrombicula autumnalis*, larva (×c.100); B, *Leptotrombidium akamushi*, larva.

Chigger bites can be prevented by treating clothing, particularly socks or stockings, cuffs and collars, with mite repellents. It should be noted however that some synthetic fabrics may be harmed by the repellents. The most efficient chigger repellent is diethyltoluamide, but dimethyl phthalate, dibutyl phthalate and benzyl benzoate are also effective. Herms (1961) notes that the following formulation is useful for the temporary relief of itching: benzocaine, 5 per cent; methyl salicylate, 2 per cent; salicylic acid, 0.5 per cent; ethyl alcohol, 73 per cent; water, 19.5 per cent. The application of this medicament to weals should give relief for an hour or more and treatment may be repeated as often as necessary.

The most important mite-borne disease of man is scrub typhus or tsutsugamushi disease, the aetiological agent being *Rickettsia tsutsugamushi*, and the role of mite vectors in the natural history of this disease has been reviewed by Audy (1949, 1958, 1968). The disease occurs in many parts of eastern and south-eastern Asia (including the Indian subcontinent and Indonesia), New Guinea, northern Australia and certain islands of the Indian and South-West Pacific Oceans, and the principal vectors are two closely related mites classified in the genus *Leptotrombidium*, namely *L. akamushi* (Brumpt) (fig. 207B) and *L. deliense* (Walch). These species rarely if ever cause trombidiosis and their bites often pass unnoticed. The mites feed only once during the larval instar and the rickettsiae are passed trans-stadially to the nymph and adult and thence transovarially to the larvae of the next generation.

Within its range, scrub typhus occurs in isolated and usually sharply delineated areas, and the factors which favour the disease are those which favour the natural hosts of the vectors (principally rodents and insectivores) without disturbing the soil in which the free-living nymphal and adult stages of the vectors live. Thus scrub typhus is generally a serious problem in areas where habitats such as neglected gardens and plantations provide an abundant food supply for rodents. In fact any subclimax vegetation with rank undergrowth will support a high population of small mammals and a correspondingly high chigger population. In particular, 'fringe habitats', that is to say habitats in intermediate zones between two major types of ecological unit, are potentially dangerous sources of infection. Fringe-habitats include a range of habitat types from bushy hedgerows to narrow forest strips as well as the scrub forest-fringe, and, under the influence of peculiar 'edge effects', they appear to be remarkable sanctuaries of small animal life. They also tend to conserve and accompany moisture, thus allowing chiggers to flourish into the dry season. Audy (1968) comments that one rule of thumb which emerged from studies in Imphal during the Second World War was that the degree of risk should increase with the number of blocks of vegetation of different types in a unit area, for this number, which could easily be determined from aerial photographs, was in effect a measure of the extent of edge or fringe-habitats.

Although the principal natural hosts of *L. akamushi* (and probably the main animal reservoirs for the scrub-typhus rickettsiae) are ground-dwelling rodents (particularly field rats) and certain insectivores, the list of recorded hosts for the mite is very extensive, and, in addition to man, includes cattle, carnivores, marsupials and birds. *L. akamushi* has a wide distribution extending from Japan and China southwards through South-East Asia to Indonesia and eastwards through the Philippines to New Guinea. It occurs also on many Indian Ocean islands, including Ceylon. In Japan this mite is found only in grassland in certain river valleys in north-western Honshu (the classical areas of tsutsugamushi fever), and in equatorial countries, although it has been recovered from scrub terrain on the margins of jungles, it occurs principally in grassland and on the hosts found in this type of habitat (such as the rice-field rat and quail in Malaya).

While not occurring so far north, *L. deliense* has a rather wider east-west distribution than *L. akamushi*. It has been recorded from China, the Indian subcontinent, Burma, Malaya, Indonesia, the Philippines, New Guinea, Australia and from many of the smaller islands in the Indian and South-West Pacific Oceans. As in the case of *L. akamushi* its principal natural hosts are rodents and insectivores. It is abundant in grassland as well as in fringe habitats and it has also been found in secondary scrub and in forest with leaf litter. It even occurs in small numbers in deep jungle.

In addition to these two major and widely distributed vectors, other species of *Leptotrombidium* are scrub-typhus vectors of local importance. For example, *L. pallida* Nagayo *et al.* is a vector in some foci near Tokyo and *L. scutellaris* Nagayo *et al.* is almost certainly the sole vector of the mild 'winter scrub typhus' (Shichto fever) of southern Honshu and the islands south of Yokohama. Many other species, while not attacking man, are important in the maintenance of enzootic infections.

BIBLIOGRAPHY

ALIFANOV, V. I., ZAKORKINA, T. N., NETSKY, G. I. & FEDOROV, V. G. 1961. [Experimental data relative to the role played by Gamasoidea in transmission of the tick-borne encephalitis and Omsk haemorrhagic fever viruses.] *Med. Parazit.* **30**: 24–26. (In Russian, English summary pp. 121–122.)

ANASTOS, G. 1950. The scutate ticks, or Ixodidae, of Indonesia. *Ent. Amer.* **30**: 1–144.

ARTHUR, D. R. 1960. *Ticks: A Monograph of the Ixodoidea. Part V. On the Genera* Dermacentor, Anocentor, Cosmiomma, Boophilus *and* Margaropus. 251 pp. Cambridge: Cambridge University Press.

—— 1962. *Ticks and Disease.* 445 pp. London.

—— 1963. *British Ticks.* 213 pp. London.

—— 1965. *Ticks of the Genus* Ixodes *in Africa.* 348 pp. London: University of London.

—— 1966. The ecology of ticks with reference to the transmission of Protozoa. pp. 61–84. In *Biology of Parasites.* Edited by E. J. L. Soulsby. New York: Academic Press.

ATKINS, J. A., WINGO, C. W., SODEMAN, W. A. & FLYNN, J. E. 1958. Necrotic arachnidism. *Am. J. trop. Med. Hyg.* **7**: 165–184.

AUDY, J. R. 1949. A summary topographical account of scrub typhus 1908–1946. *Bull. Inst. med. Res., Malaya* (N.S.) **1**: 1–86.

—— 1958. The role of mite vectors in the natural history of scrub typhus. *Proc. 10th int. Congr. Ent., Montreal 1956* **3**: 639–649.

—— 1968. *Red Mites and Typhus.* 191 pp. London.

BAERG, W. J. 1929. Some poisonous arthropods of North and Central America. *Trans 4th int. Congr. Ent., Ithaca 1928* **2**: 418–438.

—— 1959. Black widow and five other venomous spiders in the United States. *Bull. Ark. agric. Exp. Stn,* No. **608**: 1–43.

BAKER, E. W., EVANS, T. M., GOULD, D. J., HULL, W. B. & KEEGAN, H. L. 1956. *A Manual of Parasitic Mites of Medical or Economic Importance.* 170 pp. New York.

BAKER, E. W. & WHARTON, G. W. 1952. *An Introduction to Acarology.* 465 pp. New York.

BALOZET, L. 1964. Le scorpionisme en Afrique du Nord. *Bull. Soc. Path. exot.* **57**: 33–38.

BASU, U. P. 1939. Observations on scorpion-sting and snake-bite. *Am. J. trop. Med.* **19**: 385–391.

BERLAND, L. 1932. *Les Arachnides (Biologie, Systematique).* 485 pp. Paris.

BERTRAM, D. S. 1962. Rickettsial infections and ticks. *Symp. zool. Soc. Lond.* No. **6**: 179–197.

BETTINI, S. 1964. Epidemiology of latrodectism. *Toxicon* **2**: 93–102.

BOGEN, E. 1956. The treatment of spider bite poisoning. In *Venoms.* Edited by E. E. Buckley and N. Porges. Washington, D.C.: American Association for the Advancement of Science, pp. 101–105.

BOOTH, B. H. & JONES, R. W. 1952. Epidemiological and clinical study of grain itch mite. *J. Am. med. Ass.* **150**: 1575–1579.

BOUISSET, L. & LARROUY, G. 1962. Envenimations par *Scorpio maurus* et *Buthus occitanus* dans le département de Tlemcen. *Bull. Soc. Path. exot.* **55**: 139–146.

BRUMPT, E. 1933. Etude du *Spirochaeta turicatae*, n.sp., agent de la fiévrè récurrente sporadique des Etats-Unis transmise par *Ornithodorus turicata*. *C.r. Séanc. Soc. Biol.* **113**: 1369–1372.

BÜCHERL, W. 1953. Escorpiões e escorpionismo no Brasil. 1. Manutencão dos escorpiões em extracão do veneno. *Mems Inst. Butantan* **25**: 53–82.

—— 1956. Studies on the dried venom of *Phoneutria fera* Perty. In *Venoms*. Edited by E. E. Buckley and N. Porges. Washington, D.C.: American Association for the Advancement of Science, pp. 95–97.

BÜCHERL, W. & BUCKLEY, E. E. 1972. *Venomous animals and their venoms*. 3. Venomous invertebrates. 560 pp. London.

BURKS, J. W., JUNG, R. & GEORGE, W. M. 1956. Norwegian scabies. *A.M.A. Archs Derm.* **74**: 131–140.

BUXTON, P. A. 1921. On the *Sarcoptes* of man. *Parasitology* **13**: 146–151.

CAIUS, J. F. & MHASKAR, K. S. 1932. Notes on Indian scorpions. *Indian med. Res. Mem.* No. **24**: 1–102.

CANESTRINI, G. & KRAMER, P. 1899. Demodicidae und Sarcoptidae. *Tierreich*, Lief. **7**: 1–193.

CARTER, H. F., WEDD, G. & D'ABRERA, V. 1944. The occurrence of mites in human sputum and their significance. *Indian med. Gaz.* **79**: 163–168.

CASALS, J., HENDERSON, B. E., HOOGSTRAAL, H., JOHNSON, K. M. & SHELOKOV, A. 1970. A review of Soviet viral hemorrhagic fevers, 1969. *J. inf. Dis.* **122**: 437–453.

CASALS, J., HOOGSTRAAL, H., JOHNSON, K. M., SHELOKOV, A., WIEBENGA, N. H. & WORK, T. H. 1966. A current appraisal of hemorrhagic fevers in the U.S.S.R. *Am. J. trop. Med. Hyg.* **15**: 751–764.

CHAKRAVORTY, A. N., GHOSH, S. & BANERJEE, A. K. 1953. Case notes of scabies in a family transmitted from goats. *Indian med. Gaz.* **88**: 153–154.

COOLEY, R. A. 1938. The genera *Dermacentor* and *Otocentor* (Ixodidae) in the United States with studies in variation. *Natn. Inst. Hlth Bull.* No. **171**: 1–89.

COOLEY, R. A. & KOHLS, G. M. 1944. The genus *Amblyomma* (Ixodidae) in the United States. *J. Parasit.* **30**: 77–111.

——, —— 1945. *The Genus* Ixodes *in North America*. Washington, D.C.: United States Government Printing Office.

COOPER, K. W. 1946. The occurrence of *Cheyletiella parasitivorax* (Mégnin) in North America, with notes on its synonymy and 'parasitic' habit. *J. Parasit.* **32**: 480–482.

CREMER, G. & MORRIEN, J. J. 1962. Dermanyssusschurft (duiveschurft) bij de mens. *Ned. Tijdschr. Geneesk.* **106**: 520–523.

CUNNINGTON, A. M. & GREGORY, P. H. 1968. Mites in bedroom air. *Nature, Lond.* **217**: 1,271–1,272.

DAVIS, G. E. 1956. The identification of spirochetes from human cases of relapsing fever by xenodiagnosis with comments on local specificity of tick vectors. *Expl Parasit.* **5**: 271–275.

DILLAHA, C. J., JANSEN, G. T., HONEYCUTT, W. M. & HAYDEN, C. R. 1964. North American loxoscelism. *J. Am. med. Ass.* **188**: 33–36.

DUBININ, V. B., GUSELNIKOVA, M. I. & RAZNATOVSKY, I. M. 1956. [Discovery of skin ticks (*Dermatophagoides scheremetewskyi* Bogandov, 1864) in some human skin diseases.] *Byull. Mosk. Obshch. Ispyt. Prir. (Biol.)* **61** (3): 43–50. (In Russian, English summary.)

EFRATI, P. 1949. Poisoning by scorpion stings in Israel. *Am. J. trop. Med.* **29**: 249–257.

ELBL, A. & ANASTOS, G. 1966. *Ixodid Ticks (Acarina, Ixodidae) of Central Africa*. 4 volumes. Tervuren: Musée Royal de l'Afrique Central. (*Annales*, Series 8, Nos. 145–148.)

EVANS, G. O. & TILL, W. M. 1965. Studies on the British Dermanyssidae (Acari: Mesostigmata). Part I. External morphology. *Bull. Br. Mus. nat. Hist.* (Zool.) **13**: 249–294.

——, —— 1966. Studies on the British Dermanyssidae. Part II. Classification. *Bull. Br. Mus. nat. Hist.* (Zool.) **14**: 109–370.

EVANS, G. O., SHEALS, J. G. & MACFARLANE, D. 1961. *The Terrestrial Acari of the British Isles. An Introduction to their Morphology, Biology and Classification*. I. Introduction and Biology. 219 pp. London: British Museum (Natural History).

HH

FAIN, A.   1967.   Le genre *Dermatophagoides* Bogandov 1864.   Son importance dans les allergies respiratoires et cutanées chez l'homme.   (Psoroptidae, Sarcoptiformes). *Acarologia* **9**: 179–225.

FINLAYSON, M. H.   1956.   Arachnidism in South Africa.   In *Venoms*.   Edited by E. E. Buckley and N. Porges.   Washington, D.C.: American Association for the Advancement of Science, pp. 85–87.

GARNET, J. R. (Editor).   1968.   *Venomous Australian Animals Dangerous to Man*.   86 pp.   Parkville, Victoria: Commonwealth Serum Laboratories.

GILBERT, E. W. & STEWART, C. M.   1935.   Effective treatment of arachnidism by calcium salts, a preliminary report.   *Am. J. med. Sc.* **189**: 532–536.

GIROUD, P., CAPPONI, M., DUMAS, M. & RAGEAU, J.   1965.   Les *Ixodes ricinus* et *hexagonus* de France contiennent des agents rickettsiens ou proches.   *C.r. hebd. Séanc. Acad. Sci. Paris* **260**: 4,874–4,876.

——, ——, ——, —— 1965a.   Résultats concernant *Dermacentor marginatus* et *reticulatus* prélévés dans differentes régions de France et leur contamination avec des rickettsies ou des éléments proches.   *C.r. hebd. Séanc. Acad. Sci. Paris* **260**: 5,419–5,421.

GOREN, S.   1950.   Akrepler ve akrep serumu.   *Turk. Ij tecr. Biyol. Derg.* **10**: 81–93.

GRASSET, E., SCHAAFSMA, A. & HODGSON, J. A.   1946.   Studies on the venom of South African scorpions (*Parabuthus, Hadogenes, Opisthophthalmus*) and the preparation of a specific anti-scorpion serum.   *Trans. R. Soc. trop. Med. Hyg.* **39**: 397–421.

HERMS, W. B. 1917.   Contribution to the life history and habits of *Ornithodorus megnini*. *J. econ. Ent.*   **10**: 407–411.

—— 1969.   *Medical Entomology* (6th edition revised by M. T. James and R. F. Harwood). 484 pp.   New York: Macmillan.

HILL, M. A. & GORDON, R. M.   1945.   An outbreak of dermatitis amongst troops in North Wales caused by rodent mites.   *Ann. trop. Med. Parasit.* **39**: 46–52.

HINMAN, E. H. & KAMPMEIER, R. H.   1934.   Intestinal acariasis due to *Tyrophagus longior* Gervais.   *Am. J. trop. Med.* **14**: 355–362.

HIRST, A. S.   1919.   The Genus *Demodex* Owen.   44 pp.   London: British Museum (N.H.).

—— 1922.   *Mites Injurious to Domestic Animals*.   107 pp.   London: British Museum (N.H.).

HOOGSTRAAL, H.   1956.   *African Ixodoidea. I. Ticks of the Sudan*.   1,101 pp.   Washington.

—— 1966.   Ticks in relation to human diseases caused by viruses.   *A. Rev. Ent.* **11**: 261–308.

—— 1967.   Ticks in relation to human diseases caused by *Rickettsia* species.   *A. Rev. Ent.* **12**: 377–420.

—— 1967a.   Tickborne hemorrhagic fevers, encephalitis and typhus in U.S.S.R. and southern Asia.   (Theobald Smith Memorial Lecture.)   *Expl Parasit.* **21**: 98–111.

HOOGSTRAAL, H., KAISER, M. N., TAYLOR, M. A., GABER, S. & GUINDY, E.   1961.   Ticks (Ixodoidea) on birds migrating from Africa to Europe and Asia.   *Bull. Wld Hlth Org.* **24**: 197–212.

HOPLA, C. E.   1951.   Experimental transmission of tularaemia by the tropical rat mite.   *Am. J. trop. Med.* **31**: 768–782.

HUEBNER, R. J., JELLISON, W. L. & POMERANTZ, C.   1946.   Rickettsial pox—a newly recognized rickettsial disease.   IV. Isolation of a rickettsia apparently identical with the causative agent of rickettsial pox from *Allodermanyssus sanguineus*, a rodent mite.   *Publ. Hlth Rep. Wash.* No. 61: 1,677–1,682.

HUGHES, A. M.   1961.   *The Mites of Stored Food*.   Ministry of Agriculture, Fisheries and Food, Technical Bulletin No. 9.   287 pp.   London: H.M.S.O.

JADIN, J., GIROUD, P. & LE RAY, D.   1969.   Présence de rickettsies chez *Ixodes ricinus* en Belgique.   *Proc. 2nd int. Congr. Acarology, Sutton Bonington 1967*, 615–617.

JONES, B. M.   1950.   The penetration of the host tissue by the harvest mite, *Trombicula autumnalis* Shaw.   *Parasitology* **41**: 229–248.

KAISER, M. N.   1966.   Viruses in ticks.   I. Natural infections of *Argas* (*Persicargas*) *arboreus* by Quaranfil and Nyamanini viruses and absence of infections in *A.* (*P.*) *persicus* in Egypt. *Am. J. trop. Med. Hyg.* **15**: 964–975.

—— 1966a.   Viruses in ticks.   II. Experimental transmission of Quaranfil virus by *Argas* (*Persicargas*) *arboreus* and *A.* (*P.*) *persicus*.   *Am. J. trop. Med. Hyg.* **15**: 976–985.

KEEGAN, H. L., WEAVER, R. E., TOSHIOKA, S. & MATSUI, T.   1964.   *Some Venomous and Noxious Animals of East and South East Asia*.   *406th Medical Laboratory Special Report*.   United States Army Medical Command, Japan.

KRAEPELIN, K.   1899.   Scorpiones und Pedipalpi.   *Tierreich* Lief **8**: 1–265.

KUBOTA, S.   1918.   An experimental study of the venom of the Manchurian scorpion.   *J. Pharm. exp. Ther.* **11**: 379–388.

LARSON, D. G., MITCHELL, W. F. & WHARTON, G. W.   1969.   Preliminary studies on *Dermato-phagoides farinae* Hughes, 1961 (Acari) and house dust allergy.   *J. med. Ent.* **6**: 295–299.

LEVI, H. W.   1959.   The spider genus *Latrodectus* (Araneae, Theridiidae).   *Trans. Am. micros. Soc.* **78**: 7–43.

McCRONE, J. D. & NETZLOFF, M. L.   1965.   An immunological and electrophoretical comparison of the venoms of the North American *Latrodectus* spiders.   *Toxicon* **3**: 107–110.

McKEOWN, K. C.   1952.   *Australian Spiders*.   274 pp.   London: Angus and Robertson.

MAUNSELL, K., WRAITH, D. G. & CUNNINGTON, A. M.   1968.   Mites and house-dust allergy in bronchial asthma.   *Lancet* **1968**: 1,267–1,270.

MAZZOTTI, L. & BRAVO-BECHERELLE, M. A.   1963.   Scorpionism in the Mexican Republic.   In *Venomous and Poisonous Animals and Noxious Plants of the Pacific Region*.   Edited by H. L. Keegan and W. V. Macfarlane.   London: Pergamon, pp. 119–131.

MEKIE, E. C.   1926.   Parasitic infection of the urinary tract.   *Edinb. med. J.* **33**: 708–719.

MELLANBY, K.   1943.   *Scabies*.   81 pp.   Oxford: Oxford University Press.   [Reprinted 1973].

MILLOT, J.   1949.   Ordre des Araneides.   In *Traité de Zoologie*.   Edited by P.-P. Grassé.   Paris: Masson, Vol. **6**, pp. 589–743.

MILLOT, J. & VACHON, M.   1949.   Ordre des scorpions.   In *Traité de Zoologie*.   Edited by P.-P. Grassé.   Paris: Masson, Vol. **6**, pp. 386–436.

MISKJIAN, H. G.   1951.   Demodicidosis (*Demodex* infestation of the scalp).   *A.M.A. Archs Derm. Syph.* **63**: 282–283.

MIYAMOTO, T., OSHIMA, S., ISHIZAKI, T. & SATO, S.   1968.   Allergenic identity between the common floor mite (*Dermatophagoides farinae* Hughes, 1961) and house dust as a causative antigen in bronchial asthma.   *J. Allergy* **42**: 14–28.

MUNDLE, P. M.   1961.   Scorpion stings.   *Br. med. J.* No. **5231**, 1,042.

NAUMOV, R. L., LEVKOVICH, E. N. & RZHAKHOVA, O. E.   1963.   [The part played by birds in the circulation of tick-borne encephalitis virus.]   *Med. Parazit.* **1**: 18–29.   (In Russian, English translation, United States Medical Research Unit No. 3, Cairo, translation 141.)

NUTTALL, G. H. F.   1911.   On the adaptation of ticks to the habits of their host.   *Parasitology* **4**: 46–67.

NUTTALL, G. H. F. & WARBURTON, C.   1911.   *Ticks: A Monograph of the Ixodoidea*.   *Part II. Ixodidae*: 105–348.   Cambridge: Cambridge University Press.

——, —— 1915.   *Ticks: A Monograph of the Ixodoidea*.   *Part III. The Genus* Haemaphysalis: 349–550.   Cambridge: Cambridge University Press.

NUTTALL, G. H. F., WARBURTON, C., COOPER, W. F. & ROBINSON, L. E.   1908.   *Ticks: A Monograph of the Ixodoidea*.   *Part I. Argasidae*.   104+35 pp.   Cambridge: Cambridge University Press.

PARKER, R. R.   1937.   Recent studies of tick-borne diseases made at the U.S. Public Health Service Laboratory at Hamilton, Montana.   *Proc. 5th Pacif. Sci. Congr.* **5**: 3367–3374.

PARRISH, H. M.   1963.   Analysis of 460 fatalities from venomous animals in the United States.   *Am. J. med. Sci.* **245**: 129–141.

PETRUNKEVITCH, A.   1928.   Systema Araneum.   *Trans. Conn. Acad. Arts Sci.* **29**: 1–270.

PHILLIP, C. B. & KOHLS, G. M.   1952.   Elk, winter ticks and Rocky Mountain spotted fever: a query.   *Publ. Hlth Rep. Wash.* No. **66**: 1672–1675.

PIRILÄ, V., NUORTEVA, P. & KALLELA, K.   1967.   The etiologic agent of Norwegian scabies.   *Trans. a. Rep. St. John's Hosp. derm. Soc., Lond.* **53**: 80–81.

POMERANTZEV, B. I.   1950.   [Ixodid ticks (Ixodidae).]   *Fauna SSR* **4** (2): 1–224.   [In Russian, translation by Alena Elbl, 1959, Washington, D.C.: American Institute of Biological Sciences.]

REEVES, W. C., HAMMON, M. W., McCLURE, H. E. & SATHER, G. 1955. Studies on mites as vectors of western equine and St. Louis encephalitis viruses in California. *Am. J. trop. Med. Hyg.* **4**: 90–105.

ROANTREE, W. B. 1961. Scorpion stings. *Br. med. J.* No. **5236**, 1,395.

ROBINSON, L. E. 1926. *Ticks: A Monograph of the Ixodoidea. Part IV. The Genus* Amblyomma. 302 pp. Cambridge: Cambridge University Press.

RUSSELL, B. F. 1962. Some aspects of the biology of the epidermis. *Br. med. J.* No. **5281**: 815–820.

SASA, M. 1950. Mites of the genus *Dermatophogoides* Bogdanow 1864 found in three cases of human acariasis. *Japan. J. exp. Med.* **20**: 519–525.

—— 1964. Special problems of mites in stored food and drugs in Japan. *Proc. 1st int. Congr. Acarology, Fort Collins* 1963. *Acarologia* **6** (hors série): 390–391.

SCHENONE, H. 1953. Mordeduras de aranas. *Boln chil. Parasit.* **8**: 35–37.

—— 1959. Estudio de 27 casos de loxoscelismo. *Boln chil. Parasit.* **14**: 7–13.

SCHENONE, H. & PRATS, F. 1961. Arachnidism by *Loxosceles laeta*. Report of 40 cases of necrotic arachnidism. *Arch. Derm.* **83**: 139–142.

SERGENT, E. 1949. Étude comparative du venin de scorpions mexicains et de scorpions nord-africains. *Archs Inst. Pasteur Alger.* **27**: 31–34.

SHULOV, A. 1939. The venom of the scorpion *Buthus quinquestriatus* and the preparation of an antiserum. *Trans. R. Soc. trop. Med. Hyg.* **33**: 253–256.

SHULOV, A., FLESH, D., GERICHTER, C., ESHKOL, Z. & SCHILLINGER, G. 1959. The anti-scorpion serum prepared by use of fresh venom and the assessment of its efficacy against scorpion stings. *Proc. 5th Int. Congr. biol. Standard., Jerusalem* 1959, 489–492.

SIMPSON, D. I. H., KNIGHT, E. M., COURTOIS, G., WILLIAMS, M. C., WEINBREN, M. P. & KIBUKAMUSOKE, J. W. 1967. Congo Virus: a hitherto undescribed virus occurring in Africa. Part I. Human isolation-clinical notes. *E. Afr. med. J.* **44**: 87–92.

SINGH, J., KAPILA, C. C., BASU, S. M., VARMA, R. N., NARSIMHAN, D., RAO, K. N. A., SARDANA, D. N., CHOPRA, S. K. & KARANI, N. D. P. 1965. Haemorrhagic disease following tick bites suspected toxoplasmosis. *Lancet* **1965**: 834–838.

SMITH, C. E. G. 1962. Ticks and viruses. *Symp. zool. Soc. Soc. Lond.* No. **6**: 199–221.

SMITH, M. H., BLATTNER, R. J., HEYS, F. M. & MILLER, A. 1948. Experiments on the role of the chicken-mite *Dermanyssus gallinae*, and the mosquito in the epidemiology of St. Louis encephalitis. *J. exp. Med.* **87**: 119–138.

SPICKETT, S. G. 1961. Studies on *Demodex folliculorum* Simon (1842). 1. Life history. *Parasitology* **51**: 181–192.

STAHNKE, H. L. 1956. *Scorpions.* 35 pp. Tempe, Arizona: Arizona State University, Poisonous Animals Research Laboratory.

—— 1963. Some pharmacological and biochemical characteristics of *Centruroides sculpturatus* Ewing scorpion venom. *Proc. Int. pharmac. Meet., Prague* 1963, 63–70.

—— 1966. *The Treatment of Venomous Bites and Stings.* 117 pp. Tempe, Arizona: Arizona State University, Poisonous Animals Research Laboratory.

STRANDTMANN, R. W. & WHARTON, G. W. 1958. *A Manual of Mesostigmatid Mites Parasitic on Vertebrates.* College Park, Md.: University of Maryland, Institute of Acarology.

SWAN, D. C. 1934. The hay itch mite, *Pediculoides ventricosus* (Newp.) (Acarina: Pediculoididae). *J. Agric. S. Aust.* **37**: 1289–1299.

TABOADA, M. DE F. 1954. Pulmonary acariasis in Spain. An illustrative case report. *Br. med. J.* No. 4859: 437–438.

TAYLOR, R. M., HURLBUT, H. S., WORK, T. H., KINGSTON, J. R. & HOOGSTRAAL, H. 1966. Arboviruses isolated from *Argas* ticks in Egypt: Quaranfil, Chenuda, and Nyamanini. *Am. J. trop. Med. Hyg.* **15**: 76–86.

THEILER, G. 1962. *The Ixodoidea Parasites of Vertebrates in Africa South of the Sahara.* 260 pp. Report to the Director of Veterinary Services, Onderstepoort, South Africa.

THOMAS, L. A., KENNEDY, R. C. & ECKLUND, C. M. 1960. Isolation of a virus closely related to Powassan virus from *Dermacentor andersoni* collected along the North Cache la Poudre River, Colo. *Proc. Soc. exp. Biol. Med.* **104**: 355–359.

THORP, R. W. & WOODSON, W. D. 1945. *Black Widow. America's Most Poisonous Spider.* 222 pp. Chapel Hill: University of North Carolina.

TODD, C. 1909. An antiserum for scorpion venom. *J. Hyg. Camb.* **9**: 69–85.

TRAPIDO, H., VARMA, M. G. R., RAJAGOPALAN, P. K., SINGH, K. P. R. & REBELLO, M. J. A. 1964. A guide to the identification of all stages of *Haemaphysalis* ticks of South India. *Bull. ent. Res.* **55**: 249–270.

TRAVER, J. R. 1951. Unusual scalp dermatitis in humans caused by the mite *Dermatophagoides*. *Proc. ent. Soc. Wash.* **53**: 1–25.

TULGA, T. 1964. Scorpions found in Turkey and paraspecific action of an antivenin produced with the venom of the species *Androctonus crassicauda*. *Turk. Ij. tecr. Biyol. Derg.* **24**: 153–155.

VACHON, M. 1951. A propos de quelques scorpions de Turquie collectés par M. le Professeur Dr Curt Kosswig. *Istanb. Univ. Fen. Fak. Mecm.* (B) **16**: 341–344.

—— 1952. *Etudes sur les Scorpions.* Alger: Institut Pasteur D'Algérie.

—— 1953. The biology of scorpions. *Endeavour* **12**: 80–89.

—— 1966. Liste des scorpions connus en Egypte, Arabie, Israël, Liban Syrie, Jordanie, Turquie, Irak, Iran. *Toxicon* **4**: 209–218.

VARMA, M. G. R. 1962. Transmission of relapsing fever spirochaetes by ticks. *Symp. zool. Soc. Lond.* No. **6**: 61–82.

1964. The acarology of louping ill. *Proc. 1st int. Congr. Acarology, Fort Collins 1963. Acarologia* **6** (hors série): 241–254.

VELLARD, J. 1936. *Le Venin des Araignées.* 311 pp. Paris: Masson.

—— 1954. L'aranéism au Pérou et dans les régions méridionales de l'Amerique du Sud. *Trav. Inst. fr. Etud. andines* **4**: 133–196.

—— 1966. La fonction venimeuse chez les araignées. *Mems Inst. Butantan (Simp. Internac.)* **33**: 35–44.

VERCAMMEN-GRANDJEAN, P. H. 1968. *The Chigger Mites of the Far East (Acarina: Trombiculidae and Leeuwenhoekiidae)* Washington, D. C.: U.S. Army Medical Research and Development Command.

VITZTHUM, H. G. 1943. Acarina. *Bronn's Kl. Ordn. Tierreichs* **5**, Abt. 4, Buch 5, 1–1011.

VOORHORST, R., SPIEKSMA-BOEREMAN, M. I. A. & SPIEKSMA, F. T. M. 1964. Is a mite (*Dermatophagoides* sp.) the producer of the house-dust allergen? *Allergie Asthma* **10**: 329–334.

VOORHORST, R., SPIEKSMA, F. T. M. & VAREKAMP, H. 1969. *House-Dust Atopy and the House-Dust Mite* Dermatophagoides pteronyssinus (*Trouessart* 1897). 159 pp. Leiden: Stafleu's Scientific Publishing Company.

WAHBEH, Y. 1965. Scorpion stings in children. *Jordan med. J.* **1**: 57–61.

WALDRON, W. G. 1962. Notes on the occurrence, observations and public health significance of the pajaroello tick, *Ornithodoros coriaceus* Koch, in Los Angeles County. *Bull. Sth Calif. Acad. Sci.* **61**: 241–245.

WALTON, G. A. 1962. The *Ornithodoros moubata* superspecies problem in relation to human relapsing fever epidemiology. *Symp. zool. Soc. Lond.* No. **6**: 83–156.

WATERMAN, J. A. 1938. Some notes on scorpion poisoning in Trinidad. *Trans. R. Soc. trop. Med. Hyg.* **31**: 607–624.

—— 1939. Some observations on the habits and life of the common scorpion of Trinidad. *Trans. R. Soc. trop. Med. Hyg.* **33**: 113–118.

WERNER, F. 1935. Scorpiones, Pedipalpi. *Bronn's Kl. Ordn. Tierreichs* **5**, Abt. 4, Buch 8, 1–316.

WHARTON, G. W. & FULLER, H. S. 1952. A manual of the chiggers. *Mem. ent. Soc. Wash.* No. **4**: 1–185.

WHITTEMORE, F. W., KEEGAN, H. L. & BOROWITZ, J. L. 1961. Studies of scorpion antivenins. 1. Paraspecificity. *Bull. Wld Hlth Org.* **25**: 185–188.

WIENER, S.   1961.   Red back spider bite in Australia: an analysis of 167 cases.   *Med. J. Aust.*
  2: 44–49.

WILSON, W. H.   1904.   On the venom of scorpions.   *Rec. Egypt. Govt. Sch. Med.* **2**: 7–40.

WORTH, C. B. & RICHARD, E. R.   1951.   Evaluation of the efficiency of common rat ectoparasites
  in the transmission of murine typhus.   *Am. J. trop. Med.* **31**: 295–298.

YEOMAN, G. H. & WALKER, J. B.   1967.   *The Ixodid Ticks of Tanzania.*   London: Commonwealth Institute of Entomology.

ZDRODOVSKIJ, P. F.   1964.   Les rickettsioses en URSS.   *Bull. Wld Hlth Org.* **31**: 33–43.

ZEMSKAYA, A. A. & PCHELKINA, A. A.   1968.   [Infection of some species of Gamasidae with
  R. burneti in natural Q fever foci.]   *Zh. Mikrobiol. Epidem. Immunobiol.* **45**: 130–132.   (In
  Russian, English summary.)

ZUMPT, F.   1950.   Preliminary study to a revision of the genus *Rhipicephalus* Koch.   *Docum.*
  Mocambique No. **60** (1949): 57–123.

15. OTHER ARTHROPODA

by J. G. Sheals and A. L. Rice

CRUSTACEA (CRABS, etc.)

THE Crustacea include the crabs, lobsters, prawns, sandhoppers, woodlice, barnacles, water fleas, fairy shrimps and a multitude of related organisms, many of which have no common names. The class embraces such a wide range of structural variation that it is not possible to give a short definition applicable to all its members. In general they can be said to be aquatic arthropods, breathing by means of gills or through the general surface of the body, having two pairs of antennae, and having at least three pairs of postoral appendages acting as jaws, the three somites corresponding with the latter being coalesced with the head. The more highly modified members of the class, however, furnish exceptions to every statement of this definition. General systematic accounts of the class are given by Calman (1909) and Zimmer (1927).

Many Crustacea, particularly decapod species (crabs, lobsters, prawns and shrimps) are eaten by man and a number are known to cause poisoning (Holthuis, 1968). Thus, *Ucides cordatus* (L.), (family Gecarcinidae), a large land crab from the West Indies, is toxic after it has eaten the fruits or leaves of the manzanilla tree (*Hippomanes mancinella*). The natives of Yap in the Caroline Islands leave the robber crab (*Birgus latro* (L.), family Coenobilidae), for a few days before eating it so that toxic substances can leave its intestine. Other crabs known to be poisonous at times include *Atergatis floridus* (L.), *Carpilius maculatus* (L.), *C. convexus* (Forskål), *Eriphia sebana* (Shaw and Nodder), *E. norfolcensis* (Grant and McCulloch), *Lophozozymus pictor* (Fabricius), *Pilumnus vespertilio* (Fabricius), *Xantho reynaudii* (H. Milne Edwards), and *Zosimus aeneus* (L.) (all of the family Xanthidae); *Cardisoma carnifex* (Herbst) (family Gecarcinidae), *Micippa philyra* (Herbst) (family Majidae), *Parthenope longimanus* (L.) and *Daldorfia horrida* (L.) (family Parthenopidae) and *Dromidiopsis dormis* (family Dromiidae).

A number of Crustacea play a role in the transmission of human parasites. For example man (and certain carnivores) acquire infections of the lung fluke, *Paragonimus westermannii* (Kerbert) and certain allied trematodes, by eating uncooked river-crabs and freshwater crayfish (*Potamon* and *Cambarus* spp. respectively). These decapods act as second intermediate hosts for the trematode larvae (metacercariae), the first intermediate hosts being some species of the freshwater snail, *Melania*.

Cyclopoid and calanoid copepods (often referred to loosely as 'water fleas') are appropriate first intermediate hosts for well-known helminth parasites of humans. In the case of the broad tapeworm of man, *Diphyllobothrium latum* (L.), infection is acquired by eating fish that have swallowed copepods infested with the second (procercoid) stage of the developing tapeworm (fig. 208). It should be noted that further development of the procercoid of this tapeworm will not occur if man accidentally swallows infested copepods. On the other hand, man is capable of acting as an intermediate host of *Diphyllobothrium erinaceieuropaea* (Rud.) (often referred to in medical literature as

Diphyllobothrium mansoni (Cobbold)), which develops into an adult worm in carnivores. In this instance man becomes infected with the plerocercoid or sparganum stage (which causes sparganosis) by swallowing infested copepods in drinking water.

The nematode *Dracunculus medinensis* (L.) which is responsible for 'guinea worm' in man, uses many species of cyclopoid copepods as intermediate hosts, and human infection is acquired through drinking water containing the infested copepods.

A phoretic association is known to occur between the larvae and pupae of species of the black-fly genus *Simulium* which transmits human onchocerciasis, and river-crabs of the family Potamonidae, particularly between *Simulium neavei* (see p. 137) and *Potamonautes niloticus* (H. Milne Edwards), (van Someren and McMahon, 1950). The aquatic stages of *Simulium* are found attached to the carapace of the crabs and the discovery of this association has made possible the precise delimitation of the breeding areas of the insects (Williams, 1969). Other species of *Simulium* have been found in association with the prawn *Atya africana* Bouvier in West Africa.

The mosquito *Aedes pembaensis* lays its eggs in rows on the meri of the legs of a mangrove crab, *Sesarma meinerti* de Man (Goiny *et al.*, 1957) and burrows produced by crabs may provide reservoirs for larvae of a wide variety of insects, some of which may be of medical importance.

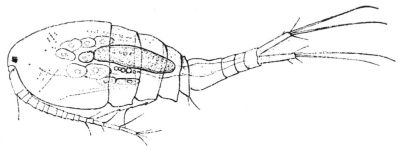

Fig. 208   *Cyclops strenuus* containing procercoids of *Diphyllobothrium latum* ($\times$35) (after Janicki and Rosen, 1918).

CHILOPODA (CENTIPEDES)

Centipedes are elongated, many-legged tracheate arthropods with a distinct head and a dorso-ventrally flattened body. The head bears a single pair of antennae and three pairs of trophic appendages: the mandibles and two pairs of maxillae. The trunk cannot be differentiated into regions and each of its segments carries a single pair of legs. The number of trunk segments varies from 15 to well over one hundred. The first pair of trunk appendages, the maxillipedes or toxognaths are provided with powerful terminal claws, at the tips of which are the orifices of ducts leading from the large venom glands. The class is made up of four* orders: Geophilomorpha, Lithobiomorpha, Scolopendromorpha and Scutigeromorpha. General systematic accounts are given by Attems (1926, 1929, 1930) and Verhoeff (1902–1925).

---

* Some authorities classify the Australian genus *Craterostigmus* in a fifth order, the Craterostigmomorpha, others classify this genus in the Lithobiomorpha.

Centipedes live in damp dark habitats, for example, under stones, in crevices of the upper layers of soil, in forest litter, under logs and under the bark of trees.   Some are cave dwellers and a few are marine.   The majority are nocturnal predators feeding mainly on insects and other small arthropods which they capture and kill with their poison claws.   The larger Scolopendromorpha have been reported to feed on mice, small birds, amphibia and reptiles.

Reports on the effects of centipede bites on man are often conflicting but it is certain that very few species living in temperate zones are able to pierce human skin with their poison claws.   Bites of *Lithobius forficatus* (L.) have been reported to be sharply painful, but the signs of injury are insignificant (Cloudsley-Thompson, 1958).   The large scolopendromorphs (fig. 209A) which are abundant in tropical and subtropical regions are generally regarded as venomous although only one human fatality resulting from centipede bite has been reported (Pineda, 1923 cited by Remington, 1950).   This case occurred in the Philippines where a seven-year-old child died 29 hours after being bitten on the head, and the species concerned was probably *Scolopendra subspinipes* Leach.   Bücherl (1946) found that the bites of *S. viridicornis* Newport and *S. subspinipes* were lethal to mice, and that, in 60 per cent of the tests, the bites of *S. viridicornis* were fatal to guinea pigs.   The bites of *S. subspinipes* were never fatal to guinea pigs.   He considered that animal experiments were sufficient to prove that the venom of the largest Brazilian centipedes was too feeble to cause intoxication in either adults or young children.

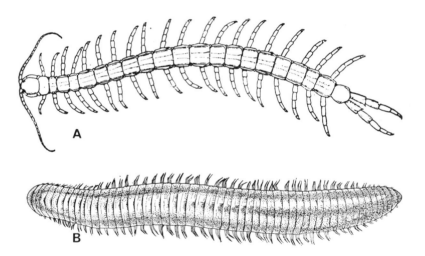

Fig. 209   A, *Scolopendra morsitans*; B, a large tropical millipede, *Rhinocricus lethifer* (both natural size).

Remington (1950) reported that *S. subspinipes* were abundant around military installations in the town of Tarragona on the island of Leyte in the Philippines.   It was difficult to keep these large centipedes out of tents and temporary buildings.   During their nocturnal hunting they occasionally crawled into the beds of soldiers and many men were bitten.   Remington, who was himself bitten in the axilla, stated that the pain caused by the bite was at first almost unbearable and it did not diminish sufficiently to permit a return to bed for about twenty minutes. The next morning the axilliary region was severely swollen and somewhat tender and this mildly painful condition persisted for about three weeks. Although numerous cases were recorded, no centipede bite caused serious effects and many suffering centipede victims were quickly relieved after local anaesthetics, of the kind used by dentists, were injected in the vicinity of the bite.

Keegan *et al.* (1964) cite an unpublished observation by Reid who stated that symptoms in one case of centipede bite seen in Malaysia were more severe than those seen in some cases of viper bite. Nevertheless these authors point out that, in spite of their spectacular appearance and occasional aggressive behaviour, the consensus is that the great majority of centipedes, regardless of size, are not dangerously venomous to man.

A number of cases of pseudoparasitism by centipedes have been reported. For example, Tartaglia (1961) reported a case of intestinal pseudoparasitism lasting nine hours in Split, Yugoslavia, where a 55-year-old woman accidently swallowed a chilopod, subsequently determined as *Scolopendra cingulata*. The symptoms included vomiting, cold perspiration, irregular heart beat and sudden discharge from the bowels.

DIPLOPODA (MILLIPEDES)

The millipedes are elongate tracheate arthropods with a distinct head and with a trunk composed of a large number of apparent segments, each of which, with the exception of the first four, bears two pairs of legs. The head bears a single pair of antennae, a pair of mandibles and only one pair of maxillae which form a broad plate known as the gnathochilarium. The detailed anatomy of the gnathochilarium is used in the characterization of the major groups down to the level of suborder. There are two subclasses: the Pselaphognatha (made up of a single order, the Polyxenida) and the Chilognatha which is made up of 11 orders. Systematic accounts of the group are given by Attems (1937, 1938, 1940), Verhoeff (1926-1932) and by Chamberlin and Hoffman (1958).

Millipedes are predominantly saprophagous and lack the poison claws of the centipedes. In the majority of the Chilognatha many of the trunk segments contain paired repugnatorial glands whose ducts open dorso-laterally. The secretion of these glands is unpalatable to predators and in the case of some large tropical species the secretion has been reported to have a strong caustic action. Some interesting observations on *Rhinocricus lethifer* Loomis, a very large spirobolid species (fig. 209B), were recorded by Loomis (1936). Whilst collecting this species in Haiti it was noticed that the millipede had the ability to eject the repugnatorial fluid several inches from the body and that the fluid caused a smarting sensation on the back of the hands and lower forearms. One large specimen ejected its secretion upwards a distance of about 45 cm (18 inches) into the face and left eye of the collector. The pain was instantaneous, intense and of a burning and smarting nature. It persisted for several hours despite immediate bathing with iced-water. Swelling of the eyelid and cheek progressed rapidly and soon the eye was closed. The following morning the pain was gone but the eyelid again was swollen shut. On the day following the attack the skin over the affected area had turned dark brown and was raised into blisters where the concentration of the secretion had been greatest. The blisters persisted for nearly a week after which the discoloured skin peeled off without leaving any scars. Another species of *Rhinocricus*, *R. latespargor* Loomis, when irritated, projected its repugnatorial fluid distances up to 82 cm (33 inches) (Loomis, 1941). The caustic action of the repugnatorial fluid of an African millipede, probably a species of *Spirobolus*, is described by Burtt (1947), who left a living specimen in his hip pocket for about an hour. A soreness developed in the skin below the pocket, and soon afterwards it was found that the skin over are area of about nine square inches had become black. No blisters were raised, and four days later the blackened skin peeled off leaving a raw wound. The scar of the injury was clearly defined more than a year later. Examples of similar injuries caused by contact with *Orthoporus* species in Mexico are cited by Halstead and Ryckman (1949).

BIBLIOGRAPHY

ATTEMS, C. 1926. Progoneata, Chilopoda. *Handb. Zool., Berl.* **4**: 1–402.

—— 1929. Myriapoda, Geophilomorpha. *Tierreich Leif.* **52**: 1–388.

—— 1930. Chilopoda, Scolopendromorpha. *Tierreich Leif.* **54**: 1–308.

—— 1937. Polydesmoidea I. *Tierreich Leif.* **68**: 1–300.

—— 1938. Polydesmoidea II. *Tierreich Leif.* **69**: 1–487.

—— 1940. Polydesmoidea III. *Tierreich Leif.* **70**: 1–577.

BÜCHERL, W. 1946. Acão de veveno dos escolopendromorfos do Brasil sôbre alguns animais de laboratório. *Mems Inst. Butantan* **19**: 181–197.

BURTT, E. 1947. Exudate from millipedes with particular reference to its injurious effects. *Trop. Dis. Bull.* **44**: 7–12.

CALMAN, W. T. 1909. Crustacea. In *A Treatise on Zoology*. Edited by R. Lankester. **7** (3): 1–346. London: Adam and Charles Black.

CHAMBERLIN, R. V. & HOFFMAN, R. L. 1958. Checklist of the millipeds of North America. *Bull. U.S. natn. Mus.* No. 212: 1–236.

CLOUDSLEY-THOMPSON, J. L. 1958. *Spiders, Scorpions, Centipedes and Mites*. London: Pergamon.

FAUST, E. C. & RUSSELL, P. F. 1964. *Craig and Faust's Clinical Parasitology*. 1099 pp. Philadelphia.

GOINY, H., SOMEREN, E. C. C. VAN & HEISCH, R. B. 1957. The eggs of *Aedes* (*Skusea*) *pembaensis* Theobald discovered on crabs. *E. Afr. med. J.* **34**: 1–2.

HALSTEAD, B. W. & RYCKMAN, R. 1949. Injurious effects from contacts with millipedes. *Med. Arts. Sci.* **3**: 16–18.

HOLTHUIS, L. B. 1968. Are there poisonous crabs? *Crustaceana* **15**: 215–222.

KEEGAN, H. L., WEAVER, R. E., TOSHIOKA, S. & MATSUI, T. 1964. Some Venomous and Noxious Animals of East and South East Asia. *406th Medical Laboratory Special Report*. United States Army Medical Command, Japan: 1–43.

LOOMIS, H. F. 1936. The millipeds of Hispaniola with descriptions of a new family, new genera, and new species. *Bull. Mus. comp. Zool. Harv.* **80**: 1–191.

—— 1941. New genera and species of millipeds from the southern peninsula of Haiti. *J. Wash. Acad. Sci.* **31**: 188–195.

MOORTHY, V. N. & SWEET, W. C. 1936. Guinea-worm infection of cyclops in nature. *Indian med. Gaz.* **71**: 568.

ONABAMIRO, S. D. 1951. The transmission of *Dracunculus medinensis* by *Thermocyclops nigerianus*, as observed in a village in south-west Nigeria. *Ann. trop. Med. Parasit.* **45**: 1–10.

REMINGTON, C. L. 1950. The bite and habits of a giant centipede (*Scolopendra subspinipes*) in the Philippine Islands. *Am. J. trop. Med.* **30**: 453–455.

TARTAGLIA, P. 1961. Intestinaler Pseudoparasitismus durch *Scolopendra cingulta* [sic]. *Z. Tropenmedizen und Parasitologie* **13**: 218–220.

VAN SOMEREN, V. D. & MCMAHON, J. P. 1950. Phoretic association between *Afronurus* and *Simulium* species, and the discovery of the early stages of *Simulium neavei* on fresh water crabs. *Nature, Lond.* **166**: 350–351.

VERHOEFF, K. W. 1902–1925. Chilopoda. *Bronn's Kl. Ordn. Tierreichs* **5**, Abt. 2, Buch 1, 1–725.

—— 1926–1932. Diplopoda. *Bronn's Kl. Ordn. Tierreichs* **5**, Abt. 2, Buch 2, 1–2084.

WILLIAMS, T. R. 1969. The taxonomy of the East African river-crabs and their association with the *Simulium neavei* complex. *Trans R. Soc. trop. Med. Hyg.* **62**: 29–34.

YOKOGAWA, S., YOKOGAWA, M. & CORT, W. W. 1960. *Paragonimus* and paragonimiasis. *Expl. Parasit.* **10**: 81–205.

ZIMMER, C. 1927. Crustacea. *Handb. Zool., Berl.* **3**: 277–1158.

16. PENTASTOMIDA
(Tongue worms)

by J. G. Sheals

MEMBERS of this enigmatic group have from time to time been classified as Cestoda, Nematoda, Acanthocephala, Hirudinea, Crustacea and Arachnida. Sambon (1922) considered them to be highly specialized Acari, but while their many arthropod-like characters are recognized, they are now generally treated as a discrete group of uncertain affinity (see for example Cuénot, 1949; Nicoli, 1963).

The Pentastomida (known also as the Linguatulida) are endoparasites which as adults live in the nose, nasal sinuses or other parts of the respiratory tract of mammals, birds or reptiles. With only one known exception the larvae are found (usually encysted in the viscera) in an intermediate vertebrate host.* The body of the adult (fig. 210) is elongate, vermiform and more or less clearly annulated. There are no appendages, although the mouth region is provided with two pairs of retractile chitinous hooks which emerge from longitudinal pits or pouches. The sexes are separate, the males being much smaller than the females. The genital opening lies on the mid-ventral line, anteriorly in the male and either anteriorly or posteriorly in the female. The anus is terminal or subterminal.

Systematic accounts are given by Sambon (1922) Heymons (1935) and Nicoli (1963) and the central African species are dealt with by Fain (1961). The group is divided into two orders: Cephalobaenida and Porocephalida. The adults of the Cephalobaenida occur in reptiles and birds and their main intermediate hosts are snakes, amphibia and probably also fish. The Porocephalida are made up of two superfamilies: Porocephaloidea and Linguatuloidea. The Porocephaloidea (made up of five families) are, as adults, parasites of reptiles, and their larvae occur in fish, reptiles, mammals and rarely in birds. The Linguatuloidea (made up of a single family) are essentially parasites of mammals, both as adults and larvae.

Amongst the Linguatulidae, adults of *Linguatula serrata* Frölich are commonly found in the upper respiratory passage of dogs, foxes and wolves, and the larvae are found encysted in the viscera of various herbivorous mammals; rabbits and rodents being commonly infested. The adult female reaches a length of 80–120 mm but the males are much smaller. Eggs are present in the nasal secretions of the definitive host and are ingested by the intermediate host with food and water. On hatching in the intermediate host the larvae migrate to the mesenteric nodes and various viscera and then become encysted. The larvae do not leave their cysts during the life of the intermediate host, and the definitive hosts become infected by eating larval-infested food. It would appear that swallowed larvae do not migrate back to the pharynx, and

---

* The only record of a pentastomid occurring in an invertebrate is that of Lavoipierre and Lavoipierre (1966), who found the larvae of *Raillietella hemidactyli* Hett (Cephalobaenida) in cockroaches (*Periplaneta americana*) in Singapore. The adults of this pentastomid parasitize geckonid lizards.

to cause infection they must cling to the mucous membrane of the mouth before being swallowed, or when vomited (Hobmaier and Hobmaier, 1940). The true definitive hosts of *L. serrata* appear to be the Canidae, for although larvae may invade the nasal cavities of other carnivorous mammals, in these hosts they rarely complete their development to become adults.

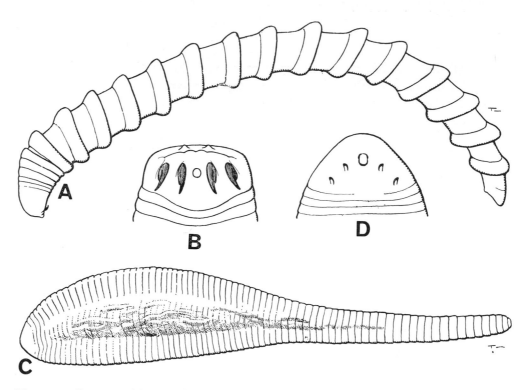

Fig. 210   Pentastomida: A, C, entire specimens (×*c.*1.5); B, D, mouth region enlarged: A, B, *Armillifer armillatus*; C, D, *Linguatula serrata*.

Several cases of human infestation with *L. serrata* (linguatulosis) have been recorded, and in man the site of parasitism can be the site in the definitive host, that is to say the upper respiratory passages. However, a survey of recorded cases suggests that adults are rarely, if ever, found in man and thus it would seem that man is an unsuitable definitive host in which the parasites fail to complete their normal developmental cycle. Papadakis and Hourmouzadis (1958) describe a case of human linguatulosis from Greece. The patient, a 29-year-old woman, developed an acute attack of coughing and sneezing, with much rhino-pharyngeal secretion and allergic symptoms. After eight hours the symptoms subsided to reappear several hours later, and minor paroxysms interrupted by amelioration of symptoms continued for 15 days. During the attacks worm-like pieces (subsequently determined as larvae of *L. serrata*) were expelled. Six months after the episode the only symptom remaining was a slight burning sensation in the region of the pharynx. A similar case from Morocco was described by Corroller and Pierre (1960). Man can also harbour the encysted larvae of *L. serrata*, but they are usually dead and calcified and of no pathological significance (reviewed Cannon, 1942; Papadakis and Hourmouziadis, 1958).

In the case of the Porocephaloidea man can function as an intermediate host; although human cases of porocephaliasis are often (but not invariably) asymptomatic and recognized only at autopsy. The porocephalids commonly found in man are species of the genus *Armillifer*,* notably *A. moniliformis* (Diesing) and *A. armillatus* (Wyman). The adults of both species live in the respiratory tracts of snakes, and larval infestations in man are generally believed to be due to the ingestion of inadequately cooked snake meat or unboiled water contaminated with the nasal secretions of snakes. Faust (1927) found an encysted larva of *A. moniliformis* in the liver of a Tibetan who had died in Peking from miliary tuberculosis, and Cannon (1942) reported the case of a 50-year-old Nigerian woman who died following an intestinal obstruction caused by the presence of a very large number of encysted larvae of *A. armillatus* in the wall of the colon. Data presented by Prathap, Lau and Bolton (1969) suggest that amongst primitive peoples the incidence of porocephaliasis can be very high. In 30 consecutive autopsies on Malaysian aborigines these authors found that 45.4 per cent of adults were infested by a species of *Armillifer*. The parasites, which occurred most frequently in the liver and lungs, were incidental findings and did not appear to have contributed to the cause of death.

BIBLIOGRAPHY

CANNON, D. A. 1942. Linguatulid infestation of Man. *Ann. trop. Med. Parasit.* **36**: 160–166.

CORROLLER, Y. E. & PIERRE, J. L. 1960. Sur un cas de linguatulose humain au Maroc. *Bull. Soc. Path. exot.* **52**: 730–733.

CUÉNOT, L. 1949. Les Pentastomides. In *Traité de Zoologie*. Edited by P.-P. Grassé. Paris: Masson, vol. **6**, pp. 61–75.

DOLLFUS, R.-Ph. 1950. *Armillifer* L.-W. Sambon 1922 tombe en synonymie de Nettorhynque H.-D. de Blainville 1824. *Annls Parasit. hum. comp.* **25**: 112–114.

FAIN, A. 1961. Les Pentastomides de l'Afrique centrale. *Annls Mus. r. Afr. cent.* (8) (Zool.) **92**: 1–115.

FAUST, E. C. 1927. Linguatulids (order Acarina) from man and other hosts in China. *Am. J. trop. Med.* **7**: 311–322.

HEYMONS, R. 1935. Pentastomida. *Bronn's Kl. Ordn. Tierreichs* **5** (4): 1–268.

HOBMAIER, H. & HOBMAIER, M. 1940. On the life-cycle of *Linguatula rhinaria*. *Am. J. trop. Med. Hyg.* **20**: 199–210.

LAVOIPIERRE, M. M. J. & LAVOIPIERRE, M. 1966. An arthropod intermediate host of a pentastomid. *Nature, Lond.* **210**: 845–846.

NICOLI, R.-M. 1963. Phylogénèse et systématique. Le phylum des Pentastomida. *Annls Parasit. hum. comp.* **38**: 483–516.

PAPADAKIS, A. M. & HOURMOUZIADIS, A. N. 1958. Human infestation with *Linguatula serrata*. Report of a case. *Trans. R. Soc. trop. Med. Hyg.* **52**: 454–455.

PRATHAP, K., LAU, K. S. & BOLTON, J. M. 1969. Pentastomiasis in Malaysian aborigines. *Am. J. trop. Med. Hyg.* **18**: 20–27.

SAMBON, L. W. 1922. A synopsis of the family Linguatulidae. *J. trop. Med. Hyg.* **25**: 188–206, 391–428.

* Nomenclature changes could be the cause of some confusion. Dollfus (1950) placed *Armillifer* Sambon, 1922 in synonymy with *Nettorhynchus* de Blainville, 1824, but his opinion does not appear to have been widely accepted. Nicoli (1963) follows Dollfus by classifying species considered to be congeneric with *Linguatula armillata* Wyman (the type–species of *Armillifer*) in *Nettorhynchus*, but having done so he improperly retains *Armillifer* for *Subtriqueta shipleyi* Hett and *Pentastoma subtriquetum* Diesing.

17. FORENSIC ENTOMOLOGY

by Kenneth G. V. Smith

AFTER death the human body is still attractive to a variety of insects and other invertebrates. Not surprisingly, the Diptera figure largely in this fauna and include many species previously considered under myiasis. The entomologist is not infrequently called upon to identify such specimens for medico-legal purposes, usually as an aid in establishing the time of death, an all-important factor in murder cases.

The classic work on this subject is that of Megnin (1894), who gives figures as an aid to identification. Leclercq (1969) has more recently published a resumé of this work brought up to date, with additional case-histories and a good bibliography, but no figures or other aids to identification are given. Easton and Smith (1970) illustrate the principal invertebrates associated with the human cadaver and give other aids to identification. Other useful literature is given at the end of this chapter. The majority of insects and arachnids invading the corpse can be identified from earlier chapters, especially Chapter 6 and the references contained therein.

The constitution of the invertebrate fauna of the cadaver varies according to whether the corpse be exposed above ground, buried, or immersed in water. Obviously many environmental factors may affect the faunal constitution, such as temperature, time of year, climate, weather conditions, altitude, etc.

In Europe, the faunal succession of exposed or shallowly buried bodies appears roughly divisible into eight waves, as given in Table 12. However, the situation is undoubtedly more complex than this elsewhere and needs further study. Bohart and Gressitt (1951) have published some observations on the fauna of a corpse exposed on a beach in Guam. As has been stressed elsewhere (Easton and Smith, 1970) the coleopterous element of the fauna is more complex than the literature reveals and an indication of this is given in a detailed study of the faunal succession on a fox corpse (Easton, 1966).

Recent studies of the fauna of pig-carrion in the U.S.A. have been carried out by Payne et al. (1965–1972).

The faunal succession in buried corpses has not been studied in detail, but as one would expect is less abundant. According to Megnin the sequence appears to be as in Table 13, for human corpses; Payne et al. (1965–1972 give data for pigs).

There is little information on corpses immersed in water, but clearly the faunal composition will depend upon whether the water is fresh or salt. In fresh water conditions caddis-flies (Trichoptera) and other aquatic insects will feed on the body. In the sea Crustacea such as barnacles, shrimps and prawns will feed on corpses.

Freshly drowned corpses should be studied for any external parasites. According to Simpson (1961) fleas take 25 hours or so to drown and body lice some 12 hours. The recovery rates of these parasites after shorter immersion periods is also indicative of time of death.

Table 12 Faunal succession of exposed human corpse

	Fauna	State of corpse	Approx. age of corpse
1st Wave	*Calliphora erythrocephala* (Dipt., Calliphoridae) *C. vomitoria* (Dipt., Calliphoridae) *Musca domestica* (Dipt., Muscidae) *M. corvina* (Dipt., Muscidae) *Muscina stabulans* (Dipt., Muscidae)	Fresh	Fresh
2nd Wave	*Sarcophaga* spp. (Dipt., Sarcophagidae) *Lucilia* spp. (Dipt., Calliphoridae) *Cynomyia* spp. (Dipt., Calliphoridae)	Odour developed	
3rd Wave	*Dermestes* (Col., Dermestidae) *Aglossa* (Lep., Pyralidae)	Fats rancid	3–6 months
4th Wave	*Piophila casei* (Dipt., Piophilidae) *Madiza glabra* (Dipt., Piophilidae) *Fannia* (Dipt., Muscidae) Drosophilidae (Dipt.) Sepsidae (Dipt.) Sphaeroceridae (Dipt.) *Eristalis* (Dipt., Syrphidae) *Teichomyza fusca* (Dipt., Ephydridae) *Corynetes* (Col., Cleridae)	After butyric fermentation protein of 'caseic' fermentation	
5th Wave	*Ophyra leucostoma* (Dipt., Muscidae) *O. anthrax* (Dipt., Muscidae) Phoridae (Dipt.) Thyriophoridae (Dipt.)	Ammoniacal fermentation Evaporation of sanious fluids	1 year

	Fauna	State of corpse	Approx. age of corpse
	Necrophorus (Col. Silphidae) Silpha (Col., Silphidae) Hister (Col., Histeridae) Saprinus (Col., Histeridae)		
6th Wave	Acari	Remaining body fluids now absorbed	
7th Wave	Attagenus pellio (Col., Dermestidae) Anthrenus museorum (Col., Dermestidae) Dermestes maculatus (Col., Dermestidae) Tineola bisselliella (Lep., Tineidae) T. pellionella (Lep., Tineidae) Monopis rusticella (Lep., Tineidae)	Completely dry	
8th Wave	Ptinus brunneus (Col., Ptinidae) Tenebrio obscurus (Col., Tenebrionidae)		3 years plus

Table 13 Faunal succession of buried human corpses

	Fauna		Approx. age of corpse
1st Wave	Calliphora and Muscina stabulans		
2nd Wave	Ophyra anthrax		
3rd Wave	Phoridae		1 year
4th Wave	Rhizophagus paralelicollis (Coleoptera Rhizophagidae) Philonthus ebeninus (Coleoptera Staphylinidae)		2 years

BIBLIOGRAPHY

BEQUAERT, J. 1942. Some observations on the fauna of putrefaction and its potential value in establishing the time of death. *New Engl. J. Med.* **227**: 856.

BIANCHINI, G. 1930. La biologia del cadavere. *Archo. Antrop. criminale Psichiat. Med. leg.* **50**: 1035–1105.

BOHART, G. E. & GRESSITT, J. L. 1951. Filth inhabiting flies of Guam. *Bull. Bernice P. Bishop Mus.* **204**, 152 pp., 17 plates. Honolulu.

BORNEMISSZA, G. F. 1957. An analysis of arthropod succession in carrion and the effects of its decomposition on the soil fauna. *Aust. J. Zool.* **5**: 1–12.

EASTON, A. M. 1966. The Coleoptera of a dead fox (*Vulpes vulpes* (L.)); including two species new to Britain. *Entomologist's mon. Mag.* **102**: 205–210.

EASTON, A. M. & SMITH, K. G. V. 1970. The Entomology of the Cadaver. *Medicine Sci. Law* **1970**: 208–215.

FULLER, M. E. 1934. The insect inhabitants of carrion: a study in animal ecology. *Bull. Coun. scient. ind. Res. Melb.* **82**: 1–62.

HAUSER, G. 1926. Ein Beitrag zum Medenfrass an menschlichen Leichen. *Dt. Z. ges. gericht. Med.* **7**: 179–192.

HOLZER, F. J. 1939. Zerstörung an Wasserleichen durch Larven der Köcherfliege. *Dt. Z. ges. gericht. Med.* **31**: 223–228.

LECLERCQ, M. 1969. *Entomological Parasitology—the relations between Entomology and the Medical Sciences.* 158 pp. Pergamon Press, Oxford, London, etc.

MEGNIN, P. 1894. La Faune des cadavres. Application de l'Entomologie à la Médicine légale. *Encycl. Léautn.*, Paris.

MOTTER, M. G. 1898. A contribution to the study of the fauna of the grave. A study of one hundred and fifty disinterments, with some additional experimental observations. *J.N.Y. ent. Soc.* **6**: 201–231.

PAYNE, J. A. 1965. A summer carrion study of the baby pig, *Sus scrofa* Linnaeus. *Ecology* **46**: 592–602.

—— 1967. A comparative ecological study of pig carrion decomposition and animal succession with special reference to the insects. Ph.D. dissertation, Clemson University, S.C. 128 pp.

PAYNE, J. A. & KING, E. W. 1969. Lepidoptera associated with pig carrion. *J. Lepid. Soc.* **23**: 191–195.

—— —— 1970. Coleoptera associated with pig carrion. *Entomologist's mon. Mag.* **105**: 224–232.

—— —— 1972. Insect succession and decomposition of pig carcasses in water. *J. ent. Soc. Georgia.* **7**: 153–162.

PAYNE, J. A., MEAD, F. W. & KING, E. W. 1968. Hemiptera associated with pig carrion. *Ann. ent. Soc. Am.* **61**: 565–567.

PAYNE, J. A., KING, E. W. & BEINHART, G. 1968. Arthropod succession and decomposition of buried pigs. *Nature, Lond.* **219**: 1,180–1,181.

PAYNE, J. A. & MASON, W. R. M. 1971. Hymenoptera associated with pig carrion. *Proc. ent. Soc. Wash.* **73**: 132–141.

REED, H. B. 1958. A study of dog carcass communities in Tennessee, with special reference to the insects. *Am. Midl. Nat.* **59**: 213–245.

SIMPSON, K. 1961. *Forensic Medicine* (4th edn.). Arnold, London.

YOVANOVITCH, P. 1888. *Entomologie appliquée à la Médicine légale.* Paris.

18. INSECTS AND HYGIENE

by Harold Oldroyd

THE reason for giving a special chapter in this book to hygiene is a practical one. The major pests such as mosquitoes and tsetse-flies are fairly easily recognized, or may be identified by means of the Key in Chapter 2, but these are only a small fraction of the insects of the world. A somewhat larger number of insect species, from a wider variety of families, may on occasion be a nuisance, or may be involved in the spread of disease, though without the direct association that exists between mosquitoes and malaria, or tsetse-flies and trypanosomiasis.

It is a mistake to imagine that the so-called 'harmful' insects form a natural group, or have anything in common biologically. Indeed only rarely is the association with man significant in the natural history of the insect. The human louse is the only insect that has evolved specifically in this way. All other insects associated with man became so accidentally, because their feeding or breeding habits caused them to intrude at some point into man's affairs. Any single species of insect might affect human health in suitable circumstances.

Obviously this indefinite number of potential pests cannot be presented in a systematic way. Any particular species that becomes a pest might be identified through one of the standard textbooks of entomology, followed by use of more detailed, specialist keys or by submission to a specialist.

The aim of the present chapter is to approach this problem from the opposite direction, by classification of habits rather than by taxonomic relationships. Ways in which insects may attract attention, and raise problems of hygiene, include:
1. By being a nuisance outdoors, flying round the head, or crawling on the skin.
2. By swarming into houses and other buildings.
3. By infesting human food and drink.

SWARMS OF INSECTS OUTDOORS

The word 'swarm' needs to be used carefully. A swarm of bees is an active aggregation, grouped around a queen, and moving off as a corporate body in search of a new nesting-site. Few so-called 'swarms' of insects have this positive coherence. Locusts in a swarm are in the gregarious phase, when the behaviour of individuals is influenced by the proximity of others, with the result that the swarm is kept together, and is carried by wind as a unit. Processionary caterpillars follow another individual, and so have collective behaviour. But generally aggregations of insects come about through a large number of individuals doing the same thing at the same time, and there is no direct influence of one upon another. The numbers may dwindle to three, two or even one insect without any change in behaviour.

Midges

'Swarms' of non-biting midges (Chironomidae) are aggregations of individuals brought together by the fact that each is 'dancing' in the air over a marker of some kind, usually a prominent object on the ground, such as the gleam of light on water, or a light patch of rock; often a swarm forms in a patch of turbulent air, as under the eaves on the lee side of a house or other building.   Swarms over markers are common in open localities with water, in which the Chironomidae can breed, natural water in rivers and lakes, and artificial water in ornamental gardens and sewage works.

'Chironomidae were a serious pest at the Nile Hotel causing intense annoyance to guests and staff.   In the mornings dead midges were swept up by the bucketful.   Nearly all lights had to be extinguished in the evenings.   Many people could not sleep owing to allergic effects, and a considerable proportion of the staff spent periods in the asthma camp.   The pest was so bad that removal of the town was discussed' (Lewis, 1956: 147).

Swarms of *Chaoborus edulis* over Lake Victoria, in East Africa, appear as a black cloud, which moves with the wind, and which also shows signs of an internal rotation. The precise mechanism by which these swarms assemble is not known, but they may have some link with the rotating columns of rising air that often give rise to waterspouts.

Besides being a nuisance, these swarms of midges can be dangerous, and even fatal to people trapped in their path; a fisherman in a small boat may be suffocated before he can escape from the flies.

Trichoptera (caddisflies) and Ephemeroptera (mayflies) (see Chapter 13) may also occur in large numbers, causing considerable nuisance and allergic effects.

Winged Ants and Termites

Ants and termites periodically send out very large numbers of winged males and females, which take part in a mating flight.   These individuals fly about in a blundering way and can be a great nuisance, but do not involve any risk to health unless they fall into food or drink, or block drainage channels.   Biologically, the remarkable aspect of this kind of swarming is the way in which it occurs simultaneously over a number of nests scattered through a wide area.

Face-Flies and Eye-Flies

Various flies congregate round the head and face of man and of domestic animals, without first forming themselves into a swarm.   In the tropics the 'eye-flies' of the family Chloropidae—*Siphunculina* of the Old World and *Hippelates* of the New World—pass the larval life in plant or vegetable debris, but the adult flies feed readily on bodily secretions, sweat, the moisture of eyes, nose and mouth, ulcers and sores.   Besides irritating the eyes, these flies may transmit infection both by superficial contact and by their droppings.

Certain Noctuid and Pyralid moths feed by night from the eyes of buffalo and man in Thailand (see Chapter 11 and pl. 8).

Some muscid flies that are technically not blood-suckers—i.e. they have no equipment for piercing and sucking blood—nevertheless feed from open wounds, sores, and the blood oozing from punctures made by primary blood-suckers such as mosquitoes or Tabanidae. *Musca domestica* will do this on occasion, as part of its general adaptability, but other species of *Musca* practise this habit more regularly, and in greater numbers: *Musca sorbens* (pl. 4), *M. vetustissima*, *M. autumnalis*, *M. vitripennis*, *M. pattoni* and *M. fasciata* in various tropical and subtropical countries. *M. crassirostris* and some other species fold back the labella of the proboscis and use the prestomal teeth to rasp through skin and so draw blood which can be sucked up; these species are placed in the subgenus (or separate genus) *Philaematomyia*.

M. *autumnalis* is particularly annoying to pasturing cattle by clustering round the head and eyes. An Old World species, *M. autumnalis* appeared in North America in 1952, and has rapidly increased in numbers to become a major pest known as the 'face-fly' (see Depner, 1969).

Biting Muscidae of the genera *Stomoxys* and *Haematobia* (inc. *Lyperosia*) have been dealt with in Chapter 5b, but non-biting Muscidae of the genus *Hydrotaea*, the 'sweat-flies' of the countryside, are very persistent in flying round the head, eyes, nostrils and mouth of man and of grazing animals. They settle relatively little, and are a nuisance rather than a danger to health (see also Chapter 5b).

INSECTS NUMEROUS INDOORS

The only truly swarming insects that penetrate indoors are the 'driver ants', 'army ants' and 'legionary ants' (Dorylinae) of the tropics. These are primitive, carnivorous ants, the communities of which consume all the food in one area and then move in a compact column to another site, attacking every living thing on the way. If they pass through a house they kill or drive out all the rats, mice, scorpions, cockroaches, bugs, fleas, etc. Provided one keeps out of their way, driver ants can be considered beneficial from a hygienic point of view.

Other invasions of houses arise either from the drifting in of a swarm that has already formed outdoors or, more often, from the accumulation of insects indoors as in a trap.

Hibernating Insects

FLIES. Certain muscoid flies—*Musca autumnalis* and *Dasyphora cyanella* (MUSCI-DAE) and *Pollenia rudis* (CALLIPHORIDAE)—overwinter as adults in sheltered places, and often congregate in the roof-spaces of houses in temperate countries. They arrive individually at the end of summer, and thus form an aggregation, not a true swarm, though they may be present in very large numbers. They remain inactive during cold weather, but easily become aroused during a mild spell, or if the building contains warm air. Then they emerge from shelter, and may be a nuisance by beating against the windows, or falling down upon people, or into food. The flies are most troublesome in spring, when the first warm spell brings them all out within one or two days, usually the males first. They then sit about on the external walls, and on shrubs in the garden. If the flies die during the winter—as they easily do if they exhaust their fat-resources by activity—they normally dry out harmlessly, but have been recorded to breed *Sarcophaga*.

Thaumatomyia notata (Meigen) is a small Chloropid, which breeds in very large numbers in the roots of grass, where its larvae are predatory on the root-aphis *Pemphigius bursarius* (L.).   The Chloropid is believed normally to overwinter as a pupa, but a certain number of adults of this generation emerge before the onset of winter.   In favourable summers a high proportion of emergences takes place, and the resulting adult flies are driven to seek shelter as soon as the nights become cooler.   These flies enter bedrooms, apparently by drifting passively in the air until trapped in an eddy under the eaves.   They walk about on the ceiling of bedrooms, and are easily killed by cold.

FLIES HIBERNATING SINGLY, AND OTHER INSECTS.   Mosquitoes and drone-flies (*Eristalis tenax*) hibernate indoors as individuals, though sometimes drone-flies cluster together in a hollow space.   A few butterflies and moths, lacewings (Neuroptera: Chrysopidae) and ladybirds (Coleoptera: Coccinellidae) are regularly found indoors, and ladybirds may be present in large aggregations.   Most of these, except the mosquitoes, are quite harmless, and form no danger to hygiene unless they contaminate food or drink: e.g. if they fall into a water-cistern from which drinking water is taken.   The house-gnat *Culex pipiens*, bites only birds, but regularly hibernates in houses.   This species, however, has a number of man-biting strains, regarded as subspecies, and as these often breed in water in dark situations indoors, complaints of biting may arise (see Chapter 5).   *Culiseta* (= *Theobaldia*) *annulata*, a conspicuously striped mosquito, regularly hibernates in houses, and remains active, biting frequently during the night.

Insects Active Indoors

ANTS.   Some live outdoors and follow trails indoors in search of food; others make nests indoors.   They are generally a nuisance rather than a danger to health.   If they locate and infest food, for example a jar or packet in the larder, the fact is obvious from their numbers, and the item can be thrown away.   Ants are pre-eminently an example of insects that are individually insignificant, but which become a difficult problem because of their large numbers and persistence.

FLEAS have been dealt with in Chapter 7.   Since their larvae do not live on the host animal, but in debris in its nest or lair, fleas breed only in association with warm-blooded animals that form some sort of regular habitation.   The house itself may harbour human-fleas (*Pulex irritans*), or the cat-flea (*Ctenocephalides felis*).   The dog-flea (*Ctenocephalides canis*) is less common.   The garden and outbuildings may harbour hedgehog-fleas (*Archaeopsylla erinacei*), mole-fleas (*Hystrichopsylla talpae*) and particularly chicken-fleas (*Ceratophyllus gallinae*) coming from abandoned nests of the common garden birds.   Any flea will bite man if it gets the chance.

FLIES.   Adults and larvae of certain families of flies are liable to cause nuisance indoors, and to be distasteful to the occupants, without directly contaminating food and drink (see next section).

HOUSE-FLIES and BLOW-FLIES have been dealt with in Chapter 5.

SEPSIDAE are shining black-flies, with rounded head, which breed, among other places, in the settlement tanks and sludge deposits of sewage works.   Normally the adult flies remain close to their breeding place, but sometimes they move away in great numbers and penetrate into houses. *Themira putris* and *Nemopoda nitidula* are common European species to do this.

COELOPIDAE, seaweed-flies, breed in very large numbers in the heaps of 'wrack', or decaying seaweed on the beaches of temperate countries, in both hemispheres. Such heaps of seaweed have the properties of a compost heap, and provide a steady temperature throughout the year by bacterial decay. The larvae live in the decomposition products of the seaweed. The adult flies normally remain near the weed, but from time to time make mass movements along the shore-line to new breeding grounds. Occasionally these swarms lose contact with the shore and are blown inland, sometimes for considerable distances. They then tend to accumulate in premises to which they have been attracted by smell.

One species, *Coelopa frigida*, is strongly attracted by the smell of trichloroethylene and similar chlorinated compounds, and congregates in any building where these substances are used: garages, small workshops, dry-cleaners', pharmaceutical chemists' shops. The flies are a nuisance to people working in such places, but have not been associated with any infection.

EPHYDRIDAE. These flies normally breed in fresh or brackish water, but *Teichomyza fusca* has been recorded in massive numbers indoors due to a faulty toilet (Theodorides & Leclercq, 1949 (see also Chapter 5).

SPHAEROCERIDAE (=BORBORIDAE). A family of acalyptrate flies breeding in dung, decaying vegetable matter, grass-clippings, compost heaps and so on. Many species prefer dark, damp places, and are found in the burrows and nests of small mammals; these flies appear in coal mines, where they may be brought together into an aggregation by the ventilation system. They also appear in cellars and subterranean rooms, outhouses, lavatories and so on. One species, *Leptocera caenosa*, is a frequent pest indoors in Europe; it appears to breed in a variety of decaying substances, especially of animal origin and is frequently associated with leaks in sewage disposal systems, septic tanks etc. (see also Chapter 5d).

DROSOPHILIDAE. The many species of *Drosophila* are of tropical origin, where they breed in all kinds of sweet, fermenting materials which produce ethers and esters, and which favour the growth of yeasts. A few species have become adapted to living outdoors in temperate countries, and many more flourish in the artificial heat of buildings, where in a uniform temperature they may breed throughout the year. The increasing speed of modern communications favours the spread of *Drosophila*, and the genus itself is notoriously plastic, and readily evolves in new directions. This plasticity, and the fact that, given suitable food and a uniform temperature, *Drosophila* colonies will breed indefinitely, are reasons why *Drosophila* became the famous experimental animal for genetical work. The same reasons make it likely that *Drosophila* will be a domestic nuisance of increasing importance.

The nuisance is constituted by the adult flies, which are attracted by the smell of wine, vinegar, fruit, trifle, custard. Usually only a single fly appears, like a tiny speck hovering over the table, but if there is a local focus of breeding the flies may be numerous. Outbreaks occur in hotels, restaurants and apartments if the disposal of food scraps is not sufficiently careful. A colony of *Drosophila* can breed in a very small amount of material. Even if garbage bins are regularly emptied there may be ample material for breeding purposes left in the bottom, or in cavities round about where scraps that have missed the bin are allowed to lie.   In the house even one rotting fruit may give trouble.

Drosophila is one of two domestic flies that infest milk bottles if these are not washed and promptly returned to the dairy (see also PHORIDAE, below).   *Drosophila* actually breeds in the bottle, if this is exposed with dregs of milk curdled round the rim and in the base.   When the larvae are fully fed they crawl up the glass and pupate, the puparium being attached to the glass so firmly that they are often not loosened by the next washing of the bottle.

Adult *Drosophila* are generally harmless, though *D. repleta* and perhaps one or two related species breed in excreta, and may be carriers of infection in hospitals (see also Chapter 5d).

PHORIDAE.   Small flies of hump-backed appearance (fig. 102), recognized in life by their peculiar 'scrabbling' movement, which is a consequence of their lateral flattening and relatively long hind legs.   The larvae feed in decaying organic matter, especially if it is of high protein content and rather desiccated.   They are particularly associated with dead animals at a mummified stage, and with cadavers that have passed beyond the stage when they are attractive to blow-flies or even to *Piophila* (see below). A number of species breed in dead snails, feeding on the shrivelled remains of the soft body and then pupating in the empty shell.

One species, *Paraspiniphora bergenstammi*, readily pupates in empty milk bottles, if these are left out of doors for long periods.   The phorid does not breed in the bottle, but its puparia are firmly attached to the glass, and may remain after the next washing (cf. *Drosophila*, above).

Window-Flies

Most flies move towards the windows of a room, even if, like the blue-bottle *Calliphora*, they make noisy excursions into dark corners, and then dash themselves against the window again.   Certain flies are so characteristically found on the windows that they attract attention as 'window-flies'.

Silvicola fenestralis (ANISOPODIDAE) breeds normally in decaying organic materials, especially in rotting debris in tree-holes, and in sewage works.   Indoors breeding occurs in damp corners, especially round sinks and wash-houses.   The adult fly is not unlike a gnat or a midge, but with conspicuously spotted wings, and indoors is usually found walking up windows (see also Chapter 3).

Scenopinus fenestralis (SCENOPINIDAE) is a small, compact, bare black-fly, with rounded head, which is seldom seen anywhere else except on windows.   Its larva is an improbable wormlike creature, almost an inch long when fully grown, bare and white in colour.   Its natural habitat is the debris of birds' nests, but the larvae are often found indoors in carpets and furnishing materials.   They do not damage the material, nor involve any risk to health; they are carnivorous, feeding upon larvae of fleas, clothes moths and other injurious insects.

Louse-Flies

Flies of the family HIPPOBOSCIDAE live entirely upon the blood of mammals and birds, and produce a single, fully grown larva viviparously.   In these respects they resemble tsetse-flies, but are much more closely tied to their host animal than are tsetse. Most Hippoboscidae have wings, and can fly actively to another host, but quite a number have reduced wings, or none at all.

One species in particular comes into houses. *Crataerina pallida* is a specific parasite of the swift, *Apus apus*. The mature larvae are dropped into the nest, which is often under the eaves of houses and other buildings. The pupa, a brown, seed-like object, remains in the nest during the winter absence of the bird, and the adult fly at a time when normally the nest is reoccupied. *Crataerina* is unable to fly, and if there is no host at hand crawls about, often entering the bedroom. It seems to have a thigmotactic instinct to press itself into narrow spaces, perhaps associated with its normal habitat among the bases of feathers. Following this instinct the fly often gets between the sheets.

The principal reaction to the fly is revulsion at its flattened, tick-like appearance, and to the scraping of its claws against one's skin. It can, and does, bite man. Though various organisms of disease have been found in one or two species of Hippoboscidae nothing has been recorded from *C. pallida*. Considering that this fly is confined to swifts which spend so much of their time on the wing, and that the fly itself is flightless, it seems most unlikely to be an effective vector of anything harmful to man.

Other Orders of Insects Indoors

THYSANURA. Wingless, primitive insects which run about, usually at night, and scurry away from the light. The silverfish (*Lepisma saccharina*) and the firebrat (*Thermobia domestica*) are to be seen near stoves and hot pipes. They feed on farinaceous debris and are harmless, although there is one rather unusual record of a Thysanuran invading the ear of a patient in Cyprus (Burgess, 1971).

PSOCIDS. Members of the Order PSOCOPTERA. The domestic ones are called 'book lice' because they feed on the moulds and fungi developing in old books. Psocids attack any slightly mouldy material in a damp atmosphere, and sometimes appear in numbers from behind wallpaper. The indoor species are mostly wingless, and are harmless, though undesirable.

Other undesirable insects such as wasps, bees and cockroaches occur indoors and are dealt with elsewhere (Chapters 10 and 12).

INSECTS INFESTING FOOD AND DRINK

Almost any insect may accidentally fall into food or drink, but a few are particularly liable to do so, and may even breed there.

THE HOUSE-FLY *Musca domestica* is the prime example. The adult fly feeds readily from most human foodstuffs, sucking and regurgitating, as well as defaecating, and so provides the optimum conditions for the rapid spread of micro-organisms, especially as the fly feeds from human and animal faeces. House-fly larvae feed readily in decaying organic material around the house (see Chapter 11). Although *Musca domestica* is not specifically associated with any particular disease it undoubtedly acts as a mechanical vector of many human infections, especially those which are contracted through the mouth.

DROSOPHILA species. These, already mentioned, are particularly associated with food and drink, since they are attracted by aromatic substances, and the breeding medium of domestic *Drosophila* is generally in decaying food material.

THE CHEESE SKIPPER. *Piophila casei* (PIOPHILIDAE) is one of a small family of flies, the larvae of which live in organic materials of high protein content: e.g. dead animals at a stage between that attractive to blow-flies and the more mummified state in which Phorids attack. The larvae of *Piophila* continue to feed for some time, and help to clean off the remains of tissue from the skeleton. Among human foodstuffs, cheese gives the larvae their common name, but they also flourish in ham, bacon and other preserved meats, where they may excavate the interior without giving any sign externally. For this reason they may easily be swallowed and give rise to myiasis (see Chapter 5).

The name cheese 'skipper' refers to the jumping habit of the larva, which curls the body into an arc, seizes with the mouth-hooks two projections at the apex of the abdomen, and then suddenly releases them. This habit is found in a number of acalyptrate larvae, and is evidently a device for moving quickly away without a tedious and dangerous crawl. Such a leap does not necessarily take the larva into safety; it seems appropriate to larvae which live on vegetation, taking them from exposure near the top of the plant to obscurity below, from whence they can later crawl up again, but it is difficult to see how it benefits larvae with the habits of *Piophila*.

Larvae of Blow-Flies

Eggs and larvae of blow-flies are often found in foodstuffs of animal origin, principally in raw or cooked meat, but sometimes in cheese. Larvae of *Calliphora* can easily be distinguished from those of *Lucilia* (Key, p. 312 and figs 136, 148A, 149F), but it is difficult to be sure about other blow-flies that may possibly be involved. The practical problem remains the same—to avoid swallowing any of them, and to throw away as much of the food as is considered to be contaminated—but in practical cases there is often a question of finding the source of the infestation, to avoid a repetition, and possibly to determine legal liability.

The first is the easier, because it is necessary only to indicate various ways in which the foodstuff might have become infested. Legal liability is a very difficult matter, and usually starts with a demand to estimate the age of the larvae from the time of egg-laying; this problem arises not only in cases of infestation of food, but in medico-legal cases where larvae are found in corpses. Textbooks quote an average duration, at a stated temperature, for the egg-stage and the various larval and pupal instars, but if these figures are traced back to the original experimental work it will be found that there are considerable differences between different experiments, even under apparently identical conditions. It is not possible to state a figure for the duration of any part of the life-cycle of an insect with the precision that can be given, say, to the coefficient of expansion of brass (to quote a famous example). Furthermore, all stages are very much affected by variations in temperature, and probably in humidity, and these quantities are not simply measured. This is especially true of food cases, in these days of refrigeration and deep-freeze. By the time a packet of sausages has reached the consumer, who has found the maggots, they have usually been manufactured for some time, and passed in and out of various degrees of refrigeration a number of times.

It is clear, therefore, that the age of maggots is a matter to be extremely cautious about.

Blow-fly larvae figure in a rather unusual problem, by sometimes invading houses by crawling under the back door, usually at night, and after very heavy rain. There is clear evidence that they come from the soil, and that they are probably driven out because the heavy rain has filled up the interstices of the soil and left the maggots deprived of air. The problem is to know how the maggots come to be in the soil in such numbers. They are usually at the back of houses, close to dustbins, and it has been suggested that they are larvae from the bin gone into the soil to pupate; but there ought then to be younger larvae in the bin, and this is usually not so. A possibility is that there is seepage from drains beneath the soil and that the larvae are feeding on organic matter from this source.

Larvae of Fruit-Flies

This section particularly relates to the Mediterranean fruit-fly, *Ceratitis capitata*, which occurs in citrus fruits (except lemons), and feeds near the centre. Hence larvae are often undetected in fruit that is introduced into temperate countries, where the Mediterranean fruit-fly does not occur. The larvae are small, white and fragile, and quickly die and turn brown or black if they are removed from the fruit. They do little damage except near the centre, and most of the fruit is still edible. If swallowed by adults the larvae seem to be digested, or at least killed, but they sometimes pass alive through the intestine of a small child. There is some risk of infestation by these larvae if a baby is given fresh orange juice squeezed from the fruit, because the juice is turbid and the larvae are easily overlooked.

BIBLIOGRAPHY

BARNES, H. F. 1933. Two further instances of flies swarming at Rothamsted Experimental Station with some references to this phenomenon. *Entomologist's mon. Mag.* **69**: 230–232.

BUSVINE, J. R. 1966. *Insects and Hygiene*, 2nd edn. 467 pp. London.

BURGESS, N. R. H. 1971. Aural infestation by Thysanura. *Trans R. Soc. trop. Med. Hyg.* **65**: 405.

BRITISH MUSEUM (NATURAL HISTORY), LONDON. ECONOMIC SERIES AND ECONOMIC LEAFLETS. [Covers several insects and arachnids of medical and public health importance.]

COLYER, C. N. 1954. 'Swarming' of Phoridae (Diptera). *J. Soc. Br. Ent.* **5**: 22–27.

DARLINGTON, A. 1969. *Ecology of refuse tips.* 138 pp. London.

DEPNER, K. R. 1969. Distribution of the face-fly, *Musca autumnalis* (Diptera: Muscidae), in Western Canada and the relation between its environment and population density. *Can. Ent.* **101**: 97–100.

GREENBERG, B. 1965. Flies and disease. *Scient. Am.* **213**: 92–99.

—— 1971. *Flies and disease*, I. 856 pp. Princeton.

GREGOR, F. & POVOLNÝ, D. 1958. Versuche einer Klassification der synanthropen Fliegen. *J. Hyg. Epid. Microbiol. Immun.* **2**: 205–216. [Other papers by these authors, 1958–1964, should be consulted—see the Zoological Record.]

——, —— 1962. Zur Chorologie und hygienisch-epidemiologischen Rolle synanthroper Fliegen in Mittel-Europa. XI. *Intl. Kongr. Entom. Wien.* **2**: 419–422.

HAINES, T. W. 1953. Breeding media of common flies. I. In urban areas. *Am. J. trop. Med. Hyg.* **2** (5): 933–940.

—— 1955. Breeding media of common flies. II. In rural areas. *Am. J. trop. Med. Hyg.* **4** (6): 1125–1130.

HAVLÍK, B. & BATOVA, B. 1961. A study of the most abundant synanthropic flies occuring in Prague. *Cas. ceské Spol. ent.* **58**: 1–11.

HOWARD, L. O. 1900. A contribution to the study of the insect fauna of human excrement. *Proc. Wash. Acad. Sci.* **2**: 541–604.

LEWIS, D. J. 1956. Chironomidae as a pest in the northern Sudan. *Acta trop.* **13**: 142–158.

LINDSAY, D. R. & SCUDDER, H. I. 1956. Non-biting Flies and disease. *Ann. Rev. Ent.* **1**: 323–346.

MACRAE, A. W. R. 1966–1967 [exhibit]. Swarming behaviour of insects in Uganda. *Proc. R. ent. Soc. Lond.* (C) **31**: 14–16.

MALLIS, A. 1969. Handbook of pest control. The behaviour, life-history and control of household pests 5th edn. 1158 pp. New York.

MENG, C. H. & WINFIELD, G. J. 1944. Breeding habits of the common West China flies. *Chin. med. J.* **62A**: 77–87.

——, —— 1951. Studies on the control of fecal-borne diseases in north China. XVI. An approach to the quantitative study of the house-frequenting fly population. D. The breeding habits of the common north China flies. *Philipp. J. Sci.* **79**: 165–192.

MIHALYI, F. 1965. Rearing flies from feces and meat, infested under natural conditions. *Acta zool. hung.* **11**: 153–164.

MINISTRY OF AGRICULTURE, FISHERIES & FOOD. ADVISORY LEAFLETS. [Cover several insects of public health importance, house-flies, cluster-flies, etc.]

NUORTOVA, P. 1963. Die Rolle der Fleigen in der Epidemiologie der Poliomyelitis. *Anz. Schädl.* **36**: 149–155.

—— 1963. Synanthropy of blow-flies in Finland. *Suomen hyönt Aikak.* **29**: 1–49.

OLDROYD, H. 1964. *The natural history of flies.* 324 pp. London.

QUARTERMAN, K. D., BAKER, W. C. & JENSEN, J. A. 1949. The importance of sanitation in municipal fly control. *Am. J. trop. Med. Hyg.* **29** (6): 973–982.

RILEY, W. A. 1918. *Drosophila* in bottled certified milk. *Rep. Minn. St. Ent.* **17**: 40–45.

SCHOOF, H. F., MAIL, G. A. & SAVAGE, E. P. 1954. Fly production sources in urban communities. *J. econ. Ent.* **57**: 245–253.

SCOTT, H. 1926. Note on the swarming of gnats or midges round lofty towers. *Entomologist's mon. Mag.* **62**: 18–19.

—— 1953. Diptera found indoors, winter season 1951–1952. *Entomologist's mon. Mag.* **89**: 87–88. [Other papers by this author on British insects found indoors are in this journal 1916–1953.]

SCOTT, H. G. & LITTIG, K. S. 1962. Flies of public health importance and their control. *Publ. Hlth Serv. Publs Wash.* **799**: 1–40.

SERVICE, M. W. 1971. Feeding behaviour and host preferences of British mosquitoes. *Bull. ent. Res.* **60**: 653–661.

SHTAKELBERG, A. A. 1956. [Synanthropic Diptera of the fauna of the U.S.S.R.] *Opred. faune SSSR* **60**: 164 pp. [In Russian.]

SMITH, K. G. V. 1972. Adult Empididae (Diptera) occurring in numbers indoors. *Entomologist's mon. Mag.* **108**: 31–32.

STEYSKAL, G. C. 1957. The relative abundance of flies, collected at human faeces. *Z. angew. Zool.* **44**: 79–83.

TESCHNER, D. 1961. Zur Dipterenfauna an Kinderkot. *Dt. ent. Z.* **8**: 63–72.

THEODORIDES, J. & LECLERCQ, M. 1949. Une invasion de *Teichomyza fusca* Macquart (Diptera, Ephydridae). *Naturalistes belg.* **30**: 116–117.

WEST, L. S. 1951. *The Housefly, its natural history, medical importance and control.* London and New York. 584 pp.

WILTON, D. P. 1961. Refuse containers as a source of flies in Honolulu and nearby communities. *Proc. Hawaii. ent. Soc.* **17** (3): 477–481.

—— 1963. Dog excrement as a factor in community fly problems. *Proc. Hawaii. ent. Soc.* **18**: 311–317.

YAO, H. W., HUAN, I. C. & HUIE, D. 1929. The relation of flies, beverages and well water to gastro-intestinal diseases in Peiping. *Nat. Med. J. China.* **15**: 410–418.

ZUSKA, J. 1966. Some instances of mass occurrence of *Thaumatomyia notata* (Meigen) (Diptera. Chloropidae). *Acta ent. bohemoslovaca* **63**: 88.

19. SUMMARY OF ARTHROPOD VECTORS

by P. F. Mattingly, R. W. Crosskey & K. G. V. Smith

THE accompanying synoptic table of insects and other arthropods involved in the transmission of organisms pathogenic to man is intended to aid the less specialized medical entomologist and students and teachers of medical parasitology and entomology.

In using a table of this kind it is important to bear in mind that vector relations are complex and rarely more than partially understood. Recovery of a pathogen from a particular vector and vertebrate host does not necessarily imply that vector and host are associated. Similarly the fact that a particular vector is incriminated within a given geographical area does not necessarily imply that it serves as a vector elsewhere. These and other points are discussed more fully in the accompanying notes, which should be read carefully and treated as essential to the correct interpretation of the table.

Assessment of vector status involves intensive long term study. Too often vectors are listed as such on the basis of wholly inadequate evidence. Neither successful experimental transmission nor recovery of a pathogen in nature is sufficient by itself to incriminate a particular arthropod as an effective vector—yet vectors are often uncritically listed on such evidence. Vector status is determined by a complex of interrelated factors. Ability to harbour a pathogen, adequate contact with the hosts, sufficient longevity for the parasite to mature, are all equally essential and none of them alone sufficient for an effective vector: the study of any one aspect in isolation is never sufficient for assessment.

Many widely used reference sources are insufficiently discriminating in their lists of vectors. Because they are so uncritical we have preferred to make our own assessments by reference to the original literature wherever possible. This has led in some cases to the omission of certain familiar names: for example, *Simulium callidum* and *S. exiguum* are omitted as there is no conclusive evidence that they transmit *Onchocerca volvulus* to man, even though they are sometimes anthropophilic and have been found infected with filariae indistinguishable from *O. volvulus*.

The original sources of our information are far too numerous to list fully, but a short bibliography of comprehensive reviews and other useful reference works is appended. The table is largely confined to systems involving regular cyclical transmission to man. At the same time an attempt has been made to include such less clearly defined systems, e.g. mechanical transmissions, as appear to be reasonably well substantiated. It has been regarded as impracticable to try to include all the organisms for which insects such as cockroaches and domestic flies have been supposed to act as carriers.

Explanatory notes on the five columns of the table, and on the accompanying distribution maps, are given below.

THE TABLE

The comments below are arranged in order of the columns, reading from left to right.

Vector

Insect vectors are given first, followed by other Arthropods, and are listed with their Latin scientific names. In each class the entries are arranged in the commonly accepted taxonomic order to the level of family; thereafter genera and species are listed alphabetically. Subgenera, where given, are indicated in brackets and are not necessarily listed alphabetically within their genera. Vectors are listed with their currently valid names; a few important synonyms are listed in parentheses under the valid names where they provide an essential link with the literature. The names of the describers of the various vectors are omitted. Vectors whose identification is considered doubtful, either because they belong to sibling species complexes or for other reasons, are cited under the names commonly applied to them followed by the letters 's.l.' (sensu lato) (this applies mainly to mosquitoes).

Regarding mosquito vectors the following points must be noted. In citing vectors of human malaria attention has been paid only to those for which there appears to be adequate epidemiological evidence of involvement. Too little is known regarding the vectors of simian malaria to permit an equally rigorous approach. In view of its potential importance as a zoonosis all the few putative vectors are included. Similar criteria have been adopted for human and simian filariasis. For the arboviruses present knowledge does not justify such an approach. Consequently all those species are listed from which viruses known or believed to involve human hosts have been recovered. This part of the table has been designed to give adequate representation to all genera and subgenera involved, but to avoid overloading it some species belonging to the larger genera and subgenera have been listed in accompanying notes.

Group of disease organism

In this column the biological group to which the transmitted pathogen belongs is indicated, as appropriate, by one of the following letters: B=Bacterium; N=Nematode; P=Protozoan; R=Rickettsia; S=Spirochaete; V=Virus.

Pathogen and disease

This column lists the names of the transmitted pathogenic organisms under their Latin scientific names or, in the case of viruses, with the standard vernacular nomenclature. Familiar names of human diseases are indicated in brackets, where appropriate, after the name of the pathogen but only in the most practicable places (e.g. in the first of a succession of entries for the same pathogen).

Present knowledge of their vector relations is not detailed enough to justify citing the four species of human malaria parasite individually. They are accordingly cited collectively as 'Plasmodium spp.' (Garnham, 1966, gives detailed information). The periodic and subperiodic forms of *Wuchereria bancrofti* and *Brugia malayi* have, in general, different vectors, as is reflected by the quite different distributions of the two forms of *W. bancrofti* (see map of Bancroftian filariasis); apart from this our knowledge is still very limited and does not justify citing these forms separately in the table (Edeson & Wilson, 1964, give further information).   Regarding the mosquito-borne viruses all that are known or believed to infect man appear at least once in the table, but no viruses are included which have not been recovered from man or associated with him on serological evidence.   Inclusion of a particular vector and a particular virus on the same line of the table implies actual recovery of virus from the vector in all cases except three. The exceptions are Dengue-6 and *Ae. aegypti*, Dengue-3 and *Ae. polynesiensis*, and Dengue-2 and *Ae. scutellaris* which are associated on epidemiological evidence only. North American viruses belonging to the California complex are listed collectively. For details see Sudia *et. al.* (1971).   Members of the Wyeomyia complex, other than the Panama strain, are not listed because this is the only member known to have been recovered from man.

Hosts

Citation of a vector and one or more hosts on the same line of the table merely implies recovery of the pathogen from each of them.   It does not necessarily imply association of the vector with any of the known hosts.   In the case of viruses names of hosts are cited in brackets if virus has been identified in them only by immune reactions; brackets are omitted when citing hosts from which virus has actually been isolated.   Sentinel hosts are cited only in those cases in which virus has not been isolated from wild hosts of the same kind in the territory concerned.

Distribution

For each entry in this column the geographical distribution shown is that of association between vector, pathogen and man.   The range of the disease, or of the vector, may be and usually is much greater than that of any one territorial area in which a particular vector transmits the pathogen.   For example, *Simulium ochraceum* occurs widely in central and northern South America but is known to transmit *Onchocerca volvulus* to man only in circumscribed pockets in the territories named.   In the case of viruses citation of more than one territory implies that the virus has been recovered from the vector in question, and from one or more of its hosts, in all of them.   Failure to cite any vertebrate host indicates that the virus has been recovered from man (and possibly other hosts) elsewhere but is known only from the vector in the territory concerned. No entries for distribution are given if such are impracticable, as in the case of pathogens transmitted by synanthropic Diptera under conditions of primitive hygiene.

THE DISTRIBUTION MAPS

Reliable distribution maps of ecosystems as complex as human endemic and epidemic diseases are notoriously difficult to produce.   We have tried to avoid misleading over-

KK

generalization, as occurs in so many text-book maps. The distribution of dynamic systems is continually changing, either (as with yellow fever) because of natural factors, or (as with malaria) because of efforts at eradication by man. In compiling the maps of these diseases we have delimited in a generalized way the areas considered by the World Health Organisation to be at risk. For the filariases, and particularly onchocerciasis, we have felt it necessary to be more conservative, so that our maps show in general a less extensive distribution for these diseases than many of the older maps. This is notably so for our mapping of Bancroftian filariasis and onchocerciasis in Africa, where we show only endemic areas that appear to be reliably substantiated (some existing maps in text-books and reviews which show the whole of equatorial Africa as an endemic area for these diseases are particularly inaccurate). Maps of the distribution of African sleeping sickness and Chagas' disease are given in the relevant chapters in the text, as important detail would be obscured on small-scale world maps; likewise, a detailed distribution map of African onchocerciasis is given in the chapter on Simuliidae. The accompanying world maps are on Mollweide's homolographic projection, which usefully tends to emphasise the tropical regions where Arthropod-borne diseases largely occur.

BIBLIOGRAPHY

(see also references under each chapter)

ALCIVAR, Z. C. & CAMPOS, R. F. 1946. Las moscas, como agentes vectores de enfermedades entericas en Guayaquil. *Revta ecuat. Hig. Med. trop.* **3**: 3–14.

ANON. 1952. *Report of the international commission for the investigation of the facts concerning bacterial warfare in Korea and China.* 665 pp. Peking.

BERTRAM, D. S. 1962. Rickettsial infections and ticks. *Symp. zool. Soc. Lond.* No. **6**: 179–197.

BOIKOV, B. V. 1932. Rol'mukh v rasprostranenii bruishnogo tifa i drugikh zheludochnoki-shechnykj zabolevanii [Role of flies in dissemination of typhoid and other stomach and intestinal diseases] *Zh. Mikrobiol. Epidem. Immunobiol.* No. **7–8**: 26–39.

CAHILL, K. M. 1965. Leishmanin skin testing in Africa and the Middle East. *E. Afr. med. J.* **42**: 213–220.

—— 1967. Clinical and epidemiological patterns of Leishmaniasis in Africa. *Trop. geogr. Med.* **20** (1968): 109–118.

CASALS, J., HOOGSTRAAL, H. ET AL. 1966. A current appraisal of hemorrhagic fevers in the U.S.S.R. *Am. J. trop. Med. Hyg.* **16**: 751–764.

COATNEY, G. R., COLLINS, W. E., WARREN, M. W., & CONTACOS, P. G. 1971. *The Primate Malarias.* Washington.

DAY, M. F. & BENNETTS, M. J. 1954. *A review of problems of specificity in arthropod vectors of plant and animal viruses.* 172 pp. Canberra. [Extensive bibliography.]

DESCHIENS, R. & YUCEL, A. 1962. Complément d'enquete sur la filariose a *Wuchereria bancrofti* en Turquie orientale. *Bull. Soc. Path. exot.* **54** (1961): 1328–1336.

EDESON, J. F. B. 1962. The epidemiology and treatment of infection due to *Brugia malayi.* *Bull. Wld Hlth Org.* **27**: 529–541.

EDESON, J. F. B. & WILSON, T. 1964. The epidemiology of filariasis due to *Wuchereria bancrofti* and *Brugia malayi.* *A. Rev. Ent.* **9**: 245–268.

FORATTINI, O. P. 1965. Chapter IV, Filariose Bancroftiana, in FORRATTINI, *Entomologia médica* **3**: 325–356. [Useful bibliography for New World.]

GARNHAM, P. C. C. 1966. Malaria parasites and other haemosporidia. 1114 pp. Oxford.

GREENBERG, B. 1965. Flies and disease. *Scient. Am.* **213**: 92–99.

[Bibliography continued on p.531.]

VECTOR	GROUP OF DISEASE ORGANISM	PATHOGEN & DISEASE	HOSTS	DISTRIBUTION
INSECTA				
PHTHIRAPTERA (LICE)				
Pediculus humanus	S	*Spirochaeta* (*Borrellia*)	Man, Animals	Cosmopolitan
„	R	*Rickettsia prowazeki* (Epidemic, exanthematic typhus)	Rats, Mice, Man	Almost cosmopolitan
HEMIPTERA (BUGS)				
Panstrongylus megistus	P	*Trypanosoma cruzi* (American trypanosomiasis or Chagas' Disease)	Man and Reservoir Animals such as Armadilloes, Opossoms, Rats, Mice, Cats, Dogs, Bats, Squirrels, etc.	Brazil
Rhodnius prolixus	P	„	„	Venezuela
Triatoma barberi	P	„	„	Mexico
„ *brasiliensis*	P	„	„	Brazil, Bolivia
„ *dimidiata*	P	„	„	Panama through Central America into Mexico
„ *infestans*	P	„	„	S. Brazil, Uruguay, Paraguay, Argentina, Chile & S. Bolivia
„ *maculata*	P	„	„	Venezuela
„ *protracta*	P	„	„ (possible vector)	Arizona, California
DIPTERA (FLIES) **PHLEBOTOMIDAE**				
Phlebotomus (*Phlebotomus* s.s.)	P	*Leishmania tropica** (Leishmaniasis)	Man & Rodents	North Africa
„ *duboscqi*				

* *Leishmania donovani* and *Le. infantum* cause human visceral leishmaniasis. The other species of *Leishmania* cause human dermal (including mucocutaneous) leishmaniasis.

,, papatasi	V	Sandfly fever	Man	Mediterranean to India
	P	Le. tropica	Man & Rodents	Mediterranean to India
(Paraphlebotomus)				
,, caucasicus	P	Le. tropica	Rodents	Central Asia
,, sergenti	P	,, ,,	Man	Mediterranean, W. Asia
(Synphlebotomus)				
,, celiae	P	,, donovani*	Man (& Animals) ?	Kenya
,, martini	P	,, ,,	,,	,,
,, vansomerenae	P	,, ,,	,,	,,
(Laroussius)				
,, ariasi	P	,, infantum*	Man, Fox & Dog	Southern France
,, kandelakii	P	,, ,,	Man (& Animals ?)	Western Asia
,, langeroni orientalis	P	,, donovani	Man, Rodents & Non-canid Carnivores	Sudan
,, longicuspis	P	,, infantum	Man & Dog	North Africa
,, longipes	P	,, tropica	Man (& Animals ?)	Ethiopia
,, major syriacus	P	,, infantum	Man & Carnivores	Western Asia
,, perfiliewi	P	,, tropica	Man	Mediterranean
,, perniciosus tobbi	P	,, infantum	Man & Carnivores	Eastern Mediterranean
(Adlerius)				
,, chinensis chinensis	P	,, donovani	Man & Dog	China
,, ,, halepensis	P	,, infantum	Man & Carnivores	Middle East
,, longiductus	P	,, infantum	Man & Carnivores	Central Asia
,, simici	P	,, infantum	Man & Carnivores	Western Asia
(Euphlebotomus)				
,, argentipes	P	,, donovani	Man	India
Lutzomyia				
(Lutzomyia)				
,, longipalpis	P	,, donovani	Man, Fox & Dog	Brazil
verrucarum-group				
,, colombiana	P	Bartonella bacilliformis (Carrion's disease)	Man	Colombia
,, verrucarum	P	Bartonella bacilliformis (Carrion's disease)	Man	Peru
(Nyssomyia)				
,, anduzei	P	Le. braziliensis*	Man & Rodents ?	Surinam
,, flaviscutellata	P	,, ,,	Rodents	Brazil
,, olmeca	P	,, mexicana	Man & Rodents	Belize

* See footnote on p. 501

CULICIDAE

Species		Virus/Disease	Hosts	Distribution
Aedeomyia catasticta	v	Alfuy	Wild Bird, (? Man, ? Fowl)	Queensland
Aedes (Ochlerotatus) caballus	v	Middelburg	(Man, Domestic Ungulates, Dog)	S. Africa
"	v	Rift Valley Fever	Man, Domestic Ungulates, (Polecat)	S. Africa
"	v	Wesselsbron	Man, Domestic Ungulates	S. Africa
" *caspius*	v	Tahyna	Man, (Domestic Ungulates, Lagomorphs, Fox, Field Mouse)	Central and S. Europe
" *dorsalis/melanimon*	v	California group	(Man, Domestic Ungulates, Lagomorphs, Rodents)	California, New Mexico, Utah
"	v	St. Louis	(Man, Rodents, Cow, Wild and Domestic Birds)	California
	v	Vesicular Stomatitis (Indiana)	Man, Cattle	New Mexico
	v	Western Equine	Man, Ungulates, Rodents, Birds, Snakes, (Lagomorphs)	Western U.S.A.
" *scapularis*	v	Ilheus	Man, Sentinel Monkey, Birds, (Monkeys, Rodents, Marsupials, Sloths, Lizards)	Brazil, Trinidad
"	v	Kairi	Rodent, Monkey, (Man, Donkey)	Brazil, Colombia, Trinidad
"	v	Lukuni	(Man)	Trinidad
"	v	Maguari	(Man, Domestic Ungulates, Monkey, Birds)	Brazil, Colombia, Trinidad
"	v	St. Louis	Man, Birds	Trinidad
"	v	Venezuelan Equine	Man, Horse (Other Domestic Animals, Opossum)	Venezuela
" *serratus*	v	Guama	Man, Sentinel Monkey, Rodents, Marsupial	Brazil
"	v	Ilheus	Man, Sentinel Monkey, Birds, (Monkeys, Rodents, Marsupials, Sloths, Lizards)	Brazil, Trinidad
"	v	Maguari	(Man, Monkey, Horse)	Trinidad
"	v	Mucambo	Man, Sentinel Monkey, Rodents, Marsupials, Bird (Edentates)	Brazil
"	v	Oriboca	Man, Sentinel Monkey, Rodents, Marsupial	Brazil
"	v	Oropouche	Man, Sloth	Brazil
"	v	St. Louis	Man, Birds	Trinidad
"	v	Una	(Man, Domestic Ungulates), Rodents	Brazil, Trinidad

Species	Virus/Parasite	V/N	Hosts	Distribution
"	Venezuelan Equine	V	Equines, Man, Rodents, Marsupial	Trinidad
" sollicitans	Cache Valley	V	(Man, Dog, Domestic Ungulates, Wild Carnivores, Rodents)	Maryland, Virginia
"	Eastern Equine	V	Man, Horses	Eastern U.S.A.
" taeniorhynchus	Cache Valley	V	(Man, Dog, Domestic Ungulates, Wild Carnivores, Rodents)	Maryland, Virginia, Jamaica
"	Eastern Equine	V	Horses, Rodents, Marsupial, Sentinel Monkey, (Man)	Brazil
"	Kairi	V	?	Colombia
"	Mucambo	V	Man, Sentinel Monkey, Rodents, Marsupials	Fr. Guiana, Brazil
"	?St. Louis	V	Man, (Birds)	Florida
"	Tensaw	V	Rodent, Dog, (Man, Cow, Raccoon)	Florida
"	Tlacotalpan	V	(Man, Cattle, Pigs)	Mexico
"	Venezuelan Equine	V	Man, Domestic Ungulates, (Rodents)	Florida, Guatemala, Venezuela
" vigilax	Wuchereria bancrofti (filariasis)	N	Man	New Caledonia
"	Kokobera	V	(Man, Domestic Ungulates, Marsupials)	Queensland
"	Ross River	V	(Man, Cattle, Bat, Marsupials)	Queensland
" spp.		V	See Note I (at end of table)	
(Finlaya) fijiensis	Wuchereria bancrofti	N	Man	Fiji
" kochi	Wuchereria bancrofti	N	Man	New Guinea
" koreicus	Japanese B	V	Man, Birds	Far-eastern U.S.S.R.
" niveus gp.	Wuchereria bancrofti (filariasis)	N	Man	Thailand
" poecilus	Wuchereria bancrofti	N	Man	Philippines
" samoanus	Wuchereria bancrofti	N	Man	Samoa
" togoi	Brugia malayi (filariasis)	N	Man	China, Japan
"	Japanese B	V	Man, Birds	China, Japan
(Howardina) arborealis s.l.*	Apeu	V	Man, Sentinel Monkey, Marsupials, Rodents	Far-eastern U.S.S.R.
"	Guama	V	Man, Sentinel Monkey, Marsupials, Rodents	Brazil

* See footnote on page 505.

,,	V	Oriboca	Man, Sentinel Monkey, Marsupials, Rodents	Brazil
,, (Skusea) pembaensis	V	Bunyamwera	Man, (Domestic Animals, Rodents, Birds)	S. Africa
,,	V	Lumbo	(Man)	Mozambique
,, (Stegomyia) aegypti	V	Chikungunya	Man, Bats, (Domestic Animals, Wild Mammals, Birds and Reptiles)	African and Asian Tropics
,,	V	Dengue-1, -2	Man, (Wild Birds, Monkeys, Bats)	Cosmotropical
,,	V	Dengue-3, -4, -5	Man	Southern Asia
,,	V	? Dengue-6	Man, (Monkey)	Thailand
,,	V	Yellow Fever	Man, Monkeys, Bat, (Apes, Lemuroids, Marsupials, Bird)	African and Central and South American Tropics
,,	V	Zika	Sentinel Monkey, (Man, Wild Monkeys, Birds)	Tropical Africa, S.E. Asia
Aedes (Stegomyia) africanus	V	Chikungunya	Man, (Monkeys)	Uganda
,,	V	Rift Valley Fever	Rodent, Calf (Man)	Uganda
,,	V	Yellow Fever	Man, Bat, (Monkeys, Lemuroids, Bird)	Uganda, Ethiopia
,,	V	Zika	Sentinel Monkey, (Man, Wild Monkey)	Uganda
,, *albopictus* s.l.†	V	Dengue-2, -4	Man, (Monkeys)	S.E. Asia
,, *polynesiensis*	N	*Wuchereria bancrofti* (filariasis)	Man	Fiji, Samoa, Tahiti, Tokelau Is.
,,	V	? Dengue-3	Man	Tahiti
,, *pseudoscutellaris*	N	*Wuchereria bancrofti*	Man	Fiji
,, *scutellaris*	V	? Dengue-2	Man	New Guinea
,, *simpsoni*	V	Yellow Fever	Man, (Monkeys, Lemuroids)	Uganda
,, *tabu*	N	*Wuchereria bancrofti*	Man	Tonga
,, *upolensis*	N	*Wuchereria bancrofti*	Man	Samoa
,, *spp.*	V	*Wuchereria bancrofti*	See Note 2 (at end of table)	

† Vectors whose identification is considered doubtful, either because they belong to sibling species complexes or for other reasons, are cited under the name commonly applied to them followed by the letters 's.l.' (sensu lato). For further details the following should be consulted. *Ae. arborealis* s.l. Berlin, O.G.W. *Contr. Am. ent. Inst.* **4** (2): 1969, *Ae. albopictus* s.l. Huang, Y-M., *Proc. ent. Soc. Wash.* **71**: 234, 1969, *An. maculipennis* s.l. Bates, M., *Am. ent. Soc. Am.* **33**: 343, 1940, *An. gambiae* s.l. Davidson, G., *Riv. Malar.* **43**: 167, 1964, *An. punctulatus* s.l. Bryan, J. H., *J. Parasit.* **56** (4.2): 40, 1970, *C. pipiens* s.l. Mattingly, P. F., *Bull. Wld Hlth Org.* **37**: 257, 1967, *C. vishnui* s.l. Reuben, R., *Bull. ent. Res.* **58**: 643, 1969, *C. caudelli* s.l. Aitken, T. H. G. et al., *Am. J. trop. Med. Hyg.* **17**: 253, 1968, *E. chrysogaster* s.l. Van Someren, E. C. C., *Proc. R. ent. Soc. Lond.*, B **18**: 119, 1949.

Species	Virus	V	Hosts	Location
" (Aedimorphus) cumminsi	Middelburg	V	(Sheep)	Senegal, Cameroun
"	Sindbis	V	Man, Birds, (Domestic Ungulates)	S. Africa
"	Spondweni	V	(Man, Domestic Ungulates)	Natal
" vexans	Eastern Equine	V	Man, Horse, Dog, Pheasant, Wild Birds, Rodents, (Bat)	U.S.A.
"	Japanese B	V	Man, Horse, Pig, Bird, Bat	Japan
"	Sagiyama	V	(Man, Horses, Pigs, Wild Birds)	Japan
"	Tahyna	V	Man, (Domestic Ungulates, Hare)	Czechoslovakia
"	Trivittatus	V	(Man, Rabbit)	U.S.A.
"	Western Equine	V	Man, Ungulates, Rodents, Birds, Snakes, (Lagomorphs) See Note 3 (at end of table)	Western U.S.A.
" spp.	Bunyamwera	V	Man, (Domestic Ungulates, Rodents, Birds)	Natal
(Neomelaniconion) circumluteolus	Middelburg	V	(Man, Domestic Ungulates, Dog)	Natal
"	Ndumu	V	(Man)	Natal
"	Pongola	V	(Man, Domestic Ungulates)	Natal, Uganda
"	Rift Valley Fever	V	Rodent, Calf (Man)	Uganda
"	Simbu	V	(Man)	Natal
"	Sindbis	V	Man, Birds, (Domestic Ungulates)	S. Africa
"	Spondweni	?	(Man, Domestic Ungulates)	Natal
"	Wesselsbron	V	Man, Sheep, (Cattle, Goats)	S. Africa
" lineatopennis	Tembusa	V	(Man)	Malaya
" sp. indet.	Middelburg	V	(Man, Domestic Ungulates, Dog)	S. Africa
"	Wesselsbron	V	Man, Sheep, (Cattle, Goats)	S. Africa
(Aedes) " cinereus	Bunyamwera group	V	(Man, Wild Mammals, Reptiles)	Wisconsin
"	California group	V	?	New York State
(Verrallina) butleri gp	Bebaru	V	(Man)	Malaya
(Cancraedes) curtipes	Bunyamwera group	V	(Man, Pig)	Sarawak
"	Japanese B	V	(Man, Pigs, Dogs, Birds, Bats)	Sarawak
spp. indet.	Uganda S	V	(Man, Monkey)	Uganda
"	Wesselsbron	V	(Squirrels, Birds)	Cameroun
Anopheles (Stethomyia) nimbus	Lukuni	V	?	Brazil
"	Maguari	V	(Man, Domestic Ungulates, Birds)	Brazil

Species		Agent	Host	Locality
(*Anopheles*) *barbirostris*	N	*Wuchereria bancrofti*	Man	? Celebes
"	P	*Plasmodium* spp. (malaria)	Man	? Celebes
"	V	Chittoor	(Man, Domestic Ungulates, Bird)	India
" *campestris*	N	*Brugia malayi* (filariasis)	Man	Malaya
"	P	*Plasmodium* spp.	Man	Thailand, Malaya
" *claviger*	P	*Plasmodium* spp.	Man	Near East
" *crucians*	V	Cache Valley	(Man, Dog, Domestic Ungulates, Rodents, Wild Carnivores)	Maryland, Virginia, Alabama, Florida
"	V	California group	(Man)	Florida, Alabama
"	V	Eastern Equine	Birds, (Man)	Alabama, Georgia
"	V	St. Louis	Man, (Birds)	Florida
"	V	Tensaw	Rodent, Dog, (Man, Cow, Raccoon)	Florida, Alabama
" *donaldi*	V	Venezuelan Equine	(Man, Horse, Rodents)	Florida
"	N	*Brugia malayi*	Man	Malaya, Borneo
" *labranchiae*	P	*Plasmodium* spp.	Man	Malaya, Borneo
" *lesteri*	P	*Plasmodium* spp.	Man	North Africa
"	N	*Brugia malayi*	Man	China
" *letifer*	P	*Plasmodium* spp.	Man	China
"	N	*Wuchereria bancrofti*	Man	Malaya
" *maculipennis* s.l.	P	*Plasmodium* spp.	Man	Malaya
"	*V	Calovo	(Man, Horse, Hare)	Czechoslovakia, Jugoslavia
" *nigerrimus*	P	*Plasmodium* spp. (malaria)	Man	Malaya
" *pseudopuncti-pennis*	P	*Plasmodium* spp.	Man	Mexico, S. America
" *punctimacula*	P	*Plasmodium* spp.	Man	Colombia, ? Panama
" *quadrimaculatus*	P	*Plasmodium* spp.	Man	? Mexico
"	V	Cache Valley	(Man, Cattle, Equines, Dog, Raccoon, Woodchuck)	Illinois, Kentucky
"	V	Tensaw	Rodent, Dog, (Man, Cow, Raccoon)	Florida, Georgia
" *sacharovi*	P	*Plasmodium* spp.	Man	Near East
" *sinensis*	N	*Wuchereria bancrofti*	Man	China
"	N	*Brugia malayi*	Man	China
" *whartoni*	P	*Plasmodium* spp.	Man	China, Korea
"	N	*Wuchereria bancrofti*	Man	Malaya

* See footnote on p. 505.

Species		Type	Agent	Hosts	Distribution
»	(Nyssorhynchus) albimanus	P	*Plasmodium* spp.	Man	New World Tropics
»	" albitarsis	V	Tlacotalpan	(Man, Cattle, Pigs)	Mexico
»	" aquasalis	P	*Plasmodium* spp.	Man	S. America
»	" "	V	Venezuelan Equine	Man, Horse, (Other Domestic Animals, Opossum)	Lesser Antilles, S. America
»	" darlingi	N	*Wuchereria bancrofti*	Man	Venezuela
»	" "	P	*Plasmodium* spp.	Man	Brazil, Guyana
»	(Anopheles) numeztovari	P	*Plasmodium* spp.	Man	C. and S. America
»	" oswaldoi	P	*Plasmodium* spp.	Man	Venezuela
»	(Kerteszia) bellator	P	*Plasmodium* spp.	Man	Panama, ? Colombia
»	" cruzii	P	*Plasmodium* spp.	Man	Brazil, ? Venezuela
»	" "	P	*Plasmodium* spp.	Man	Brazil, ? Venezuela
»	" "	P	*Plasmodium simium*	Monkeys, ? Man	Brazil
»	" "	V	Tacaiuma	Sentinel Monkey, (Man, Horse, Rodent)	Brazil
»	" neivai	V	Guaroa	Man	Colombia
»	(Cellia) acomitus	V	Yellow Fever	(Man, Monkeys)	Panama
»	" amictus	P	*Plasmodium* spp.	Man	Thailand, Java
»	" "	V	? Getah	(Man, Fowl, Cattle, Birds)	Queensland
»	" annularis	V	Kowanyama	(Man, Fowl, Horse, Kangaroos)	Queensland
»	" annulipes	P	*Plasmodium* spp.	Man	India, Burma
»	" "	V	Kowanyama	(Man, Fowl, Horse, Kangaroos)	Queensland
»	" "	V	Trubanaman	(Man, Fowl, Domestic Ungulates, Wallabies, Kangaroo)	Queensland
»	" balabacensis	P	*Plasmodium* spp.	Man	Assam, Thailand, Vietnam, N. Borneo
»	" culicifacies	P	*Plasmodium* spp.	Monkeys	Malaya
»	" "	P	*Plasmodium* spp.	Man	W. Pakistan, India, Ceylon
»	" "	V	Chittoor	(Man, Domestic Ungulates, Bird)	India
»	" fluviatilis	P	*Plasmodium* spp.	Man	Iran, W. Pakistan, India
»	" funestus	N	*Wuchereria bancrofti*	Man	Tropical Africa, Madagascar
»	" "	P	*Plasmodium* spp.	Man	Tropical Africa, Madagascar
»	" "	V	Bwamba	(Man)	Uganda
»	" "	V	Nyando	(Man)	Kenya
»	" "	V	Onyongnyong	(Man)	Kenya, Uganda
»	" "	V	Semliki Forest	(Man)	Kenya
»	" "	V	Tanga	(Man)	E. Africa

Species		Parasite/Virus	Host	Distribution
"	gambiae s.l.*	N	Wuchereria bancrofti	Man	Africa, Malagasy Region
"	"	P	Plasmodium spp.	Man	Tropical Africa, Madagascar
"	"	V	Bwamba	Man	Central Africa
"	"	V	Chikungunya	Man, Bats, Monkey, Galago, Squirrel	Senegal
"	"	V	Onyongnyong	Man	Kenya, Uganda
"	"	V	Tataguine	Man	West Africa
"	hackeri	P	Plasmodium spp.	Monkeys	Malaya
"	jeyporiensis	N	Wuchereria bancrofti	Man	Vietnam, ? China
"	"	P	Plasmodium spp.	Man	Vietnam, China
"	leucosphyrus	P	Plasmodium spp.	Man	S. Borneo, ? Sumatra
"	ludlowae	P	Plasmodium spp.	Man	Celebes
"	maculatus	N	Wuchereria bancrofti	Man	Malaya
"	"	P	Plasmodium spp.	Man	Malaya, Sumatra, ? Java
"	minimus	N	Wuchereria bancrofti	Man	Philippines, China
"	"	P	Plasmodium spp.	Man	Assam, Thailand, Vietnam, Philippines
"	moucheti	P	Plasmodium spp.	Man	Uganda, Congo, Cameroun, ? Nigeria
"	nili	P	Plasmodium spp.	Man	Congo, W. Africa
"	pharoensis	P	Plasmodium spp.	Man	Egypt
"	"	V	Sindbis	Man, (Wild Birds and Mammals)	Egypt
"	philippinensis	P	Plasmodium spp.	Man	Bengal, E. Pakistan, ? Burma
"	"	V	Tembusu	(Man)	Malaya
"	pujutensis	P	Plasmodium spp.	Monkeys	Malaya
"	pulcherrimus	P	Plasmodium spp.	Man	Afghanistan
"	punctulatus s.l.*	N	Wuchereria bancrofti	Man	New Guinea, Solomon Is.
"	"	P	Plasmodium spp.	Man	Moluccas, Melanesia
"	sergentii	P	Plasmodium spp.	Man	? N. Africa, ? Arabia
"	stephensi	P	Plasmodium spp.	Man	Persian Gulf to India, ? Vietnam
"	subpictus	P	Plasmodium spp.	Man	? Indonesia
"	"	V	Chittoor	(Man, Domestic Ungulates, Bird)	India
"	sundaicus	P	Plasmodium spp.	Man	Bengal, Malaya, Java, Borneo
"	superpictus	P	Plasmodium spp.	Man	Near East
"	tessellatus	N	Wuchereria bancrofti	Man	Maldive Is.
"	"	P	Plasmodium spp.	Man	Maldive Is.
"	"	V	Chittoor	(Man, Domestic Ungulates, Bird)	India

* See footnote on p. 505.

Vector		Organism / Virus	Host	Distribution
" varuna	P	Plasmodium spp.	Man	India
" spp.	V	Tahyna	See Note 4 (at end of table)	S. France
Culex (Barraudius) modestus	V	West Nile	(Man, Horse, Rabbit)	S. France
" (Neoculex) rubinotus	V	Banzi	Man, Horse	E. and S. Africa
" "	V	Germiston	Man, (Cattle, Sheep)	Uganda, S. Africa
" "	V	Witwatersrand	Man, Rodents (Domestic Ungulates)	Uganda, S. Africa
" (Lophoceraomyia) spp. indet.	V	Bakau, Bebaru, Ketapang	(Man)	Malaya
" (Culiciomyia) nebulosus	V	Ntaya	(Man)	Cameroun
" (Culex) annulioris	V	Kamese	?	Uganda
" "	N	Ntaya	(Man)	Uganda
" " annulirostris	V	Wuchereria bancrofti	Man	New Guinea
" "	V	Corriparta	Birds, (Fowl, ? Man)	Queensland
" "	V	Kokobera	(Man, Domestic Ungulates, Marsupials)	Queensland
" "	V	Kunjin	(Man, Fowl, Cattle)	Queensland
" "	V	Murray Valley	Man, (Wild and Domestic Birds and Mammals)	Australia
" "	V	Ross River	(Man, Cattle, Bat, Marsupials) (Man, Cattle, Fowl, Dog, Wallaby)	Queensland
" bitaeniorhynchus	N	Wuchereria bancrofti	Man	New Guinea
" "	V	Chittoor	(Man, Domestic Ungulates, Bird)	India
" "	V	? Getah	(Man, Fowl, Cattle, Birds)	Queensland
" " gelidus	V	Sindbis	?	Philippines
" "	V	Batai	(Man)	Malaya
" "	V	Chikungunya	Man, (Monkeys, Domestic Ungulates, Dogs, Rabbits, Bats)	Thailand
" "	V	Dengue-1	Man, (? Monkeys)	S.E. Asia
" "	V	Getah	(Man)	Malaya
" "	V	Japanese B	Man, Domestic Ungulates, (Dogs, Birds, Bats)	S.E. Asia
" "	V	Sindbis	?	Sarawak
" " nigripalpus	V	Tembusu	(Man, Fowl)	Malaya, Sarawak
" "	V	California group	(Man)	Florida
" "	V	? Caraparu	Rodents	Trinidad
" "	V	Eastern Equine	Birds, (Man)	Alabama, Trinidad
" "	V	Ilheus	Man, Birds, (Rodents)	Panama
" "	V	St. Louis	Man, Birds	Florida, Jamaica, Trinidad
" "	V	Tensaw	Rodent, Dog, (Man, Cow, Raccoon)	Florida

Species	Virus / Agent	V/N	Hosts	Location
"	Venezuelan Equine	V	Man, Domestic Ungulates, Rodents, Marsupial	Florida, Guatemala, Trinidad
"	Western Equine	V	Birds	Alabama
"	Wyeomyia complex	V	Man	Panama
" pipiens fatigans	Wuchereria bancrofti	N	Man	Cosmotropical
"	Chikungunya	V	Man, (Monkeys, Domestic Animals, Bats)	Tanzania, Thailand
"	St. Louis	V	Man, Wild and Domestic Birds, Bat	Arizona, Texas
"	Sindbis	V	?	Philippines
"	Venezuelan Equine	V	Man, Rodents, Birds, (Horse, Mule, Cow, Dog, Fowl, Opossums, Bat)	Panama
"	Western Equine	V	Man, Ungulates, Rodents, Birds, Snakes, (Lagomorphs)	Southern U.S.A.
" var. molestus	West Nile	V	Man	India
"	Wuchereria bancrofti	N	Man	Egypt
" s.l.*	West Nile	V	Man, Horse, Bats, Birds	Egypt, Israel
"	Wuchereria bancrofti	N	Man	China, Japan
"	Japanese B	V	Man, Horse, Pigs, Birds, Bat	Far-eastern U.S.S.R., Japan
"	St. Louis	V	Man, Birds, Bat, (Domestic Ungulates, Poultry, Lagomorphs, Rodents, Carnivores)	U.S.A.
"	Sindbis	V	Man, Birds, (Domestic Ungulates)	S. Africa
"	Western Equine	V	Man, Ungulates, Rodents, Birds, Snakes, (Lagomorphs)	U.S.A.
" tarsalis	California group	V	(Man, Domestic Ungulates, Rodents)	California, Utah
"	Lokern	V	Lagomorphs, (Man, Domestic Ungulates, Rodents, Fowl, Wild Birds)	California
"	St. Louis	V	Man, Birds, Bat, (Domestic Ungulates, Poultry, Lagomorphs, Rodents, Carnivores)	Western U.S.A.
"	Western Equine	V	Man, Ungulates, Rodents, Birds, Snakes, (Lagomorphs)	Western U.S.A.
" telesilla	Middelburg	V	?	Cameroun
"	Ntaya	V	?	Cameroun
"	Wesselsbron	V	(Squirrels, Birds)	Cameroun

* See footnote on p. 505.

Genus	Species		Virus	Hosts	Distribution
"	theileri	V	Rift Valley Fever	Man, Domestic Ungulates, (Polecat)	S. Africa
"	"	V	Sindbis	Man, Birds, (Domestic Ungulates)	S. Africa
"	"	V	West Nile	Man, Wild Birds, (Domestic Ungulates, Fowl, Monkeys, Rodents)	S. Africa
"	tritaeniorhynchus	V	Chikungunya	Man, Monkeys, (Domestic Ungulates, Dogs, Rabbits, Bats)	Thailand
"	"	V	Dengue-3	Man	Philippines
"	"	V	Japanese B	Man, Horse, Pig, Birds, Bat, (Wild Rodents, Mongoose)	Southern and Eastern Asia
"	"	V	Sagiyama	(Man, Pigs, Birds, Horses)	Japan, Okinawa
"	"	V	Sindbis	?	Malaya, Sarawak
"	univittatus	V	Tembusu	(Man, Fowl)	Malaya, Sarawak
"	"	V	Mossuril	(Man, Baboon)	S. Africa
"	"	V	Sindbis	Man, Birds, (Domestic Ungulates)	Egypt, S. Africa
"	"	V	Spondweni	(Man, Domestic Ungulates)	Natal
"	"	V	Wesselsbron	Man, Sheep, (Cattle, Goats)	S. Africa
"	"	V	West Nile	Man, Domestic Ungulates, Wild Birds, (Fowl, Monkeys, Bats, Rodents)	Egypt, S. Africa, Israel
"	vishnui s.l.*	V	Ganjam	Man, (Sheep, Goats)	India
"	"	V	Japanese B	(Man, Wild Birds, Fowl, Domestic Ungulates, Mongoose)	India
"	"	V	West Nile	Man	India
"	spp.	V	Venezuelan Equine	See Note 5 (at end of table)	
"	(Melanoconion) aikenii	V	Venezuelan Equine	Man, Rodents, Birds, (Horse, Mule, Cow, Dog, Fowl, Opossums, Bat)	Panama
"	crybda	V	Bussuquara	Man, Rodents	Panama
"	epanastasis	V	Catu	Rodents	Trinidad
"	"	V	? Guama	?	Panama
"	"	V	Venezuelan Equine	Man, Equines, Rodents, Marsupial	Trinidad
"	iolambdis	V	Venezuelan Equine	Equines, Opossum, Bat (Cattle, Pigs, Chickens, Rabbit, Wild Carnivores and Rodents)	Mexico

* See footnote on p. 505.

				Hosts/Vectors	Location
„	*opisthus*	Venezuelan Equine	V	Equines, Opossum, Bat (Cattle, Pigs, Chickens, Rabbit, Wild Carnivores and Rodents)	Mexico
„	*portesi*	Bimiti	V	Rodents, (Man, Donkey)	Trinidad
„	„	? Caraparu	V	Rodents	Trinidad
„	„	Catu	V	Rodents	Trinidad, Fr. Guiana
„	„	Guama	V	Rodents	Trinidad, Fr. Guiana
„	„	Mucambo	V	?	Fr. Guiana
„	„	Murutucu	V	?	Fr. Guiana
„	„	Oriboca	V	?	Trinidad, Fr. Guiana
„	„	Restan	V	Man	Trinidad
„	„	Venezuelan Equine	V	Man, Equines, Rodents, Marsupial	Trinidad
„	*spissipes*	Caraparu	V	Man, Sentinel Monkey, Rodents	Brazil
„	„	? Caraparu	V	Rodents	Trinidad
„	„	Guama	V	Man, Sentinel Monkey, Rodents, Marsupials	Brazil
„	„	Itaqui	V	Man, Sentinel Monkey, Marsupial, Rodents	Brazil
„	„	Kairi	V	(Man, Monkey, Donkey)	Trinidad
„	„	Oriboca	V	Man, Sentinel Monkey, Marsupial, Rodents	Brazil
„	„	St. Louis	V	Man, Birds	Trinidad
„	„	Venezuelan Equine	V	Man, Equines, Rodents, Marsupial	Trinidad
„	*taemiopus*	Eastern Equine	V	Horses, Rodents, Marsupial, Sentinel Monkey, (Man)	Trinidad, Brazil
„	„	Guama	V	Man, Sentinel Monkey, Rodents, Marsupials	Brazil
„	„	St. Louis	V	Man, Birds	Trinidad
„	„	Venezuelan Equine	V	Man, Equines, Rodents, Marsupials, Birds, (Cow, Dog, Fowl, Bat)	Panama, Trinidad
„	*vomerifer*	? Caraparu	V	Rodents	Trinidad
„	„	? Guama	V	?	Panama
„	„	Ilheus	V	Man, Birds, (Rodents)	Panama
„	„	Venezuelan Equine	V	Man, Equines, Rodents, Marsupials, Birds, (Cow, Dog, Fowl, Bat)	Panama, Trinidad
„	spp.		V	See Note 6 (at end of table)	

		Virus	Hosts	Locality
,, (Aedinus) accelerans	V	? Caraparu	Rodents	Trinidad
,, ,,	V	Nepuyo	(Man)	Trinidad
,, ,,	V	Venezuelan Equine	Man, Equines, Rodents, Marsupial	Trinidad
,, amazonensis	V	? Caraparu	Rodents	Trinidad
,, ,,	V	Catu	Rodents	Trinidad
,, ,,	V	Guama	Rodents	Trinidad
,, ,,	V	Venezuelan Equine	Man, Equines, Rodents, Marsupial	Trinidad
,, (Mochlostyrax) caudelli	V	Caraparu	Man, Sentinel Monkey, Rodents	Brazil
,, ,,	V	Murutucu	Man, Sentinel Monkey, Rodents, Marsupial, Sloth	Brazil
,, ,,	V	Oriboca	Man, Sentinel Monkey, Rodents, Marsupial	Brazil
,, ,, caudelli s.l.*	V	Ilheus	Man, Birds	Trinidad
,, ,,	V	St. Louis	Man, Birds	Trinidad
Culex spp.			See Note 7 (at end of table)	
Culiseta (Culiseta) annulata	V	Tahyna	Man, (Domestic Ungulates, Hare)	Czechoslovakia
,, impatiens	V	California Group	Wild Rodents	Montana
,, inornata	V	Cache Valley	(Horses)	N. Dakota, Utah
,, ,,	V	California group	(Man, Domestic Ungulates, Lagomorphs, Rodents)	Alberta, Western U.S.A.
,, ,,	V	Western Equine	(Man, Domestic Ungulates, Birds, Rodents)	Washington State
,, tonnoiri	V	Whataroa	(Man, Birds, Opossum)	New Zealand
,, (Climacura) melanura	V	Eastern Equine	Man, Birds	South-east U.S.A.
,, ,,	V	Western Equine	Birds, (Horse)	South-east U.S.A.
Deinocerites pseudes	V	St. Louis	Man, (Birds, Rodent)	Panama
,, ,,	V	Venezuelan Equine	Equines, Opossum, Bat (Cattle, Pigs, Chickens, Rabbit, Wild Carnivores, and Rodents)	Mexico
Eretmapodites chrysogaster s.l.*	V	Middelburg	?	Cameroun
,, ,,	V	Okola	(Man, Birds)	Cameroun
,, ,,	V	Rift Valley Fever	Rodent, Calf, (Man)	Uganda
,, ,,	V	Simbu	?	Cameroun
,, silvestris	V	Spondweni	(Squirrel, Birds)	Cameroun
,, ,,	V	Spondweni	(Man, Domestic Ungulates)	Natal
Haemagogus (Longipalpifer) equinus	V	Yellow Fever	Man, Monkeys	Guatemala, Panama

* See footnote on p. 505.

" (Stegoconops) mesodentatus	V	Venezuelan Equine	Equines, Opossum, Bat (Cattle, Pigs, Chickens, Rabbit, Wild Carnivores and Rodents)	Mexico
" (Stegoconops) mesodentatus	V	Yellow Fever	Man, Monkeys	Guatemala
" (Stegoconops) spegazzinii	V	Ilheus	Man, Birds, (Rodents)	Panama
" (Stegoconops) spegazzinii	V	Yellow Fever	Man, Monkeys, (Marsupials)	Panama, Trinidad, Brazil, Colombia
" (Haemagogus) lucifer	V	Yellow Fever	Man, Monkeys	Panama
" (Conoepstegus) leucocelaenus	V	Ilheus	Man, Sentinel Monkey, Birds, (Monkeys, Rodents, Marsupials, Sloths, Lizards)	Brazil
" "	V	Maguari	(Man, Domestic Ungulates, Birds)	Brazil
" "	V	Una	(Man, Rodents, Horses, Cows)	Brazil
" "	V	Yellow Fever	Man, Monkeys, (Marsupials)	Brazil, Colombia
" sp. indet.	V	Mayaro	Man, (Monkeys, Rodents, Marsupials, Birds, Sloth)	Brazil
" "	V	Mucambo	Man, Sentinel Monkey, Rodents, Marsupials, Birds, (Bat)	Brazil
" "	V	Tacaiuma	Sentinel Monkey, (Man, Horses, Rodents, Birds)	Brazil
Limatus durhamii	V	Guama	Man, Sentinal Monkey, Rodents, Marsupials	Brazil
" flavisetosus	V	Venezuelan Equine	Man, Rodents, Marsupial	Trinidad
Mansonia (Coquillettidia) fuscopennata	V	Chikungunya	Man, (Monkeys)	Uganda
" "	V	Rift Valley Fever	Rodent, Calf, (Man)	Uganda
" " metallica	V	Sindbis	Man	Uganda
" perturbans	V	West Nile	Man, (Birds)	Uganda
" "	V	Eastern Equine	Birds, (Man)	Alabama
" "	V	Tensaw	(Man)	Alabama
" richiardii	V	Tahyna	(Man)	Austria
" (Mansonioides) africana	V	Bunyamwera	Man, (Domestic Animals, Rodents, Birds)	Nigeria, S. Africa
" "	V	Chikungunya	(Man, Monkeys)	Nigeria
" "	V	Middelburg	?	Cameroun
" "	V	Pongola	(Man, Monkeys)	Nigeria, Uganda, Kenya

LL

	Species		Virus/Disease	Hosts	Locality
"	"	V	Rift Valley Fever	Rodent, Calf, (Man)	Uganda
"	"	V	Sindbis	Man, Birds, (Domestic Ungulates)	S. Africa
"	*annulata*	V	Spondweni	(Man, Domestic Ungulates)	Natal
"	*annulifera*	N	*Brugia malayi*	Man	Thailand, Malaya, Indonesia
"	"	N	*Brugia malayi*	Man	Ceylon, India, Malaya
"	*bonneae*	N	Sindbis	?	Sarawak
"	"	N	*Brugia malayi*	Man	Thailand, Malaya, Indonesia, Philippines
"	*dives*	N	*Brugia malayi*	Man	Malaya, Indonesia
"	*indiana*	N	*Brugia malayi*	Man	Ceylon, India, Thailand
"	*uniformis*	N	*Brugia malayi*	Man	Ceylon, India, Thailand, Malaya
"	"	N	*Wuchereria bancrofti*	Man	New Guinea
"	"	V	Bunyamwera	?	Sudan
"	"	V	Ndumu	(Man)	Natal
"	"	V	Spondweni	(Man, Domestic Ungulates)	Natal
"	*africana/uniformis*	V	Wesselsbron	Man, (Sheep, Cattle, Goats)	S. Africa
"	"	V	Chikungunya	Man	Uganda
"	"	V	Pongola	(Man, Domestic Ungulates)	S. Africa
"	"	V	Rift Valley Fever	Rodent, Calf, (Man)	Uganda
"	sp. indet.	N	*Brugia malayi*	Leaf Monkeys	Malaya
"	(*Mansonia*) *titillans* s.l.	V	Guama	Man, Sentinel Monkey, Rodents, Marsupials	Brazil
"	"	V	Tlacotalpan	(Man, Cattle, Pigs)	Mexico
"	"	V	Venezuelan Equine	Man, Domestic Ungulates, Rodents, Marsupial	Guatemala, Trinidad
"	(*Rhynchotaenia*) *arribalzagai*	V	Oriboca	Man, Sentinel Monkey, Rodents, Marsupial	Brazil
"	*venezuelensis*	V	Catu	Man, Sentinel Monkey, Rodents, Bat	Brazil
"	"	V	Guama	Man, Sentinel Monkey, Rodents, Marsupials	Brazil
"	"	V	Guaroa	?	Panama
"	"	V	Mayaro	Man, (Monkeys)	Trinidad
"	"	V	Mucambo	Man, Sentinel Monkey, Rodents, Marsupials, Birds, (Bat)	Fr. Guiana, Brazil
"	"	V	Oriboca	Man, Sentinel Monkey, Rodents, Marsupials	Brazil
"	"	V	Oropouche	Man, (Monkeys)	Trinidad

Species			Virus	Hosts	Location
,,		V	Una	?	Fr. Guiana
,,	,,	V	Venezuelan Equine	Man, Rodents, Marsupial	Trinidad
,,	sp. indet.	V	Guama	Rodents	Trinidad
,,		V	Mucambo	Man, Sentinel Monkey, Rodents, Marsupials, Birds, (Bat)	Brazil
Psorophora (Psorophora) cilipes		V	Venezuelan Equine	Man, Domestic Ungulates, (Rodent)	Guatemala
,,	howardii	V	California group	(Man)	Alabama
,,	(Janthinosoma) albipes	V	Ilheus	Man, Sentinel Monkey, Birds, (Monkeys, Rodents, Marsupials, Sloths, Lizards)	Brazil
,,	ferox	V	Una	(Man, Horses, Cows, Rodents)	Panama, Brazil
,,	,,	V	Cache Valley	(Man, Cattle, Dog, Raccoon)	Illinois
,,	,,	V	Ilheus	Man, Sentinel Monkey, Birds, (Monkeys, Rodents, Marsupials, Sloths, Lizards)	Panama, Trinidad, Brazil, Colombia
,,	,,	V	Kairi	(Man, Monkey, Donkey)	Trinidad
,,	,,	V	Mayaro	?	Panama
,,	,,	V	Oriboca	Man, Sentinel Monkey, Rodents, Marsupial	Brazil
,,	,,	V	St. Louis	Man, Birds	Trinidad
,,	,,	V	Una	(Man, Horses, Cows, Rodents)	Panama, Trinidad, Brazil
,,	,,	V	Venezuelan Equine	Man, Rodents, Marsupial	Trinidad
,,	lutzii	V	Wyeomyia complex	Man	Panama
,,	,,	V	Guama group	Man, Sentinel Monkey, Rodents, Marsupials, Bat	Brazil
,,	albipes/ferox	V	Ilheus	Birds	Panama
,,	,,	V	Mayaro	(Man)	Colombia
,,	sp. indet.	V	Venezuelan Equine	Man, Horses, (Monkeys)	Colombia
,,	(Grabhamia) cingulata	V	Wyeomyia complex	Man	Panama
,,	confinnis	V	? Cache Valley	?	Panama
,,	,,	V	Cache Valley	?	Indiana
,,	,,	V	Tensaw	Rodent, Dog, (Man, Cow, Raccoon)	Florida
,,	,,	V	Venezuelan Equine	Man, Domestic Ungulates, (Rodent)	Guatemala
Sabethes (Sabethes) belisarioi	sp. indet.	V	Ilheus	Man, Birds	Trinidad
		V	St. Louis	Sentinel Mouse, Marsupial, Birds	Brazil
,,	(Sabethoides) chloropterus	V	Ilheus	Man, Birds	Panama

Species		Virus/Disease	Host	Locality
"	V	St. Louis	Man	Panama
"	V	Yellow Fever	Man, Monkeys	Guatemala, Panama
" sp. indet.	V	St. Louis	Man	Panama
Trichoprosopon sp. indet.	V	Bussuquara	Man, Rodents	Panama
"	V	Ilheus	Man, Birds	Panama
"	V	Triniti	(Man)	Trinidad
"	V	? Caraparu	Rodents	Trinidad
Wyeomyia (Wyeomyia) medioalbipes	V	Ilheus	Man, Birds	Trinidad
"	V	Venezuelan Equine	Man, Rodents, Marsupial	Trinidad
" mitchellii	V	Venezuelan Equine	Equines, Opossum, Bat (Cattle, Pigs, Chickens, Rabbit, Wild Carnivores and Rodents)	Mexico
" (Dendromyia) aporonoma	V	Kairi	(Monkey, Donkey)	Trinidad
" ypsipola	V	Kairi	(Monkey, Donkey)	Trinidad
" sp. indet.	V	Kairi	Monkey, Rodent, (Man)	Brazil
"	V	Maguari	?	Fr. Guiana
Undetermined sabethine spp.	V	Caraparu	Man, Sentinel Monkey, Rodents, (Marsupials)	Brazil
"	V	Guama gp	Man, Sentinel Monkey, Rodents, (Marsupials), Bat	Brazil
"	V	Kairi	Monkey, Rodent, (Man, Donkey)	Trinidad, Brazil
"	V	Mayaro	Man, (Monkeys, Rodents, Sloth)	Brazil
"	V	Mucambo	Man, Sentinel Monkey, Marsupials, Rodents, (Monkeys, Sloth)	Brazil
"	V	Murutucu	Man, Sentinel Monkey, Rodents, Marsupials, Sloth	Brazil
"	V	Oriboca	Man, Sentinel Monkey, Rodents, Marsupials	Brazil
"	V	Yellow Fever	Man, Monkeys, (Marsupials)	Brazil

CERATOPOGONIDAE

Species		Disease	Host	Locality
Culicoides furens	N	Mansonella ozzardi	Man	St. Vincent, W.I.
" grahami	N	Dipetalonema perstans	Man	Cameroun
"	N	Dipetalonema streptocerca	Man	Cameroun
" inornatipennis	N	Dipetalonema perstans	Man	Cameroun
" milnei (syn. austeni)	N	Dipetalonema perstans	Man	Cameroun
"	N	Dipetalonema streptocerca	Man	Cameroun
" sp.	V	Shuni	Man, Sheep, Cattle	Nigeria

SIMULIIDAE*

Simulium damnosum	N	*Onchocerca volvulus* (Onchocerciasis)	Man	W. Africa, Sudan, Ethiopia, Congo Basin, locally E. Africa, Fernando Po, Yemen
" *metallicum*	N	*Onchocerca volvulus*	Man	Venezuela, minor extent Mexico and Guatemala
" *neavei*	N	*Onchocerca volvulus*	Man	Congo, Uganda, formerly Kenya
" *ochraceum*	N	*Onchocerca volvulus*	Man	Mexico, Guatemala
" *woodi*	N	*Onchocerca volvulus*	Man	Tanzania (one small focus)

TABANIDAE

Chrysops centurionis	N	*Loa loa* (Loiasis)	Man, Monkeys	Sierra Leone to Angola and the Congo, and eastwards to Uganda and Bahr el Ghazal
" *dimidiata*	N	*Loa loa*	Man, Monkeys	Sierra Leone to Angola and the Congo and eastwards to Uganda and Bahr el Ghazal
" *langi*	N	*Loa loa*	Man, Monkeys	Sierra Leone to Angola and the Congo, and eastwards to Uganda and Bahr el Ghazal
" *silacea*	N	*Loa loa*	Man, Monkeys	Sierra Leone to Angola and the Congo, and eastwards to Uganda and Bahr el Ghazal
Tabanus discalis	B	*Pasteurella* (Tularaemia)	Rodents, Man	U.S.A.
Tabanus spp.	B	*Pasteurella* (Tularaemia)	Rodents, Man	Eurasia
Tabanid spp.	V	Vesicular stomatitis	Horses, Bovines, Pig, Mouse, Man	U.S.A.
Tabanid spp.	B	*Bacillus anthracis* (Anthrax)		
	P	Trypanosomes (mechanical transmission)		

CHLOROPIDAE

Hippelates collusor	S	Conjunctivitis ('pink-eye')		South West U.S.A. and Mexico
" *flavipes* (=*pallipes*)		*Treponema pertenue* (yaws)		Jamaica
" *pusio*		conjunctivitis		Southern U.S.A.
Siphunculina funicola		epidemic conjunctivitis		India, Ceylon, Java

MUSCIDAE

Musca domestica and possibly other domestic Diptera	B	*Shigella* (Dysentery)		Samoa (Satchell & Harrison) and certainly elsewhere in tropics
	B	*Salmonella* (Typhoids)		

* See fourth paragraph on p. 497.

Vector		Disease / Pathogen	Hosts	Distribution
Musca domestica and possibly other domestic Diptera	S	Yaws		Samoa (Satchell & Harrison) and certainly elsewhere in tropics
Musca domestica and possibly other domestic Diptera	B	*Bacillus anthracis* (Anthrax)		Samoa (Satchell & Harrison) and certainly elsewhere in tropics
Musca domestica and possibly other domestic Diptera	B	? *Vibrio comma* (Cholera)		Samoa (Satchell & Harrison) and certainly elsewhere in tropics
Musca sorbens s.l. (including *vetustissima*)	V	? Poliomyelitis		
	S	Yaws		Samoa (Satchell & Harrison) and certainly elsewhere in tropics
	V	Trachoma		Old World tropics
	B	? *Bacillus anthracis* (Anthrax)		
GLOSSINIDAE				
Glossina morsitans	P	*Trypanosoma rhodesiense* (Rhodesian sleeping sickness)	Man, Game (potentially important: common duiker, eland, reedbuck, bushbuck), Cattle (potentially important)	West, Central and East Africa north of forest belt
" *pallidipes*	P	*Trypanosoma rhodesiense*	Man, Game (proven: bush-buck; potentially important: reedbuck), Cattle (potentially important)	East Central and East Africa
" *palpalis*; *G. fuscipes*	P	*T. gambiense* (Gambian sleeping sickness)	Man, Domestic Pigs (potentially important)	Mainly West and Central Africa
" *fuscipes*	P	*T. rhodesiense*	Man, Cattle (proven), Game (potentially important)	Central Africa and northern part of East Africa
" *swynnertoni*	P	*T. rhodesiense*	Man, Game (potentially important: eland, impala), Cattle (potentially important)	East Africa
" *tachinoides*	P	*T. gambiense*	Man, Cattle and Domestic Pigs (potentially important), Game Animals (potentially important)	Mainly West Africa north of forest belt
SIPHONAPTERA				
Xenopsylla cheopis	R	*Rickettsia prowazeki* (epidemic, exanthematic typhus)	Rats, Mice, Man	Almost Cosmopolitan

ERRATA

pp. 521–3. Some unfortunate errors (of classification) and omissions (e.g. scrub-typhus, louping ill) have occurred in the Arachnida entries. The editor accepts full responsibility and the reader is referred to Chapter 14 for rectification of these. The following specific corrections should be made:

p. 521, entry   4, for ARACHINIDA read ARACHNIDA

    ”   5, delete —METASTIGMATA

    ”   6, delete IXODIDAE and brackets

    ” 19, for fever read typhus

    ” 20, for *japponica* read *japonica*

    ” 21, for *leporis palustris* read *leporispalustris*

p. 522, entry 14, for *spinipalpus* read *spinipalpis*

p. 523, entry   1, delete ACARI-MESOSTIGMATA and brackets

    ”   2, replace LELAPTIDAE with DERMANYSSIDAE

    ”   5, for Sinbis read Sindbis

		Infectious disease	Vertebrate Hosts	Geographical Distribution
”	X	(Siberian tick fever) Russian spring-summer encephalitis (RSSE)	Small rodents, birds mammals, Man	Russia, Asia, Europe
japponica douglasi				
leporis palustris	V	Colorado tick fever	Leporids, Rodents, Man	Canada, U.S.A.
neumanni	V	Russian spring-summer encephalitis (RSSE)	Small Rodents, Birds, Mammals, Man	Russia, Asia, Europe
papuana kinneari	V	Kyasanur Forest disease (KFD)	Monkeys, Ground-living Birds, Rodents, Cattle, Man	India (Mysore State)

Species		Disease	Hosts	Distribution
,,	*spinigera*	V	Kyasanur Forest disease (KFD)	Monkeys, Ground-living Birds, Rodents, Cattle, Man	India (Mysore State)
,,	*turturis*	V	Kyasanur Forest disease (KFD)	Monkeys, Ground-living Birds, Rodents, Cattle, Man	India (Mysore State)
Hyalomma spp.		R	*Rickettsia conori* (Boutonneuse fever)	Leporids, Dogs, Rodents, other Mammals, Man	Africa, European Mediterranean, Middle East and S.E. Asia
,,	*anatolicum*	V	Crimean-Congo haemorrhagic fever (CHF-Congo)	Small Mammals, Domestic Mammals, Birds (including migrants), Man	Steppes of S.E. Europe, Asia, W. Pakistan, Uganda, Kenya, Nigeria and Congo
Ixodes sp.		R	*Rickettsia conori* (Boutonneuse fever)	Dogs, Rodents, Man	Indies, European Asiatic and African Mediterranean zones, Africa
,,	*cookei*	V	Powassan encephalitis	Wild Rodents, etc., Man	Canada, U.S.A.
,,	*granulatus*	V	Langat encephalitis	Rodents, Man	Malaysia
,,	*lividus*	V	Russian spring-summer encephalitis (RSSE)	Small Rodents, Birds, Mammals, Man	Russia, Asia, Europe
,,	*marxi*	V	Powassan encephalitis	Wild Rodents, etc., Man	Canada, U.S.A.
,,	*persulcatus*	V	Russian spring-summer encephalitis (RSSE)	Small Rodents, Birds, Mammals, Man	Russia, Asia, Europe
,,	,,	V	Omsk haemorrhagic fever (OHF)	Rodents, other small Mammals, Man	S.W. Siberia, possibly elsewhere
,,	,,	V	Kemerovo fever	Rodents, Man	W. Siberia
,,	*ricinus*	V	C. European encephalitis (aestivo-autumnal) (TE)	Small Rodents, Birds, Mammals, Man	W. Russia, C. Europe, Scandinavia, Balkans, Austria, Germany
,,	*spinipalpus*	V	Powassan encephalitis	Wild Rodents, Man	Canada, U.S.A.
Ornithodoros sp.		S	*Spirochaeta (Borrellia) duttoni* (relapsing fever)	Man	Central and Southern Africa
,,		S	*Spirochaeta (Borrellia) hispanica* (relapsing fever)	Man, Pig	Mediterranean, Asia Minor
,,		S	*Spirochaeta (Borrellia) persica* (relapsing fever)	Man, Rodents, Bats	Central Asia
,,		S	*Spirochaeta (Borrellia) turicatae, S. parkerii, S. hermsi* (relapsing fever)	Man	U.S.A.
,,		S	*Spirochaeta (Borrellia) venezuelensis* (relapsing fever)	Man	Central and South America
Otobius lagophilus		V	Colorado Tick fever	Leporids, Rodents, Man	U.S.A., Canada
,,	,,	R	*Rickettsia conori* (Boutonneuse fever)	Leporids, Dogs, Rodents, other Mammals, Man	Africa, European Mediterranean, Middle East and S.E. Asia

ACARI-MESOSTIGMATA (MITES)

TROMBICULIDAE, LELAPTIDAE

"	V	Haemorrhagic nephroso-nephritis	Rats, Man	Siberia, Manchuria, Korea, Hungary
"	V	Tacaribe group-Junin virus	Wild Rodents, Man	Argentina
"	V	Slight jaundice vesicles Sinbis virus (febrile disease)	Wild Birds, Man	Africa, India, Malaya, Philippines, N. Australia

DERMANYSSIDAE

Liponyssoides sanguineus	R	Rickettsia akari (rickettsial-pox)	Mice, Man	North East U.S.A., U.S.S.R. Korea

NOTES

1. Includes the following: PANAMA—*Ae. angustivittatus* (Ilheus, Venezuelan Equine), *Ae. (Ochl.)* pool (Guaroa, Ilheus, Una, Venezuelan Equine). BRAZIL—*Ae. fulvus* (Yellow Fever). COLOMBIA—*Ae. angustivittatus* (Maguari). AUSTRALIA—*Ae. normanensis* (Murray Valley, Sindbis). N. AMERICA—*Ae. aberratus* (California gp), *Ae. atlanticus/tormentor* (Eastern Equine, Tensaw, Venezuelan Equine), *Ae. canadensis* (California gp), *Ae. communis* gp (Bunyamwera gp), *Ae. fitchii* (California gp), *Ae. flavescens* (Western Equine), *Ae. fulvus* (Eastern Equine), *Ae. infirmatus* (California gp, Western Equine), *Ae. michellae* (Eastern Equine, Tensaw), *Ae. nigromaculis* (Western Equine), *Ae. sticticus* (California gp, Eastern Equine), *Ae. stimulans* gp (California gp), *Ae. trivittatus* (Bunyamwera gp, California gp, Cache Valley). CENTRAL EUROPE—*Ae. cantans* (Tahyna), *Ae. flavescens* (Tahyna), *Ae. punctor/communis* (Tahyna). FINLAND—*Ae. punctor/communis* (Tahyna).

2. Includes the following: UGANDA—*Ae. dendrophilus* (Rift Valley). SENEGAL—*Ae. luteocephalus* (Chikungunya). CAMEROUN—*Stegomyia* pool (Wesselsbron).

3. Includes the following: SENEGAL—*Ae. irritans* (Chikungunya). CAMEROUN—*Ae. domesticus* gp (Bunyamwera), *Ae. simulans* (Middelburg), *Ae. tarsalis* gp (Wesselsbron). UGANDA—*Ae. abnormalis* gp (Semliki Forest), *Ae. tarsalis* gp (Rift Valley). KENYA—*Ae. dentatus* gp (Pongola). ETHIOPIA—*Ae. dentatus* (Yellow Fever). MOZAMBIQUE—*Ae. argenteopunctatus* (Semliki Forest). NATAL—*Ae. albocephalus* (Middelburg), *Ae. marshalli* (Middelburg), *Ae. minutus* (Wesselsbron), *Aedimorphus* pool (Middelburg).

4. Includes the following: U.S.A.—*An. (An.) freeborni* (Cache Valley, California gp, Western Equine), *An. (An.) punctipennis* (Cache Valley). PANAMA—*An. sp. indet.* (Guaroa, Venezuelan Equine). JAMAICA—*An. (An.) grabhamii* (Cache Valley). TROPICAL AMERICA—(Territory not stated.) *An. (An.) neomaculipalpus* (Venezuelan Equine). VENEZUELA—*An. (N.) triannulatus* (Venezuelan Equine). ISRAEL—*An. (An.) coustani* (West Nile).

5. Includes the following: U.S.A.—*Cu. erythrothorax* (California gp), *Cu. peus* (St. Louis, Western Equine), *Cu. restuans* (Eastern and Western Equine), *Cu. salinarius* (Eastern and Western Equine), *Cu. tarsalis* (Eastern and Western Equine). MEXICO—*Cu. coronator*, *Cu. thriambus* (Venezuelan Equine). TRINIDAD—*Cu. coronator* (? Caraparu, St. Louis), *Cu. declarator* (Catu, ? St. Louis). EGYPT—*Cu. antennatus* (West Nile). SENEGAL—*Cu. thalassius* (? Ntaya). CAMEROUN—*Cu. guiarti/ingrami* (Ntaya). KENYA—*Cu. zombaensis* (Pongola). MOZAMBIQUE—*Cu. sitiens* gp (Mossuril). SARAWAK—*Cu. pseudovishnui* (Sindbis, Tembusu). TAIWAN—*Cu. annulus* (Japanese B), *Cu. fuscocephalus* (Japanese B). QUEENSLAND—*Cu. squamosus* (Kunjin). NEW ZEALAND—*Cu. pervigilans* (Whataroa).

6. The following viruses have been recovered from unidentified species of *Culex (Melanoconion)*: U.S.A.—California gp, Venezuelan Equine. MEXICO—Nepuyo. GUATEMALA—Venezuelan Equine. PANAMA—Guama gp, Venezuelan Equine. BRAZIL—Eastern Equine, Guama, Guama gp, Mucambo, Oriboca.

7. The following have been recovered from *Culex* spp. identified only to the genus: MEXICO—Nepuyo. TRINIDAD—Bimiti, Venezuelan Equine. FR. GUIANA—St. Louis. BRAZIL—Apeu, Caraparu, Catu, Eastern Equine, Itaqui, Mayaro, Mucambo, Murutucu, Oriboca. CAMEROUN—Ntaya. UGANDA—Germiston, Witwatersrand. SARAWAK—Japanese B.

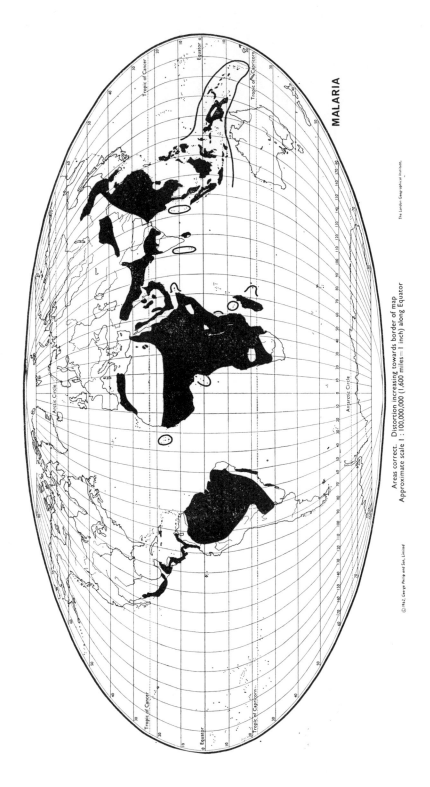

MALARIA

Areas correct.   Distortion increasing towards border of map
Approximate scale 1 : 100,000,000 (1,600 miles = 1 inch) along Equator

© 1962, George Philip and Son, Limited

The London Geographical Institute.

Fig. 211   Areas actually or potentially malarious.   Areas in the 'consolidation' phase omitted (30th June 1969, modified from World Health Organisation Chronicle 24).   Transmission by mosquitoes (Culicidae).

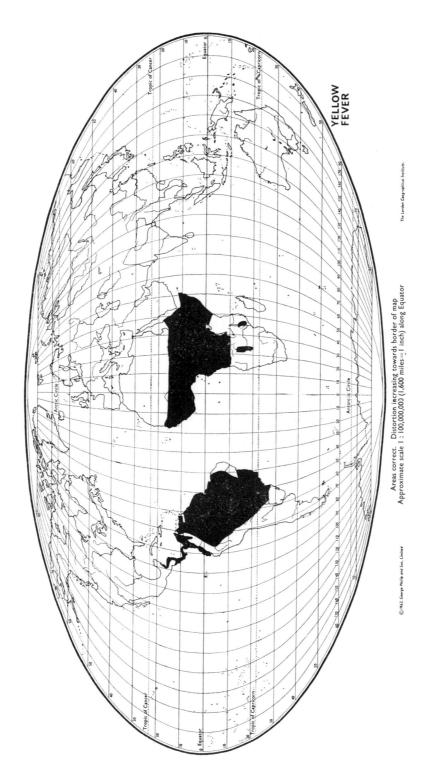

YELLOW
FEVER

Areas correct.   Distortion increasing towards border of map
Approximate scale 1 : 100,000,000 (1,600 miles = 1 inch) along Equator

© 1962, George Philip and Son, Limited

The London Geographical Institute.

Fig. 212   Areas actually or potentially subject to yellow fever (after Mattingly, 1969).   Transmission by mosquitoes.

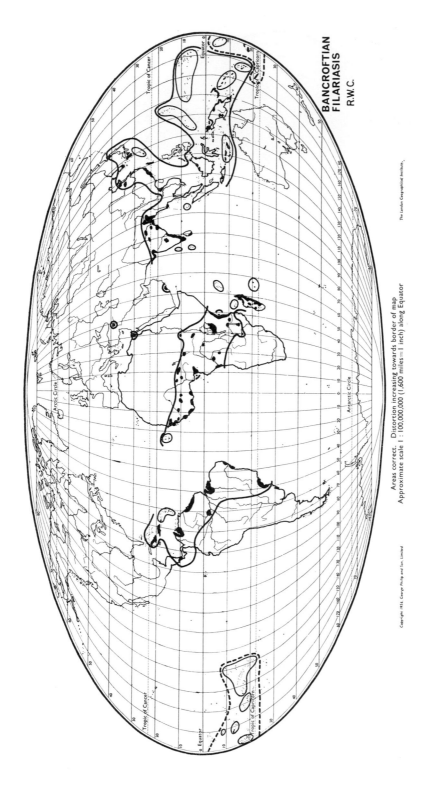

Fig. 213   Geographical distribution of Bancroftian filariasis.   The extent of known foci is shown in black and the overall range of possible distribution is suggested by the thick lines (continuous line for the limits of the periodic form of *Wuchereria bancrofti* and broken line for limits of the superiodic form).   Island groups where the disease is endemic are circled.   Transmission by mosquitoes (Culicidae).

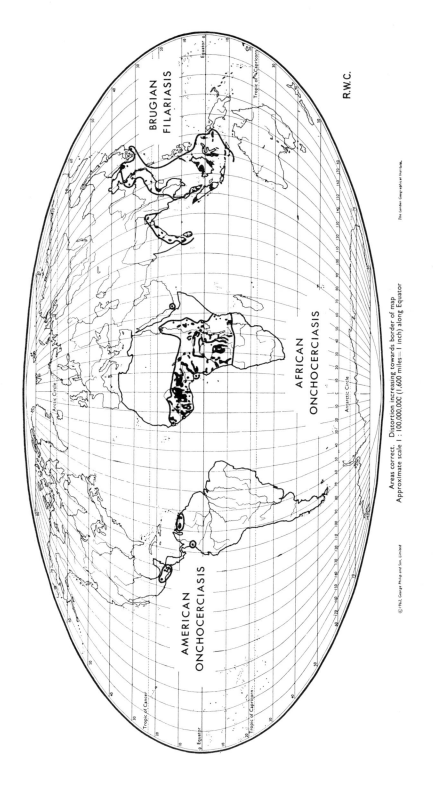

Fig. 214 Geographical distribution of onchocerciasis and Brugian filariasis in man. The approximate extent of known foci is shown in black, and the thick lines are given as a rough indication of the possible overall limits. Transmission by black-flies (Simuliidae) and mosquitoes (Culicidae) respectively.

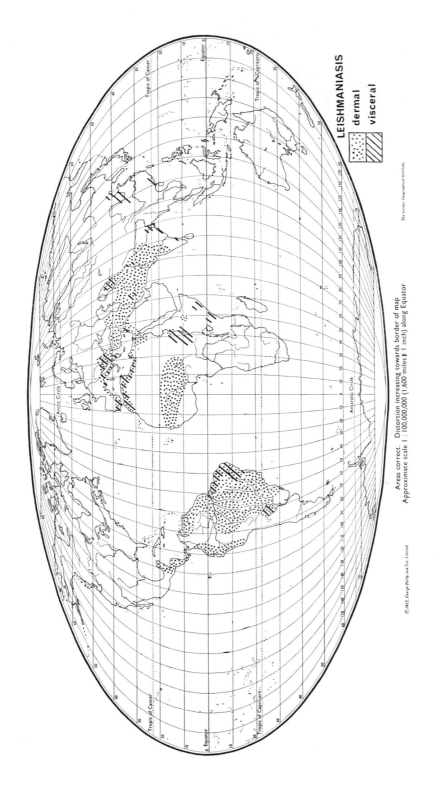

Fig. 215 Approximate distribution of established human leishmaniasis; dermal (Oriental sore, espundia etc.) and visceral (kala-azar). Transmission by Phlebotomidae.

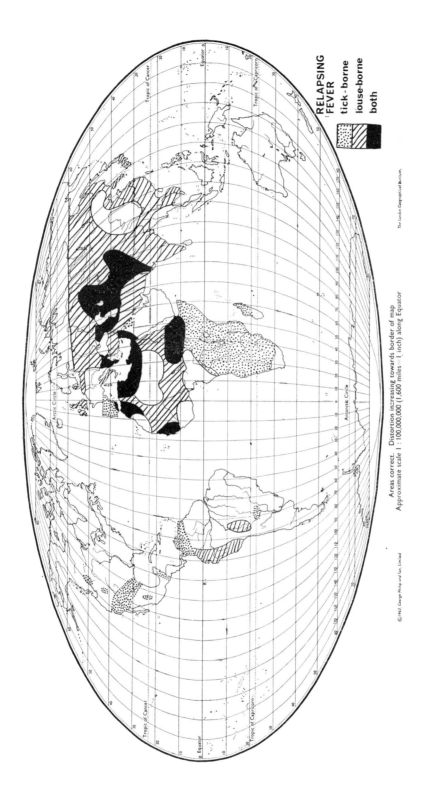

Fig. 216 The geographical distribution of relapsing fever, borne by ticks (dotted), lice (striped) or both (black).

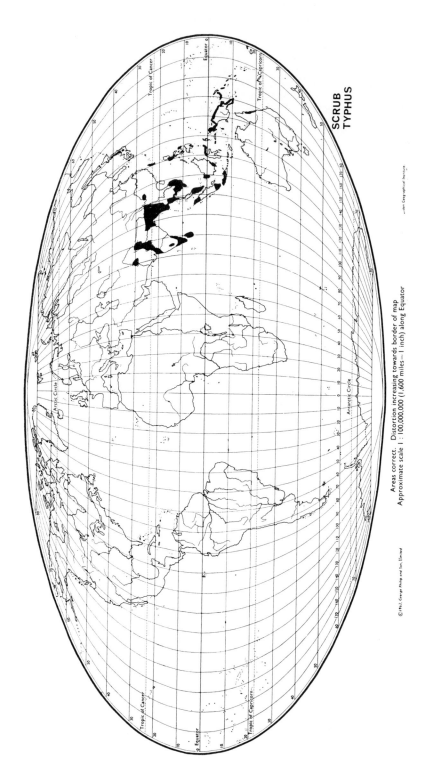

Fig. 217 The geographical distribution of scrub typhus (after Mackie, Hunter & Worth, 1954). Transmission by mites (Acari).

HARRIS, A. H. & DOWN, H. A. 1946. Studies of the dissemination of cysts and ova of human intestinal parasites by flies in various localities on Guam. *Am. J. trop. Med.* **26**: 789–800.

HAWKING, F. 1957. The distribution of Bancroftian filariasis in Africa. *Bull. Wld Hlth Org.* **16**: 581–592.

HOOGSTRAAL, H. 1967. Ticks in relation to human diseases caused by *Rickettsia* species. *A. Rev. Ent.* **12**: 377–420.

IYENGAR, M. O. T. 1965. Epidemiology of filariasis in the South Pacific. *Tech. Pap. S. Pacif. Commn*, No. 148. x+183 pp.

JAMES, M. T. & HARWOOD, R. F. 1969. *Herms's medical entomology.* 484 pp. London.

KESSEL, J. F. 1965. Filarial infections of man. *Am. Zool.* **5**: 79–84.

KUMM, H. W. 1935. The natural infestation of *Hippelates pallipes* Loew, with the spirochaetes of yaws. *Trans. R. Soc. Trop. Med. Hyg.* **29**: 265–272.

KUMM, H. W. & TURNER, T. B. 1936. The transmission of yaws from man to rabbits by an insect vector, *Hippelates pallipes. Am. J. trop. Med.* **16**: 245–262.

LECLERCQ, M. 1948. La transmission de la poliomyèlite par les insectes. *Revue méd. Liège* **3** (7): 154–156; **3** (8): 197; **3** (11): 279–281.

—— 1969. *Entomological Parasitology—the relations between entomology and the medical sciences.* 158 pp. London, Oxford, etc.

LINDSAY, D. R. & SCUDDER, H. I. 1956. Non-biting flies and disease. *A. Rev. Ent.* **1**: 323–346.

MACKIE, T. T., HUNTER, G. W. & WORTH, C. B. 1954. *A manual of tropical medicine.* 907 pp. Philadelphia and London.

MANSON-BAHR, P. H. 1966. *Manson's tropical diseases.* 16th edn. 1068 pp. London.

MATTINGLY, P. F. 1969. *The biology of mosquito-borne disease.* 184 pp. London.

MOUCHET, H., GRJÉBINE, A. & GRENIER, P. 1965. Transmission de la filariose de Bancroft dans la région éthiopienne. *Cah. Off. Rech. Sci. Tech. Outre-Mer.* (Ent. méd.) No. **3–4**: 67–90. [Useful bibliography.]

NASH, T. A. M. 1969. *Africa's bane, the tsetse fly.* 224 pp. London.

NUORTOVA, P. 1959. Studies on the significance of flies in the transmission of poliomyelitis. *Suom. hyönt. Aikak.* [*Ann. ent. Fennica*] **25**: 1–14.

OHANESSIAN, A. & ECHALIER, G. 1967. Multiplication du virus *Sindbus* chez *Drosophila melanogaster* (Insecte diptère), en conditions experimentales. *C.r. hebd. Séanc. Acad. Sci., Paris* **264**: 1356–1358.

PAUL, J. R., HORSTMANN, D. M., RIORDAN, J. T., OPTON, E. M., NIEDERMAN, J. C., ISACSON, E. P. & GREEN, R. H. 1962. An oral poliovirus vaccine trial in Costa Rica. *Bull. Wld Hlth Org.* **26**: 311–329.

PAVLOVSKI, E. N. 1946–1948. Manual of parasitology of man with discussion of the theory of the vectors of transmissible diseases. *Iz. Akad. Nauk SSSR.* [In Russian.]

PHILIP, C. B. 1939. Ticks as vectors of animal diseases. *Can. Ent.* **71**: 55–65.

POVOLNY, D. & PRIVORA, M. 1961. Kritische Bewertung mikrobiologischer Befunde bei synanthropic Fliegen in Mitteleuropa. *Angew. Parasit.* **2**: 66–74.

PROVOST, M. W. 1972. Environmental hazards in the control of disease vectors. *Environ. Ent.* **1**: 333–339.

RADVAN, R. 1960. Persistence of bacteria during development in flies. I–III. *Folia microbiol., Praha* **5**: 50–56, 85–92, 149–156.

RAGHAVAN, N. G. S. 1957. Epidemiology of filariasis in India. *Bull. Wld Hlth Org.* **16**: 553–579.

—— 1961. The vectors of human infections by *Wuchereria* species in endemic areas and their biology. *Bull. Wld Hlth Org.* **24**: 177–195.

RHODES, A. J. & VAN ROOYEN, C. E. 1968. *Textbook of Virology.* 5th edn. Baltimore.

RICHARDS, C. S., JACKSON, W. B., DE CAPITO, T. M. & MAIER, P. P. 1961. Studies on rates of recovery of *Shigella* from domestic flies and from humans in southwestern United States. *Am. J. trop. Med. Hyg.* **10**: 44–48.

MM

SATCHELL, G. H. & HARRISON, R. A. 1953. Experimental observations on the possibility of transmission of yaws by wound-feeding diptera, in western Samoa. *Trans. R. Soc. trop. Med. Hyg.* **47** (2): 148–153.

SEN, S. K. 1944. Experiments on the transmission of anthrax through flies. *Indian J. vet. Sci.* **14**: 149–158.

STEINHAUS, E. A. 1966. *Insect microbiology.* 763 pp. New York.

STRODE, G. K. (ED.). 1951. *Yellow fever.* 710 pp. New York.

SUDIA, W. D., NEWHOUSE, V. F., CALISHER, C. H. & CHAMBERLAIN, R. W. 1971. California group arboviruses: isolations from mosquitoes in North America. *Mosquito News* **31**: 576–600.

TAYLOR, R. M. 1967. Catalogue of arthropod-borne viruses of the world: a collection of data on registered arthropod-borne animal viruses. *Publ. Hlth Serv. Publs*, Washington No. **1760.** 898 pp.

TOOMEY, J. A., PIRONE, P. P., TAKACS, W. S. & SCHAEFFER, M. 1947. Can *Drosophila* flies carry poliomyelitis virus? *J. Infect. Dis.* **81**: 135–138.

VARMA, M. G. R. 1962. Transmission of relapsing fever spirochaetes by ticks. *Symp. zool. Soc. Lond.* No. **6**: 61–82.

—— 1964. The acarology of louping ill. Proc. 1st int. Congr. Acarology Fort Collins 1963. *Acarologia* **6** (hors série): 241–254.

WORLD HEALTH ORGANISATION. 1962. WHO Expert Committee on Filariasis. First Report. *Wld Hlth Org. Tech. Rep. Ser.* No. **233**, 49 pp. Geneva.

—— 1966. WHO Expert Committee on Onchocerciasis. Second Report. *Wld Hlth Org. Tech. Rep. Ser.* No. **335**, 96 pp. Geneva.

——1967. Inter-regional travelling seminar on Leishmaniasis.

——1970. Malaria eradication in 1969. *Wld Hlth Org.* Chronicle **24**: (9) 395–403.

——1972. Vector control in international health. 144 pp. Geneva.

INDEX TO AUTHORS CITED

SUBJECT INDEX

Entries should be sought under both common and scientific names. Where many species are repeatedly included in a few pages only the generic name and page range is given. As an aid to diagnosis, the subject index includes diseases, symptoms, reservoir hosts, etc. Since the vector table (pp. 501–523) is largely a summary of information already included in the text of the relevant chapters, the entries are not included in the index.

Plate 2   *Glossina morsitans* female ($\times 6\frac{1}{2}$), tsetse-fly.

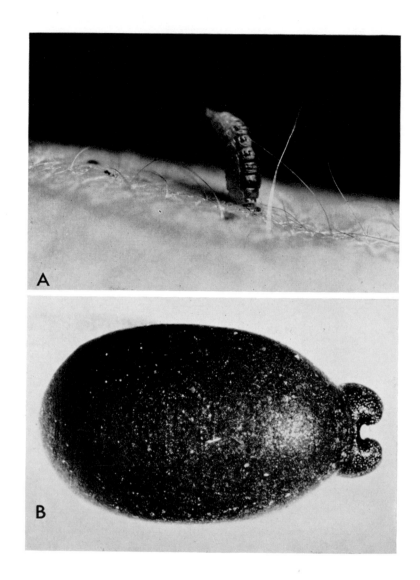

Plate 3   A, larva of *Auchmeromyia luteola*, the Congo floor maggot on human arm (photo S. A. Smith); B, puparium of tsetse-fly *Glossina palpalis* (×15).

Plate 4   *Musca sorbens* is completely ignored by the Masai but is the principal vector of certain eye infections (photo Raymond Lewis and Denys Dawnay).

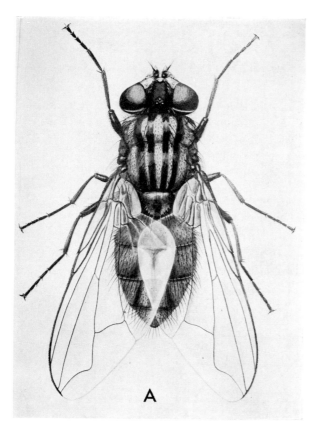

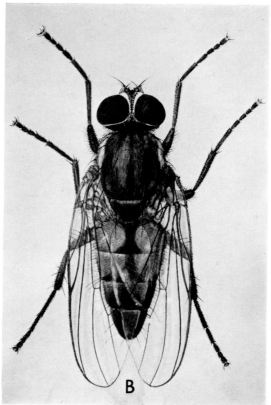

Plate 5   A, *Musca domestica*, female, the house-fly ($\times 7\frac{1}{2}$); B, *Fannia canicularis*, male, the lesser house-fly ($\times 7\frac{1}{2}$).

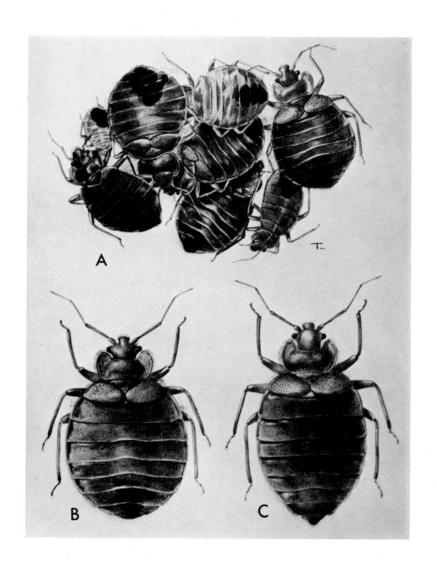

Plate 6 *Cimex lectularius*, the bed-bug: A, cluster of nymphs and adults; B, female; C, male, (×10).

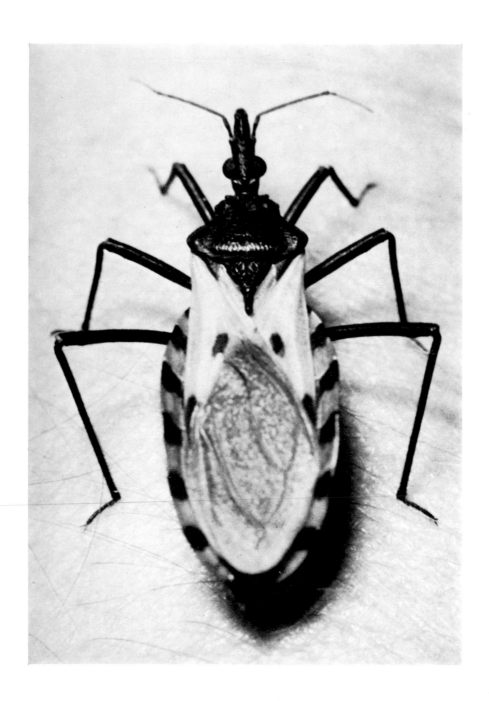

Plate 7   *Triatoma dimidiata*, adult bug found infected with *Trypanosoma cruzi* in Cayo district of British Honduras (photo J. R. Coura and W. B. Petana) (×4).

Plate 8   *Filodes fulvodorsalis* (Pyralidae) feeding at human eye (photo H. Bänziger).

Plate 9   A, *Calyptra eustrigata* (Noctuidae) re-ingesting blood that has been regurgitated a few seconds earlier.   The proboscis still remains in the wound (photo H. Bänziger); B, larva of South American Saturniid moth which injects a powerful anticoagulant if handled, causing serious bleeding.

Plate 10   Some dangerous scorpions (Buthidae): A, *Androctonus crassicauda* (Oliver) (×1); B, *Leiurus quinquestriatus* (Hemprich & Ehrenberg) (×1); C, *Centruroides suffusus* Pocock (×¾); D, *Tityus trinitatis* Pocock (×1).

Plate 11   Two dangerous spiders: A, the black widow, *Latrodectus mactans mactans* (Fabricius) (×4); B, the Sydney funnel web spider, *Atrax robustus* O.P.-Cambridge (×1).

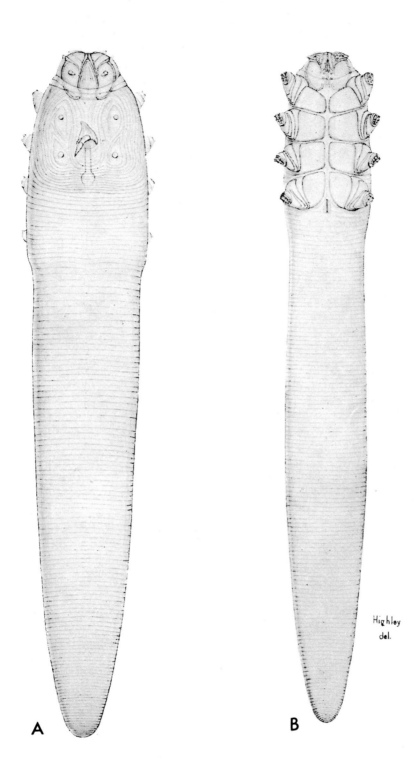

A

B

Plate 12   The follicle mite, *Demodex folliculorum* (Simon): A, Dorsum of male; B, venter of female. ($\times$500).